P9-AFZ-602

METHODS in MICROBIOLOGY

METHODS in MICROBIOLOGY

Edited by

J. R. NORRIS
Milstead Laboratory of Chemical Enzymology,
Sittingbourne, Kent, England

D. W. RIBBONS
Department of Biochemistry,
University of Miami School of Medicine,
and Howard Hughes Medical Institute,
Miami, Florida, U.S.A.

Volume 1

1969

ACADEMIC PRESS
London and New York

ACADEMIC PRESS INC. (LONDON) LTD
Berkeley Square House
Berkeley Square,
London, W1X 6BA

U.S. Edition published by
ACADEMIC PRESS INC.
111 Fifth Avenue,
New York, New York 10003

Library of Congress Catalog Card Number: 68–57745
SBN: 12–521501–0

PRINTED IN GREAT BRITAIN BY
ADLARD AND SON LIMITED
DORKING, SURREY

LIST OF CONTRIBUTORS

N. Blakebrough, *Chemical Engineering Department, Indian Institute of Technology, Delhi, India*

R. Brookes, *Bacteriological Bioengineering Research Group, Department of Bacteriology, Karolinska Institutet, Stockholm* 60, *Sweden*

C. T. Calam, *Imperial Chemical Industries Ltd, Pharmaceuticals Division, Alderley Park, Macclesfield, Cheshire, England*

R. C. Codner, *School of Biological Sciences, Bath University of Technology, Bath, England*

H. M. Darlow, *Microbiological Research Establishment, Porton, Salisbury, Wilts., England*

E. C. Elliott, *Unilever Research Laboratory, Colworth House, Sharnbrook, Beds., England*

R. Elsworth, *Microbiological Research Establishment, Porton, Salisbury, Wilts., England*

D. L. Georgala, *Unilever Research Laboratory, Colworth House, Sharnbrook, Beds., England*

Charles E. Helmstetter, *Radiation Physics Section, Roswell Park Memorial Institute, Buffalo, New York, U.S.A.; Department of Biophysics, State University of New York, Buffalo, New York, U.S.A.*

K. I. Johnstone, *Department of Bacteriology, The School of Medicine, Leeds, England*

H. E. Kubitschek, *Argonne National Laboratory, Argonne, Illinois, U.S.A.*

M. F. Mallette, *Department of Biochemistry, Pennsylvania State University, University Park, Pennsylvania, U.S.A.*

J. J. McDade, *Biohazards Department, Pitman-Moore Division, The Dow Chemical Company, Zionsville, Indiana, U.S.A.*

J. G. Mulvany, *Millipore (U.K.) Ltd, Wembley, Middlesex, England*

G. B. Phillips, *Biological Safety and Control, Becton, Dickinson and Company, Cockeysville, Maryland, U.S.A.*

J. R. Postgate, *University of Sussex, Falmer, Sussex, England*

Louis B. Quesnel, *Department of Bacteriology and Virology, University of Manchester, Manchester, England*

K. Sargeant, *Microbiological Research Establishment, Porton, Salisbury, Wilts., England*

H. D. Sivinski, *Planetary Quarantine Department, Sandia Laboratories, Albuquerque, New Mexico, U.S.A.*

A. H. Stouthamer, *Botanical Laboratory, Microbiological Department, Free University, De Boelelaam* 1087, *Amsterdam, The Netherlands*

G. Sykes, *Boots Pure Drug Co. Ltd, Nottingham, England*

W. J. Whitfield, *Planetary Quarantine Department, Sandia Laboratories, Albuquerque, New Mexico, U.S.A.*

ACKNOWLEDGMENTS

For permission to reproduce, in whole or in part, certain figures and diagrams we are grateful to the following publishers—

Blackwell Scientific Publications; Bradley and Son, Ltd; Cambridge University Press; H. K. Lewis & Co., Ltd; Macmillan and Company, Ltd; Maskinaktiebolaget Karlebo; Oliver & Boyd Ltd; Royal Microscopical Society; University of Texas; U.S. National Academy of Sciences; Verlag S. Karger A. G.; Weed Society of America; John Wiley & Sons, Inc; Williams and Wilkins Company; Year Book Publishers.

Detailed acknowledgments are given in the legends to figures.

PREFACE

The rapid expansion of microbiology, both as a separate discipline in itself, and as a means of providing research material in allied fields, such as those of biochemistry and genetics, during the past twenty years has led to a proliferation of technical methods. These appear in scientific literature but are nowhere collected together to form a comprehensive reference source for the laboratory scientist. "Methods in Microbiology" is an attempt to fill this gap in the contemporary literature. Its intent is to describe research techniques in microbiology at the practical level, and it is aimed at the advanced student and the research worker who studies, or uses, micro-organisms. The field covered is a large one and when we first conceived the idea of editing such a manual, some two-and-a-half years ago, it was immediately obvious to us that we would need to call on the services of a considerable number of our colleagues specializing in the various branches of microbiology. Our first, tentative, approaches to potential contributors produced an enthusiastic response and we quickly decided to go ahead with the project. The scope of the manual has increased progressively and the final publications will contain the contributions of some 150 authors. We soon realized that we were, by the nature of our backgrounds and experience, not competent to deal with the mycological aspects of microbiology and it was with great pleasure that we were able to enlist the help of Dr. Colin Booth of the Commonwealth Mycological Institute to edit that section of the work.

This kind of publication, involving the participation of a large number of contributors must, inevitably, lack the structural homogeneity of a work produced by a small number of authors. We have attempted to maintain as uniform a style as possible and much of our work as editors has been concerned with the prevention of duplication and omission where the contributions of different authors have approached one another in content. Beyond this we have imposed little restriction, taking the view that the specialist in a particular field is the best person to judge what should be presented in the way of practical information and the best way in which the details should be expressed. Our main theme has been to provide sufficient practical detail to ensure that sorties into the original literature may be avoided. Within this framework, however, we have avoided areas which are already well documented. For example, microbiological assay is very well treated by F. Kavanagh in his book "Analytical Microbiology", and in this area we have dealt only with a few selected topics. Similarly, we have not dealt with the methodology of applied aspects of microbiology, partly because the methods employed in the applied areas of the subject are usually standard techniques extensively documented and, partly, because these areas are gradually being covered at a practical level, in the Techniques Series of the Society for Applied Bacteriology. Their inclusion in "Methods in Microbiology" would necessarily mean extensive, pointless duplication. In spite of this limitation of scope, however, it is obvious that there must be a certain amount of overlap between "Methods in Microbiology" and other publications such as "Methods in Enzymology", "The Bacteria", "Methods in Virology" and "Methods in Immunology and Immunochemistry". Where our coverage borders on that of such publications we have tried to present methods and approaches which are important for the microbiologist and so, although our selection of topics has to a certain extent been arbitrary, we believe that we have selected areas which are particularly relevant to the readers of our publication.

Although the contributions are intended primarily to be of a practical nature, it is our belief that the publication of a "cookery book" would be an inadequate approach to research technology at this level and we have encouraged authors to provide a certain amount of theoretical background as a context for their technical detail. Furthermore, some of the contributions are of a general nature and are intended primarily for the orientation of a new worker entering the field for the first time. Although each Volume has its own theme we recognize that it is impossible to compartmentalize a subject of this type and we have tried, by the use of extensive cross-references, to simplify the location of information in this Series as much as possible. It has been a great source of pleasure to us that the vast majority of contributions have been presented on, or before, the date for which they were commissioned. We have considered that we were not justified in holding up the publication of any portion of the work because of the late appearance of a small number of manuscripts. The subsequent inclusion of these manuscripts has meant a certain rearrangement of later Volumes but it has meant that we have been able to adhere reasonably closely to our original publication schedule.

The willingness with which authors have responded to requests for articles and the pleasant way in which they have co-operated with our various, and at times naive, suggestions for improving their manuscripts has made our task a most pleasant one. We would like to take this opportunity of thanking all our contributors for their co-operation. We would also like to thank the publishers for their meticulous and prompt handling of the manuscripts and proofs which has done much to lighten our task. We are also extremely grateful to those of our colleagues in the microbiological and biochemical fields who have freely given their time to read and comment on various manuscripts, the contents of which were unfamiliar to us. Among these we would like particularly to thank Dr. Ella M. Barnes, Dr. R. P. M. Bond, Dr. G. Hamer, Dr. D. J. D. Hockenhull, Dr. M. Knight, Professor S. J. Pirt, Professor J. R. Postgate, Professor R. Y. Stanier, Dr. D. W. Tempest and Mr A. H. Walters. We are also grateful to Shell Research Limited, the University of Miami and the Howard Hughes Medical Institute for the generous provision of secretarial assistance without which our work as editors would have been much more difficult.

We hope that "Methods in Microbiology" will provide a useful source of reference for the research worker today, and a valuable stimulus for future developments in the techniques by which micro-organisms are handled and examined.

J. R. Norris

D. W. Ribbons

April 1969

CONTENTS

CHAPTER I

Sources, Handling and Storage of Media and Equipment

E. C. ELLIOTT AND D. L. GEORGALA

Unilever Research Laboratory, Colworth House, Sharnbrook, Beds., England

I. LAYOUT OF MEDIA SERVICE AREA

A. General considerations

The layout of a new service area requires careful planning if the unit is to function smoothly. Services and equipment should be sited so that a continuous flow of glassware and media is obtained with the minimum of handling.

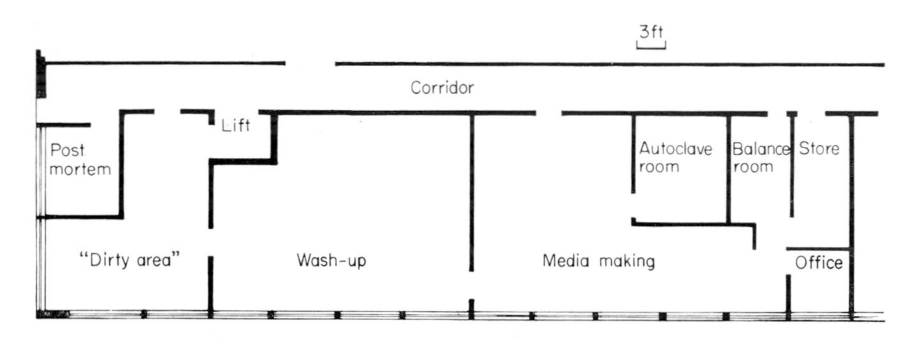

FIG. 1. Plan of total floor area.

A special unit was designed for this purpose when the Colworth bacteriology building was constructed. A plan of the service area is shown in Fig. 1, and the unit is described in detail in this Chapter. Forty-six microbiologists and research assistants are supplied from this service unit. The services include washing of all equipment, glassware, etc., used in the laboratories, and provision of media, sterile glassware and special equipment.

A large variety of culture media is prepared by this service unit, either from basic ingredients or from commercial dehydrated products. The research in this laboratory encompasses both long- and short-term investigations, and work is always in progress with food-spoiling micro-organisms (e.g., *Pseudomonas*, *Microbacterium*), food-poisoning bacteria (*Salmonella*, *Staphylococcus aureus* and *Clostridium botulinum*), sporing bacilli, and others, such as streptococci and lactobacilli.

B. Dirty area

If the laboratories have more than one floor it is a great advantage to have a lift to serve the dirty area. Contaminated equipment or glassware can be transferred in trays, baskets or buckets via the lift direct to the dirty area.

The dirty area, shown in Fig. 2, is fitted with floor drains so that the whole area can be hosed down with water. Electrical fittings and sockets are water proof.

FIG. 2. View shows the dirty area leading through to the wash-up room. The two autoclaves are shown built into the partition wall and the drying ovens can be seen at the rear of the wash up room.

Contaminated equipment and glassware is loaded direct from trolleys into either of the two double-door autoclaves (Fig. 2). After autoclaving, the sterile equipment is unloaded from the autoclaves from the other doors situated in the wash-up room (Fig. 3).

C. Wash-up room

Figure 3 shows the wash-up room, which like the dirty area is fitted with floor drains. Large extraction hoods over the sinks also supply warm air for winter heating. The area was designed for a wash-up staff of four.

The centre island consists of two identical suites of four stainless-steel sinks. A similar suite is shown on the left adjoining two porcelain "acid" sinks. The remaining sinks and equipment on the right were designed for pipette washing. De-mineralized water is supplied via the overhead Pyrex glass lines to one sink in each suite. Hot and cold soft water is supplied via a flush side vent to the other sinks. This flush feed enables large baskets to be used in the sinks without fouling taps.

FIG. 3. Wash-up room showing: 1, piped de-ionized water supply; 2, auxillary de-ionized water supply; 3, bottle-brushing machine, 4, pipette washers; 5, foot-operated water jet; 6, extraction hoods.

The sinks nearest to the autoclave in each suite are fitted with a steam heater so that the sink can be used for boiling equipment in detergent solutions.

A foot-operated jet, as shown in Fig. 3, is fitted to the next sink. Sufficient flat bench space has been left at this spot to accommodate a mechanical bottle-brushing machine (Fig. 3). This is used in conjunction with the water pressure jet.

A syphoning device is built into the third sink, so that a basket of bottles can be automatically rinsed for an indefinite time.

The fourth sink as previously mentioned is supplied with de-mineralized water for the final rinse before drying.

A low-pressure air tap (40 p.s.i.) is sited behind the pipette sink, adjacent to which are sited two automatic pipette-rinsing machines. The compressed air is used to blow wool cotton plugs out of pipettes.

D. Media-preparation room

Except for a swing door, the whole of the wall between the wash-up room and the media-preparation room consists of a built-in drying oven (Fig. 2). The oven is steam-heated and has doors on both sides. The wet glassware is loaded into the oven from the wash-up room and unloaded directly into the media-making room where it is going to be used. Although the capacity may appear to be large (175 cu. ft), in our experience anything less than this can produce a bottleneck in the system. If the full capacity is not required for short periods, it is very useful for storing clean glassware.

The media-making room was designed to accommodate 3–4 people. It consists of a centre bench, a dispensing bench plus media-making machine,

FIG. 4. Media-making room and autoclave room, showing: 1, the dual autoclave control panel; and 2, the Bikini recording control instrument.

Fig. 5. The media preparation room and supervisor's office.

a hooded bench for hot-air ovens, and storage cupboards. Two views of it can be seen in Figs. 4 and 5.

Figure 6 shows the media-making machine plus dispensing bench and filamatic dispensing machine. The media-making machine has been described elsewhere (Elliott and Hurst, 1964). It has now been in use for 5 years in this laboratory with complete success.

E. Autoclave room

The autoclave room as shown in Fig. 4 leads directly off the preparation room. It is equipped with a separate extract system and is vented straight to atmosphere. With the control panel for the two autoclaves sited in the media-making room, the heat generated does not unduly effect the temperature of the media room, provided the sliding door leading to the autoclave room is kept closed.

FIG. 6. Media-making unit showing: 1, stainless-steel boiling vessel; 2, filtration chamber; 3, filamatic dispenser.

F. Office and storage space

To complete the service area, the supervisor's office, shown in Fig. 5, and storage cupboards for prepared media are sited at the end of the media-making room. This enables the supervisor to control the whole operation from dirty glassware to completed media by maintaining contact easily with all his staff.

II. STAFF REQUIREMENTS

It is necessary to have an experienced technician in overall charge of a media-supply centre. He or she should have had a wide experience of bacteriology, be capable of controlling staff and mature enough to "handle" requests and complaints from senior graduates.

The practice of employing young technicians, direct from college, for making media, has much to recommend it. They do acquire useful basic knowledge of sterilization techniques and formulation of culture media, before practising bacteriology in the laboratories. A disadvantage is that it is difficult to maintain their interest for much longer than three months. At this stage in their career it is important to maintain their initial enthusiasm for bacteriology. In practice they are generally transferred to a laboratory at an early stage. This inevitably leads to a rapid turnover of staff in the media room. Good consistent batches of media cannot be produced by a "floating staff".

We have found over several years that to maintain continuity of staff, successful results are obtained by training older persons who do not intend taking qualifications. They are more interested in media making, accept responsibility more readily than junior technicians, and their liaison with the wash-up staff, an important factor, is better than the newcomers.

New laboratory technicians should still have some experience in a media-preparation room, but they should be looked upon as supernumary rather than as part of the permanent staff.

The type of individual employed strictly on washing-up is largely governed by the area in which the laboratory is located. It is difficult to find men or women who are prepared to do this type of work for several years. A good "washer-up" is someone to be nurtured and treasured.

III. EQUIPMENT FOR MEDIA-PREPARATION ROOM AND HANDLING

A. Equipment required for washing-up

Large baskets, made to fit easily into the wash-up sinks are required in quantity. They should be fitted with a clip-on wire-mesh lid, to retain the contents when rinsing or boiling. In our experience, although the initial cost is high, it is worthwhile having these baskets made of monel metal wire. They will withstand boiling in detergents, and autoclaving for years, without showing any signs of wear or rust.

Storage trays made in sheet aluminium and designed to fit the storage cupboards are also required in large numbers. In practice, trays holding about 150 $\frac{1}{4}$-oz, 100 1-oz, and 50 6-oz bottles, seem to be a convenient size for handling. A few holes should be drilled in the base of the tray to allow condensation water to escape when the trays are autoclaved.

The pipette washers shown in Fig. 3 are supplied by Laboratory Thermal Equipment Ltd, Greenfield, Nr Oldham. They are essentially a continuous-syphoning device with a removable stainless-steel inner basket. Large numbers of pipettes can be rinsed with water with the minimum of effort.

An ideal pipette washer would be one that automatically gives three rinses, hot detergent, tap water, and distilled water. We have not as yet seen a washer designed to do this.

An automatic bottle-brushing machine is an essential piece of equipment. Numerous models are available that can be fitted with a different brush for each size of bottle.

There are several washing machines on the market now, designed specifically for laboratory glassware. When designing a new laboratory they must be considered as an alternative to the large number of sinks shown in Fig. 3. Installation of washing machines should go hand in hand with a careful assessment of staffing requirements.

B. Equipment required for sterilization

Most autoclaves on the market have been designed specifically for hospital use. To ensure adequate sterilization, the probe controlling the temperature of the autoclave is normally sited in the chamber. Providing the autoclave is functioning correctly the contents must have reached the temperature of the probe, but they could be hotter.

In sterilizing culture medium, it is important that the medium is sterilized but not overheated. To avoid this, the tip of the probe should just enter the bottom of the chamber. An even better control is to fit a "wandering" probe inside the chamber. This probe can be immersed in any size of bottle with different loads. A calibration chart can then be made for different volumes of medium.

It is worth noting here, that the time taken for a large autoclave to get up to temperature, and come down to a safe temperature for unloading bottles of medium, is longer than that taken with a small autoclave. Thus, although a large autoclave may be required for routine use, an additional small autoclave is very useful for heat-sensitive media.

The choice of autoclaves is largely dictated by the amount of money available. One can have a fully automatic system fitted with sliding doors, and recorders giving a visual record of the process. At the other end of the scale is the manually operated autoclave fitted with a pressure–temperature gauge.

Whatever type of autoclave is chosen, we consider that the most essential accessory is a device to control and record the temperature and time of the process. Equipment that does this and can be fitted to any autoclave is supplied by Fielden Electronics Ltd, Paston Road, Wythenshawe, Manchester.

A steamer of some sort is required for melting agar, maintaining medium hot, and for sterilizing at 100°C. The Koch-type of steamer heated by gas, electricity or steam is available from commercial suppliers. Unfortunately, they are very slow in heating up and require regular topping up with water. These disadvantages are readily overcome by fitting a constant-head device on the water supply and a timer on the heater so that it is switched on about an hour before it is required.

If mains steam is available, a much more useful piece of equipment is a simple copper cabinet, with front opening doors. Steam at 5–15 p.s.i. is fed direct to the chamber. Excess steam can be re-circulated or vented to atmosphere. The cabinet in our laboratory measures $4' \times 3' \times 2'\ 6''$.

For dry sterilization, an oven fitted with a circulating fan and capable of being maintained at 180°C is required. If the oven is fitted with a 24 h timing control clock, all dry sterilization can be done overnight.

C. Equipment required for media preparation

Containers can be made of either stainless steel or heat-proof glass. Handling large amounts of molten agar in glass containers can be quite dangerous when breakages occur. Stainless-steel containers are therefore recommended despite the higher initial cost. A range of graduated jugs with lids, capacity 2, 4 and 6 litres, and cylindrical containers with lids, capacity 9 and 13 litres, cover most demands. These can be obtained from the Taw Manufacturing Co. Ltd, Campsbourne Works, High Street, Hornsey, London N.8.

Graduated measuring cylinders in polypropylene complete the measuring equipment.

Large Buchner funnels in either glazed porcelain or stainless steel are used for filtering media through paper pulp.

A Lovibond comparator or indicator solutions can be used for controlling pH values. However, a pH meter is preferred for accurate control. The pH meter should be fitted with one of the combined electrodes with reinforced ends, such as the type used for soil testing (Pye 401/EO7) supplied by W. G. Pye and Co. Ltd, York Street, Cambridge. The advantage of this type of electrode in a media room, is that the pH of agar media can be measured after they have been allowed to solidify. Molten agar inevitably blocks an electrode.

Dispensing large quantities of medium accurately is an important part of media preparation. There are several hand dispensers on the market, but they are not completely satisfactory for handling large quantities of medium.

The dispenser we have used for some years is a Filamatic Vial Filler Model DAB6 supplied by National Instrument Co. Ltd, Baltimore 15, Md. U.S.A. This electronically controlled dispenser can be fitted with two stainless-steel syringes of desired capacity. The syringes are set at the required volume to be dispensed with a vernier fitting and the speed of delivery can be adjusted from 0 to 32 deliveries/min. With all types of media, including agars, the cut off between deliveries is positive, and the amounts dispensed are accurate to $\pm 1\%$. If desired, the syringes can be autoclaved and sterile solutions dispensed aseptically. The filler is seen being used in Fig. 6.

For most purposes screw-capped bottles are more convenient than tubes for both use and storage of culture medium. These are available in various sizes in either a round or flat shape. United Glass Bottle Co., Kingston Road, Staines, Middlesex, supply them specially cleaned for immediate use without any further treatment.

The sizes we find most useful are the $\frac{1}{4}$ oz, 1 oz (round) and the 4 oz, 6 oz, 20 oz flat bottles. As the 1 oz bottle is supplied with either a narrow neck (McCartney bottle) or a wide neck (Universal bottle) it is as well to decide from the beginning which to standardize on. The Universal bottle is more easily cleaned, but the cap is more difficult to manipulate when inoculating media.

Cleaning screw caps is time consuming and the rubber wads are difficult to clean. We buy the complete caps in bulk from United Glass Bottle Co. and use them as disposables, which is both economical and certainly safer to do than re-using cleaned caps. As some rubbers contain inhibitory substances to bacteria, it is essential to specify bacteriological-grade rubber wads when ordering.

Glass tubes are still required for some of the special media, e.g., where columns of media are required, and for Durham tubes. Heat resistant, rimless hard-glass tubes of bacteriological standard should be specified. The minimum selection one requires in a media preparation room should include sizes $2 \times \frac{1}{4}$ in, $2 \times \frac{3}{8}$ in, $3 \times \frac{3}{8}$ in., $3 \times \frac{1}{2}$ in., $6 \times \frac{1}{2}$ in and $6 \times \frac{5}{8}$ in.

These can be plugged with non-absorbent cotton wool or one of the numerous test-tube caps made of metal or polypropylene. In our experience, with one exception, the contamination rate obtained with these caps is no greater than that obtained with cotton-wool plugs. The exception is the large cap required for a 250 ml conical flask.

The medium-making unit previously mentioned speeds up production, reduces fatigue and risk. With this unit, which has been in use for 5 years in this laboratory, large or small quantities of medium can be produced both safely and economically.

IV. PROCEDURES FOR CLEANING GLASSWARE

A. Choice of detergent

The choice of detergents is almost unlimited, but whichever one is chosen should satisfy the following criteria—

(a) From a hardness point of view it must soften the local water supply.
(b) It must be capable of removing organic soil quickly at a temperature of 60°C.
(c) After normal rinsing, the glassware should be neutral.
(d) No antimicrobial factors should be left on glassware after normal rinsing. Before using a new detergent, a bacteriological test should be carried out to confirm this. A standard inoculum of about 100 cells of a suitable test organism is added to 10 ml of liquid culture medium in three treated bottles and three untreated bottles. Viable counts done on the six bottles after overnight incubation should be approximately the same.

B. Treatment of new glassware

Any free alkali present in new glassware must be removed. It is good practice with all new glassware, with the exception of bottles that have been treated by the manufacturer, to neutralize by soaking in 1% HCl solution overnight. Neutralization can be checked by filling the containers with neutral water, autoclaving and checking the pH of the water after autoclaving.

With few exceptions, new glassware is contaminated with spores from packing material. Autoclaving at 121°C for 20 min should render them sterile.

C. Cleaning dirty bottles, flasks, tubes, etc.

It is not possible to define a washing routine that is fully effective in all situations. Using the wash-up facilities described previously, the following routine gives a high standard of clean glassware.

(a) Soon after autoclaving, while the contents such as agar are still fluid, the bottles or flasks are drained to waste. The containers are then stacked in the wire baskets designed to fit the sinks.
(b) The basket and its contents are placed in the boiling sink with a suitable detergent and boiled for 20–30 min.
(c) The bottles are drained by inverting the basket. Each bottle is then individually rinsed in the foot-operated jet, brushed on the bottle brushing machine, re-rinsed on the foot jet and stacked into another basket.
(d) The basket is transferred to the rinsing sink and the bottles are allowed to rinse until the next basket is ready.

(e) After draining by inverting, the basket is transferred to the next sink containing de-mineralized water. The bottles are rinsed in the de-ionized water before the basket and its contents are put in the drying oven.

(f) After drying, the basket of clean glassware is unloaded from the oven direct into the media-preparation room. All glassware is inspected at this stage and dirty, cracked or chipped glassware is rejected. The bottles are stacked into the aluminium dispensing trays ready to receive media.

D. Cleaning pipettes

1. *Pipettes less than* 1 *ml in volume*

As it is unlikely that many small pipettes are used these are better cleaned individually. After rinsing with water, or a suitable solvent for the material they have been used with, they should be rinsed in distilled water and finally in acetone. This routine can be done very rapidly using a water-suction pump where the rinsing fluid is diluted and goes direct to waste.

2. *Larger size pipettes*

These are often blocked with solidified agar or cotton-wool plugs. They require some treatment before the normal pipette washer can be used.

When received in the wash-up room, from the laboratory, these pipettes have generally been standing in a polythene container with a suitable disinfectant solution overnight. The choice of disinfectant will depend on the organisms being used. For most organisms a commercial hypochlorite solution (e.g., Chloros) used at a concentration of 300 p.p.m. of free chlorine, is satisfactory. Pipettes that have been contaminated with dangerous pathogens should be autoclaved before washing. After draining, the cotton-wool plugs are blown out with compressed air. The pipettes are then boiled in a suitable detergent solution before being transferred to the pipette washer. A final rinse in distilled water and the pipettes can then be dried off in the drying oven.

After drying the pipettes are plugged with cotton wool, and dry-sterilized in copper pipette canisters.

V. SOURCES OF CULTURE MEDIUM

A. General considerations

Culture media can be either obtained from commercial suppliers or made in the laboratory from basic ingredients. Whether to use one of these sources or a combination of the two depends on several factors. The strictly routine laboratory may be able to get along with media from commercial suppliers, but most laboratories will probably find that they still need to make some media from basic ingredients. In larger laboratories, the type of media and components to be used will be decided, following close consultation between research staff and the media-preparation supervisor.

The following are some of the factors that should be considered.

B. Commercial media

Depending on the supplier and the type of medium, this is available either as a powder, granules or tablets. Preparation normally involves soaking in cold water and dissolving by heating, either before or during sterilization. Medium that will not withstand further sterilization is supplied as poured plates, slopes, or sterilized and ready to pour.

The advantages of using commercially prepared dehydrated media are fairly obvious. The preparation and testing of the medium is such that the final product as received should give consistent results. We particularly emphasize, as received, because these products can be abused by excessive heat, inadequate soaking before heating, and possible deterioration on storage.

In laboratories that carry out standard routine tests for control purposes, consistency requirements may be sufficient to justify using commercial dehydrated media.

Another advantage that the commercial suppliers claim is the saving in time and equipment involved in making media in the laboratory. This point is much more difficult to evaluate and depends on the amount and type of medium used. If small amounts of media are required it is much more convenient to rely on commercial sources. In this laboratory, where we use 200 litres of a particular medium per month, the work and equipment involved in dissolving and dispensing a commercial product, is very little less than that required to make the medium from basic ingredients.

C. Laboratory-prepared media

Even when commercial media are in general use, there is always some medium that has to be made up. It may be one which is not available from commercial sources, or a "pet" formula of the laboratory concerned, or a new formula just published.

D. Conclusions

It can be concluded therefore that most larger laboratories must have the equipment and the staff (perhaps limited) with the expertise for making and testing media. Such laboratories will usually use a mixture of "home-produced" and commercially dehydrated media. The commercial suppliers of media do of course accept this situation, and supply all the basic ingredients required for making culture medium.

Commercial dehydrated media have been included in various inter-laboratory test programmes. In the future, increasing international standardization of bacteriological techniques could lead to very widespread acceptance of particular commercial media.

The main commercial suppliers listed below issue free manuals containing a wealth of information on formulae and purity of basic ingredients, together with methods for growing and isolating organisms—

(a) Oxoid Ltd, Southwark Bridge Road, London S.E.1.
(b) The Difco Laboratories, Detroit 1, Michigan, U.S.A.: British Agents, Baird & Tatlock (London) Ltd, Freshwater Road, Chadwell Heath, Essex.
(c) Baltimore Biological Laboratory Inc., Baltimore 18, Maryland, U.S.A.

VI. MEDIA PREPARATION

A. Coding of media

Various systems of colour-labelling or coding media have been used in the past. Coloured cotton wool, cellulose paint on bottles or caps, and coloured beads have been used.

We prefer to use a range of coloured beads, or a combination of different sized coloured beads. Use of beads allows permanent identification of a medium from the time it is dispensed until it is discarded. One must test that any dye that could be released from the beads during autoclaving is not toxic to micro-organisms. Rocaille beads, supplied by Ellis and Farrier Ltd, 5 Princes Street, Hanover Square, London W.1, are used in our laboratory and have proved to be non-toxic.

B. Recorded details of media

A complete record should be prepared for each batch of medium as it is being produced. It should contain details of the manufacturer's batch number of all the ingredients used, the steps of preparation, e.g., pH adjustment, filtration, and the duration and temperature of sterilization. This information can easily be recorded on the front of an 8×5 in. card. The results of the tests done after sterilization, can be entered on the reverse of the card.

C. Treatment of ingredients

Very often the formula used dictates the method of preparation. Only a few general points of technique will be made here.

Ingredients are always dissolved in either de-mineralized or distilled water. De-mineralized water seems to be quite satisfactory for most media, but there is some evidence to suggest that with media for clostridia distilled water gives better growth of some of the more fastidious organisms.

When dissolving ingredients, as a general principle it is better to dissolve one completely, before adding the next ingredient. Where trace quantities of chemicals have to be added these are best incorporated as concentrated solutions.

A good quality agar does not alter the pH of a medium. It is therefore more convenient to add the agar after the other ingredients have been dissolved and any adjustment of pH done. To decide when agar has been completely dissolved, dip a glass rod into the medium. On withdrawing it, undissolved globules of agar are easily seen on the rod.

The minimum quantity of heat should be used at all times in producing a culture medium. Re-sterilization, or holding a medium for hours at highish temperatures all tend to reduce its growth-supporting potential.

Complete dehydrated media from commercial manufacturers must always be soaked in cold water for at least 15 min before being heated or sterilized. This time is necessary to allow complete rehydration of the product before mixing, dispensing, and sterilizing.

D. Filtration of media

Small quantities of medium can be filtered through filter papers, but for quantities above 500 ml filter pulp is recommended. The type of pulp employed is that used by breweries for beer filtration, and is made of pure cotton. It is supplied in sheets (Baird & Tatlock (London) Ltd, Freshwater Road, Chadwell Heath, Essex) and is prepared for use by soaking in cold distilled water.

The thickness of filter pulp required will depend on the volume to be filtered. A thickness of $\frac{1}{2}$–1 in. will cope easily with 25 litres. The wet pulp is poured into a Buchner funnel, or in the case of the media-making unit, the filtration chamber. Negative pressure is applied to remove the excess water. Using a flat surface, such as the base of a polypropylene measuring cylinder, the pulp is packed flat. The pulp is now covered with a Whatman No. 54 filter paper to prevent the pulp breaking up when the medium is poured onto it.

This type of filter will rapidly clear all types of culture media. The small amount of water remaining in the pulp is probably not of any importance, but if desired the first few ml through can be run to waste. The usual filtration technique should of course be observed, i.e., use the minimum negative pressure required until the filter material shows signs of blocking.

With some media it is necessary to add small quantities of solutions, such as carbohydrates, to the media after bottling and sterilization. These solutions are conveniently sterilized by filtration through either a Sietz or membrane filter.

E. Adjustment of pH of culture media

With some media, precipitation of phosphates occurs on cooling after autoclaving. Some bacteriologists prefer to leave these phosphates in the medium, but as they can very easily mask light growth of organisms, they are more often removed before autoclaving. This is done by adjusting the reaction of the completed medium to a pH of 8·0 before filtration. After filtration the pH is re-adjusted to the desired figure.

Most laboratories will have a pH meter available, which may or may not have a temperature-compensating control. Other laboratories will use a Lovibond comparator. When testing media not containing agar it is generally more convenient to cool the sample before testing. With agar media, this can also be done if one of the combined electrodes previously mentioned is used. With the Lovibond comparator, the indicator can be mixed through the molten agar and read after the agar has set. It is the pH of the medium at room temperature that is of importance.

The pH of a medium very often falls after autoclaving. This depends on the formulation of the medium and can only be ascertained by trial or previous experience. The pH of most media falls by about 0·2 units during autoclaving. This must be allowed for when making the final pH adjustment after filtration.

The simplest method for adjusting the pH of a medium is by the addition of M HCl or M NaOH. However, the formulation of the medium dictates what can be used, and is normally included in the recipe. Whatever is used to make the final adjustment, mix well after its addition, check the pH, mix again and recheck the pH.

F. Dispensing culture medium

Medium should be dispensed and autoclaved as soon as possible after production. If the coloured-bead system of coding is used, the beads can be put into the bottles in advance. Any other system of coding should be done immediately after dispensing. The trays containing the bottles, or baskets if tubes are used, should be labelled with the batch number before dispensing, using an autoclave-proof ink.

The method of dispensing will be governed largely by the amount to be dispensed. If the quantities are large, a machine, such as the Filamatic previously mentioned, is almost essential. Whatever equipment is used, immediately after use it should always be given a thorough rinsing in hot water when agar has been dispensed. Trace amounts of inhibitory factors or carbohydrates are almost impossible to detect in a subsequent medium dispensed with the same equipment.

When agar medium is dispensed, its temperature should be maintained at about 70°C. At this temperature, particularly when using automatic dispensers, the agar remains fluid long enough to allow time for adequate cleaning of valves, etc.

Bottles should be capped if possible as the medium is being dispensed. With bottles up to and including 1 oz, the caps should be screwed finger tight. Above this size they should be left loose until after autoclaving. These larger bottles may burst if autoclaved with tightened caps, and the bursting can be particularly dangerous if it occurs while the bottles are being removed from the autoclave. Before autoclaving, autoclave indicator tape No. 1222 should be attached to each tray, to give a visual indication that it has been autoclaved. The tape is supplied by 3M Company, 3M House, Wigmore Street, London W.1.

G. Sterilization of culture medium

As already stated, all culture media are heat labile to some extent, but some are more so than others. When sterilizing medium by heat, one has to reach a compromise between complete destruction of spores, and possible damage to the growth-supporting potential of the medium.

Most types of spores are destroyed when autoclaved for 15 min at 121°C. If a culture medium will withstand this treatment it is the method of choice. If the medium is heat sensitive, then a reduced temperature must be applied for an extended period. Examples of this are 115°C for 20 min or 109°C for 30 min.

Some media cannot be autoclaved; these are steamed in free steam for 20 min on three successive days. In this process, called Tyndallization, the spores produce vegetative forms that are killed off with the succeeding days' treatment.

Some media containing serum or egg fluid are sterilized by heating to 80°–85°C for 2–4 hours, depending upon the likely level of contamination of the serum or egg fluid.

For fuller details of the different sterilization methods see Sykes (this Volume p. 77.

VII. MEDIA TESTING

A. Principles of media testing

In designing suitable test procedures, the limits of these tests and the effort one is prepared to allocate must be clearly defined. To test fully all aspects of every medium produced could conceivably employ most of the staff in any laboratory!

Two simple checks recommended for every batch of medium is to check pH after autoclaving, and sterility at different incubation temperatures.

Broadly speaking, each medium is designed for a specific purpose. It either supports growth of a range of organisms or of a specific organism. It could inhibit a particular organism or it could contain a specific carbohydrate for a fermentation test. The tests performed should confirm that the medium does specifically what it is designed to do. They should also detect any undue variation between successive batches of medium produced.

B. Recommended procedures for testing media

It is virtually impossible to give procedures to cover all media, but the examples chosen, nutrient broth and nutrient agar will serve as models.

1. *Nutrient broth*

This would be expected to support growth of many different types of organisms. One should choose three organisms, representative of the types expected to be isolated, from the specimens handled in the laboratory concerned.

After sterilization, nine bottles of medium should be chosen at random from the trays as they are being removed from the autoclave. They should be chosen from different parts of the trays, and from trays at different sites of the autoclave. The bottles should be allowed to cool to room temperature before being tested.

Two bottles are incubated for 72 h at room temperature, 30° and 37°C. If no growth is visible after incubation, wet preparations are made from them and examined microscopically. A negative result on all six bottles indicates effective sterilization. A check pH measurement is done on three other bottles.

Suspensions of 18 h cultures of the three test organisms are made in 0·1% peptone water. Three concentrations of each organism are required, approximately 5×10, 5×10^2 and 5×10^3 organisms/ml. One millilitre of the 5×10 suspension of each culture is added to 9 ml of the nutrient broth. The three bottles are incubated at the optimum temperature for 18 h, before a viable count is made by using the Miles–Misra technique (Miles and Misra, 1938). A control broth of a previous batch is treated in the same way. The results obtained will show whether the new medium is capable of supporting growth of small inoculums and also how it compares with the previous batches.

2. *Nutrient agar*

For the sterility tests the agar is melted and poured into Petri dishes, then incubated for 72 h and examined. pH tests are done using the combined electrode.

The growth-supporting potential is checked by adding 1 ml of the 5×10^2 suspensions of each organism to 19 ml of 0·1% peptone water. This is repeated with the 5×10^3 suspensions and duplicate pour plates are made. After 72 h, the colonies are counted and compared with a control agar.

This sort of test procedure can be modified to suit any medium. If freeze-drying facilities are available, it is good practice to maintain the test organisms as freeze-dried cultures.

VIII. STORAGE OF CULTURE MEDIUM

It is difficult to lay down a maximum storage life for prepared medium, and the onus for doing this must rest on the bacteriologist who is going to use it.

The media-making supervisor must ensure that media are stored under the correct conditions and are routinely examined at fixed intervals. The ideal temperature for storing media is 2°–4°C, but very few laboratories would have the cold space available to store all their culture media and ingredients at this temperature.

Inspissated medium, sera and unsterilized medium, such as Rogosa's agar, are best stored in the cold room. Cupboards with doors to exclude light, situated in a coolish room, provide the best alternative accommodation.

Monthly checks on the visual appearance of all batches of medium should be made. With liquid medium, contamination, fading of indicators, or precipitation, may be a warning sign that the batch as a whole needs re-examining. In the case of agar medium, particularly slopes, any sign of drying indicates that the batch needs replacing.

Six months is probably about the maximum safe shelf life of most media stored at room temperature. If media over this age are going to be used they should be fully re-tested.

REFERENCES

Elliott, E. C., and Hurst, A. (1964). *J. appl. Bact.*, **27**, 134.
Miles, A. A., and Misra, S. S. (1938). *J. Hyg., Camb.*, **38**, 732.

CHAPTER II

Properties of Materials Suitable for the Cultivation and Handling of Micro-organisms

R. BROOKES*

Bacteriological Bioengineering Research Group, Department of Bacteriology, Karolinska Institutet, Stockholm 60, Sweden

I. INTRODUCTION

In general, microbiological work on the laboratory scale presents few problems as far as the choice of materials for equipment is concerned.

* Present address: Imperial Chemical Industries Ltd., Mond Division, Runcorn, Cheshire.

On a larger, industrial scale, economics becomes the primary factor and the best choice is dependent on capital costs of the equipment, possible corrosion effects of the process, depreciation and a host of other factors, the interaction of which can be estimated. An evaluation then makes possible a rational choice of materials giving the lowest cost. In the laboratory these factors are of lesser importance compared with other factors, connected with the biological behaviour of micro-organisms.

Basic parameters restricting the choice of materials

Certain factors, inherent in the nature of laboratory microbiology, govern and restrict the choice of materials, except in isolated circumstances, to a narrow range.

1. *Range of micro-organisms*

In most laboratories a number of micro-organisms are cultivated in a variety of media, under a wide range of conditions, either simultaneously or in rapid succession. This applies to teaching, routine control or research laboratories. A cheap steel used in the construction of a fermenter for the non-sterile production of a yeast could be entirely useless for the growth of another fastidious micro-organism, sensitive to trace metal contamination. Thus it is a false economy to choose materials for specialized equipment suitable only for a few micro-organisms.

2. *Low absolute materials costs*

Because of the small physical size of laboratory equipment, the capital costs of the materials used in the construction of equipment are usually of low significance in comparison with the machining costs and other laboratory costs.

As a result of these two constraints, the choice of materials for most routine work becomes limited to a handful, of which glass, certain plastics and stainless steel predominate. The plastics are an interesting example of increasing application. Most industrialized countries have experienced a steady rise in the cost of unskilled labour for the washing and autoclaving of equipment. Simultaneously, the costs of simple plastic items has fallen dramatically over the same period of time, so that the consumption of disposable plastic equipment for most routine handling operations is rapidly increasing.

It is not only the physical properties of a material that are the limiting factors governing its use for a particular purpose. This is often true, but cases do arise where chemical or surface properties interfere with the microbial system under study. Such factors are also of importance when a process is being prepared for transfer from the laboratory scale to production. Possible interference of the fermenter material with the process is one of the factors used in the evaluation of the most economic choice of construction material.

II. GLASS

Glass is the most widely used material in laboratory-scale equipment for both chemical and microbiological work. Until recent years it was almost exclusively used for this purpose but gradually other materials, such as plastics and metals, have taken an ever increasing proportion of the expanding production of equipment. A summary of the properties of glass illustrate the background to its popularity.

Glass owes most of its properties to the fact that it does not exhibit crystallinity (Charles, 1967). When most liquids are cooled, the transition from the liquid to the solid state occurs abruptly at a specific temperature, heat being evolved simultaneously. The heat output is an expression of the ordering of the atoms or molecules as they take up their positions in the crystal. If impurities are present the extra degree of ordering necessary to form crystals results in a depression of the freezing point. However, this freezing process is dependent on some form of seed, which may emanate from molecules attached to foreign particles or to irregularities in the surface of the vessel. Alternatively, small numbers of atoms aggregate spontaneously to form a crystal nucleus having a high surface energy when the temperature falls below the thermodynamic freezing temperature, i.e., when the liquid becomes supercooled.

The ease with which crystal nuclei can be formed on supercooling depends on the nature of the molecules, and since most common liquids contain atoms or molecules of roughly spherical shape, their sliding movement into an ordered crystalline configuration is easily accomplished. Certain liquids do, however, become viscous as the freezing point is approached, the formation of the nuclei is therefore hindered so that the molecular configuration at any point in the cooling cycle corresponds to that expected for a higher temperature. Since the nuclei are not formed, the liquid remains supercooled until a solid state has been formed. Glass has this property.

The ordering of the molecules in the glass is strongly dependent on the rate of cooling so that the physical properties are not only dependent on the chemical composition, but also on the thermal pattern of its manufacture. This explains the importance of correct annealing during the manufacture of glassware.

Chemically, glasses consist of chain or network molecules made up mainly from silicon and elements of group VI of the periodic table—

oxygen, selenium and tellurium. However, the properties of these molecules are greatly affected by the inclusion of elements of groups I and II, so that the melting point and viscosity characteristics are dependent upon these inclusions.

A. Chemical properties

The approximate composition of three common types of glass is given in Table I. These types are by no means the only ones available, but they show the variation available in clear glass.

TABLE I

Approximate chemical composition of some glasses

Type of glass	Composition, %					
	SiO_2	Na_2O	K_2O	CaO	B_2O_5	Al_2O_3
96% Silica	96·3	0·2	0·2	..	2·9	0·4
Soda-lime	72·0	13–16†	..	10–13	..	1·5–2·5
Borosilicate	80·5	3·8	0·4	..	12·9	2·2

† Sum of Na_2O and K_2O.

Glass possesses the desirable quality of almost perfect resistance to acids of all strengths, with the exception of hydrofluoric acid. This inertness is extremely valuable in microbiology because so many organisms are affected by even small traces of certain ions, particularly metal ions. Strong alkalis do have a corrosive effect on glass but it is an extremely slow process and at the low concentrations usually employed in pH control systems is of no importance.

All these glasses behave in a similar manner chemically.

B. Physical properties

The physical property most relevant to the application of different varieties of glass is the thermal coefficient of expansion, which varies widely, as can be seen from Table II.

Silica glass does not have particular application in microbiology but is included as it is a high temperature glass. It is relatively expensive and its high melting point makes it difficult to work, but with its chemical purity it is feasible that it could be useful under special circumstances. Soda-lime glass is the normal glass used in bottles. Its application is limited by the high value of its coefficient of thermal expansion. This prevents the glass from surviving the type of thermal shock encountered in autoclaving, and so it cannot be used for microbial cultures.

TABLE II
Physical properties of some glasses

Type of glass	Specific gravity	Coefficient of thermal expansion 10^{-5}°C	Modulus of elasticity 10^5kg/cm^2	Annealing Temperature °C
96% Silica	2·18	0·44	6·7	910
Soda-lime	2·47	4·70	7·0	550
Borosilicate	2·23	1·70	6·4	570

Borosilicate glass is the material of choice for laboratory glassware. It is known under a variety of trade names, such as Pyrex, Jena and Corning and has a characteristically low thermal expansion coefficient. Its working properties are excellent, so it is easily fabricated into any type of special equipment as long as care is taken with proper annealing.

There is no doubt that glass is the most popular material for the submerged culture of micro-organisms in liquid media in the laboratory. Its chemical inertness and smooth surface facilitate washing, either with detergents or soaking in acid, so that contamination from this source is at a minimum. There is no doubt, however, that the cost of this washing is allowing plastics to compete more favourably in many of the mass produced items.

Glass lining is sometimes used in fermenters on a production or pilot plant scale, but it is uneconomic for small equipment where corrosion resistant metals are more suitable.

C. Boundary surface effects

All construction materials used in conjunction with liquid media have one factor in common—an interface between the solid boundary and the culture medium. Interface reactions are complicated phenomena, difficult to analyse in the context of complex solutions used in cultivating micro-organisms, but the gross effect of such interactions has been observed in many systems, particularly where glass forms the solid boundary.

1. *Growth at low nutrient concentration*

When an aqueous electrolyte forms a boundary surface with an ionogenic material such as glass, some cations (e.g., H^+) are split off from the silicic acid of the glass and go into solution, leaving the silicate groups in the solid phase. The surface of the solid now bears a negative charge and is bounded by a diffuse double layer of mobile cations. This can lead to a concentration of specific ions in the region of the boundary surface and its related double layer.

Under the special conditions of low nutrient concentration, the concentration of ions becomes greater than the critical level required for growth and cell multiplication is facilitated under conditions normally unfavourable to microbial growth. ZoBell (1937) observed the correlation of increasing bacterial population in stored sea water and the size of the receptacle used for its storage. In a later work (1943) he showed that other inert particulate materials had a beneficial effect on the growth of bacteria, but only in very dilute nutrient solutions. This was confirmed by Heukelekian and Heller (1940), who inoculated washed *Escherichia coli* cells into glucose or peptone media of concentrations varying from 0·5 to 100 ppm in flasks of 250 ml capacity, each containing 20 ml medium. To one series, 50 g of 4 mm glass beads were added. No growth could be observed in the absence of glass beads in glucose concentrations up to 0·5 ppm and peptone concentrations up to 2·5 ppm. An enhanced growth effect in the presence of glass beads was noticeable up to 25 ppm of either glucose or peptone, but above these levels growth was similar in both series of experiments.

These results illustrate the importance of boundary surfaces under specific conditions which may have significance in certain cases. One example is the sampling of water supplies for later examination of micro-organism content. Storage in vessels of large surface/volume ratio can increase microbial counts under otherwise unfavourable growth conditions.

2. *Adhesion of Tissue Cells*

The attachment of tissue cells to solid surfaces, and in particular to glass, is a phenomenon fundamentally affecting their growth behaviour. For example Coman (1944) has suggested that the reduced adhesiveness of carcinoma cells may have an important influence on their ability to metastasize and invade.

The importance of the subject has led to various studies quantitating the adhesiveness of cells by using standardized techniques. Taylor (1961), after inverting a chamber with cells attached to one glass surface, subjected it to a standard stroke with an electric hammer so that cells not firmly attached were released. Centrifugation under controlled conditions has been employed by Berwick and Coman (1962) and Easty *et al.* (1960). Gravity was used by Nordling *et al.* (1965a, 1965b) to study the effects of different sterilization procedures and the addition of a number of anionic polymers on cell attachment and growth behaviour. Weiss (1965) used a fluid shear force when studying the influence of anti-sera on cell detachment.

The results of such experiments reveal our incomplete knowledge of the very complex biological and physical factors involved in the interaction between cells and solid phases in the vicinity of cells. This is illustrated by the observations of Nordling *et al.* (1965a). Cells adhered much more readily to glass sterilized with dry heat after alcohol-ether drying than to steam-sterilized glass, where attachment was poor. On the other hand, cell attachment proved to be remarkably good if the glass surface had been autoclaved under water at 120°C rather than in steam at the same temperature. These considerable differences were noted after washing in detergents, followed by 10 rinses in tap water and 5 rinses in pyrogen-free, twice distilled water.

3. *Attachment of other micro-organisms*

Anyone familiar with continuous culture of micro-organisms will be aware that cells adhere to the boundary surfaces. This probably has little effect on the results obtained from short batch culture experiments but can become a factor in the evaluation of results from experiments involving cultures over long periods of time.

Nordin *et al.* (1967), realizing the importance of adhesion in process equipment, have studied the phenomenon with an alga as test organism, correlating the degree of cell adhesion with the magnitude of the charges on the algal and glass surfaces. The technique used in the experiments was to observe the algae in a small, temperature controlled, glass electrophoresis cuvette where the charges could be determined. Algal adhesion was measured by allowing the cells to settle by gravity to the base of the cuvette. The force necessary to remove them from the glass was then measured arbitrarily by directing a stream of liquid at a controlled rate past the adhering cells for a given time. The data thus obtained was measured as per cent cells detached against flow rate for a range of flow rates.

It was found that the charges on both the algae and glassware varied considerably, both in sign and magnitude, with the concentration of salt in the suspending medium when either NaCl or $FeCl_3$ was employed. These charges, as might well be expected, led to differences in the algal adhesion to glass. It was found that at very low $FeCl_3$ concentrations of up to 5×10^{-6} M, when both the *Chlorella* and the glass exhibited negative charges, very little adhesion of the algae to the glass occurred. As the concentration of $FeCl_3$ increased to 3×10^{-5} M the adhesion was very strong, as the charge on the *Chlorella* had disappeared and that on the glass had become slightly positive. At 10^{-4} M the cells had also become positively charged thus reducing the adhesion, but a further increase to 0·05 M once again resulted in a negative charge on the *Chlorella* and the adhesion became so strong it could not be measured with the apparatus. In conclusion, it could be said that adhesion was strongest under those conditions which produced the greatest difference in zeta potentials between organism and boundary surface.

It has been popular to perform continuous culture experiments in small volumes, of the order of 50 ml, where the ratio of vessel volume to surface area is relatively small. In this case a monolayer of cells attached to the walls of the vessel would influence the output of cells to a considerable extent, as has been demonstrated by Larsen and Dimmick (1964). These workers used a cylindrical vessel of 27 mm diameter and 25 mm depth to study wall effects with a variety of bacteria, most experiments being performed with *Serratia marcescens* as test organism. It was found that the dilution rate could be increased to 7 h^{-1}, far beyond the maximum specific growth rate of the organism, without achieving washout of the cells. On subsequently stopping the medium flow the population density returned to its original level within a very short time. Removing the cells from the walls by washing with saline and a sterile brush released $2{\cdot}8 \times 10^9$ colony forming units. By adding glass wool to the vessel, which was agitated by a stream of air bubbles, the output of cells from boundary surfaces could be increased by a factor of 30.

In one particular experiment, when the cell count in the outflow was reduced to a low level by a high medium flow rate, the resultant growth curve showed that as many as 98% of the cells in the liquid culture originated from the walls of the vessel in a one hour period! On the other hand experiments on *Bacillus cereus* and *Bacillus subtilis* failed to reveal such artefacts.

In conclusion it must be said that it is not possible to predict the behaviour of a particular micro-organism in a specific environment without the results from careful experiments only warranted in special circumstances, because of the extreme complexity of interaction of the factors involved. Nordin *et al.* (1967) believe that it is not necessary for the charges on the cells and boundary surface to be of opposite sign in order for cell adhesion to occur. The charges may only have to differ in magnitude, or the cells or boundary surfaces show some roughness property to reduce the electrostatic barrier which must be surpassed before adhesion occurs. The situation is further complicated by the fact that different culture conditions result in varying zeta potentials of the cells. Silicone treatment of glass has been used as a method of preventing adhesion (Larsen and Dimmick, 1966; Holmström, unpublished work) but although it is possible that the onset of cell adhesion is delayed, sooner or later adhesion occurs.

III. METALS

The mechanical strength and low cost of steel makes it the natural choice as a construction material in engineering. Unfortunately, ordinary mild steel has a poor resistance to corrosion when in contact with aqueous solutions. This makes it unsuitable for use in the laboratory, although it has been used considerably in the past for the construction of production fermenters, where there is often a scale build-up to protect the metal. Alternatively large tanks have been lined with glass or resins.

Stainless steel is therefore the natural choice for laboratory scale equipment and fundamentally it is the corrosion resistance of stainless steels that is their desirable property. Not only does this property manifest itself in less deterioration of equipment, but the biochemical effects of corrosion are also minimized, if not always eliminated. These effects are a result of ions released from metals during corrosion and it is therefore worthwhile considering what is meant by corrosion.

A. Corrosion

1. *Nature of corrosion*

The process of corrosion may be used to cover all reactions in which a metal is converted from the elemental state to a chemically combined form, whether this combination is in the solid state or not. It must first be realized that all metals, even if corrosion resistant, undergo some oxidation when exposed to oxygen or oxidizing systems. The changes brought about by oxidation may fall into one of two types, depending on the ability or non-ability of the oxidation products to produce a firmly bound film. This film-forming property of some metals is of vital importance to their corrosion resisting qualities. Film-forming reactions usually show a decreasing rate of progress as the film thickness increases, whereas in the presence of all reactants a non-film-forming reaction can continue indefinitely (Evans, 1948).

Obviously then the corrosion-resistant metals are those forming a film; however film formation is not the only criterion for this desirable property. Metals have been divided into two groups by Pilling and Bedworth (1923) according to whether the oxide occupies a volume smaller or larger than that of the metal destroyed in producing it. Obviously in the latter case the oxide is likely to contain voids and fail to isolate the metal from the oxidizing phase. This is the case for the lighter metals, such as sodium and magnesium, so that if these metals are heated in air they can ignite and generate enough heat to maintain the burning temperature.

The heavier metals, together with beryllium and aluminium, yield compact oxide layers which, under certain conditions, are able to protect the metal. Even a dense oxide layer is not sufficient in itself to guarantee a rapidly decreasing oxidation rate. Films are subject to lateral stress and, if brittle, are liable to crack, allowing further oxidation to occur. This type of oxidation occurs with aluminium at room temperatures, as demonstrated by Vernon (1923, 1927), but in this case the process is usually so slow that it does not render the aluminium completely unserviceable. Similarly, iron and steel are not immune to oxidation since, although the oxide layer is dense, mechanical flexing and the effects of temperature cycling combine to disrupt an otherwise homogeneous film.

The cultural conditions under which micro-organisms are grown in stirred fermenters present difficult corrosion problems, particular where aerobic organisms are concerned. Not only is deterioration of the equipment an important consideration, but also minute traces of dissolved metals can drastically alter the course of a biochemical reaction.

Corrosion is naturally enhanced by an increased concentration of oxygen in turbulent conditions, precisely those conditions desirable in aerobic cultures. In addition, if dead spaces occur in the flow pattern of an aerated fermenter, it is quite feasible for an electrolytic oxygen cell to be formed with resultant corrosion. Secondly, the salts added to media assist in the dissolution of ionic materials, thus accelerating the process.

Other factors play a part in corrosion, such as stress and other types of electrochemical action, but these are of greater importance with industrial plants than in the laboratory.

2. *Corrosion guide*

Table III is a brief corrosion guide for some of the metals applicable in microbiology. The corrosive compounds include salts and carbon sources commonly found in media, together with acids and alkalis used for cleaning equipment.

For several reasons the data must be treated with some caution. In the first place 10% concentration is obviously an arbitrary choice and media do not contain only one salt, but a combination of several salts. Secondly, even if an alloy is classed as good, meaning that little metal goes into solution, the ions becoming soluble may well represent a toxic factor to the process under study.

Handbooks giving detailed information on corrosion have been produced by Rabald (1951) and Uhlig (1948) and these should be consulted for special cases.

3. *Microbiological effects of corrosion*

Obviously it is undesirable that corrosion should cause too frequent replacement of mechanical parts, but in the laboratory the main concern is to prevent contamination and interference with the microbiological

TABLE III

Corrosion guide for metals used in microbiology

Adapted from Parker (1967)

Compound	Aluminium	Monel	Nickel	Inconel	Hastelloy	Stainless Steels BSS En 60 AISI 430	BSS En 58A AISI 301	BSS En 58J AISI 316	Titanium
Acetic acid	G	G	F	G	E	E	E	E	E
Ammonium chloride	P	G	G	G	E	F	G	G	E
Ammonium sulphate	P	G	G	G	G	F	F	G	E
Butyric acid	G	F	F	F	E	F	G	G	E
Chromic acid	P	P	P	G	G	G	G	G	E
Citric acid	G	G	G	G	E	G	G	G	E
Ethyl alcohol	G	E	E	E	E	G	G	E	E
Ethylene oxide (gas)	E	G	G	G	E	G	G	G	E
Ferric chloride	P	P	P	F	E	P	P	P	E
Ferrous chloride	P	P	F	P	G	P	P	P	E
Ferrous sulphate	E	.	P	G	G	G	G	G	E
Formaldehyde	G	E	E	E	G	E	E	G	E
Hydrochloric acid (aerated)	P	P	P	P	G	P	P	P	G
Lactic acid	G	P	G	G	G	P	G	G	E
Potassium hydroxide	P	E	E	G	G	G	G	G	E
Sodium chloride	F	E	E	E	G	G	G	G	E
Sodium hydroxide	P	E	E	E	E	E	E	E	E
Sodium phosphates	P	G	G	G	G	G	G	G	..
Sulphuric acid (aerated)	P	F	F	P	E	F	P	F	G
Sulphuric acid (100%, aerated)	P	F	F	P	E	P	G	G	P

Unless stated otherwise the concentration of corrosive fluid is 10% in water. Code: E, excellent; G, good; F, fair; P. poor. AISI = American Iron and Steel Institute. BSS = British Standards Specification.

system under study. Bowen (1966) has reviewed the toxic effect of metals on micro-organisms and in the field of biochemistry in general.

Contamination is a serious consideration in the case of alloys since, although the quantities of alloying elements are small some are highly toxic to micro-organisms, copper and tin for example, being toxic to green algae at concentrations of 0·01 ppm and 0·002 ppm respectively (Young, 1935; Wiessner, 1962). Somers (1959), evaluated the toxicity of divalent metals on *Alternaria tenuis* and found the order of decreasing toxicity to be—

$$\mathrm{Hg} > \mathrm{Cu} > \mathrm{Pb} > \mathrm{Pd} > \mathrm{Ni} > \mathrm{Co} > \mathrm{Be} > \mathrm{Zn} > \mathrm{Mn} > \mathrm{Sr} > \mathrm{Mg}, \quad \mathrm{Ca} > \mathrm{Ba}.$$

The toxic action of metallic ions differs according to the mode of action on the organisms, so although the pattern is similar for most micro-organisms there is a fair degree of interchange in the order of toxicity.

The most common mode of attack is the inhibition of enzyme action, particularly in the cases of copper, silver and mercury. However arsenate acts as an antimetabolite, occupying sites normally filled by phosphate ions. Iron may form a stable precipitate with ATP and several elements including copper, mercury and lead combine with the cell membrane and affect its permeability. These are just some of the factors which may be brought into play by corrosion effects, which may not in themselves be sufficient to show on the metal in the form of mechanical deterioration.

The complex interplay of reactions occurring when a medium containing metals, salts, carbohydrates, amino-acids and other components is autoclaved in an alloy culture vessel, such as one made of stainless steel, together with the unknown sensitivity of the micro-organisms to be cultivated in the medium precludes any general prediction of interference in growth. That interference does occur has been documented by a number of workers. For example, Fåhreus and Reinhammar (1967) growing *Polyporus versicolor* in a defined liquid medium containing salts, glucose and amino-acids at pH 5·0 in a stainless steel pilot plant fermenter of 100 litre capacity, were forced to sterilize a concentrated solution of medium and add this to sterilized water in the fermenter. If the complete medium was sterilized in the fermenter, the growth rate of the fungus was reduced considerably.

Similarly, Hernandez and Johnson (1967) found it necessary to add a chelating agent to a synthetic minimal medium when growing *E. coli* anaerobically in a 2·5 litre fermenter having a stainless steel agitator and sparger. No such difficulty was experienced when the same strain was grown in the same medium under the same conditions but in a glass vessel.

The problem is usually of relatively minor importance in the laboratory since it can be overcome by modifications in the sterilization technique, such as a separate autoclaving or filtration of the solution causing the change in composition. The changes in the medium normally take place when the temperature is elevated to 121°C during autoclaving.

When laboratory studies are performed with a view to eventual scale-up for production, due consideration must be taken of the possibility that this problem may arise, and it is wise to carry out a laboratory programme of investigation aimed at evaluating the effects of likely metals of construction on the process to be studied. Since enzyme reactions are the most sensitive to this type of disturbance, particular note must be taken in the case of processes involving their participation for specific product formation.

Some reports have been published on the toxic effects of metals during growth of micro-organisms, not only inhibition of growth due to sterilization effects. Dyer and Richardson (1962) using two algae in a medium having a pH of 7·5 found, as might be expected, that copper both in the pure state and in a number of alloys, was strongly inhibitory. However it is of interest to note that bearing bronze, an alloy containing copper, tin and zinc, had no such effect. Other metals, including aluminium, nickel, silver, solder, titanium and several grades of austenitic stainless steel were also without effect. Rosenwald *et al.* (1962) tested a number of materials on *Pasteurella tularensis*, noted for extreme sensitivity to toxic metallic ions, by incubating the organism on agar plates for 72 h, following which a circular sample of test material was placed in the centre before incubating the plates for a further period. Zones of inhibition were measured using the replica plate technique of Lederberg and Lederberg (1952). Not only was inhibition detectable but the diameter of the inhibition zone gave an indication of the order of toxicity. In confirmation of the results of Dyer and Richardson (1962) zinc, brass and copper showed inhibition of increasing magnitude in that order. Again aluminium and stainless steel were without detrimental effect.

Tissue cells are also noted for extreme sensitivity to inhibitors and Giardinello *et al.* (1958) have shown that copper is no exception in the case of L cells. In addition, aluminium, brass, mild steel and monel alloy all showed considerable toxicity, indicated not only by a reduction in cell number but also by morphological changes. Of particular interest was the fact that one grade of stainless steel, AISI 304, showed some degree of toxicity, whilst two other grades, AISI 316 and AISI 347, were inocuous, even though all three have the same components in different concentrations.

Titanium has been used by Molin and Hedén (1968) as a surface for monolayer cultivation of human diploid cells in order to avoid all trace metal contamination. Several workers have grown a variety of tumor cells in suspended culture in stainless steel fermentors, including ERK and L cells (Pirt and Callow, 1964), BHK cells (Telling *et al.*, 1967) and HeLa cells (Holmström, 1968), all apparently without inhibition problems.

B. Properties and uses of metals

1. *Aluminium and associated alloys*

Small pilot plant scale fermenters have been constructed from aluminium, where the suitability of the metal for the process under investigation had been established. In such cases the use of the metal presents a considerable financial saving over other possible choices.

However, the limited corrosion resistance properties of the metal do not make it a wise choice for general purpose laboratory equipment. The oxide film, mentioned previously, is easily rendered soluble when the pH is far removed from neutral whether acidic or alkaline. Since commercial aluminium is over 99% pure and aluminium salts are inocuous to most biochemical reactions, there is no toxicity problem, but the equipment can rapidly deteriorate and accidental spillage of pH control liquids, for example, causes serious corrosion. In addition the pure metal is very sensitive to chloride ions, although certain alloys containing magnesium and manganese have been developed which overcome this problem.

All aluminium alloys are weldable, but some require special techniques and it is always advisable to carry out annealing afterwards. Resistance welding has some disadvantages for microbiological purposes since metallurgical changes occur in the heat-affected zones of such welds, and these changes not only weaken the metal, but also adversely affect its corrosion resistance.

2. *Copper*

Copper is used for many yeast fermenters in the brewing industry. However, although the metal is fairly corrosion resistant there is no doubt that even in this case some copper finds its way into solution. This precludes any possibility of copper being used for general microbiological purposes, since so many enzyme reactions are inhibited by minute quantities of copper. This is a pity as copper has good working properties and a high thermal conductivity. For the same reason brass components, like impeller bushes in fermenters, should be avoided, particularly as new materials having more suitable properties are available.

3. *Nickel and associated alloys*

Nickel is a very stable metal, easily worked and with good strength and resistance to abrasion, even in the presence of steam. The compositions of high purity nickel and some of its alloys with copper and iron are given in Table IV.

TABLE IV

Chemical composition of nickel and nickel alloys

Nickel or alloy	Chemical composition %							
	Ni	Cu	Fe	Mn	Si	Cr	Mo	C
Nickel	99	0·25	0·40	0·35	0·35	..	..	..
Monel	63–70	26–33	2·50	2·0	1·0	..	..	0·3
Inconel	72	0·50	6–10	1·0	0·50	14–17	..	0·15
Hastelloy B	62	..	5·0	..	..	..	28	0·12

Monel is primarily an alloy with copper, inconel contains both iron and chromium, and hastelloy B contains iron and molybdenum. These alloys have been developed to cater for different environments. Nickel and monel are most resistant to alkalis, whereas inconel and hastelloy B are suitable for use with acids. In general the corrosion resistance of the nickel alloys can be said to cover the range offered by stainless steel and most have the same disadvantage of being sensitive to chlorides.

Since the cost of nickel and its alloys is approximately twice that of the most expensive standard grades of stainless steel its application is limited by economic considerations. It is, however, useful for threaded parts in combination with stainless steels, in order to prevent the binding which occurs when two identical metals are screwed together. This is usually only necessary with the pilot plant size of equipment, small threads on laboratory fermenters presenting no difficulty in this respect.

Machining and polishing of nickel and its alloys are easily accomplished. Similarly nickel alloys can be welded, brazed and soldered satisfactorily with compatible alloys using standard techniques.

4. *Stainless steels*

As has been indicated, stainless steels are easily the most popular group of metals in use in the laboratory. The addition of relatively small quantities of other elements to form alloys with iron, retains the desirable mechanical properties of the latter, whilst imparting a vastly improved resistance to the effects of many corrosive compounds, as is shown in Table III. The stainless steels can be divided into three groups by virtue of their chemical composition and physical properties, these groups being ferritic, martensitic and austenitic steels. Their application in biochemical plants has been described by Cooke and Williams (1967).

(a) *Martensitic and ferritic steels.* These steels are both iron-chromium alloys, exhibiting magnetic properties. The martensitic group have mechanical properties of a similar nature to the low alloy steels but the addition of 12–14% chromium imparts a degree of corrosion resistance. In the case of the ferritic group, the chromium is further increased from 14–18% and in some cases up to 27%. These steels are the cheaper grades and have good casting properties, but find little application as general purpose stainless steels because of their limited corrosion resistance.

(b) *Austenitic steels.* Undoubtedly the most useful and most commonly encountered group of stainless steels available for use in microbiological laboratories are the austenitic types, easily distinguished from the martensitic and ferritic types by being non-magnetic. They are iron alloys including both chromium and nickel, often in the concentrations of 18% and 8% respectively, which has led them to be known as 18 : 8 stainless steels. However there is such a wide variety of alloys having specialized properties for different applications that a more detailed specification is required. Each country has its own specification system, so that it is not possible to define exact equivalents. However a list of some common British (BSS) and American (AISI) equivalents is given in Table V together with martensitic and ferritic types. The figures are somewhat meaningless to a microbiologist but briefly it may be said that carbon is included to ensure malleability, silicon, sulphur and manganese confer the necessary desirable machining properties, whilst molybdenum improves the corrosion resistance even further than is possible with chromium and nickel. It is in fact this addition of molybdenum which makes the BSS En 58J steel, and its American equivalent AISI No. 316, the natural choice for the construction of equipment for laboratory microbiological operations. It has the highest corrosion resistance of the standard grades and is therefore likely to stand up to most service requirements. This grade is used, for example, in the laboratory and pilot plant fermenters produced by Taylor Rustless Fittings Ltd., Leeds, England, New Brunswick Scientific Co., Inc. New Brunswick, N.J., U.S.A., and Biotec AB, Stockholm, Sweden. Should a process require development beyond the laboratory stage and large scale equipment become necessary, then the most economical material of construction can be evaluated with the aid of laboratory experiments For example, it has been found possible to produce beer barrels in En 58E steel and this grade is also popularly used in dairy equipment and other branches of the biochemical industry, where the high cost of the En 58J steel can be avoided.

Stainless steel does have one limitation from the corrosion point of view, which can be seen from Table III; a sensitivity to hydrochloric and sulphuric acids, and to some chloride salts in addition. The sulphuric acid presents no problem microbiologically but chloride ions are particularly important physiologically.

Composition and relative cost of some wrought stainless steels

BSS no.	AISI no.	Composition limits % C (max)	%P (max)	% S (max)	% Mn (max)	% Si (max)	% Cr	% Ni	Other components	Relative cost of sheets
Austenitic steels										
	201	0·15	0·06	0·03	(5·5–7·5)	1·0	16·0–18·0	3·5–5·5	N = 0·25% (max)	
	302	0·15	0·045	0·03	2·0	1·0	17·0–19·0	8·0–10·0	..	1·0
En 58A		0·16			2·0	0·2 (min)	17·0–20·0	7·0–10·0	..	
	304	0·08	0·045	0·03	2·0	1·0	18·0–20·0	8·0–12·0	..	
	304L	0·03	0·045	0·03	2·0	1·0	18·0–20·0	8·0–12·0	..	
	316	0·08	0·045	0·03	2·0	1·0	16·0–18·0	10·0–14·0	Mo = 2·0–3·0%	
	316L	0·03	0·045	0·03	2·0	1·0	16·0–18.0	10·0–14·0	Mo = 2·0–3·0%	
En 58J		0·12			2·0	0·2 (min)	17·0–20·0	8·0–12·0	Mo = 2·5–3·5%, Ti and Nb optional	
En 58B		0·15			2·0	0·2 (min)	17·0–20·0	7·0–10·0	Ti = 4 × C	
	321	0·08	0·045	0·03	2·0	1·0	17·0–19·0	9·0–13·0	Nb + Ti = 5 × C	1·25
En 58E		0·08			2·0	0·2 (min)	17·5–20·0	8·0–11·0	Ti = 4 × C (min)	
	347	0·08	0·045	0·03	2.0	1·0	17·0–19·0	9·0–13·0	Nb + Ti = 10 × C	1·55
	348	0·08	0·045	0·03	2·0	1·0	17·0–19·0	9·0–13·0	Nb = 10 × C, Ti = 0·1% (max)	
En 58F		0·15			2·0	0·2 (min)	17·0–20·0	7·0–10·0	Ni = 8 × C	
Martensitic steels										
	414	0·15	0·045	0·03	1·0	1·0	11·5–13·5	1·25–2·5		
En 56C		0·25			1·0	1·0	12·0–14·0	1·0		
	431	0·20	0·045	0·03	1·0	1·0	15·0–17·0	1·25–2·5		
En 57		0·25			1·0	0·10–1·0	15·5–20·0	1·0–3·0		
	501	0·10 (min)	0·045	0·03	1·0	1·0	4·0–6·0	..	Mo = 0·4–0·65%	0·70
Ferritic steels										
En 56A		0·12			1·0	1·0	12·0–14·0	1·0		
	405	0·08	0·045	..	1·0	1·0	11·5–14·5	..	Al = 0·1–0·3%	
	430	0·12	0·045	0·03	1·0	1·0	14·0–18·0	..		0·8
En 60		0·12			1·0	1·0	16·0–18·0	0·5		

From Cooke and Williams (1967), and Parker (1957).

Studies of halophilic micro-organisms, where salt concentrations of 20% or even 30% are required, cannot be carried out in stainless steel equipment, as corrosion is severe during autoclaving with such high salt concentrations.

When En 58J steel, containing molybdenum is used, the quantities of chloride usually present in media for microbial cultures do not cause difficulty, particularly if the pH is kept above 6·0 during autoclaving. Care must also be taken to ensure no metallurgical changes during welding, otherwise the chloride causes rapid local pitting.

(c) *Fabrication of stainless steel equipment*. Stainless steels are not easy metals to fabricate, but their wide range of uses has acted as a stimulant for the development of suitable techniques for both forming and welding.

For large production series in austenitic steels, cold forming can be employed for vessels having large radii of curvature in relation to the metal thickness, using either stretching or spinning. When a severe working procedure is carried out, such as forming dished ends for fermenters, hot working at about 870°C is required, with careful heating and cooling cycles.

The manufacture of most laboratory equipment involves a considerable proportion of machining. Stainless steels are strongly work-hardening, so it is essential for the cutting tool to be given a constant and heavy load. Carbide tipped tools offer an advantage in that higher cutting speeds can be used.

Because of the stainless nature of these steels, welding processes are restricted to those that are compatible with corrosion resistance requirements. Acceptable processes include the inert gas method, the use of flux-covered electrodes and electrical resistance. The austenitic grades are most amenable, but the high thermal expansion coefficient can easily lead to distortion. In addition they are prone to form grain boundary cracks near the melting temperature. Intergranular corrosion can then occur in the zone of carbide precipitation. For the other grades, preheating to 200°–300°C is essential if cracking of the weld is to be avoided.

Brazing can be used for joining stainless steel if silver brazing alloys are used. Other alloys penetrate along grain boundaries and make corrosion much more possible, especially in stressed assemblies manufactured from cold worked steel. However, even silver brazing alloys give rise to much greater risks of corrosion and distortion than argon welding, and the latter method is preferable for these reasons.

The surface roughness of stainless steel sheet in the raw state is quite coarse. This would enable both micro-organisms and nutrients to become lodged in crevices. Such micro-organisms surrounded by substances of poor thermal conductivity can withstand autoclaving. In addition, washing of equipment would be rendered difficult. For these reasons polishing of stainless steel equipment is normally carried out to eliminate these crevices. Because of the work-hardening nature of stainless steel the procedure must be carried out in stages, each giving a successively smoother finish. Initially grinding wheels are used but the final stage is performed with a buffing wheel and fine carborundum paste.

5. *Titanium*

The extremely high corrosion resistance of titanium is illustrated in Table III. At the same time the commercially pure metal, containing over 99% titanium and only traces of nitrogen, carbon, hydrogen, iron and oxygen, has thermal expansion, heat transfer and strength characteristics roughly equivalent to those of austenitic stainless steels. Thus, in a metal having mechanically excellent qualities, the problem of trace metal contamination in cultures, always threatening with alloys, can virtually be eliminated. Halophilic micro-organisms could be satisfactorily grown in titanium equipment. The only situation in microbiology where a reagent would conceivably cause deterioration to titanium occurs with hydrochloric acid. Fermenter components of titanium in contact with hydrochloric acid of greater than 5% concentration would be corroded, the rate increasing with concentration and temperature. Hence pH control systems cannot be completely free from corrosion problems even with this material.

There are two drawbacks to the use of the metal for sophisticated laboratory equipment, such as fermenters. One is the high cost of the metal, which would double the cost of a fermenter, and the second is technical, namely the difficulties involved in welding, due to the high reactivity of the metal. Special inert gas techniques have been developed using closed chambers, but as yet, according to Parker (1967) there are no generally accepted methods.

Titanium has certainly been successfully used as a surface on which to grow large quantities of human diploid cells, by Molin and Hedén (1968), using a multi-disc arrangement. Fermenters and tissue culture propagators in this metal are now available from Biotec AB, Stockholm, Sweden.

6. *Other alloys and related materials*

Modern technology is continually developing new alloys to serve a variety of specialized purposes and some of these are finding application in the microbiological laboratory. One example is tungsten carbide, from which it is possible to produce lubricant-free bearings. These have been used for an ingenious impeller system in the laboratory fermenters of Biotec AB, Stockholm, Sweden. By combining these bearings with a magnetic drive through multi-polar ceramic magnets, a hermetically sealed fermenter has been made possible, requiring no steam seal or lubricant.

The corrosion problems associated with hydrochloric acid addition for pH control of fermenters can be entirely eliminated by using tantalum for specific details in contact with the acid, such as hypodermic needles. This metal is remarkably inert but its high cost—about 25 times that of titanium—precludes its use for fermenter construction for other than minor components.

IV. PLASTICS

A. The scope of plastics

The term plastics covers a heterogeneous collection of chemicals which have however one basic chemical characteristic in common—a polymeric structure produced by polymerization of a raw material of low molecular weight. They play an ever increasing role in all types of applications as can be understood from the exapanding production figures over the last few decades. In 1939, world production was approximately 350,000 tons, whereas in 1950 it had increased to 1,300,000 tons and by 1964, the figure of 10,000,000 tons had been reached and this rate of increase shows no signs of decrease as new types come into production (Salevid, 1966).

The increase in production in general has been marked also in the field of microbiology, with essentially two approaches to the demand for sterility for microbiological operations. Either the plastic chosen is able to withstand autoclaving at 120°C in a moist atmosphere, or another method of sterilization is used for mass produced disposable items. Usually these plastics are chosen for cheapness and for their transparency.

It is clear that disposable plastics are increasing their share of the production of routine handling equipment in the microbiological laboratory. Petri dishes and syringes are often completely the prerogative of this type of item and even disposable pipettes are now available.

The more temperature resistant plastics are usually more expensive and are used for special applications. Examples are polycarbonate centrifuge tubes available from all major centrifuge manufacturers, and details manufactured from polytetrafluoroethylene, such as sealing gaskets for small fermenters.

B. General properties

1. *Physical and mechanical properties*

The physical and mechanical properties of plastics span such a range that they can take the place of materials as widely different as glass or stainless steel, depending on the requirements and type of polymer chosen. A complete description of all standard tests would be out of place in this review, but a summary of the most important characteristics of plastics and their comparison with those of common metals is of value in deciding on specific applications. Most of the data have been taken from the works of Salevid (1966) and Parker (1967).

The test methods for plastics have been developed independently within different countries. Usually the aim of each test is the same, but the standard conditions vary, so that the test method must be specified. Here, unless otherwise stated, all data are given according to the test methods specificed by the American Society for the Testing of Materials (ASTM) for the sake of convenience.

Tables VI and VII give the general physical properties of thermoplastic resins and thermosetting resins respectively, such as specific gravity and thermal conductivity, which are similar for all test methods. Table XII of Section VII gives the permeability of some plastics to water and gases and is often of significance. Tables VIII and IX give the mechanical properties of the two types of plastics, where the margin for variation between the test methods is wide. This may be because of the difference in dimensions of the test pieces, or because the tensile properties of a plastic may alter markedly by changes in strain rate or temperature.

The variations can be better understood if the nature of polymeric materials is considered. According to Mark (1967) the three factors which govern the physical and mechanical properties of a plastic are the degree of crystallinity, the degree of chemical cross-linking and the degree of chemical stiffening in the molecules themselves.

Examples of crystalline plastics in order of increasing crystallinity are polyethylene, polypropylene and polystyrene. Their melting points, 130°C, 175°C and 230°C respectively, reflect this characteristic. Vulcanized rubber and phenol-formaldehyde plastics consist of cross-linked polymers which increase their rigidity. The higher the degree of cross-linking, the less elastic the polymer becomes. The interplay between these two factors at various temperatures can produce some anomalies. Chemical stiffening is achieved by attaching heavy groups of atoms on the backbone structure to inhibit flexing. In the case of acrylic thermoplastics, methyl (CH_3) and methacrylate ($COOCH_3$) groups are used to give the polymer its hardness, combined with optical clarity. Tensile strength data must on this account be treated with some caution, since a plastic may be deformed elastically up to a characteristic strain, above which the stress-strain curve deviates from linearity (Parker, 1967). The softer types of plastics have lower elastic limits and greater elongations, some even to the extent of having no elastic behaviour, as in films from parylene thermoplastic polymers.

TABLE VI

Physical properties of some thermoplastic resins.

Type of plastic and common trade names	Variant	Specific gravity D 792	Light transmission % D 791–54	Thermal expansion 10^5/°C D 696–41	Thermal conductivity kcal/m h°C C 177–45	Operating temperature maximum °C	Water absorption % D 570–57T
Acetal							
Delrin (Du Pont, U.S.A.)	..	1·4	Opaque	8	0·11	130	0·01
Alkon (ICI, England)							
Celcon (Celanese, U.S.A.)							
Hostaform C (Hoechst, West Germany)							
Acrylics							
Lucite (Du Pont, U.S.A.)	Normal	1·17–1·20	91–92	5–9	0·12–0·18	80	0·3–0·4
Plexiglas (Röhm and Haas, West Germany)	Heat resistant	1·18	87–92 Transparent	7–8	0·16	95	0·3
Perspex (ICI, England)							
Fluorocarbons							
Teflon (Du Pont, U.S.A.)	PTFE (polytetra fluoroethylene)	2·10–2·20	Opaque	6–7	0·21	260–280	0
Fluon (ICI, England)							
KEL-F (Minnesota Mining and Mfg., U.S.A.)	FEP (fluorinated ethylene propylene)	2·0–2·20	Transparent	6–7	0·21	..	0

Polyamides, Nylons							
Zytel (Du Pont, U.S.A.)	Nylon 6/6	1·14	Opaque	10–15	0·26	110	0·3–1·5
Maranyl (ICI, England)							
Durethan BK (Bayer, West Germany)	Nylon 6/10	1·09	,,	10	0·26	110	0·3–0·4
Ultramid (BASF, West Germany)							
Rilsan (Organico, France)	Nylon 6	1·13	,,	8–13	0·26	110	1·9–3·3
Plaskon (Allied Chem., U.S.A.)	Nylon 11	1·05	,,	13	0·30	100	1·0–1·9
Polycarbonates							
Lexan (G.E.C., U.S.A.)	. .	1·20	80	7	0·14–0·20	130–135	0·3
Makrolon (Bayer, West Germany)			Translucent				
Polyethylenes							
Alathon (Du Pont, U.S.A.)	Low density	0·91–0·93	Opaque	15–30	0·29	80	<0·01
Marlex (Phillips, U.S.A.)							
Alkathene (ICI, England)	Medium density	0·93–0·94	Opaque	15–30	0·29	90	<0·01
Hostalen (Hoechst, West Germany)							
Lupolen (BASF, West Germany)	High density	0·94–0·96	,,	15–30	0·29	95	<0·01
Fertene (Montecatini, Italy)							
Polypropylenes							
Hostalen PP (Hoechst, West Germany)	. .	0·90–0·91	Normally opaque, transparent films can be made	11	0·10–0·11	95–120	<0·01
Luparen (BASF, West Germany)							
Moplen (Montecatini, Italy)							
Escon (Esso, U.S.A.)							

TABLE VI—*continued*

Type of plastic and common trade names	Variant	Specific gravity D 792	Light transmission % D 791–54	Thermal expansion $10^5/°C$ D 696–41	Thermal conductivity kcal/m h°C C 177–45	Operating temperature maximum °C	Water absorption % D 570–57T
Polystyrenes							
Styron (Dow, U.S.A.) Distrene (Brit. Resin Prod., England)	Ordinary	1·04–1·06	78–90 Transparent	8	0·14	65–75	0·03–0·05
Trolitul (Dynamit Nobel, West Germany) Afcolene (Pechney St. Gobain, France)	Heat and chemical resistant	0·05–1·10	78–87 Transparent	8	0·14	75–105	0·10–0·40
Stiroplasto (Montecatini, Italy)	Impact resistant	0·98–1·10	Opaque	7	0·15	60–80	0·05–0·25
Tyril 767 (Dow, U.S.A.) Luran 52 (BASF, West Germany)	ANS (acrylonitril-styrene)	1·04–1·10	80 Transparent	8	0·14–0·15	50–90	0·05–0·30
Kralastic (U.S. Rubber, U.S.A.) Cycolac (Borg-Warner, England) Novodur (Bayer, West Germany)	ABS (acrylonitril-butadiene-styrene)	0·99–1·12	Opaque	8	0·14–0·15	50–100	0·05–0·30
Vinyl polymers							
Trovidur (rigid) (Dynamit Nobel, West Germany)	Rigid	1·35–1·45	Transparent to opaque	8	0·14	70	0·1
Mipolam (flexible) (Dynamit Nobel, West Germany) Solvic (Solvic, Belgium) Geon (British Geon, England) Pevikon (Stockholms Superfosfat, Sweden)	Flexible	1·16–1·35	Transparent to opaque	15	0·14	50	0·2
Methylpentene polymers			90			At least	

TABLE VII

Physical properties of some thermosetting resins

Type of plastic and common trade names	Type of filler and construction	Specific gravity D 792	Thermal expansion 10^{-5}/°C D 696–44	Thermal conductivity Kcal/cm h°C DIN 52612	Water absorption % D 570
Epoxy Resins	Short glass-fibre,	1·85–1·9	3	0·2	0·03–0·05
Epikote (Shell, England)	moulded				
Araldite (CIBA, Switzerland)	Woven glass-fibre,	1·65	1·5	0·2	0·2–0·3
Epotuf (Reichhold, U.S.A.)	laminated				
Phenol-formaldehydes and Phenol-furfurals	Asbestos, moulded	1·5–1·9	1·5–3	0·20–0·50	0·1–0·3
Bakelite (Union Carbide, U.S.A.)	mica, moulded	1·6–1·9	1·5–3	0·20–0·45	0·8
Bakelite (Bakelite, England)	Short glass-fibre,	1·7–2·0	1·2	0·15–0·30	0·05
Resinit (Bakelite, W. Germany)	moulded				
Fluosite (Montecatini, Italy)	Pulp, laminated	1·4	2·0	0·20–0·25	1·5–8
Polyesters	Short glass-fibre,	1·8–2·3	2–2·5	0·15	0·1–0·2
Polylite (Reichhold, U.S.A.)	moulded				
Marcon (Scott Bader, England)	Woven glass-fibre,	1·5–2·1	1·5–3	0·12	0·2–0·9
Leguval (Bayer W. Germany)	laminated				
Palatal (BASF, W. Germany)					
Laminac (Cyanamid, U.S.A.)					
Silicones	Woven glass-fibre,	1·6–1·7	1·2	0·2	0·35
MS-Silicone (Midland Silicones England)	laminated				
DC-Silicone (Dow Corning, U.S.A.)					
GE-Silicone (G.E.C., U.S.A.)					

TABLE VIII

Mechanical and electrical properties of some thermoplastic resins

Type of plastic	Variant	Compression strength kg/cm^2 D 695–54	Flexure strength kg/cm^2 D 790	Tensile strength kg/cm^2 D 638	Tensile elongation % D 651–58 T	Izod impact energy ft lb/in D 256–65	Volume resistivity ohm cm D 257	Dielectric strength kV/mm D 149
Acetal	..	1100–1200	900–1000	600–700	15–75	1·2–1·7	10^{14}	19
Acrylics	Normal	840–1300	900–1200	500–800	2–10	0·5	10^{12}–10^{15}	10–40
	Heat resistant	1050–1260	1300–1450	750–800	3–4	0·4–0·6	10^{16}	22
Fluorocarbons	PTFE	120	180–200	100–250	300	2·5–4	10^{15}–10^{19}	16–20
	FEP	..	..	130–150	290	No break	10^{18}	16–20
Polyamides-Nylons	Nylon 6/6	500–1100	550–1000	500–800	60–300	0·9–2	10^{10}–10^{15}	14–18
	Nylons 6/10	500	650	420–600	85–320	0·6–1·6	10^{14}	19
	Nylon 6	400–850	550–1100	700–850	90–320	1·2–4·0	10^{11}–10^{15}	12–20
	Nylon 11	..	..	350–560	30–300	4·5	10^{12}–10^{13}	9–16
Polycarbonates	..	800–850	700–900	600–650	65–80	12–16	10^{16}–10^{17}	16–27
Polyethylenes	Low density	Not determinable	Not determinable	70–140	300–1000	Not determinable	10^{15}–10^{20}	17–20
	Medium density	Not determinable	Not determinable	85–350	50–350	Not determinable	10^{15}–10^{16}	19–28
	High density	170–370	95–115	300–400	10–300	1–10	$>10^{10}$	18–48
Polypropylenes	..	400–550	Not determinable	300–400	300–750	1·0–15	10^{16}–10^{17}	30–32
Polystyrenes	Ordinary	800–1000	500–1000	350–600	1–3	0·25–0·6	10^{16}–10^{21}	20–65
	Heat and chemical resistant	800–1000	600–1100	450–800	1–3	0·3–0·6	10^{13}–10^{17}	16–24
	Impact resistant	300–600	350–1100	250–450	5–50	0·7–3	10^{12}–10^{17}	12–25
	ANS	1000–1200	800–1300	650–850	1–2	0·8–4	10^{13}–10^{17}	12–45
	ABS	200–700	250–900	200–600	10–45	0·7–4·5	10^{11}–10^{17}	12–26
Vinyl Polymers	Rigid	560–900	700–900	480–600	20–40	0·4–1 (High impact 3–15)	10^{12}–10^{16}	17–52
	Flexible	Not determinable	Not determinable	160–180	370–400	Not determinable	10^{13}	24–26

TABLE IX
Mechanical and electrical properties of some thermosetting resins

Type of plastic	Type of filler and construction	Compression strength kg/cm² D 695-54	Flexure strength kg/cm² D 790-58 T	Tensile strength kg/cm² D 638/D 651	Izod impact energy ft lb/in D 256-56	Volume Resistivity ohm cm D 257	Dielectric strength kV/mm D 149-55T
Epoxy resins	Short glass-fibre, moulded	1500	900–1200	500–700	0·5–1·5	10^{13}–10^{15}	10–20
	Woven glass-fibre, laminated	700–900	2800–3500	..	5–7	10^{12}	10–20
Phenol formaldehyde and phenol-furfurals	Asbestos, moulded	1050–2100	450–850	200–500	1–3	10^{5}–10^{13}	6–12
	Mica, moulded	1050–1750	500–800	200–450	0·4–1	10^{6}–10^{14}	6–18
	Short glass-fibre, moulded	1150–1900	1100–1400	500	2–5	10^{12}	5–16
	Pulp, laminated	1500	1300–1500	1200	0·55–0·66	10^{9}–10^{10}	5–20
Polyesters	Short glass-fibre, moulded	700–2000	200–2000	400–700	3–5	10^{10}–10^{15}	6–24
	Woven glass-fibre, laminated	1900–4000	800–2500	500–2000	8–9	10^{12}	10–16
Silicones	Woven glass-fibre, laminated	1500	1250–1400	..	5·5–6·5	10^{12}	32

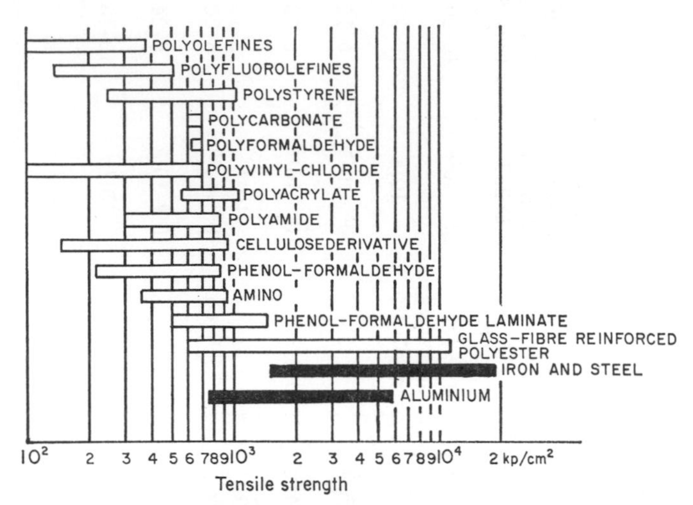

FIG. 1. Tensile strength of some plastics and metals at 20 °C. Adapted from Salevid (1966) and reproduced by permission of the author.

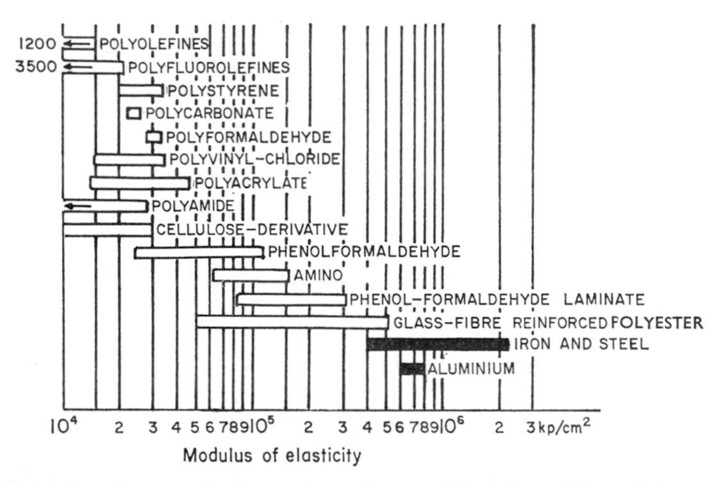

FIG. 2. Elasticity of some plastics and metals at 20 °C. Adapted from Salevid (1966) and reproduced by permission of the author.

However, the tests do serve as guide lines for the comparison of plastics for a given purpose. There are many more tests than those listed here, for which, apart, from the standards specifications, references are available when the design of plastic articles is contemplated (Klein, 1959; Kluckow, 1963; Lever and Rhys, 1957).

Figures 1 and 2 show in diagrammatic form the tensile strength and modulus of elasticity of some polymers as compared to certain metals. It can

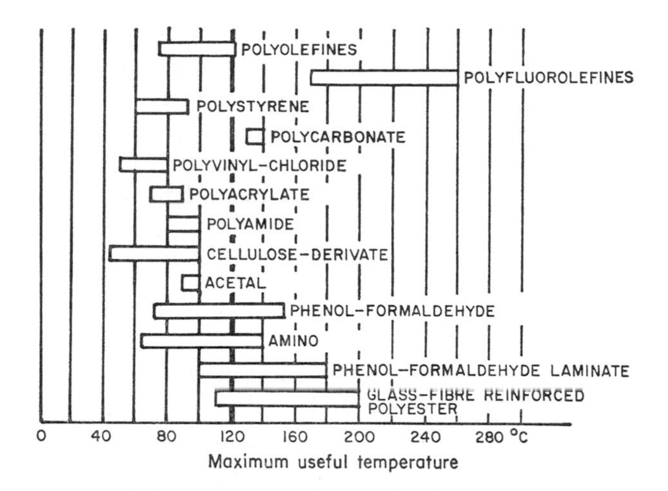

FIG. 3. Maximum useful temperature of some plastics. Adapted from Salevid (1966) and reproduced by permission of the author.

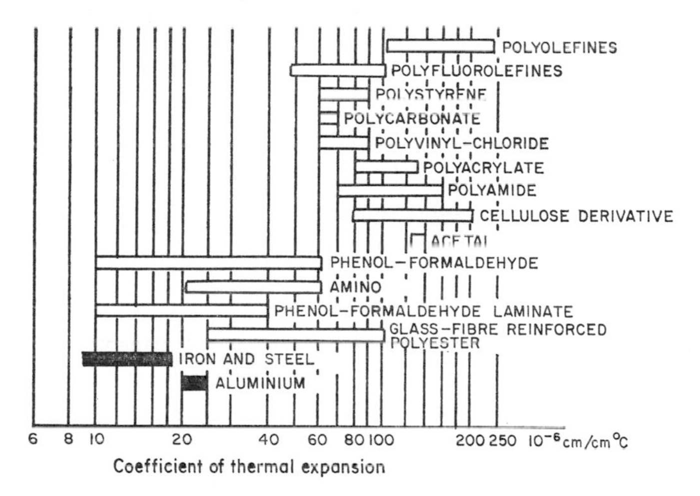

FIG. 4. Coefficient of thermal expansion for some plastics and metals at 20 °C. Adapted from Salevid (1966) and reproduced by permission of the author.

be seen that, in general, plastics have a tensile strength and modulus of elasticity one order of magnitude lower than the metals. As with metals there is a wide variation and in the case of glass-fibre reinforced polyester, tensile properties approaching that of steel are obtained, this property being used to advantage in many applications.

The temperature properties of plastics must also be carefully borne in mind, particular in mixed plastic-metal constructions requiring autoclaving.

Figure 3 shows the maximum useful temperature of various plastics, which is considerably lower than the softening point since plastics become deformed under stress at much lower temperatures. Stresses are easily encountered from the large differences in thermal coefficient of expansion and thermal conductivity between plastics and metals as shown in Figures 4 and 5. If the combined effects of strain rate and temperature are not taken into account, failures will most certainly occur.

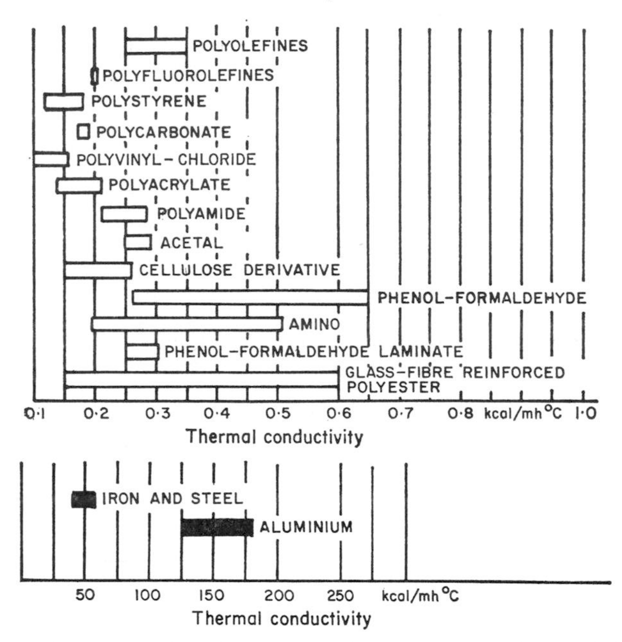

FIG. 5. Thermal conductivity of some plastics and metals. Adapted from Salevid (1966) and reproduced by permission of the author.

All those plastics shown in Fig. 3 as having a maximum useful temperature greater than 121°C may be autoclaved repeatedly without deterioration. In addition some of the plastics having a nominal temperature limit lower than this figure may also withstand autoclaving once or even many times. Such plastics include nylons and acetal polymers, provided the components are of small physical size and are very lightly stressed both mechanically and chemically. However, because of the complex stress-temperature relationship of plastics, it is advisable to make use of those having a higher useful maximum temperature where autoclaving is to be undertaken.

2. *Chemical resistance*

Plastics in general are characteristically inert to the effects of aqueous inorganic solutions at neutral pH. Their resistance to strong acids and alkalis varies but they commonly resist low concentrations at room temperature. Organic solutions often react with polymers and Table X gives a summary of the compatibility of a number of common plastics with a range of both organic and inorganic solutions.

As in the case of metals, some caution should be exercised in its interpretation since a number of plasticizers can be used in combination with each polymer. It is often the former which is affected by the solution, not the polymer molecules themselves.

For almost complete chemical inertness the fluorinated polymers stand in a class of their own, the limitations to their wider application being primarily cost and also some constructional difficulties.

C. Thermoplastic and thermosetting resins

1. *Basic definitions*

Broadly speaking plastics can be divided into thermoplastic resins and thermosetting resins. The former are rigid at lower temperatures and soften with increasing temperature, the process being reversible within certain limits. Thermosetting resins, on the other hand, obtain their final chemical form in an irreversible hardening process when the shape is simultaneously obtained. This grouping however is not rigid and modern techniques can modify properties of plastics to give a wide range of properties with the same basic chemical structure. However all the resins useful because of their transparent or translucent properties are to be found among the thermoplastic types. For this reason no values for light transmission are given in Table VII.

2. *Common thermoplastic resins and their uses*

(a) *Polythene.* Polyethylene (or polythene) is formed by either a high or low pressure polymerization of ethylene. It is a crystalline substance with the degree of crystallinity proportional to the density, which it is possible to control to a certain degree to give different properties. As a result polyethylene is defined as low density or high density, with the low density plastic having a lower melting point and being softer (Renfrew and Morgan, 1957).

Some laboratory containers, particularly reagent bottles, beakers and measuring cylinders, not requiring sterilization, are often made from polyethylene of high density type since it withstands most chemicals and boiling water. It is also often used for the membrane covering electrodes for the measurement of dissolved oxygen, in which case chemical sterilizing agents must be used, such as ethylene oxide. The limited temperature range in which polyethylene can be used, together with the fact that only thin films can be made transparent, would appear to restrict its area of applicability in the microbiological field.

TABLE X

Chemical resistance of various plastics

Type of plastic	acetone		ethyl ether, etc.	ethanol, propanol, etc.		benzene, toluene, etc.		ethyl acetate		vegetable fats		ammonia, 30%	
°C	20	60	20	20	60	20	60	20	60	20	60	20	60
Acetal	+	+	+	+	+	+	O	+	+	+	+	+	–
Acrylic	–	–	–	O	–	–	–	–	–	+	O	+	+
Acrylonitril-butadiene styrene	–	–	–	+	O	–	–	–	–	+	+	+	+
Epoxy	O	–	+	+	O	+	+	O	O	+	+	O	O
Melamine-formaldehyde	+	+	O	+	+	+	+	+	+	+	+	O	O
Phenol-formaldehyde	O	O	O	+	+	+	+	+	+	+	+	O	–
Polyamide	+	O	+	+	+	+	+	+	+	+	+	+	O
Polycarbonate	–	–	–	+	+	O	O	–	–	+	+	–	–
Polyester	–	–	O	O	O	O	O	O	–	+	+	O	O
Polyethylene, high density	+	O	O	+	+	O	O	O	O	+	O	+	+
Polyethylene, low density	O	–	O	+	O	–	–	O	–	O	–	+	+
Polypropylene	+	O	O	+	+	O	–	O	O	+	O	+	+
Polystyrene	–	–	–	+	+	–	–	–	–	O	–	+	+
Polytetrafluorethylene	+	+	+	+	+	+	+	+	+	+	+	+	+
Polyvinyl chloride, flexible	–	–	–	O	–	–	–	–	–	O	O	+	O
Polyvinyl chloride, rigid	–	–	O	+	O	–	–	–	–	+	+	+	O
Silicone	–	–	–	–	–	–	–	–	–	+	+	O	O
Urea-formaldehyde	+	+	O	+	+	+	+	+	+	+	+	O	O

Table X.—*continued*

Type of plastic		sodium hydroxide				hydrochloric acid 30%		sulphuric acid				nitric acid 10%		acetic acid 50%	
		10%		conc.				10%		conc.					
	°C	20	60	20	60	20	60	20	60	20	60	20	60	20	60
Acetal		+	+	+	O	–	–	–	–	–	–	–	–	–	–
Acrylic		+	+	+	O	O	O	+	+	–	–	+	O	–	–
Acrylonitril-butadiene styrene		+	+	+	+	+	O	+	+	–	–	O	–	O	–
Epoxy		O	O	O	O	+	+	+	+	–	–	+	O	+	O
Melamine-formaldehyde		O	–	–	–	–	–	O	–	–	–	–	–	O	–
Phenol-formaldehyde		–	–	–	–	+	O	+	+	–	–	–	–	+	+
Polyamide		+	+	O	O	–	–	–	–	–	–	–	–	–	–
Polycarbonate		–	–	–	–	+	+	+	+	–	–	+	+	–	–
Polyester		O	O	–	–	+	+	+	+	–	–	+	O	+	O
Polyethylene, high density		+	+	+	+	+	+	+	+	+	O	+	+	+	+
Polyethylene, low density		+	+	+	+	+	O	+	+	O	O	+	+	+	+
Polypropylene		+	+	+	+	+	+	+	+	+	–	+	+	+	+
Polystyrene		+	+	+	+	+	+	+	+	O	O	+	+	+	+
Polytetrafluorethylene		+	+	+	+	+	+	+	+	+	+	+	+	+	+
Polyvinyl chloride, flexible		+	O	O	–	+	O	+	O	–	–	O	O	+	O
Polyvinyl chloride, rigid		+	+	+	O	+	O	+	+	+	O	+	+	+	+
Silicone		+	+	–	–	O	O	+	+	–	–	–	–	–	–
Urea-formaldehyde		O	–	–	–	–	–	O	–	–	–	–	–	O	–

\+ completely resistant.
O partially resistant
– attacked.

(b) *Polypropylene.* Polypropylene is a more recent addition to the polyolefines, having a degree of crystallinity which gives the polymer a high softening temperature. This is coupled with a low density and very good chemical resistance, even against the effects of fats and organic solvents where polyethylene is less resistant (Haines, 1963; Kressler, 1960). Earlier, polypropylene was noted for becoming brittle at low temperatures and the impact strength reduced drastically at approximately 0°C, but by the incorporation of small quantities of rubber or by forming a copolymer with other olefines, it has proved possible to produce varieties with good qualities in this respect at temperatures as low as −40°C. Like polyethylene, it has a very low permeability to water whilst allowing gas molecules to pass and as such very thin films can be used for autoclavable dissolved oxygen electrodes (Brookes, 1968). This property of allowing autoclaving has permitted its use for the production of a wide range of products, including sterilizable centrifuge tubes. Another property which may increase its usefulness in the laboratory is its high fatigue strength under constant flexing and at least one fraction collector (LKB Instruments Ltd., Stockholm, Sweden) incorporates hinged test tube holders of this plastic, where the hinge is simply a thin section of polypropylene.

(c) *Polystyrene.* As would be expected polystyrene is produced mainly by the polymerization of styrene, but copolymers with acrylonitrile and butadiene are also produced in large quantities. Acrylonitrile-styrene in particular has a greatly increased chemical resistance so that substances which release stress cracks in unmodified polystyrene have very little effect (Teach and Kiessling, 1960).

Polystyrenes are cheap and can be produced glass clear, so they have found wide application as disposable items in the microbiological laboratory, particularly as they are easily injection moulded. Since the maximum usable temperature lies below 100°C the articles produced from these polymers cannot normally be re-used. The consumption of polystyrene in general has been increasing by as much as 25% per annum and there is no doubt that within the laboratory field equally rapid expansion is taking place. In addition to Petri dishes, sample cups and syringes, other precision items such as pipettes are manufactured and it has even been possible to develop such optical qualities that disposable cuvettes are now used in some optical instruments.

The growing cost of service facilities, such as glassware washing and autoclaving, will no doubt increasingly favour the use of such disposable equipment. Its future in the microbiological laboratory is assured.

(d) *Polyvinyl chloride (PVC).* Amongst the vinyl plastics only polyvinyl chloride has much application in the laboratory field. It is formed by polymerization of vinyl chloride in an emulsion or suspension in water with the help of initiators. The resulting small particles can be spray-dried for further processing. In pure form it has a specific gravity of 1·38 and a softening point of 80°C. Its usefulness depends on its ability to accept different softening substances in a wide range of concentrations, so that the end product can have varying characteristics (Penn, 1962).

It can be extremely hard and in this form is used for wastepipes in laboratory furniture and aerating devices in biological effluent treatment, since it is resistant to a wide range of chemicals, whilst being easy to manufacture in a module system. Softer variations are used for the manufacture of floor tiling, whilst even more soft versions are used to coat protective gloves.

The most useful application of a vinyl chloride based polymer in the laboratory is in the form of tubing. The clear, flexible tubing, manufactured under the trade name of Tygon (U.S. Stoneware, Inc., Akron, Ohio, U.S.A.) commonly used for the transfer of many biochemical fluids, is a copolymer of vinyl chloride and vinylidene chloride. In its basic form it is very resistant to a wide variety of chemicals and there are different formulations for specific requirements. The disadvantages compared with some rubber tubings are a somewhat limited elasticity, connections requiring both careful sizing and a clamping device, together with only a moderate resistance to steam in the standard grades. It will, however, withstand autoclaving at least once.

(e) *Polyamides—nylon.* Nylon is a generic term covering long chain polymeric amides with the recurring amide groups forming part of the main chain, separated by a variable number of methylene groups (Parker, 1967; Floyd, 1958). It is possible to manufacture over 3000 different types of polyamides but only a few are commercially produced. The nomenclature used to define their composition is the number of carbon atoms in the diamine followed by the number of carbon atoms in the di-basic acid used in their formation. For example, nylon 6/6 is formed from hexamethylene diamine and adipic acid as in the following equation—

$$NH_2(CH_2)_6NH_2 + HOOC(CH_2)_4\,COOH \rightarrow$$
$$-NH(CH_2)_6NHCO(CH_2)_4CO- + 2H_2O$$

Similarly, nylon 6 is formed from caprolactam with the aid of a catalyst—

$$NH(CH_2)_5CO \xrightarrow{\text{catalyst}} -NH(CH_2)_5CO-$$

Polyamides, apart from their fibre-forming properties which are of value in other spheres, have good mechanical and fair chemical characteristics, except that their maximum useful temperature is of the order of 110°C. They will, however, withstand a few repeated autoclavings at 120°C, but absorb considerable quantities of water vapour which results in dimension changes. They will not withstand strong acids, oxidizing agents and chlorine-containing bleaches. Phenols dissolve polyamides. These factors limit the usefulness of nylons in the laboratory, but they find applications in small equipment.

(f) *Acrylic plastics*. These are based on 90% methyl methacrylate with supplementation by an ester of acrylic or methacrylic acid. Their main useful properties are transparency and ease of machining, which make them of use in the field of modelling for design of instruments. They do not withstand high temperatures nor most organic solvents, so do not find direct use in the microbiological laboratory.

(g) *Polycarbonates*. Outstanding toughness characterizes polycarbonates, together with dimensional stability, heat resistance and transparency. The one which has found most large-scale use is based on bisphenol A with the structure—

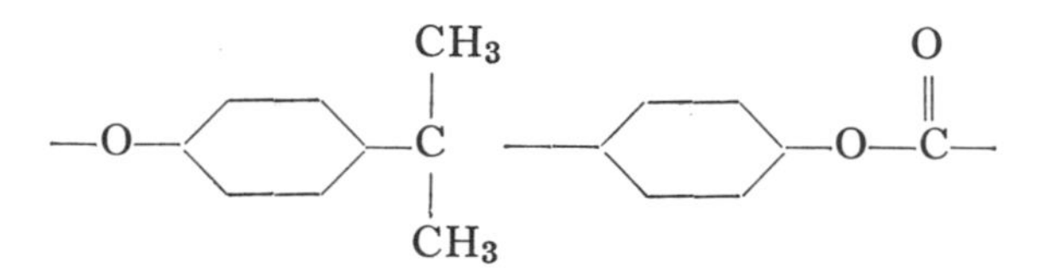

Since the plastic can withstand a continuous temperature of 120°C it is a well used plastic in the microbiological laboratory, for such items as centrifuge tubes. Its area of usefulness is somewhat limited by a sensitivity to organic solvents and alkalis although it can withstand weaker acids. It can also be deformed, if autoclaved under mechanical stress. Other articles produced from polycarbonate are syringes and Petri dishes.

(h) *Fluoro plastics*. Fluorine can replace hydrogen atoms in some polymers with straight carbon chains having 10,000 to 100,000 units (Rudner, 1958). This replacement results in a remarkable series of plastics such as polytetrafluoroethylene (PTFE), fluorinated ethylene-propylene (FEP) and polyvinylidene (VF_2).

They are uniquely resistant to attack by nearly all chemicals and have operating temperatures from −200°C to +250°C, combined with excellent electrical properties and negligible water absorption. They have a very low coefficient of friction making it difficult for other materials to adhere and it has been suggested that micro-organisms should not attach themselves to PTFE surfaces, although no direct evidence has been published to support this hypothesis. The low friction quality finds its use in lubricant-free bearings in small fermenters, but since PTFE is not dimensionally stable with changes in temperature it is not a good material for large items. FEP is produced as a film with good gas transfer qualities, at the same time having extremely low water transmission, and has been used as a membrane for autoclavable dissolved oxygen electrodes by Johnson *et al.* (1964).

(i) *Cellulose derivatives.* The natural polymer cellulose forms the basis for several plastics produced by chemical substitution into the molecule. These are the cellulose esters and cellulose ethers, together with cellulose nitrate.

After dissolving these polymers in a mixture of methylene chloride and methanol, the resultant solutions may be cast into membranes. It is these membranes which are of importance in microbiology, because they have varying degrees of porosity, making them suitable for dialysis or differential filtration.

(j) *Methylpentenes.* Methylpentene polymers are relative newcomers to the list of thermoplastic resins of potential importance to microbiology. They are almost as clear as acrilics but have the big advantage of being able to withstand a virtually unlimited number of steam sterilizations without loss of transparency or shape. Chemically they are resistant to inorganic salts and most inorganic acids and alkalis. Resistance to many organic chemicals is good, although there are exceptions, and, in addition their permeability to gases is higher than that of the olefines, for example.

3. *Common thermosetting resins and their uses*

Most thermosetting polymers are of only indirect use to the microbiologist, since the final product must be formed during the manufacturing process. This means that they are most suitable for series production of specific items such as instrument cases, and are not normally available in sheet form, for example, for further machining.

There are exceptions to this generalization, in the case of laminated plastic protective sheets for use as laboratory bench tops and polymer kits of thermosetting resins for manufacture of special items or coatings for protective purposes.

(a) *Phenol-formaldehyde resins.* One of the earliest plastics to be developed was Bakelite, from the research carried out by Baekeland on the reaction between phenol and formaldehyde. Phenol resins now comprise a wide range of plastics produced from not only phenol and formaldehyde but also cresol, resorcinol, alkylphenols and furfural (Megson, 1958).

Although used for the production of a tremendous range of household and industrial articles, their application in the laboratory is mainly confined to bench tops. Phenol-formaldehyde resins permit high pressure lamination techniques, the final product, sometimes having a melamine resin finish, being extremely resistant to impact and the effects of chemicals. Thin aluminium layers are sometimes incorporated to disperse heat and reduce the effects of its sensitivity to heat.

(b) *Epoxy resins.* Epoxy resins have ethylene oxide groups as the common chemical factor, although they are formed of rather complicated chains having molecular weights of the order of 1,000,000 (Parker, 1967). The epoxide groups can react with a wide range of hardeners, such as amines, alcohols, phenols, carboxylic acids and acid anhydrides (Skeist, 1958). This reactivity combined with the ability to accept xylene or dibutyl phthalate as diluents, makes it possible to create a range of resins with properties tailor-made to suit requirements.

Their general qualities are good mechanical properties, with low shrinkage, resistance to chemical attack and excellent adhesive properties. This adhesiveness makes epoxy resins excellent glues for specialized construction work of the type encountered in advanced laboratory operations and, with certain preliminary treatment, even PTFE can be bonded to other common materials such as rubber or metals.

Since no bubbles are produced during the hardening process, the low shrinkage quality is useful for the construction of instruments with encapsulation of electrical connections, or coating equipment with a protective covering.

(c) *Polyesters.* Polyesters are usually formed through an esterification condensation between glycols and dicarboxylic acids (Boenig, 1964). As with most resins, the structure can be modified considerably to give desirable properties which in this case may be strength, flexibility or curing behaviour. For example, if the glycol used is based on bisphenol A, the resin has excellent resistance to acids and alkalis.

Their most widely used property is an ability to accept reinforcement by glass fibre. This increases the tensile strength of the plastics enormously and enables them to be used for a wide number of applications where metals would be the only other choice. Biochemical storage tanks (Reid, 1967) and even parts for centrifuge rotors have been made from reinforced polyester. Their low density gives them a weight advantage over steel and their chemical stability enables them to replace stainless grades.

(d) *Silicones.* Silicones are organopolysiloxanes, consisting of alternating silicon and oxygen atoms located in the backbone of the molecule as follows—

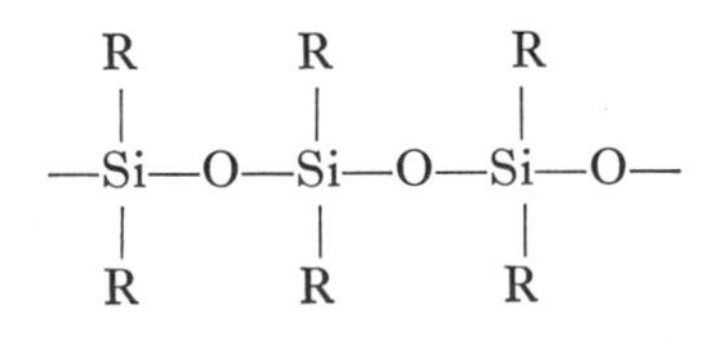

The sidegroups are simple hydrocarbon radicals (Mansfield and Magnusson, 1966), usually methyl, phenyl or vinyl radicals. The length of the chain can be varied readily so that the resultant polymers vary from low viscosity liquids to semi-solids. Branched polymers may also be produced.

Silicones have certain outstanding characteristics including thermal and oxidative stability at temperatures of 200°C and in some cases as high as 400°C, flexibility at low temperatures, general chemical inertness and excellent electrical properties.

In themselves silicones are of little value, but their ability to incorporate a wide variety of fillers and dissolve in a number of solvents makes their range of applications enormous. This covers antifoams to cold-setting plastics to silicone rubbers. The plastics have not as yet been much used in the microbiological laboratory but they do have potential for incorporation into special equipment where they can be used for sealing in components, such as metal items if there is ionic interference.

4. *Methods of fabricating plastic articles*

Manufacture of plastic articles has become an industry of its own with specialization even within the industry, so only the briefest outline is given here.

Thermoplastics can be processed by extrusion for tubing and injection moulding is used extremely commonly for the production of many plastic articles. These methods provide most of the plastic items produced from thermoplastics. Other methods include blow moulding and vacuum moulding. In the case of thermosetting resins there are fewer alternatives. Compression moulding is used for the series production of reinforced polyester components of small size, but commonly the layers are added by hand. Special techniques are used for the mass production of laminates used in bench tops.

As far as the individual laboratory worker is concerned, epoxy resins are particularly suitable for casting special electrodes and other small items, small kits of resin and hardener being available having a wide range of properties. Otherwise machining from the solid is the only alternative available.

Many plastics are available in rod, tube or sheet form and ordinary wood or metal-working machines are suitable for machining as long as care is taken in choosing the most suitable cutting speed. This varies widely but a useful rule is to start with a low cutting speed and increase this steadily until burning occurs in the case of a thermosetting plastic or softening in the case of a thermoplastic resin. A speed 10–20% lower than this limit is the usual optimum. Heat removal can be a problem but sharp tools, possibly combined with an air stream over the work, can minimize heating of the plastic. In particular, care must be taken in machining polytetrafluoroethylene since it decomposes to evolve a highly toxic gas when burnt. The fluorinated polymers also present some problems in construction. Processing of the raw powder is achieved by compression moulding or extrusion, followed by sintering at 325°C. If this is not carefully carried out the product remains porous and for most applications the articles are machined from rods, sheets or tubes. In precision work machining presents difficulties as a result of dimension changes following temperature variations, which make close tolerances impossible to achieve for items which are autoclaved. Screw threads can usually be machined in holes and on rods, but holes should be made slightly larger than normal to avoid excessive heat on cutting the thread.

D. Future development of applications of plastics

There is no doubt that the range of applications of plastics in the microbiological laboratory will continue to be extended steadily during the next few years. Already routine handling equipment, except for shake flasks, is available as disposable or autoclavable plastics and is replacing glassware. It is for more sophisticated equipment, such as fermenters, that plastics have not as yet made a break-through, although there is no reason why they should not be used for such purposes. Series production would undoubtedly make them considerably cheaper than autoclavable stainless steel fermenters, although their poor heat transfer characteristics would probably make heating and cooling fingers of metal necessary. Polypropylene, for example, tolerates 120°C satisfactorily, has sufficient rigidity without reinforcement in small sizes and is insensitive to virtually all medium components and fermentation products.

Reinforced plastics are becoming widely used as storage vessels in the biochemical industries as has been pointed out by Reid (1967), but for smaller scale tanks this reinforcement would in most cases be unnecessary, unless autoclaving under pressure is required, when reinforcement is fully capable of providing the strength necessary to counteract the required pressure. Their ease of construction compared with stainless steel is another element in their favour. This type of development will certainly be the next phase in the increasing use of plastics in microbiology.

V. RUBBER

A. Types and properties of rubber

As with plastics, there is a wide range of different types of rubber available, having varying properties which affect their usefulness as materials for use in the field of microbiology. In the laboratory there is a limited number of applications for rubbers. They are primarily used for connecting tubing, fermenter gaskets and enclosure gaskets for sample or injection bottles.

Table XI summarizes the properties of the main types of rubber that may be of use to the microbiologist, together with some trade names that help to identify them. In many cases, such as the silicones, there is no real distinction between rubbers and plastics, as their properties overlap each other.

B. Applications in microbiology

Tubing is a common item in the fermentation laboratory, for the various connections to small cultivation tanks including air and medium supplies. Various types of tubing may be employed, latex or natural rubber tubing being the cheapest. However, this type of tubing cannot withstand 120°C for more than a short period, the actual time being somewhat dependent on the quality, so that it requires either replacement after each autoclaving, or at best after 3 or 4 such treatments.

A second drawback is that screw clamps used for sealing the tubing during autoclaving have a strong tendency to weld the walls of the tubing together.

These disadvantages are not present in the case of silicone rubber tubing, such as that manufactured by Esco (Rubber) Ltd., 25 Seething Lane, London, E.C.3., England, which is extremely inert chemically and biologically, being the choice for all tissue culture applications. It has no tendency to adhere to itself or any other material. On the other hand, it has a considerably lower tensile strength, elasticity and resistance to abrasion. It is easily cut accidently and it readily becomes kinked under its own weight, preventing the flow of a fluid. On the positive side it withstands enumerable autoclavings and can be used in some types of peristaltic pumps, although if the tubing is required to be in tension, high quality latex tubing is often more satisfactory. Tube life for silicone tubing under continuous operation is usually quoted as a week or more.

Certain fluorinated rubbers, however, do have a high tensile strength and combine the advantages of both natural and silicone rubbers, but their price does not normally warrant their use.

In the case of anaerobe cultures, silicone rubber tubing is unsuitable for an entirely different reason. It is highly permeable to a number of gases, including oxygen as can be seen from Table XII of Section VII. The ingression of oxygen, at the rates permitted by silicone tubing, almost invariably inhibits the growth of such organisms in fermenters on the laboratory scale.

Somewhat similar mechanical problems occur in the choice of membrane seals for flasks and sampling ports. In this case natural rubber does not withstand the stresses of autoclaving and ages on storage. Silicone rubbers

TABLE XI

Properties of common types of rubber.

Type (ASTM code) Trade names	Chemical structure	Mechanical properties		Temperature limit, °C		Chemical resistance			Approximate price U.S. $ per kg (1965)	Autoclaving qualities
		Tensile strength kg/cm²	Abrasion resistance	high, dry atmosphere	low	steam, weak acids and alkalis	strong and oxidizing acids	oils		
Natural rubber (*NR*) "latex" "smoked sheets" and numerous other trade names	Cis-polyisoprene	150–300	good	100	−70	excellent	fair	poor	0·6	Can be autoclaved once or twice
Isoprene rubber (*IR*) "*Synthetic natural rubber*" Natsyn (Goodyear, U.S.A.) Shell IR (Shell, U.S.A.) SKI (USSR)	Cis-polyisoprene	100–250	good	100	−70	excellent	fair	poor	0·6	Can be autoclaved once or twice
Butadiene rubber (*BR*) Ameripol CB (Goodrich-Gulf U.S.A.) Astyr (Montecatini, Italy) Budene (Goodyear, U.S.A.) Cis-4 (Phillips, U.S.A.) Cisdene (ASR, U.S.A.) Diene (Firestone, U.S.A.) Duragen (General Tire, U.S.A.) Europrene Cis (Anic, Italy) Taktene (Polym. Corp. Canada SKB, SKV, SKBM (USSR)	Cis-polybutadiene	100–200	excellent	120	−80	excellent	fair	poor	0·6	Suitable for repeated autoclaving
Ethylene-propylene rubber (*EPM, EPDM*) Dutral (Montecatini, Italy) Enjay EPT (Enjay, U.S.A.) Nordel (Du Pont, U.S.A.) Royalene (U.S. Rubber, U.S.A.)	Copolymer of 60–70% ethylene, 30–40% propylene and small amounts of a diene e.g. dicyclopentadiene	100–225	good	140	−80	excellent	good	poor	0·8	Suitable for repeated autoclaving
Styrene rubber (*SBR*) *emulsion polymerized* Ameripol (Goodrich-Gulf, U.S.A.) Austrapol (Australian Synth. Rubber) Buna (IG Farben, Germany) Cariflex (Shell, U.S.A.) Europrene (Anic, Italy) FR-S (Firestone, U.S.A.) Intol (IRC, England) Kralex (Czechoslovakia) Naugapol (Naugatuck, U.S.A.) Philprene (Phillips, U.S.A.) Plioflex (Goodyear, U.S.A.) Polysar S, Krylene, Krynol (Polymer Corp. Canada) SKS (USSR) Synpol (US Rubber, U.S.A.)	copolymer of 20–30% styrene and 70–80% butadiene	100–250	good	120	−60	excellent	fair	poor	0·4	Suitable for repeated autoclaving

Styrene rubber (SBR) solution polymerized Duradene (Firestone, U.S.A.) Solprene (Phillips, U.S.A.)	Copolymer of 20–30% styrene and 70–80% butadiene	100–250	good	120	−70	excellent	fair	poor	0·6	Suitable for repeated autoclaving
Nitrile Rubber (NBR) Breon (Goodrich, England) Buna N (Germany) Butakon (ICI, England) Butacril (Ugine, France) Butaprene (Firestone, U.S.A.) Chemigum N (Goodyear, U.S.A.) Hycar (Goodrich, U.S.A.) Paracril (Naugatuck, U.S.A.) Perbunan N (Bayer, Germany) Krynac (Polymer Corp, Canada) SKN (USSR)	Copolymer of 20–60% acrylonitrile and 60–80% butadiene	100–250	good	130	−60	good	fair	good	1·0	Suitable for repeated autoclaving
Carboxylic rubber Hycar 1072 (Goodrich, U.S.A.)	Polymer of butadiene acrylonitrile and acrylic acid derivative	100–250	good	130	−70	good	fair	good	1·4	Suitable for repeated autoclaving
Vinylpyridine rubber Philprene VP (Phillips, U.S.A.) SKMVP (USSR)	Copolymer of 15–35% 2-methyl-5-vinyl-pyridine and 65–85% butadiene	50–200	good	130	−60	good	fair	good	..	Suitable for repeated autoclaving
Chloroprene rubber (CR) Neoprene (Du Pont, U.S.A.) Perbunan C (Bayer, Germany) Sovpren, Majrit (USSR) Svedopren (Sweden)	Polychloroprene	100–250	good	130	−50	excellent	fair	fair	1·0	Suitable for repeated autoclaving
Butyl rubber (IIR) GR-I (U.S.A.)	Copolymer of isobutylene and 0·5–3% isoprene	100–180	fair	140	−50	excellent	good	poor	0·6	Suitable for repeated autoclaving
Brominated butyl rubber Hycar 2202 (Goodrich, U.S.A.)	Butyl rubber plus 1–3·5% bromine	100–150	fair	140	−40	excellent	good	poor	1·6	Suitable for repeated autoclaving
Chlorinated butyl rubber Butyl HT 10–66 and 10–68 (Enjay, U.S.A.)	Butyl rubber plus 1·2–1·3% chlorine	100–180	fair	140	−40	excellent	good	poor	0·8	Suitable for repeated autoclaving
Acrylic rubber Acrylon (Borden Co., U.S.A.) Cyanacryl (Cyanamid, U.S.A.) Hycar 4021 (Goodrich, U.S.A.) Thiacril (Thiokol Corp. U.S.A.)	Polymer of an acrylate with small quantities of other ingredients	75–125	fair	170	−20	fair	poor	good	3·0	Suitable for repeated autoclaving
Chlorosulphonated polyethylene Hypalon (Dupont, U.S.A.)	Chlorosulphonated polyethylene containing 27·5% chlorine and 1·5% sulphur	100–200	good	150	−55	good but swells in steam	good	fair	1·4	Not recommended for autoclaving

TABLE XI—*continued.*

Type (ASTM code) Trade names	Chemical structure	Mechanical properties: Tensile strength kg/cm²	Mechanical properties: Abrasion resistance	Temperature limit, °C: high dry atmosphere	Temperature limit, °C: low	Chemical resistance: steam, weak acids and alkalis	Chemical resistance: strong and oxidizing acids	Chemical resistance: oils	Approximate U.S. $ price per kg. (1965)	Autoclaving qualities
Urethane rubber Vulkollan Desmophan Urepan (Bayer, Germany) Adiprene (Du Pont, U.S.A.) Daltoflex (ICI, England) Elastothane (Thiokol Corp. U.S.A.) Estane (Goodrich, U.S.A.) Genthane S. (General Tire, U.S.A.) Multrathane, Texin (Mobay, U.S.A.)	Polyester or polyether joined with polyisocyanates and other cross-linking reagents	300–500	excellent	120 but shrinks badly over 80°C	−60 but stiffens at −20°C	poor	poor	good	2·4	Cannot be autoclaved
Silicone rubber (Si) Rhodorsil (Rhone Poulence France) Silastic (Dow Corning, U.S.A.) Silastomer (Midland Silicones, England) Siloprene (Bayer, Germany) SKT (USSR)	Poly-(methyl-, phenyl- or vinyl-) siloxanes	40–100	poor	275	−100	fair	poor	fair	8·0	Suitable for repeated autoclaving
Nitrile-(cyan)-silicone rubber NSR (General Electric, U.S.A.) (Union Carbide, U.S.A.)	Silicone rubber with cyano-groups in the side chains	40–70	poor	225	−70	fair	poor	fair	35·0	Suitable for repeated autoclaving
Fluoro-silicone rubber Silastic LS (Dow Corning U.S.A.)	Silicone rubber with fluorinated side groups	50–70	poor	175	−70	fair	poor	good	30·0	Suitable for repeated autoclaving
Fluorine rubber Kel-F Elastomer 5500 and 3700 (Minnesota Mining, U.S.A.)	Copolymer of trifluoro-monoc loroethylene with small quantities of vinylidene fluoride	150–200	good	200	−55	excellent but shrinks in steam	excellent	excel-lent	40·0	Suitable for repeated autoclaving
Fluorine rubber Silastic LS (Dow Corning, U.S.A.)	Silicone rubber with fluorinated side groups	50–70	poor	175	−70	fair	poor	good	40·0	Suitable for repeated autoclaving
Fluorine rubber Viton (Du Pont, U.S.A.) Fluorel (Minnesota Mining, U.S.A.)	Copolymer of hexa-fluoropropylene and vinylidene fluoride	150–200	good	250	−45	excellent but shrinks in steam	good	excel-lent	25·0	Suitable for repeated autoclaving

This table was prepared with data supplied by H. Palmgren, Trelleborgs Gummifabriks Aktiebolag, Trelleborg, Sweden.

do not normally have the required tensile strength to be self-sealing after puncture by a syringe needle. Qualities are being developed which are better in this respect, but the rubber of choice seems to be vulcanized types containing various fillers. These have the necessary temperature and ageing qualities, combined with sufficient elasticity to prevent a hole being punched through the rubber on puncturing with a syringe needle. The rubbers are self-sealing in that the subsequent hole is also sealed on removal of the needle.

Gaskets do not normally present a problem since, apart from the additional stresses of autoclaving, they do not require such a high degree of elasticity as other applications. Synthetic rubbers are most suitable, since if they resist the medium chemicals then there is little likelihood of extraction from the rubber. Chloroprene in particular is widely used, having excellent resistance to temperatures of 120°C, steam and moderately strong acids and alkalis. Other types are also possible such as chlorinated butyl rubber, styrene rubber and nitrile rubber, the latter also having good resistance to oils.

Silicone rubbers do not have the required tensile strength. For certain extreme conditions a fluorine rubber, consisting of a copolymer of hexafluoropropylene and vinylidene fluoride known as Viton or Fluorel, although expensive, can provide a satisfactory solution. These types, however, show a tendency to shrink slightly when in contact with steam, without deterioration in any other sense.

C. Chemical contamination of solutions by pharmaceutical grades

Because of its importance during the storage of filled vials having rubber stoppers, the leaching effect of solutions in contact with rubber has received some attention. One of the problems involved in such studies is the wide variety of rubbers produced with quite complex compositions. Reznek (1953) reports studies on a particular pharmaceutical grade of rubber containing approximately 40% natural rubber, 50% barium sulphate, together with small quantities of zinc oxide, stearic acid, sulphur, mercaptobenzole, thiuram monosulphide, paraffin wax and titanium oxide. Zinc in particular was partially soluble when in contact with aqueous solutions, increasing temperature greatly accelerating the leaching effect. After two days contact, the quantity of zinc in solution at 45°C was twice that of an ampoule stored at room temperature.

Reznek also studied the effect of the quantity of zinc in the rubber and the proportion released, since zinc is often used in rubber in considerable quantities as a stiffener. He found that increasing zinc oxide content beyond 0·6% did not significantly affect the tensile strength but beyond this concentration the proportion of zinc leached from the rubber by 0·1 N HCl in a given time increased abruptly, so that as much as 40% of the zinc oxide became soluble at 70°C after 56 days.

In a comprehensive study of 26 samples of natural, synthetic or silicone rubbers, suitable for pharmaceutical enclosures, subjected to autoclaving in the presence of water, Steiger and Dolder (1954) found several chemical changes in the water. These included pH variations of ± 2 pH units, redox potential changes, evolution of reducing agents, heavy metals, ammonia, amines, aldehydes, ketones and organic sulphur. No one type was satisfactory in all respects and attempts to manufacture a new grade showing negative reaction were also unsuccessful.

Wing (1958), in a review of the subject, states that solutions autoclaved in the presence of rubber enclosure caps almost invariably show the presence of reducing substances, according to titrations with 0·01 N permanganate. They also conclude that for pharmaceutical solutions no one special type is suitable. Similarly in microbiological experiments, cases of interference with a biological system by contact with rubber can arise. Prediction is impossible, it is a matter of bearing the possibility in mind when difficulties in cultivation or product formation arise.

VI. INTERACTION OF MICRO-ORGANISMS AND NON-METALS

A. Microbiological deterioration of common non-metals

Although it is known that mildew, dry rot of wood and rotting of fibrous materials are caused by the action of micro-organisms, it is not so widely realized that more rigid materials such as rubber and plastics are on occasions susceptible to microbiological attack. It has even been reported by Jones (1945) and Hutchinson (1948) that fungi can grow on glass surfaces of optical instruments in the tropics, producing metabolites capable of etching the glass. This presupposes, of course, a very high relative humidity.

1. *Rubber*

There is no doubt that, although there is conflicting evidence on specific effects, micro-organisms are able to attack rubber. As early as 1913, Söhngen and Fol (1914) isolated pure cultures of natural rubber-consuming micro-organisms from an enrichment culture in which pure rubber hydrocarbon was the main carbon source. Since then there have been many reports which have been reviewed and summarized by Greathouse *et al.* (1951). ZoBell *et al.* (1942, 1944) developed a technique to measure oxygen consumption, carbon dioxide production, multiplication of micro-organisms and deterioration of the rubber, as a means of evaluation, whereas Blake *et al.* (1949, 1950) used electrical resistance measurement tests of wire insulated with various rubbers. Agreement is complete that natural rubber is readily attacked by a number of micro-organisms including moulds, actinomycetes and bacteria. In the case of vulcanized rubber there is no such general agreement since there is a wide range of additives which may be capable of supporting growth. For example Blake *et al.* (1955) found that raw styrene rubber exposed to soil or inoculated water was rapidly coated with fungi. After acetone extraction the raw material did not support visible growth, but fungi did grow profusely on the evaporated extract.

It is in the field of soil microbiology that rubber attack by micro-organisms becomes of real significance, partly because the types of micro-organisms which possess the ability to attack rubbers are normally found in soil, and partly because the destructive action is slow in nature and only observed in long term experiments. Rook (1955) used a technique devised by Spence and van Niel (1936), employing latex-agar plates to isolate specific rubber decomposing micro-organisms. Clear zones indicated solubilization of rubber from which the organism responsible could be isolated. He also used thin strips of vulcanized rubber maintained in tension in which holes developed after attack by micro-organisms. Several *Streptomyces* strains were able to attack rubber hydrocarbon and one strain appeared to be capable of attacking vulcanized rubber.

It must be borne in mind that the experiments involving vulcanized rubber were carried out over a period of 12 months, so that it is really only in soil microbiology or when rubbers are tested for use in underground cables or tropical service that particular consideration of the problem must be taken.

2. *Plastics*

Biodeterioration of plastics has become an important area of study with the ever increasing range of new polymers and additives for specific purposes. Although it has not been conclusively proved for most synthetic polymers, as in the breakdown of natural polymers, the action of micro-organisms is presumed to be enzymatic in nature. A distinction must be made between microbial attack on the polymer itself and utilization of the plasticizer as a carbon source by micro-organisms. In general it is the latter that is susceptible. According to Greathouse and Wessel (1954) only phenol-aniline formaldehyde, melamine formaldehyde and casein formaldehyde, of the plastics which are likely to be found in the microbiological laboratory, have been shown to be capable of undergoing microbiological deterioration.

On the other hand a number of studies on the growth of micro-organisms on the array of plasticizers now used in the plastics industry have been carried out. The earlier work was reviewed by Brown (1945) and was mainly concerned with deterioration of equipment in the tropics. Of more direct interest are the studies of Stahl and Pessen (1953) and more recently of Berk *et al.* (1957) on the range of plasticizers most prone to attack by micro-organisms and their molecular configuration. These articles and other relevant literature have been reviewed by Wessel (1964).

B. Inhibitory effects of rubbers and plastics on micro-organisms

There have been few reports on the effects of plastics on microbiological environments and most of those published indicate little interference. Such negative results have been noted by Mackenzie (1951) and Walter *et al.* (1958) in experiments involving fluids in contact with a wide variety of plastics. Tiedeman and Malone (1955), however, did observe that rubber modified polystyrene greatly increased the microbial count of water stored in contact with the polymer.

Inhibitory effects of both rubbers and plastics on the growth of micro-organisms have been reported by Dyer and Richardson (1962). They tested several polymers for inhibition of growth of two algae, *Synechococcus lividus* and *Chlorella pyrenoidosa.* Of the rubbers tested, a nitrile rubber and some chloroprene rubbers showed strong inhibition, together with one type of cold-setting silicone rubber. The toxicity of the latter was almost certainly due to the catalyst used for the room temperature vulcanization. A sample of butyl rubber was shown by Rosenwald *et al.* (1962) to be highly toxic to *Pasteurella tularensis*, although this rubber is commonly used for gaskets.

Natural rubbers in particular seem to be common offenders in giving rise to problems in cultivation of micro-organisms. Davies *et al.* (1953) found no inhibitory effects of polyvinyl chloride tubing on the growth of *Chlorella*, but if latex rubber tubing was used in a peristaltic pump in the cultivation equipment, growth was stopped. A similar phenomenon occurred in the study of the growth of *Neisseria gonorrhoeae* by Brookes and Hedén (1967), where it was found that if the medium was autoclaved in the presence of one brand of latex tubing, growth was completely inhibited, whereas another brand was entirely without effect. The brand showing toxicity was without inhibitory effect on several other bacteria, both aerobic and anaerobic.

In the case of plastics similar anomalies occur. The results of Dyer and Richardson (1962) showed that the choice of plasticizer played the major role in determining the toxicity of a specific plastic. One sample each of polyvinyl chloride and polypropylene were highly toxic, whereas other types having the same basic polymer were entirely free from toxic effects. No other group of plastics among polyethylenes, polytetrafluoroethylenes, acrylics, polyamides, polyesters or epoxy resins showed any inhibitory effect.

The conclusion that can be drawn is that if one formulation gives rise to problems, it is not evidence that all other plastics incorporating the same basic polymer will exhibit the same effect. Factors such as purity of constituents are almost certainly involved in some instances.

VII. MEMBRANES

As was mentioned earlier, it has proved possible to produce membranes having more or less controlled porosity, some of which can tolerate autoclaving, thus making them particularly useful for microbiological studies. Other membranes may be useful for gas holders in specialized cultures such as methane oxidizing or hydrogen oxidizing organisms.

TABLE XII

Gas transmission rates of plastic films

Film	Oxygen ml/24 h 100 in^2 atm	Carbon Dioxide ml/24 h 100 in^2 atm	Water Vapour† g/24 h 100 in^2
Cellulose acetate	110	560	90·0
Opaque high density polyethylene	142	348	0·25
Polypropylene	187	639	0·7
Clear high density polyethylene	226	1030	—
Low density polyethylene	573	1742	1·2
Polytetrafluoroethylene	1100	3000	0·32
Silicone rubber	98,000	519,000	170·0

All films 0·001 in. thick.

† Measured with one side of film exposed to 90% relative humidity at 40°C and the other side maintained at 0% relative humidity with the aid of calcium chloride.

Compiled with the aid of manufacturers' data.

The permeability of some of these membranes to oxygen, carbon dioxide and water is shown in Table XII.

A. Membranes relatively impervious to water

Certain of the artificial polymers have an extremely low permeability to water, particularly polyethylene, polypropylene and fluorine olefines. On the other hand, they do permit relatively rapid transfer of gas molecules. These combined properties are extremely useful for one purpose, namely as a membrane covering in electrodes used for the measurement of dissolved oxygen. The ability of such membranes to prevent the passage of larger ionic molecules enables stable electrodes to be made with a very low rate of poisoning of the protected electrode metal.

Polyethylene cannot be autoclaved, but polypropylene and the fluorine derivates, polytetrafluoroethylene and fluorinated ethylene-propylene, withstand this treatment. Lampi (1967), after very careful experiments, found only negative results in attempts to obtain growth across polypropylene membranes. The technique used was to employ a membrane as the boundary surface of a vessel containing growing bacteria in liquid medium, the outer face of the membrane being in contact with a nutrient agar surface. A second technique, using a dye, was developed to detect the presence of pinholes. The results of the two tests could be correlated to show the exact location of any microbial penetration and dye leakage. In a large series of experiments there was no case of bacterial penetration without simultaneous detection of a pinhole at the same location. Pinholes themselves were rare unless the film had previously been mishandled deliberately. This result confirms the suitability of this type of membrane for use in dissolved oxygen electrodes, where it is essential that bacteria do not penetrate the protective film.

On occasions micro-organisms need to be grown under a controlled atmospheric environment. If balloon type reservoirs are used, then butyl rubber is the material of choice since it has an extremely high resistance to gas diffusion. As illustrated in Table XII, silicone rubber in particular has a very high permeability to gases, and natural rubbers are also offenders in this respect, but to a lesser degree. In fact the permeability of oxygen through silicone rubber may lead to specific uses of such membranes for oxygenators in enclosed equipment.

B. Dialysis-type membranes

The growth of micro-organisms in a relatively small environment, with liquid contact, by means of a dialysis membrane, to a much larger liquid volume can in certain circumstances be advantageous. Some of these advantages have been described by Gallup and Gerhardt (1963) and include an ability to produce a virtually unlimited population density—these workers producing cultures containing 10^{12} viable cells/ml, and the possibility of concentrating the cell suspension even further after growth by osmotic dehydration. An additional advantage for dialysis cultures was shown to be a retention of viability of the micro-organisms subsequent to the termination of the growth phase, compared with certain control cultures having a fully synthetic medium.

Regenerated cellulose and Cellophane are the membranes most commonly used for this purpose. They have remarkably fine pores of less than 3 nm diameter, but have a reasonably high porosity in spite of this small pore size. Macromolecules, such as proteins, are prevented from crossing the membrane and this factor may or may not be an advantage. These types of membranes have not been shown to be toxic, but all cellulosic membranes are naturally susceptible to microbial attack.

Newer types of synthetic membranes are now being manufactured from polyion complex resins—a mixture of two highly ionized linear polyelectrolytes of opposite charge (Amicon Corp., Cambridge, Mass, U.S.A.), for dialysis or ultrafiltration. They are claimed to be non-porous, the liquid transfer being accomplished by diffusion, and as such are not so prone to clogging as conventional cellulosic membranes, whilst at the same time allowing much higher liquid transfer rates. Chemically they are rather inert and will withstand a wide pH range, so chemical sterilization is used when necessary as the maximum tolerated temperature is in the region of 60°C. Their use in dialysis cultures has not as yet been reported.

C. Controlled porosity membranes

In addition to the dialysis type membranes, techniques have been developed over the last decade to produce thicker membranes from cellulose esters having a controlled pore size. These are produced, under a number of trade names, (e.g. Millipore, Bedford, Mass, U.S.A.), from cellulose nitrate, cellulose acetate and mixtures of the two polymers. Pore sizes are available ranging from 14 μm to 10 nm with quite close tolerances and a total porosity such that only 20% of the filter volume is composed of the polymer. The membranes withstand autoclaving, but do become brittle following this treatment, and they are normally pre-sterilized by the manufacturer

A wide range of specific uses has been devised for this type of membrane filter, the most common being the microbiological examination and sterile filtration of fluids (see Mulvany, this Volume, p. 205). Most applications are variations on these themes. They are not suitable for separating microorganisms from dense suspensions because clogging of the pores occurs very rapidly.

It might be thought that this type of membrane would be ideal for a variant of dialysis culture where macro-molecules can be separated from cells. Herold *et al.* (1967) attempted to use several types of membranes in such cultures of staphylococci. In all cases they were only partially successful as it was impossible to prevent the bacteria from crossing the membrane after a maximum period of approximately 100 h. Bacteria can exert tremendous osmotic pressures and it is likely that some cells divide in the region of a pore and destroy the continuity of the membrane.

Cells can be attracted to membranes and the adherence may or may not be advantageous. It has been used by Helmstetter and Cummings (1963) as a technique for obtaining synchronous cultures (see Helmstetter, this Volume, p. 327). When a culture of *E. coli* was filtered through a cellulose anion exchange membrane, a proportion of the cells became bound. By using a flow technique the daughter cells could be removed from the adhered cells, immediately subsequent to division. Consequently portions of the eluted culture could be used for synchronous growth experiments.

Other fully synthetic polymers are now becoming available, produced from such widely different resins as polyvinyl alcohol, polyvinyl chloride, polytetrafluoroethylene, nylon and vinylidene fluoride. It remains to be seen whether they have qualities superior to the cellulosic type of membrane.

VIII. LUBRICANTS

Modern microbiology usually involves the use of equipment incorporating mechanical motion. This in turn implies the use of some form of lubrication for the mutual separation of sliding parts.

Unfortunately, lubricants, usually of mineral oil composition, are not necessarily inert with respect to the microbial system under investigation. In most cases the problem can be avoided by designing the equipment so that lubricated parts lie outside the system; shaker tables are an example where this instruction is easy to follow. Normal lubrication technique can then be applied since no sterilization is required. This is undoubtedly the best solution wherever applicable, but naturally there are specific circumstances where this alternative is not available. The most obvious case where this advice cannot be strictly followed occurs in the stirred fermenter. Here bearings for the impeller require lubrication, unless use is made of tungsten carbide bearings, as mentioned in Section III B.6. Again the advice is to avoid contact between the lubricated bearing and the fermentation broth if at all possible.

The usual technique adopted to avoid contact is to design the impeller such that support bearings are mounted in the lid, then a cup can be attached to the impeller shaft to collect any possible grease. If this solution is adopted normal lubrication procedures can be followed, apart from the fact that a lubricant capable of withstanding 120°C should be selected. It should be sufficiently viscous at this temperature to prevent flow and in addition should be compatible with the materials it is intended to lubricate.

Occasionally small equipment can incorporate plastics and the polyamides in particular have excellent mechanical properties without the aid of a lubricant. However even the heat stabilized varieties cannot withstand an indefinite number of autoclavings, which restricts their use in manufactured equipment.

Fluorocarbons, which have superb temperature characteristics combined with the lowest coefficient of friction of any known solid, have poor mechanical properties and as a consequence wear very rapidly. Development work aimed at improving them in these respects by the use of reinforcements may improve their range of applications.

REFERENCES

Berk, S., Ebert, H., and Teitell, L. (1957). *Ind. Engng. Chem.*, **49**, 1115–1124.

Berwick, L., and Coman, D. R. (1962). *Cancer Res.*, **22**, 982–987.

Blake, J. T., and Kitchin, D. W. (1949). *Ind. Engng. Chem.*, **41**, 1633–1641.

Blake, J. T., Kitchin, D. W., and Pratt, O. S. (1950). *Elec Engng.*, **69**, 782–787.

Blake, J. T., Kitchin, D. W., and Pratt, O. S. (1955). *Appl. Microbiol.*, **3**, 35–39.

Boenig, H. V. (1964). "Unsaturated Polyesters. Structure and Properties". Elsevier, Amsterdam.

Bowen, H. J. M. (1966). "Trace Elements in Biochemistry". Academic Press, London and New York.

Brookes, R. (1968). Unpublished report.

Brookes, R., and Hedén, C.-G. (1967). *Appl. Microbiol.*, **15**, 219–223.

Brown, A. E. (1945). "The Problem of Fungal Growth on Synthetic Resins, Plastics and Plasticizers". U.S. Office of Scientific Research and Development, Report No. OSRD 6067.

Charles, R. J. (1967). *Scient. Am.*, **217**, No. 3, 127–136.

Coman, D. R. (1944). *Cancer Res.*, **4**, 625–629.

Cooke, F., and Williams, N. T. (1967). *Process Biochem.*, **2**, No. 7, 19–24.

Davies, E. A., Dedrick, J., French, C. S., Milner, H. W., Myers, J., Smith, J. H. C., and Spoehr, H. A. (1953). *In* "Algal Culture" (ed. J. S. Burlew), pp. 105–153. Carnegie Inst. Wash. Publ. No. 600, Washington, D.C.

Dyer, D. L., and Richardson, D. E. (1962). *Appl. Microbiol.*, **10**, 129–131.

Easty, G. C., Easty, D. M., and Ambrose, E. J. (1960). *Expl Cell Res.*, **19**, 529–548.

Evans, U. R. (1948). "An Introduction to Metallic Corrosion". Arnold, London.

Fåhreus, G., and Reinhammar, B. (1967). *Acta chem. scand.*, **21**, 2367–2378.

Floyd, D. (1958). "Polyamide Resins". Reinhold, New York.

Gallup, D. M. and Gerhardt, P. (1963). *Appl. Microbiol.*, **11**, 506–512.

Giardinello, F. E., McLimans, W. F., and Rake, G. W. (1958). *Appl. Microbiol.*, **6**, 30–35.

Greathouse, G. A., and Wessel, C. J. (1954) (Joint eds). "Deterioration of Materials: Causes and Preventive Techniques". Reinhold, New York.

Greathouse, G. A., Wessel, C. J., and Shirk, H. G. (1951). *A. Rev. Microbiol.*, **5**, 333–358.

Haines, H. W. (1963). *Ind. Engng. Chem.*, **55** : **2**, 30–37.

Helmstetter, C. E., and Cummings, D. J. (1963). *Proc. natn. Acad. Sci. U.S.A.*, **50**, 767–774.

Hernandez, E., and Johnson, M. J. (1967). *J. Bact.*, **94**, 991–995.

Herold, J. D., Schultz, J. S., and Gerhardt, P. (1967). *Appl. Microbiol.*, **15**, 1192–1197.

Heukelekian, H., and Heller, A. (1940). *J. Bact.*, **40**, 547–558.

Holmström, B. (1968). *Biotechnol. Bioengng.* **10**, 373–384.

Hutchinson, W. G. (1948). "The Frosting and Etching of Glass by Fungi". Final Report under U.S. Office of Naval Research, Contract No. N5-ORI-122 with the Univ. of Pennsylvania.

Johnson, M. J., Borkowski, J., and Engblom, C. (1964). *Biotechnol Bioengng.*, **6**, 457–468.

Jones, F. L. (1945). *J. Am. Ceram. Soc.*, **28**, 32.

Klein, M. (1959). "Einführung in die DIN-Normen". Teubner, Stuttgart.

Kluckow, P. (1963). "Rubber and Plastics Testing". Chapman and Hall, London.

Kressler, T. O. J. (1960). "Polypropylene". Reinhold, New York.

Lampi, R. A. (1967). Resistance of Flexible Packaging Materials to Penetration by Microbial Agents". Technical Report 67-62-GP. U.S. Army Natick Laboratories, Natick, Mass., U.S.A.
Larsen, D. H., and Dimmick, R. L. (1964). *J. Bact.*, **88**, 1380–1387.
Lederberg, J., and Lederberg, E. (1952). *J. Bact.*, **63**, 399–406.
Lever, A. E., and Rhys, J. (eds.) (1957). "The Properties and Testing of Plastics Materials". Temple Press, London.
Mackenzie, E. F. W. (1951). *J. Instn Wat. Engrs*, **5**, 596–604.
Mansfield, R. A., and Magnusson, J. A. (1966). *Mod. Plast.*, **44**, 284–286.
Mark, H. F. (1967). *Scient. Am.*, **217**, No. 3, 149–156.
Megson, N. J. L. (1958). "Phenolic Resin Chemistry". Butterworth, London.
Molin, O., and Hedén, C.-G. (1968). *Prog. immunobiol. Standard.*, **3**, (in press).
Nordin, J. S., Tsuchiya, H. M., and Fredrickson, A. G. (1967). *Biotechnol. Bioengng.*, **9**, 545–558.
Nordling, S., Penttinen, K., and Saxén, E. (1965a). *Expl. Cell Res.*, **37**, 161–168.
Nordling, S., Vaheri, A., Saxén, E., and Penttinen, K. (1965b). *Expl. Cell Res.*, **37**, 406–419.
Parker, E. R. (1967). "Materials Data Book". McGraw-Hill, New York.
Penn, W. S. (1962). "PVC Technology". Maclaren, London.
Pilling, N. B., and Bedworth, R. E. (1923). *J. Inst. Metals*, **29**, 529–582.
Pirt, S. J., and Callow, D. S. (1964). *Expl Cell Res.*, **33**, 413–421.
Rabald, E. (1951). "Corrosion Guide". Elsevier Publishing Co., New York.
Reid, I. W. (1967). *Process Biochem.*, **2**, No. 7, 27.
Renfrew, A., and Morgan, P. (1957). "Polyethylene". Iliffe, London.
Reznek, S. (1953). *J. Am. pharm. Ass., Sci. Ed.*, **42**, 288–293.
Rook, J. J. (1955). *Appl. Microbiol.*, **3**, 302–309.
Rosenwald, A. J., Hodge, H. M., Metcalfe, S. N., and Hutton, R. S. (1962). *Appl. Microbiol.*, **10**, 345–347.
Rudner, M. A. (1958). "Fluorocarbons". Reinhold, New York.
Salevid, I. (1966). "Perstorpsboken, Plastteknisk Handbok". Maskinaktiebolaget Karlebo, Stockholm.
Skeist, I. (1958). "Epoxy Resins". Reinhold, New York.
Söhngen, N. L., and Fol, J. G. (1914). *Zentbl. Bakt. ParasitKde.*, II Abt., **40**, 87–98.
Somers, E. (1959). *Nature, Lond.*, **184**, 475–476.
Spence, D., and van Niel, C. B. (1936). *Ind. Engng. Chem.*, **28**, 847–850.
Stahl, W. H., and Pessen, H. (1953). *Appl. Microbiol.*, **1**, 30–35.
Steiger, K., and Dolder, R. (1954). *Pharm. Acta Helv.*, **29**, 311–337, 341–351.
Taylor, A. C. (1961). *Expl. Cell Res.*, Suppl., **8**, 154–173.
Teach, W. C., and Kiessling, G. C. (1960). "Polystyrene". Reinhold, New York.
Telling, R. C., Radlett, P. J., and Mowat, J. N. (1967). *Biotechnol. Bioengng.*, **9**, 257–265.
Tiedeman, W. D., and Malone, N. A. (1955). "A Study of Plastic Pipe for Potable Water Supplies". National Sanitation Foundation, School of Public Health, University of Michigan, Ann Arbor, Michigan.
Uhlig, H. H. (ed.) (1948). "Corrosion Handbook". Wiley, New York.
Vernon, W. J. H. (1923). *Trans. Faraday Soc.*, **19**, 839–900.
Vernon, W. J. H. (1927). *Trans. Faraday Soc.*, **23**, 113–183.
Walter, W. G., Beadle, B., Rodriguez, R., and Chaffey, D. (1958). *Appl. Microbiol.*, **6**, 121–124.

Weiss, L. (1965). *Expl. Cell. Res.*, **37**, 540–551.
Weissner, W. (1962). *In* "Physiology and Biochemistry of Algae" (ed. R. A. Lewin), p. 267, Academic Press, New York and London.
Wessel, C. J. (1964). *S.P.E. Trans.*, July, 193–207.
Wing, W. T. (1958). *I.R.I. Trans. Proc.*, **5**, 67–72.
Young, R. S. (1935). *Mem. Cornell Univ. agric. Exp. Stn*, No. 174.
ZoBell, C. E. (1937). *J. Bact.*, **33**, 86.
ZoBell, C. E. (1943). *J. Bact.*, **46**, 39–56.
ZoBell, C. E., and Beckwith, J. D. (1944). *J. Am. Wat. Wks Ass.*, **36**, 439–453.
ZoBell, C. E., and Grant, C. W. (1942). *Science, N.Y.*, **96**, 379–380.

CHAPTER III

Methods and Equipment for Sterilization of Laboratory Apparatus and Media

G. SYKES

Boots Pure Drug Co. Ltd., Nottingham, England

I. METHODS OF STERILIZATION

Sterilization is fundamental to all microbiological procedures. It features at some stage or other in every method employed. It is essential therefore to have some knowledge of the processes involved and of the reliability and applicability of the methods available.

Sterilization is an absolute term and implies the total inactivation of all forms of microbial life in terms of their ability to reproduce: it does not necessarily imply the destruction of all of their constitutive enzymes or of their metabolic side products, toxins, etc. The word is often erroneously used where *disinfection* is really meant, the difference being that disinfection implies only the process of destroying infection, that is, of reducing the numbers of contaminating bacteria to a "safe" level, but not necessarily their total elimination. Nevertheless, it will be appreciated that in some situations, such as occur when only non-sporing bacteria are involved, disinfection can be synonymous with sterilization. In parenthesis, the term *sterilant* is also used, especially in the food industries, but again it is misleading. A *sterilant* does not sterilize, in the strict sense of the word; it only disinfects.

Sterilization can be achieved through the agency of—

(a) moist heat,
(b) dry heat,
(c) irradiation with γ- or X-rays,
(d) certain chemicals in solution,
(e) certain gases or vapours,
(f) filtration.

Ultraviolet irradiation is also described as a sterilizing agent, but its limitations as a lethal agent to micro-organisms, and especially viruses, indicate that it should not be placed in this category. Its characteristics are discussed later in this Chapter. Freezing and drying are also ineffective: both processes can cause some deaths in a bacterial population, but in some situations both are methods of preserving micro-organisms (see Lapage *et al.*, this Series, Vol. 3A). Neither can ultrasonic or other methods of cellular disintegration by physical means be included. According to some reports they can inactivate some microbial cells, as can also high-voltage discharges between electrodes, but they are far from absolute.

II. THE MECHANISMS OF STERILIZATION

The detailed mechanism of the death process in micro-organisms differs according to the method employed. The effect, however, is the same, namely, one or more of the essential enzymes or proteins of the cell is inactivated or immobilized, thus interfering with the metabolic cycle of the cell and blocking its ability to develop and reproduce. Information concerning the exact enzymes, or even the groups of enzymes, or proteins concerned is still very fragmentary.

With heat, for instance, it has been known since the time of Koch that the mechanism of the destruction of organisms by dry heat is not the same as that by moist heat: the difference is generally attributed to the greater heat stability of proteins in the dry state. Rahn (1945) has stated: "Death by dry heat is primarily an oxidation process; death by moist heat is due to coagulation of some protein in the cell", and in expansion of this thesis Hansen and Riemann (1963) explained that proteins in the presence of moisture release free SH groups and give rise to smaller peptide chains. These chains are mobile and realign themselves to form new complexes within themselves. In the absence of water, the number of polar groups on the peptide chains becomes less and more energy is required to open the molecules; hence the apparent increased stability of the protein, and the consequent increased stability of the organism.

The mechanism of the lethal action of irradiation is less clear. The general impression is that it is centred on the deoxyribonucleic acids which are variously subject to ring fissure, dephosphorylation and other reactions (Pratt *et al.*, 1950; Scholes and Weiss, 1954), although disruption of the enzyme systems of the cell is also a possibility. This can arise from the rupture of the internal cellular membranes and the consequent release of essential enzymes (Bacq and Alexander, 1961), or it can be the result of direct inactivation of certain of the enzymes, and its action can be extended into the post-irradiation period.

The action is initiated by the energy of the radiating beam raising molecules in its path to a state of internal excitation, the extent of which depends on the sensitivities of the molecules concerned and the total energy of the radiation. The final effect is the total disruption of the molecule. At low radiation energies only the outer electrons of the molecules are affected. They are induced into wider orbits, and this causes the molecule to become excited and therefore more reactive. At higher radiation energies, electrons are actually discharged from the molecule, thus leaving it in a positively charged state, and therefore highly reactive, and so setting up, through its released electrons, a chain reaction involving adjacent molecules.

The chemical and gaseous sterilizing agents, in contrast to the simply disinfecting ones, are all highly reactive compounds such as either oxidizing or alkylating agents. As such, therefore, they might be expected to affect profoundly the basic constituents of the cell. But not all oxidizing or alkylating agents are sterilizing agents; those that are must be available in adequate quantity or concentration to achieve the desired effect.

In contrast to all other methods, sterilization by filtration is unique, in that it is the only process by which all of the cells, both viable and non-viable, are actually removed from the gas or liquid being sterilized. Even so, it does not remove the metabolic products, the toxins, etc., from the menstruum. The complete mechanism of filtration is still not fully understood.

A. The death rate of micro-organisms

In any lethal process, the death rate of a microbial population usually follows a logarithmic path, but there are exceptions. Thus, in some processes there may be an initial lag phase and in others, where the survivor–time curve is steep, a direct, rather than logarithmic, relationship appears to occur. Much more important from the sterilization point of view, however, is the situation, not infrequent, where there is a flattening of the slope towards the end of the process, thus resulting in a more prolonged or intense treatment being required to deal with the so-called "occasionally resistant survivors". The reason for the occurrence of such cells is unknown, but clearly they are more likely to be present in an initially large microbial population than in a smaller one.

Because of the logarithmic form of the survivor–time curve, it follows that the larger the number of cells to be destroyed the more prolonged or intense must be the sterilization treatment. Thus, in theory it takes twice the "energy", in terms of heat or intensity of radiation or time of treatment, etc., to kill 99·9% of a bacterial population than it takes to kill 99%, and twice as much again to kill 99·99%, and so on. Taken to its limit the treatment required to kill extremely large numbers of bacteria would approach infinity, but in practice certain levels of kill have become accepted. Thus, for steam sterilization the reduction factor is of the order of 10^{15}–10^{20}, but for other processes, e.g., sterilization with ethylene oxide, a lower level of 10^{8}–10^{9} is recognized.

B. Resistances of different organisms

Needless to say, all micro-organisms do not respond in the same way to adverse conditions, and these include all sterilization treatments. Differences occur not only between types, but also between strains of the same species, and even between cells within a given culture according to their age and individual metabolic states. Difference can also arise from varied cultural conditions, the nature of the suspending fluid and other physical considerations. And these can all be important in assessing the efficacy of a sterilization process.

In general the naturally occurring organisms appear to be more resistant than are the cultivated ones, but the evidence available shows this to be attributable not to the higher resistances of the cells concerned but to the greater protection afforded by their surrounding menstruum. The basis for this attribution is that when an organism, which exhibits apparently abnormal resistance in the environments in which it is found, is cultured even for one generation under laboratory conditions its resistance is of the same order as that of the species or strain as a whole. Nevertheless, the phenomenon must be taken into account in practice, because in the main it is the naturally occurring organisms that have to be dealt with, and not those in artificial culture.

There is no real evidence that significantly increased resistance to sterilization can be induced by selection or by any other means. The one exception to this is with radiation sterilization, where in some conditions resistance is said to be increased twofold. In this respect the process of sterilization is quite different from that of disinfection where there are numerous examples of induced resistance amongst non-sporing bacteria.

No known relationship exists between the resistances or susceptibilities of micro-organisms, including viruses, to the various methods of sterilization. Resistance of a particular culture or strain in one direction does not necessarily imply resistance in any other, nevertheless it is generally accepted that bacterial spores as a group exhibit greater resistances than do vegetative cells; mould spores do not possess the same high resistance. Bacterial spores, therefore, must be the first choice for monitoring any sterilization procedure. But this has its disadvantages, and other methods are commonly used.

C. Sterilization indicators and controls

It should be axiomatic that each individual sterilization, made by whichever method may be chosen, should be suitably controlled to ensure that the treatment has been adequate. This is not always easy or convenient to carry out, and in practice it is usually confined to those methods that are used most frequently and that are known to be subject to various difficulties and to misuse, and this, in fact, restricts the list to sterilization by heat and by ethylene oxide.

As indicated in the preceding Section, bacterial spores are the first and natural choice, but such a choice has two principal disadvantages: first, even with a selected and recommended strain the resistances of its spores vary according to the culture conditions employed, and special conditions are needed to obtain preparations of satisfactorily high resistance; secondly, the result cannot be read immediately but must be delayed until any possible surviving spores have had an opportunity to grow out. Nevertheless, spore preparations are quite commonly used, either alone or in conjunction with one or other of the control methods.

For heat sterilizations, the spores of *Bacillus stearothermophilus* (a thermophile), and for ethylene oxide and radiation sterilizations, those of *Bacillus subtilis* var. *niger* (sometimes known as *B. globigii*) have been found to be the most resistant. They are usually dried from an aqueous or methanolic suspension onto small strips of filter paper (Beeby and Whitehouse (1965) have recommended aluminium foil), the inoculum being of the order of 10^5–10^6 spores per test piece. Several such pieces should be used for each sterilizer load, the pieces being placed at strategic points where the heat or other treatment is least likely to penetrate. After the treatment the test pieces must be suitably cultured, remembering that it will take at least several days, and perhaps even weeks or months, before the occasional surviving spore recovers and grows out: the usual 24 h or "overnight" period is quite inadequate. It is also possible that organisms may be recovered better at temperatures below their normal optima for growth.

Next to the direct biological test the thermocouple is the most reliable indicator for heat sterilizations. This is a direct method and simply involves placing the probes at significant points in the load to be sterilized and recording the temperature. Such observations enable the course of the sterilization to be controlled directly and so its efficacy to be assessed on the spot without delay.

For steam sterilization, pure chemicals of known and well-defined melting points can also be used. The substance selected must be one that melts at or just below the required sterilizing temperature; for treatments at 121°C, succinic anhydride (m.p. 120°C) is recommended, and for treatment at 115°C, sulphur (m.p. 115°C) or acetanilide (m.p. 116°C) are suitable. If such compounds are simply sealed, for example, in small tubes they merely show that the temperature has been reached, but if incorporated in a device described by Brewer and McLaughlin (1954) the time factor is also controlled. This device is based on the hour-glass principle and depends on the flow of the molten substance through the constriction of the glass. The quantity of the substance is adjusted so that it all passes through in the specified sterilizing time. The device is limited in its usefulness because its size prevents it from being inserted easily into packaged materials or individual closed containers.

Other chemical indicators are those that change colour after a given heat treatment. The most reliable of these are Browne's tubes (A. Browne Ltd, Leicester, England). These are small sealed glass tubes containing a red fluid that changes colour through amber to green on heating. They are available in three types: type I responds to sterilization at 121°C for 15 min or 115°C for 25 min; type II at 115°C for 15 min; type III is for dry heat treatment at 160°C for 1 h. They have been reported upon favourably by several investigators, but in the opinion of the author and others (Kelsey, 1959; Medical Research Council Working Party, 1959) they are not entirely reliable because, being dependent on a chemical reaction, they can change colour by heating at a lower temperature for a longer period—in fact, it is recommended that they be stored in a cold place before use. Some additional check is necessary, therefore, to confirm at least that the required temperature has been actually attained.

Besides the Browne's tubes there are several strip or patch indicators that also show a colour change, they include the Klintex paper (Robert Whitelaw Ltd, Newcastle-upon-Tyne), Diacks, Tempilpellets and a type of Scotch tape (Minnesota Mining & Manufacturing Co. Ltd (3M)). None responds in the same quantitative manner as do the other types of indicators, and probably the most accurate is the 3M tape. They serve very well as a label to indicate that a package or other material has been processed.

For dry heat sterilizations, the choice of indicator controls is more limited. It is restricted to spore strips, thermocouples and Browne's tubes (type III).

Besides the spore paper or aluminium foil strips of *B. subtilis* var. *niger*, the only device produced for controlling ethylene oxide sterilizations is the Royce sachet (Royce and Bowler, 1959). This is a solution of $MgCl_2$ and HCl, with indicator, sealed in a thin polythene sachet measuring about 1×2 in. When the ethylene oxide gas diffuses through the polythene, it is hydrolysed rapidly and ultimately, if enough of the gas diffuses, all of the acid is neutralized. The diffusion and reaction velocity varies with temperature, gas concentration and time, and the variations closely parallel the corresponding variations in the sterilization rate.

III. STERILIZATION BY HEAT

A. Moist heat

Heat, and particularly moist steam under pressure, has proved to be the most reliable and universally applicable method of sterilization, and as far as possible should be the method of choice. The only occasions when it is not suitable are with materials that are adversely affected by moisture or unable to withstand the elevated temperature. Temperature and time of treatment are interconnected, a higher temperature needing a shorter time. Rahn (1945) has calculated that, for most spores, Q_{10}, the temperature coefficient for each 10°C increase, lies between 8 and 10 in the range 100°–135°. Equivalent sterilizing times calculated on this basis by Thiel *et al.* (1952) are—

Temperature, °C	100	110	115	121	125	130
Time	20 h	2·5 h	51 min	15 min	6·4 min	2·4 min

These figures however, should only be taken as guides. Table I quotes some figures taken from the literature on the recorded destruction times of different bacterial spores. There have been occasional reports of even greater degrees of resistance, but usually this has been traced to inaccurate temperature recording or to an exceptional, abnormal environment. Oils and fats, for instance, can apparently enhance resistance, but this is because the fats prevent access of moisture to the cell and so create local conditions approaching those of dry heat treatment.

TABLE I

Some quoted destruction times of bacterial spores by moist heat

Organism	Destruction times, min, at 100°C	105°C	110°C	115°C	120°C	125°C	130°C	134°C
B. anthracis	2–15	5–10	..	..	..	..	..	..
B. subtilis	Many hours	..	..	40	..	..	..	..
A putrefactive anaerobe	780	170	41	15	5·6	..	..	..
Cl. tetani	5–90	5–25	..	..	..	..	..	..
Cl. welchii	5–45	5–27	10–15	4	1	..	..	..
Cl. botulinum	300–530	40–120	32–90	10–40	4–20	..	..	..
Soil bacteria	Many hours	420	120	15	6–30	4	..	1·5–10
Thermophilic bacteria	..	400	100–300	40–110	11–35	3·9–8·0	3·5	1
Cl. sporogenes	150	45	12	..	..	..	..	..

1. *Steam sterilization*

(a) *Factors governing the efficacy of steam sterilization.* The basic essential in a steam sterilization is that the whole of the material to be sterilized shall be in contact with saturated steam at the required temperature for the necessary length of time. Each of these three points is important. First, saturated steam must be used, otherwise the process becomes virtually a dry heat treatment for which different temperature–time relationships hold. This implies that superheated steam must be avoided, and that normally all free air must be removed, although there are certain conditions where air–steam mixtures can be used (see p. 88). Secondly, as in all heat sterilization processes, there is a well established temperature–time relationship that must be observed if reliable sterilization is to be achieved. For sterilizing liquids, and for most other purposes, 30 min at 115°C (10 lb steam pressure/sq. in.) or 20 min at 121°C (15 lb steam pressure/sq. in.), as is specified in the British Pharmacopoeia, is used. Under normal circumstances, these treatments provide a substantial margin of safety, so that if the level of contamination is known to be low they can be reduced. Thus, for many years 20 min at 115°C or 15 min at 121°C was used in the author's laboratory for sterilizing media. and other workers use even shorter periods. Boiling alone will sterilize, of course, in some situations.

Implicit in the specifications for steam sterilization is the need for sufficient time to be allowed for the load to reach the required temperature before the actual sterilizing period commences. This varies considerably with the nature and size of the load and the size of the sterilizer, and two examples indicate the extent of the variation to be expected. In the first experiment different sized containers of water were heated in a mains-operated sterilizer and the time was recorded for each container to reach 100°C, as indicated by thermocouples in the containers; in the second, a single 100 ml bottle, a crate of bottles and two stacked crates were used, and similar times recorded. In the first experiment, a 100 ml bottle required about 16 min to be heated, against 42 min for a 5 litre flask; in the second experiment, the single 100 ml bottle required about 12 min, against 19 min for those in the single crate and 30 min for the stacked crates.

(b) *The design and operation of the autoclave.* Fundamentally an autoclave is simply a chamber in which the necessary stages of a steam sterilization can be carried out. It must be of sufficient strength to withstand steam

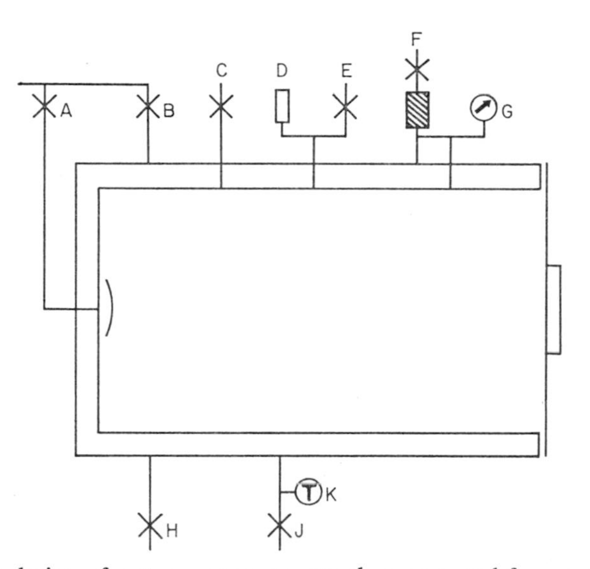

FIG. 1. Basic design of a steam pressure autoclave operated from a steam main.

A Steam valve to chamber
B Steam valve to jacket
C Vacuum line
D Safety valve
E Exhaust cock
F Air filter
G Pressure valve
H Jacket drain cock to open waste
J Chamber drain cock with non-return valve to open waste
K Temperature recorder

pressures up to 20 lb/sq. in., and for some purposes as high as 30 lb/sq. in., as well as possibly high vacuum.

The steam can be from either a mains supply or it can be generated within the chamber by gas or electricity. The latter is only suitable for the smaller "vertical" models ranging from the bench cooker to those measuring up to about 18 in. dia × 3 ft deep: they are always manually operated. The mains-supply types can measure up to 6 × 6 × 20 or 30 ft long and they can be operated manually or be fully automated. A rectangular design is the more economical in terms of useful loading space, but it is more

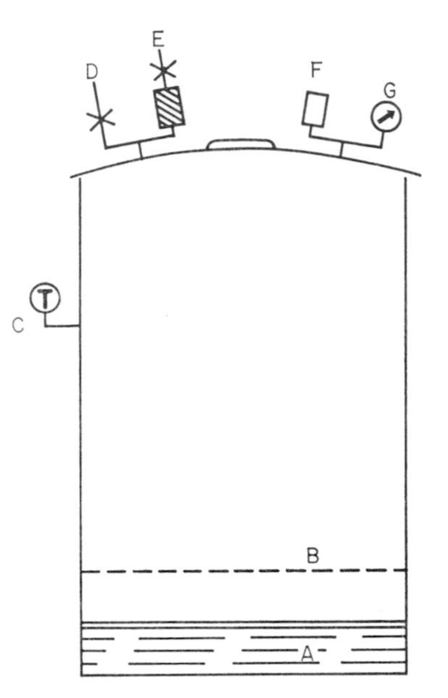

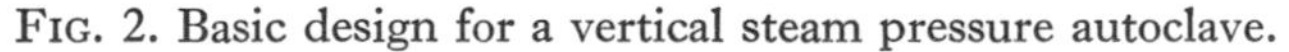
FIG. 2. Basic design for a vertical steam pressure autoclave.

A Water reservoir
B Support tray
C Temperature recorder
D Exhaust cock
E Air filter
F Safety valve
G Pressure gauge

expensive because it requires heavier gauge metal and additional reinforcement to withstand the pressure involved.

The design of a typical mains operated sterilizer (with jacket) is illustrated in Fig. 1 and that of a sterilizer generating its own steam in Fig. 2. The jacket and vacuum port in the mains-operated type are required only for the "high prevacuum–high temperature" cycle used for sterilizing dressings, rubber gloves and certain apparatus (see p. 89). Except for these special purposes a non-jacketed model, lagged sufficiently to prevent undue heat losses, is quite satisfactory. The mains-operated type is much more complicated in design and operations, and its principal features are—

(a) Separate valves must control the steam supplies to the chamber and the jacket.
(b) The steam as it reaches the chamber must not be dry or superheated to any degree and it should enter the chamber either centrally or towards the top through a baffle or sparge.
(c) A thermometer or other temperature-recording device should be sited in the steam drain on the chamber side of the control valve. Thermocouples with flexible leads should also be available to check as needed the temperatures at various points in the load.
(d) Steam- and air-escape cocks must be fitted at the top and bottom of the chamber.
(e) A safety valve is essential.
(f) The air inlet and vacuum cocks should each carry an air filter.
(g) All joints and seals, especially the door gasket, must be air-tight.

These and other details are incorporated in two British Standard Specifications (1966) for different types of hospital sterilizers. Incidentally, all sterilizers, being pressure vessels, are subject to regular inspection in Great Britain under Board of Trade regulations.

The sterilizer can be operated by either the standard "downward displacement" method or by the "high prevacuum" cycle. The former is the more commonly used and is the only one suitable for sterilizing liquids. The procedure in outline is—

(1) Having loaded the sterilizer and sealed the door, turn on the steam flow, gently at first, with both top and bottom escape cocks open.
(2) When all of the contents have become heated to 98°–100°C and all of the air displaced (as indicated by the temperature recorded in the bottom steam drain), close the top and bottom exhaust valves and allow the pressure in the chamber to rise slowly to the required level.
(3) Hold the pressure at this level for the specified length of time.
(4) Turn off the steam flow and allow the pressure to fall slowly and evenly to atmospheric. If the fall in pressure is too rapid or uneven, loss from boiling may result.
(5) Open the bottom escape valve and the air filter inlet and allow the contents to cool naturally to 80°C or less before opening the sterilizer.

When the containers in the load are sealed so as to exclude the access of water, such as in ampoules or screw-capped bottles, the last stage of the cooling process can be speeded up by means of a fine non-wetting water spray directed uniformly throughout the sterilizer (Wilkinson *et al.*, 1960). The droplets must not be too large otherwise they may crack the glass containers. It is necessary also to replace the steam pressure with air to avoid creating a sudden vacuum.

Mention was made earlier (p. 84) that it is not always necessary to eliminate all of the air from the sterilizing chamber. Such a situation always obtains, of course, whenever a sealed container of an aqueous fluid is sterilized. It also obtains in the counterpressure or "pulsating" sterilizer (see below). The reasons for eliminating air in a normal sterilization are twofold: first, there is a tendency for air–steam mixtures, unless kept in a constant state of agitation, to separate and so give an uneven temperature distribution in the load and possibly less than saturated moisture conditions in some areas (this applies more in large sterilizers); secondly, it is common practice to assess the temperature in the sterilizer by the steam pressure registered. This is acceptable when one is dealing with pure steam, but is wholly unsuitable with air–steam mixtures; the temperature–pressure ratios are quite different as the following figures show—

Pressure gauge reading, lb/sq. in.	Temperature, °C, with pure steam	half air removed	no air removed
5	109	94	72
10	115	105	90
15	121	112	100

If a sterilization is controlled by temperature (as it ought to be) instead of from the pressure gauge (as is so often the case), residual air in the chamber is of little concern, provided it does not accumulate in pockets and so prevent full access of the steam. The following experiment illustrates how easily this can occur. Two crates of 1 litre sealed bottles of saline were put through the normal sterilization cycle in a large autoclave. The crates each measured about 2 × 3 ft and were 10 in. deep: one had the normal open-mesh base so that air and steam could flow through it freely, the other was lined completely with brown paper to inhibit such flow. The bottles in the open crate reached properly the sterilizing temperature of 115°C and were satisfactory; in those in the lined crate the maximum temperature achieved was 85°C.

In the counterpressure sterilizer deliberate advantage is taken of an air–steam mixture operating at pressures considerably greater than normally used to counter the pressure developed when sealed containers are heated to the sterilizing temperature, and thereby reduce bursting of glass bottles or distortion of certain types of plastic containers now used for dispensing transfusion solutions. In this system the sterilization is carried out at 115°C, and at 30 lb/sq. in. pressure (instead of the equivalent 10 lb/sq. in.) and these conditions are maintained by alternately pulsated jets of steam and air, the amounts of each being carefully controlled within narrow limits by automatic temperature and pressure gauges. A basic essential also is a fan system to keep the air and steam in constant and uniform mixture.

In the preceding paragraphs attention has been drawn to the salient features for the satisfactory operation of a sterilizer, but it is well also to be reminded of the possible reasons for failure. They include—

(1) Blocked exhaust cocks or valves.
(2) Reliance on faulty instruments to measure temperature and steam pressure.
(3) Reliance on the pressure in the sterilizer as an indicator of temperature.
(4) Overloading or incorrect packing of the sterilizer, thus preventing the free access of steam to all of the material to be sterilized.
(5) Failure to remove all of the air.
(6) Insufficient time allowed for the materials to be preheated to the required temperature.
(7) Attempts to sterilize materials impervious to steam; these include powders in deep layers, oils and sealed packages.
(8) Recontamination during cooling and drying, during storage and due to faulty wrapping.
(9) The presence of a rare organism of unexpectedly high resistance.

The problem of air pocketing and removal can occur just as frequently and hazardously with surgical dressings and any wrapped materials as with equipment, apparatus and containers to which the only access is through a narrow orifice closed with, say, an air filter. If such material or apparatus is assembled wet, the situation is mitigated considerably, but generally they are put together and wrapped dry. Thus, because steam and air do not mutually diffuse easily, it is not difficult to imagine that even after a fairly prolonged pre-steaming period some air will still remain trapped in the package or container, so that when the pressure in the chamber is raised to the sterilizing level the air will be compressed within the package and steam will not have full access to the centre of the package or to some part of the apparatus.

To meet such contingencies the high pre-vacuum cycle, developed in the first place for sterilizing surgical dressings, was devised. Because of its applicability to equipment as well as to dressings it justifies a brief description: it is not applicable to liquids.

In this process (Medical Research Council Working Party, 1959, 1961), now incorporated in a British Standard (B.S. 3970 : 1966), the loaded chamber is first evacuated to give a residual pressure not exceeding 20 mm absolute. This is a critical level. The vacuum is held for a sufficiently long period for " absorbed" air to be extracted, but not long enough to remove the natural moisture of the fibres and other materials. Steam is then let into the evacuated chamber and the pressure taken immediately to the level required. After the appropriate holding time, the steam is turned off, the pressure in the chamber is immediately released and vacuum applied to a final residual pressure of 50 mm or less (Penniket *et al.*, 1959). Within a few minutes all of the condensed moisture will have been evaporated and the load can be taken out quite dry.

This process can be applied equally effectively to gloves and other rubber equipment without damage to the rubber. Contrary to earlier opinions, it has now been shown (Fallon and Pyne, 1963) that it is not the high temperature but the presence of air which spoils rubber, so that in this context the pre-vacuum cycle is advantageous. The same applies to the repeated sterilization of cotton materials (Henry, 1964).

From the evidence available (Bowie, 1961: Henfrey, 1961; Wilkinson and Peacock, 1961), steam penetration in this process is almost instantaneous, so that there is virtually no preliminary heating period. Because of this, the holding times can be reduced somewhat. The temperatures employed are also generally higher, and the times correspondingly shorter, than in downward displacement method. The treatments recommended by the Medical Research Council Working Party (1959) are 3 min at 134°C (30 lb steam pressure/sq. in.), 10 min at 126°C (20 lb steam pressure/sq. in.) or 15 min at 121°C (15 lb steam pressure/sq. in.), and these include adequate safety periods.

2. *The steam–formaldehyde process*

This process, described by Alder *et al.* (1966), depends on the combined lethal effects of steam and formaldehyde. It is useful for sterilizing all textiles as well as for plastics and other equipment that would be damaged at normal steam sterilizing temperatures.

The treatment is carried out in an autoclave at 80°C, and this is achieved by the use of steam at $\frac{1}{3}$ atmosphere pressure. The loaded autoclave, the steam lead of which is fitted with a control valve to maintain the chamber at the required temperature and pressure, is first evacuated to a residual pressure not exceeding 20 mm Hg, then formaldehyde is introduced via a steam-heated vaporizer at the rate of 6 ml of Formalin (a 38% w/v formaldehyde solution) followed by steam for 30 min, after which the steam pressure is released and vacuum applied to dry off the load. It is claimed that there is practically no residual smell of formaldehyde. The main drawback with the method is that the vapour does not apparently penetrate readily into narrow tubes, such as catheters.

3. *Tyndallization*

This process, named after its discoverer, is an old established method for sterilizing liquids or gels and similar semisolid materials. It consists of heating the material at 80°C for 30 min on three successive days, with intermediate storage at normal temperature. Nowadays it has several variants, with temperatures ranging up to 100°C (in either boiling water or flowing steam) and times up to 1 h.

It is essentially a fractional method of sterilization, the theory being that vegetative cells and some spores are killed at the first heating and that the more resistant spores subsequently germinate and are killed at either the second or third heating. This implies that the menstruum is sufficiently nutrient to allow bacterial spores to germinate in the intervening 24 h periods. As such, therefore, it is useful only for sterilizing heat-sensitive culture media containing materials such as carbohydrates, egg or serum. The initial contamination level should obviously be low, and the method can easily be shown to be ineffective when resistant spores are present in non-nutrient solutions such as drug injections.

4. *The heat-plus-bactericide process*

Although this process has little, if any, application in the normal laboratory, it justifies a brief reference here as an established method of sterilizing solutions, especially those used for injection. The method relies on the enhanced lethal action of a bactericide at elevated temperature, and consists simply of heating the solution to 100°C for 30 min. It can be carried out with substances such as brilliant green, $HgCl_2$ and formaldehyde, but is generally confined to phenol, *p*-chloro-*m*-cresol and the phenylmercuric salts, the recommended concentrations being 0·5%, 0·2% and 0·002%, respectively.

The major disadvantage with the process, from the bacteriological standpoint, is that the antibacterial agent remains in the sterilized solution. It cannot be used, therefore, with culture media or in any situation in which bacteria are intended to grow or survive. It also breaks down if the initial spore contamination level is high. Heat, acids and alkalis can act in the same rôle and, in particular, solutions of pH 4·5 or less (Coulthard, 1939) can be sterilized by this method. This is, in fact, the principle of the present method for sterilizing canned fruits. By the same rule, a boiling 2% solution of sodium carbonate will sterilize surgical and other small instruments in 10 min (Anon, 1943), even those carrying the most resistant spores.

B. Dry heat

Sterilization by dry heat requires higher temperatures and longer heating periods than does sterilization with steam. For this reason it has comparatively limited applications, its use being confined mainly to sterilizing oils, certain powders, surgical instruments and glassware. It is not suitable for sterilizing cotton, wool or certain plastic materials, neither can it be used when water, including water of crystallization, is present. Paper is rendered brittle and tends to become discoloured, glass syringes with metal attachments may be damaged or broken because of the different coefficients of expansion and care is needed in handling powders because of their low heat conductivities.

The basic requirement for the process is that the *whole* of the material to be sterilized shall be treated at the specified temperature for a sufficient length of time. The only possible reasons for failure, therefore, are (i) faulty temperature control, (ii) lack of heat penetration or (iii) the presence or organisms of abnormal resistance.

The factors governing the efficacy of dry heat sterilization have been discussed by Quesnel *et al.* (1967), and from these it is clear that the lethal action results from the heat conveyed from the material with which the organisms are in contact and not from the hot air surrounding them. These observations underline the importance of uniform heating of the whole of the material to be sterilized.

Opinions are not unanimous concerning the amount of heating required. The commonly accepted treatment is heating at 150°C for 1 h and this is the accepted procedure described in the British Pharmacopoeia. There is a tendency, however, to prefer 160°C for 1 h, and in the United States the treatment is at 180°C for 2 h. The first mentioned is certainly adequate for most purposes—140°C for 4 h can also be used equally well—but there are situations when the more extreme treatment is needed. The writer, for example, has encountered soil samples containing organisms which survived 170°C, but not 180°C, for 1 h. Some recorded killing times of various organisms at different temperatures are quoted in Table II.

Higher temperatures than those recorded in Table II, up to 300°C and greater, for much shorter times have also been studied, especially for sterilizing air: adiabatic compression is also effective for the same purpose (see p. 94).

1. *The hot-air oven*

The apparatus most commonly used for dry heat sterilizing is the hot-air oven. Such ovens can be heated by either electricity or gas. The heating of the contents takes place mainly by radiation from the walls of the oven, and this means that unless the appropriate steps are taken the heating will

TABLE II

Some quoted killing times of bacterial spores by dry heat

Organism	Destruction times, min, at						
	120°C	130°C	140°C	150°C	160°C	170°C	180°C
B. anthracis	..	..	up to 180	60–120	9–90	..	3
Cl. botulinum	120	60	15–60	25	20–25	10–15	5–10
Cl. welchii	50	15–35	5	..	..	..	..
Cl. tetani	..	20–40	5–15	30	12	5	1
Soil spores	..	..	..	180	30–90	15–60	15

not be even and uniform throughout the load. To meet this, the oven must be adequately heat-insulated to avoid excessive losses, it must not be overloaded, ample air spaces must be left between individual items and the air within the chamber should be kept in circulation by means of an enclosed fan. A British Standard (1961) specifies closely the performance characteristics of a hot-air oven, and these include (*i*) a maximum heating up time for a given load, (*ii*) adequate temperature control to restrict the drift to within 1°C and (*iii*) a variation in temperature of not more than 5°C between parts of the load.

The greatest problem is undoubtedly in ensuring that the heating within the load is uniform, and for this purpose thermocouples are needed. Gas-heated ovens are said to be more variable than are electrically heated ones and differences of up to 40°C have been recorded in ovens heated by gas to a nominal 160°C (Darmady and Brook, 1954). Similar differences (148°–189°C) have also been found by the writer between shelves of an electrically heated oven with elements in the base only. Clearly these examples illustrate the importance of design of such ovens.

The significance of the nature and size of the load in relation to the rate of heating is illustrated in the following experiences. In the first, a laboratory oven, heated from the base only but fitted with a fan, was packed with Petri dishes, bottles and flasks on both shelves and the rate of heating in different parts of the load was measured by means of thermocouples. In general the temperature at each point rose uniformly, but whereas on the lower shelf it took only about 45 min to reach 150°C, on the upper shelf it took nearly 100 min. In the second experiment, 4 oz jars of powder were found to take 55 min longer to reach 160°C than did the same material spread in thin layers in open dishes.

2. *Infrared heat*

The radiant heat from infrared units can be used effectively for sterilizing small pieces of equipment, syringes, pipettes, etc., and a device for this purpose has been described by Darmady *et al.* (1957). The items to be sterilized are passed on a slowly moving belt through a tunnel in which are mounted a series of heating units. The conditions are so adjusted that each item in its passage though the tunnel is heated at 180°C for 15 min, a treatment which, it is stated, even the most resistant spores do not survive. The total heating time is only 22½ min, a much shorter period than with the more traditional hot-air oven, but clearly the method is only suitable where large numbers of small items are involved.

3. *Other dry-heat methods*

At temperatures greater than about 200°C, the minimum sterilizing time becomes progressively shorter, requiring from minutes down to seconds, so that sterilization takes place as the temperature is being reached. Such is the ease in the everyday "flaming" of platinum loops and needles, but it is doubtful whether the equivalent flaming of the necks of culture tubes or the so-called "burning off" with alcohol has the same effect. In the latter instance it is doubtful whether the surfaces (where the organisms are) reached anywhere near the temperature needed to sterilize them, but the alcohol (*per se*) certainly contributes.

(a) *Sterilization of air*. Apart from the instances just quoted, temperatures above 200°C are not used because, amongst other considerations, the amount of heat energy required is too great to make it practicable. The one possible exception to this is in the sterilization of air, but again it is extremely uneconomic on anything but a very small scale. The actual heating times at different temperatures vary according to the equipment and conditions used, but the highest times quoted in the literature range from 24 sec at 218°C to 8 sec at 272°C and 3 sec at 302°C. Using a different approach, and with figures similar to these, Elsworth *et al.* (1961) estimated the level of kill of spores of *B. subtilis* after an exposure of 1 sec and found values of 1×10^{-3} at 210°C, $3{\cdot}6 \times 10^{-3}$ at 270°C and 7×10^{-7} at 330°C.

In all of these treatments, death of the cells is by destructive oxidation.

(b) *Adiabatic compression*. A special example of the sterilization of air by dry heat is in the adiabatic compression process. This was first discovered by Stark and Pohler (1951) and later confirmed by Sykes and Carter (1954), both of whom obtained their results using single-stage reciprocating compressors. The minimum working pressure needed to effect sterilization is about 40 lb/sq. in. Turbo-compressors, even though working at the same pressures, only effect a partial destruction of airborn organisms.

The mechanism of the action is not clear. The cylinder in a reciprocating compressor working at 60 lb/sq. in. reaches a temperature of about 150°C, but during compression the temperature of the air may actually be much greater than this; nevertheless the flow rate is such that the treatment can only be for a small fraction of a second. Some factor, or factors, other than temperature therefore appear to be involved. These might include the suddenness of the heating and the rate of increase of the relative humidity of the air due to compression.

IV. RADIATION STERILIZATION

A. Ionizing radiations

1. *Characteristics*

The radiations of most practical value for sterilization purposes are the electromagnetic X- and γ-rays, and the particulate cathode rays, of which the former originate from radioactive elements either directly (γ-rays) or through generating machines (X-rays) and the latter from radioactive elements. γ-Rays are of short wavelength and are akin to hard X-rays. Cathode rays can be subjected to increased energy by passage through electron accelerators, of which the van de Graaff accelerator is the best known. The Capacitron is another type of electron-accelerator machine, and emits pulsed radiations of extremely high intensity and short duration. The emission can be as great as 1 Mrad in 1 μsec, and because of this the effect might be expected to be different from the more even emissions at lower intensities. Reports indicate, in fact, that much lower doses are required in this form of treatment. The commonest source of cathode rays is cobalt-60 (^{60}Co), but substantial quantities of caesium-137 (^{137}Cs) are also available.

From the sterilization point of view, penetrating power is one of the important characteristics of ionizing radiations. Cathode rays, because of their particulate nature, have the greater intrinsic energy, and so the greater power of penetration, but X- and γ-rays have the greater relative powers of penetration. With cathode rays, penetration is related directly to the energy of the electron beam and inversely to the density, in relation to water, of the material being irradiated, and this is expressed simply—

$$\text{Penetration} \propto \frac{\text{Electron energy}}{\text{Material density}}$$

A somewhat similar, though rather more complex, relationship also holds for γ-radiations.

Because of the nature of the reaction mechanisms involved, optimum activity never occurs at the surface of the material being treated. With γ-rays it occurs just below or inside the surface, and with cathode rays a few centimetres deeper. The comparative patterns of penetration are illustrated in Fig. 3, which also shows the advantage of irradiation with two beams in opposite directions.

The original unit of measurement is the röntgen (R) which is defined in terms of the number of ionizations produced in a unit volume of air and is equivalent to an energy absorption of about 83 ergs/g of air. The rep is also used, but the more popular term is the rad, which is similar to the röntgen, except that it is based on an energy absorption of 100 ergs/g of air. For practical purposes the Mrad, equal to 10^6 rad, is commonly used, because this is the dose required for sterilization. These units can be measured electrometrically, calorimetrically or chemically, the last being the simplest and most useful. All chemical measurements are based on oxidative changes, of which the oxidation of the ferrous to the ferric ion is typical.

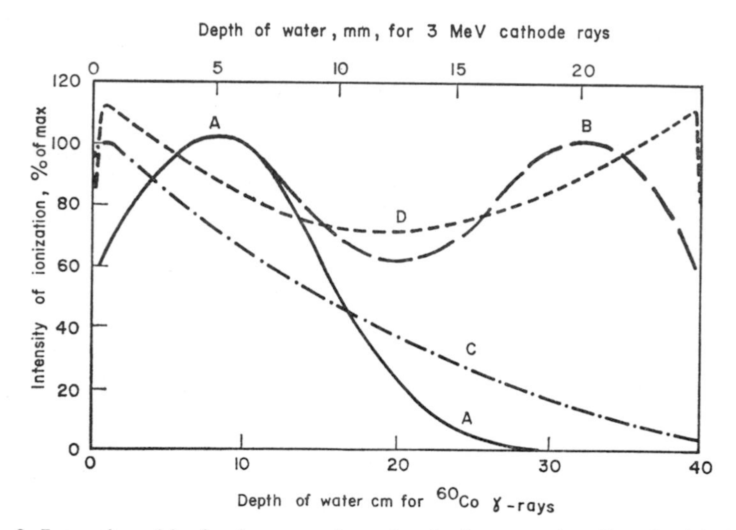

FIG. 3. Intensity of ionization at various depths in water irradiated with 3 MeV cathode rays and ^{60}Co γ-rays: A, curve for cathode rays irradiated from one side; B, curve for cathode rays irradiated from opposite sides; C, curve for γ-rays irradiated from one side; D, curve for γ-rays irradiated from opposite sides.

2. *Lethal doses*

As is to be expected from other fields, the bacterial spores are the most resistant, with *Clostridium botulinum* standing significantly high in the list. The non-sporing bacteria exhibit the least resistance, and moulds and fungi tend to occupy intermediate positions. Gram-positive non-sporing bacteria are generally more resistant than are the Gram-negative ones, and within this latter group the pseudomonads are amongst those most readily killed. It has been estimated (Thornley, 1963) that the difference between the lethal dose for sensitive pseudomonads and a highly resistant *Cl. botulinum* can be as great as 50 fold. The ranges of lethal doses quoted in the literature

TABLE III

Lethal doses of different radiations

Type of organism	Lethal doses, Mrad, from Cathode rays (a)	Cathode rays (b)	γ-Rays (c)	X-rays (d)
Vegetative, non-pathogenic	0·1–0·25	..	..	..
Vegetative, pathogenic	0·45–0·55	0·1–0·25	0·15–0·25	0·03–0·5
Bacterial spores	0·5–2·1	0·2–0·4	c 1·5	0·5–2·0
Moulds	0·25–1·15	0·35–0·4	0·2–0·3	0·25–1·0
Yeasts	0·5–1·0	..	0·3	0·25–1·5

(*a*) From van de Graaff accelerators (various authors).
(*b*) From Capacitron pulsed beam (Huber and colleagues, quoted by Hannan, 1955).
(*c*) From ^{60}Co.
(*d*) From 3 MeV source.

TABLE IV

Percentage kills of various micro-organisms with different radiation doses

Organism	Percentage kill by 35,000 rad	100,000 rad	500,000 rad
E. coli	92·86	99·97	100
Ps. fluorescens	99·97	99·999	100
Staph. aureus	94·6	99·998	100
B. stearothermophilus	..	99·0	99·99
B. thermoacidurans	..	95·41	99·997
Torulopsis rosea	67·3	97·98	99·999

for the different groups of organisms are summarized in Tables III and IV, and, bearing in mind the widely differing conditions under which the figures were obtained, the agreement, is remarkably close. The rather lower values obtained with the Capacitron are due to the different form of emission from this type of machine.

Several notable examples of unusually resistant organisms merit particular mention, including an *Achromobacter* sp. isolated from chicken meat by Thornley (1962) and several strains of *Diplococcus pneumoniae* (Koh *et al.*, 1956), but outstanding is the organism, aptly named *Micrococcus radiodurans*, which can withstand doses of up to 6 Mrad (Anderson *et al.*, 1956). This organism fortunately does not occur with any frequency: it was isolated originally from minced beef and pork. Other examples could be quoted, but they all involve organisms as they occur in the natural state where so-called "resistance" is attributable to the protection afforded by the environment, especially foodstuffs, rather than to an intrinsic property of the organism itself.

It is important to note that bacterial toxins are inactivated at about the same rate as are bacterial spores (Wagenaar *et al.*, 1959), but that many enzymes, including those associated with food spoilage, can survive doses up to as much as 10 Mrad. The same applies to the pyrogens, which may require up to 20 Mrad for inactivation.

The range of resistances of the viruses is about the same as that for bacterial spores, but again there are important exceptions. Thus, whereas vaccinia virus is inactivated by 2–2·5 Mrad, the Lansing polio virus requires 3·5–4 Mrad and the encephalitis viruses 4–4·5 Mrad (Jordan and Kempe, 1956). Pulsed radiations from a Capacitron machine inactivated at lower doses. There is one example in the literature (Kaplan, 1960) of a freeze-dried vaccinia virus being substantially inactivated by 3–4 Mrad, but with occasional viable particles surviving 11 Mrad. This is an example of the "tailing effect" already referred to in earlier pages.

(a) *Factors affecting lethal activities.* The principal factors influencing the lethal activities of ionizing radiations are the "oxygen" effect, the presence of reducing compounds, sensitizing agents, freezing and drying. Of these, the most significant is the oxygen tension during, and, to a less extent, post irradiation.

Reduced oxygen tension always increases the lethal dose level, but it does not become operative until the oxygen concentration falls below 10 mg/litre, after which it becomes progressively more significant. In the presence of oxygen the oxidized radicals of water, namely $^{\bullet}OH$ and $^{\bullet}HO_2$ are produced, with the consequent formation of peroxides, all of which tend to enhance the lethal action. In the absence of oxygen, such radical formation is restricted and so the radiation is less effective.

Reducing the tension increases the necessary radiation dose by some 3 or 4 times, and between anaerobic and aerobic organisms the factor is said to be even greater. The most remarkable difference was reported by Tallentire (1958) who found that the lethal dose for *B. subtilis* spores in a spray-dried powder sealed under vacuum was 100 fold greater than when sealed in air.

The presence of any substance capable of reacting competitively with the oxidized water radicals naturally affords some protection to the organisms concerned. Amino-acids, proteins and other nutrient substances act in this way, but it is most marked by the natural sulphydryl compounds, such as glutathione, cysteine and thiourea, as well as by hydrosulphites, metabisulphites, nitrites and other reducing compounds. Quoted concentrations for the sulphydryl compounds range up to 0·04% and of other reducing compounds up to 0·001 M. Catalase is another natural protective agent, and alcohols and glycols also act in the same rôle by virtue of their ability to modify the metabolic processes of the organisms, but much higher concentrations, up to 3·5%, are needed.

There is no real evidence of any substance other than oxygen being able to sensitize micro-organisms to ionizing radiations, but preheating to sub-lethal temperatures can sensitize vegetative cells, but not spores. On the other hand pre-irradiation renders bacterial spores much more sensitive to subsequent heat treatment. The degree of heat sensitization is a function of the pre-irradiation dose, the relationship being logarithmic. By this means the heating time required to sterilize various meats has been reduced to one-quarter of the original.

Within reasonable limits the temperature at which an irradiation is carried out has no material effect on its efficiency, but freezing seems to have an adverse effect, although this is not universally agreed. According to some workers, frozen organisms require a 50% greater dose than those in the liquid or unfrozen state. Others could find no difference between resistances at −80° and +80°C, and others have simply reported genus or species differences.

Much more important is the effect of moisture, the removal of which always increases the apparent resistance of all types of organisms, including the viruses. The reasons for this are threefold: first, the reduced availabilities of the H• and •OH radicals and the consequent absence of their reaction products; secondly, the reduction in the mobilities of the free radicals that are present; thirdly, the greater protection afforded to the organisms by substances when dried on them.

The change in the order of resistance is significant. Thus, Tallentire (1958) reported that the irradiation dose which inactivated just over 99% of spores dried normally on an inert powder was only 90% effective against the same spores when hard dried. Similarly, a factor of 10^3 has been reported for vaccinia virus (Kaplan, 1960).

(b) *The rate of kill.* From the evidence available the rate, or intensity, of irradiation is of little importance in determining the level of kill; it is the total dose administered that matters. The exception to this, as already noted, is the very high intensity radiation from the Capacitron. In most instances the death-rate curve follows the usual exponential form, although in some situations it tends to be more sigmoidal. The slope of the time–survivor curve is determined by the intensity of the irradiation, but in terms of dose against percentage kill the relationship is always logarithmic, except on occasion, towards the end of the process, when the "tailing" effect may become prominent. In this connection the figures in Table IV are interesting. They illustrate unmistakably the importance of ensuring that the full dose has been applied.

The sterilizing dose in any particular situation is also associated with the level of contamination to be dealt with, the larger numbers of organisms requiring larger doses to kill them. It is for this reason that with some materials which are known to be bacteriologically clean one occasionally reads of lower sterilizing doses being recommended.

3. *Practical applications*

The main fields of application of ionizing radiations for sterilization purposes are in pharmacy and medicine, and to a much smaller extent in the food industries. But even in these fields, apart from the very high capital cost of installing suitable units, there are restrictions, all of which stem from the many side reactions produced during the course of irradiation. Thus, it cannot be used for sterilizing solutions containing proteins, because of the denaturation that occurs, neither can it be applied to substances or materials that are subject to oxidation. Some of the antibiotics, e.g., benzylpenicillin, streptomycin and polymyxin, can be sterilized in this way, but not others, e.g., neomycin and bacitracin. In line with the known theory of the mechanism involved substances in solution are more sensitive than when in the solid state. The first sign of deterioration is usually discoloration; this is most marked in soda glass, which gradually takes on a dirty brown colour.

There is no problem with surgical instruments or plastic disposable equipment—syringes, Petri dishes, catheters, etc.—or even with rubber gloves and tubing, but cotton gauzes, bandages and adhesive dressings need careful attention, because of the damage that can occur to their fibres. In this respect, vacuum-packed materials always have the advantage, although as explained earlier, such conditions might require a higher sterilizing dose.

For pharmaceutical and medical products, the accepted sterilizing dose is 2·5 Mrad. This level has been agreed as a result of experiment and on the assumption that the materials do not carry an initially high bacterial contamination.

With foods, the sterilizing dose for most normal situations is about 2 Mrad, but there are numerous exceptions as well as some disadvantages. The exceptions arise when frozen foods are to be treated, when there is a high initial bacterial content or when specific organisms of high resistance are known to be present. In these situations, doses of 4 Mrad or greater may be needed. The disadvantages arise from the side effects produced by the irradiation and take the form of off-flavours ("radiation flavours") and discoloration—the latter especially with red meats. In this type of work it is especially important to remember that sterilization, while destroying at the same time bacterial toxins, does not necessarily inactivate spoilage enzymes which may require up to 10 Mrad.

B. Ultraviolet irradiation

Although undoubtedly possessing bactericidal and fungicidal properties, ultraviolet radiation cannot be classed as a sterilizing agent because of its many uncertainties. The energy of the radiations, unlike those of the ionizing radiations, is very low, and this means that its power of penetration is also low. In fact, penetration into solids is virtually nil, and into liquids is only slight, depending on their opacities. In clear water over 50% of the radiation energy is lost at a depth of less than 2 in., and in river water it may be within 1 cm; in milk significant penetration is limited to 1 or 2 mm.

Because of this lack of penetration only direct ultraviolet rays are effective; they have no action against organisms shielded or otherwise protected from the incident beam.

The main applications of ultraviolet irradiation are in the field of air disinfection, as a vector in controlling cross-infection, for disinfecting, but not sterilizing, enclosed spaces and certain equipment and to a limited extent in water treatment.

The most effective lethal wavelength is in the range 240–280 nm, with the optimum at 253·7 nm. This is probably not a fixed optimum, but varies with the different types of organism. In fact, other values of 254·0, 265·2 and 280·4 nm have been quoted; nevertheless, 253·7 nm is convenient because it is the wavelength at which most of the emission occurs from the lamps most commonly available. The unit of radiation energy is measured in terms of ergs or microwatts/unit area/unit of time.

Ultraviolet radiation is equally affective against Gram-positive and Gram-negative bacteria, the lethal doses falling in the range 1000–6000 μW/sec/sq. cm. Bacterial spores require up to 10 times this dose and mould spores up to 50 times. Opinions concerning the susceptibilities of viruses vary considerably, and ratios of inactivating doses for viruses and bacteria in the range 10 : 1 to 100 : 1, with extremes of 1 : 1 and 200 : 1, have been reported.

The nature of the environment determines to a large extent the efficacy of the irradiation. Thus, to quote a simple example, two types of virus that were susceptible in Ringer's solution in 15 min were still active in nutrient broth after 1 h owing to the protection afforded by the peptones and amino-acids. In the dried state some organisms can experience total protection.

A similar argument obtains with organisms on exposed surfaces. In the first place they may be protected to a greater or less extent by the materials causing them to adhere to the surface, and in the second place they might be expected to be more protected by virtue of only one "face" of the organisms being exposed to the direct radiation. For these reasons airborne organisms are more susceptible than those on surfaces.

V. STERILIZATION BY GASES AND VAPOURS

All of the gaseous sterilizing agents, almost by definition, are highly reactive chemically. They include ethylene oxide and other heterocyclic ring compounds, formaldehyde and ozone, of which the principal ones are ethylene oxide and formaldehyde. They can be used for sterilizing both surfaces and any materials containing micro-organisms to which the gas or vapour can obtain direct access.

A. Ethylene oxide

The most effective gaseous sterilizing agents are found amongst the small heterocyclic molecules, including those containing oxygen in the ring, e.g., ethylene oxide, propylene oxide, β-propiolactone, epichlorohydrin and glycidyl derivatives; those containing sulphur in the ring, e.g., ethylene sulphide; and those containing nitrogen, e.g., ethylene imine and its homologues. Several of these are potentially more active than is ethylene oxide, but they have a number of significant disadvantages, in terms of low vapour pressure, high boiling point or toxicity. β-Propiolactone is an example of the last, being substantially more active than ethylene oxide, but it is also a carcinogen. Propylene oxide, although carrying less hazards than ethylene oxide, is also less active and does not vaporize easily, having a boiling point of 35°C. It is used mainly as a disinfectant for treating toilet and cosmetic powders.

Ethylene oxide has the structural formula—

$$\begin{array}{c} CH_2\text{—}CH_2 \\ \diagdown \ \diagup \\ O \end{array}$$

It boils at 10·8°C, it is violently explosive in nearly all mixtures with air, it is chronically toxic at concentrations not detected by smell and is a skin irritant producing erythema and odoema with potential hypersensitization. The explosive hazard can be eliminated by mixing the gas with carbon dioxide or nitrogen, and the skin problem only arises under conditions where it is in intimate contact with the skin and can be absorbed by it. Ethylene oxide is also hydrolysed slowly in water to form ethylene glycol, but in the presence of chloride ions the action is rapid with the formation of ethylene chlorohydrin, which is allegedly highly toxic.

Concentrations of ethylene oxide are quoted either in terms of mg/litre of air or as percentages by volume, the latter always being related to the gases at normal pressure. The former is the more accurate because in the end it is the amount of ethylene oxide gas present that matters. At normal atmospheric pressure a 10% concentration in air is equal to approximately 200 mg/litre.

1. *Lethal properties*

The sterilizing action of ethylene oxide is not rapid—the fastest method used requires an exposure time of 3 h—but it is almost as active against spores as it is against vegetative cells, the ratio being not greater than 5, and in most instances nearer 2. It is also a virucide, its activity, at least against the larger viruses, being about the same as that against vegetative bacteria. The minimum time of exposure necessary to ensure sterilization at a fixed temperature varies inversely with the gas concentration. The temperature

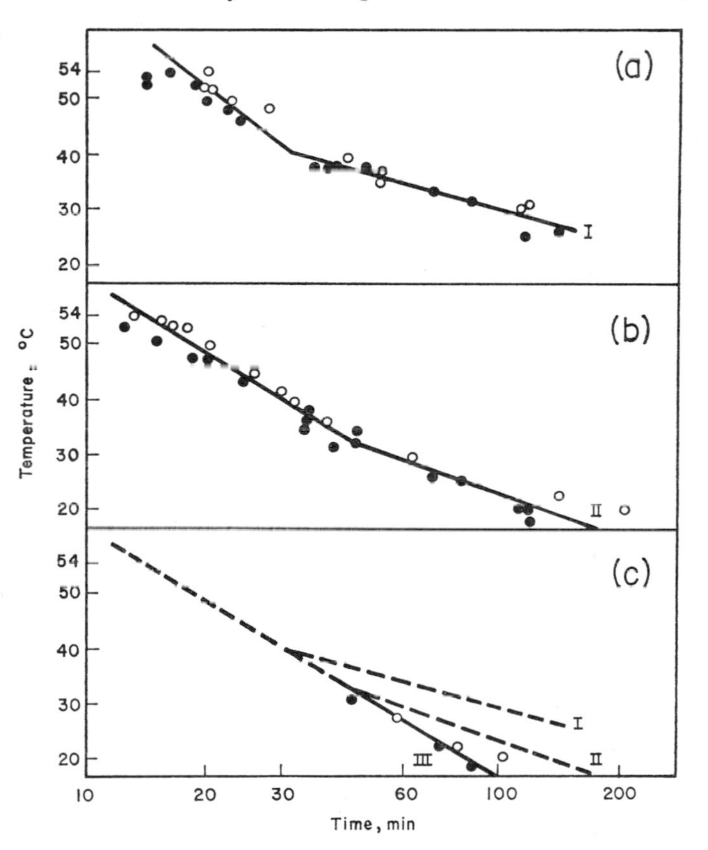

FIG. 4. Thermochemical death–time curves with ethylene oxide for spores of *Bacillus subtilis* var. *niger* dried on glass beads; (a), 440 mg/litre; (b) 880 mg/litre; (c) 1,500 mg/litre. ○, sets sterile; ●, sets not sterile.

coefficient for each 10°C range up to about 37°C and with gas concentrations of up to 880 mg/litre is about 2·7; above these limits the coefficient increases abruptly and the slope of the temperature–concentration curve becomes more even and equal to that for a concentration of 1500 mg of gas/litre, which gives a straight line throughout (see Fig. 4).

2. *Influence of humidity*

All workers are agreed concerning the importance of humidity in ethylene oxide sterilizations, but opinions differ concerning the optimum. According to Phillips (1961) it occurs at about 30% relative humidity, and certainly the efficacy falls away rapidly below this level; Mayr (1961) put it as high as 95% in some instances. The difference, as explained by Royce and Bowler (1961), is that whereas Phillips was concerned with naked washed spores, Mayr was dealing with organisms as they occurred in the natural state and so were protected to some extent. It appears, therefore, that with washed organisms it is only necessary to have enough moisture to allow the reaction to proceed, but with protected ones additional moisture must be available to dissolve the protecting substances. In this respect, and contrary to expectations, serum offers less protection than does salt. In practice, most sterilizations appear to proceed quite satisfactorily in the lower relative humidity range, provided, as already stated, it is not less than about 30%. Because of this it is not possible to sterilize substances that are hygroscopic, even to only a small degree. In such situations the relative humidity of the immediate atmosphere in and near the substance is reduced by the substance itself below the safe level.

At the other end of the scale there are situations where even at high relative humidity values the treatment is ineffective. Thus, organisms totally enclosed in crystals or other particulate substances remain untouched because the ethylene oxide is unable to gain access to them. Sodium chloride, glucose and the sulphonamides are examples. Raising the relative humidity to levels greater than 95% has other physical disadvantages.

3. *Absorption and diffusion*

A range of materials, including rubber, plastic, paper and board, as well as finely divided powders, such as french chalk and kaolin, all absorb certain amounts of ethylene oxide, and this must be taken into account in attempting to sterilize such materials. Failing this, the concentration of ethylene oxide gas in the sterilizing chamber will be below the intended level with a possible resultant sterilization failure. There is, however, a compensating feature because the gas remains absorbed for some time after the end of the normal sterilization period and so continues to exert its action, but to a diminishing extent.

The amount of ethylene oxide absorbed depends to some extent on the initial concentration of gas, but from a 10% v/v concentration Royce and Bowler (1959) have reported absorption for different materials as shown in Table V.

TABLE V

The amounts of ethylene oxide absorbed by different materials

Material	Amount, mg, of ethylene oxide absorbed/g of material
Polythene	2
PVC	19·2
Bakelite	Nil
Brown paper & board	6–10·4
Wood	18·4
Cotton wool	3·5–4
Rubbers (various)	5·5–15·2
Starch glove powder	10·5
French chalk	0·15
Sulphanilamide	0·8

These figures were obtained after the materials had been in a 10% gaseous concentration of ethylene oxide for 18h at 20°C.

The rate of desorption depends largely on the area of material exposed to the atmosphere. Thus, rubber gloves hung in free air will become degassed in about 30 min, but if wrapped and sealed in boxes, several days may be required.

Diffusion of the gas in terms of its sterilizing ability presents no problem unless narrow orifices or lengths of tubing are involved. It also passes fairly readily through film wraps of some plastics, but not all. Pre-vacuation helps the gas to penetrate more easily, and this is common practice, especially in large-scale operations.

4. *Practical applications*

Ethylene oxide is now widely used in hospitals and in industry for sterilizing heat-sensitive materials, such as pieces of equipment, sampling apparatus and pipettes, as well as some chemicals. It is also used for disinfecting chemical respirators, heart–lung machines, ophthalmoscopes, blankets and bedding.

The treatments given fall into two broad categories: (*i*) the low concentration–long period treatment requiring exposure to a minimum gas concentration of 10% (200 mg/litre) for 18 h at not less than 20°C; and (*ii*) the high temperature–short period treatment that requires a concentration of 800–1000 mg/litre at 55–60°C for 3–4 h. In the latter process it is essential that the material to be sterilized shall all be at the same elevated temperature.

Several variations of these treatments have been used. Thus, a concentration of 500 mg/litre for 4 h at room temperature has been found successful for sterilizing surgical instruments; a concentration of 200 mg/litre for 6 h at 60°C, with pre-vacuum treatment, has also been used for the same purpose, and an 80% concentration has been reported to sterilize soil in 6 h at a room temperature.

Ethylene oxide sterilizers range in style from a simple chamber, into which the gas can be introduced, with or without a pre-vacuum, to the more complicated and fully automated units. For the former, converted steam sterilizers are often used, and an autoclave unit has been described (Winge-Hedèn, 1963) that can be used with both steam and ethylene oxide. In general a pre-vacuum stage is desirable, the gas being introduced into the evacuated chamber, to encourage penetration. With wrapped or packaged material, or for sterilizing powder bulks, it is essential. Post-sterilization evacuation should also be included in the cycle, again to assist in removing the residual gas, but it does not appear to have a significant effect in extracting absorbed gas from rubber and plastics. A suitable degassing period should be allowed for all treated materials. Such a period may range from minutes to several days.

A sealed screen technique has been described by Royce and Sykes (1955) that permits difficult aseptic manipulations to be carried out without the risk of contamination. It is particularly useful in carrying out sterility tests, its only limitation being that it cannot be used where ethylene oxide comes into direct contact with the samples concerned. The principle of the method is that having loaded all of the necessary materials, samples, media and other equipment into the screen, which is constructed of sheet metal, it is sealed and ethylene oxide gas is introduced to a concentration of about 15%. The sterilization period is for 18 h at not less than 20°C, after which the ethylene oxide is flushed out with sterile filtered air. The manipulations are then carried out through long gauntlet gloves sealed to suitably positioned, flanged armholes. A sketch of the screen is shown in Fig. 5.

An even simpler method is to pack the materials to be sterilized in a gas-tight bag, fill it with an ethylene oxide–carbon dioxide or –Freon mixture, seal it and leave it at room temperature for a given period of time. Terylene fabric proofed with Neoprene and a three-foil laminate of paper, aluminium and polythene are suitable gas-impermeable materials. The method has been used successfully in sterilizing equipment as well as for disinfecting bedding. Whichever method may be employed the relative humidity should not be allowed to fall below about 50%. Where such a possibility exists, compensating water or water vapour should be introduced, but generally the natural moisture content of the materials being sterilized is adequate.

Sterilization indicators, in the form of spore strips or Royce's sachets, should be included in each sterilization load as a matter of routine to check that the treatment has been adequate. Such evidence is more important

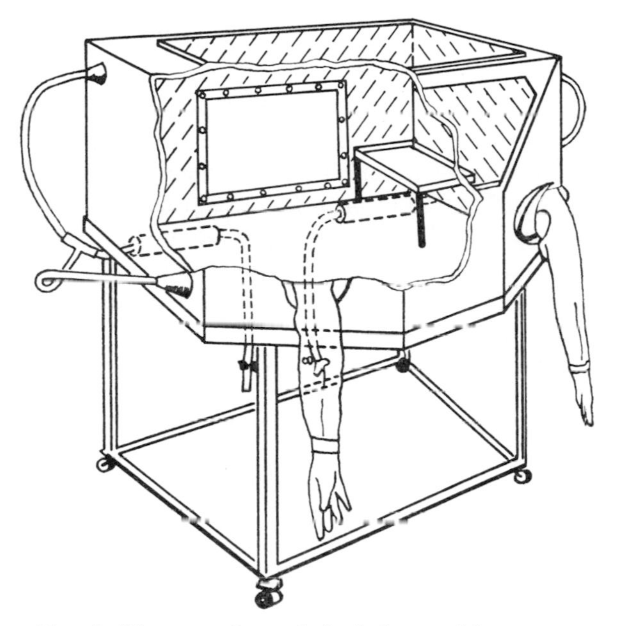

FIG. 5. Diagram of a sealed ethylene oxide screen.

with ethylene oxide sterilization than with steam or dry heat methods, because the margin of safety is much smaller. The level of kill is acknowledged to be only of the order of about 10^9, in contrast to 10^{15} or greater for an efficient steam sterilization.

5. *Sterilization of liquids*

Besides being active in the gaseous state, ethylene oxide can also be used in solution for sterilizing some aqueous liquids. To do this, the liquid is cooled to below 10°C, and then 1% (v/v) of liquid ethylene oxide is added and the container closed, but not sealed. It is allowed to stand in this state for an hour or so and then transferred to a warm place and left there for a few hours, or overnight, until all the ethylene oxide has been volatilized and removed.

The method has been recommended (Judge and Pelczar, 1955) for sterilizing media containing carbohydrates, and it has been used also to sterilize milk and serum. In the writer's experience, however, media treated in this way are rather less capable of supporting bacterial growth. This is to be expected from the high chemical reactivity of ethylene oxide.

B. Formaldehyde

The usefulness of formaldehyde gas or vapour is severely restricted because of its intense pungency, its lack of diffusibility and penetration and its tendency to polymerize and form a white film over all surfaces. This does not occur, however, at concentrations below about 3 mg/litre of air or at relative humidity values of >60%. Nevertheless, it is an effective bactericidal and sporicidal as well as virucidal agent. Spores are only 2–15 times more resistant than are vegetative bacteria (i.e., the ratio is about the same as with ethylene oxide), and tubercle bacilli and viruses are about as susceptible as the cocci.

The properties of formaldehyde were examined extensively by Nordgren (1939), and later a Committee of the Medical Research Council (Report, 1958) extended the work. Nordgren established a relationship between gas concentration and time of kill, and demonstrated increased activity at higher temperatures and the effect of humidity. One essential requirement is that the gas shall have easy access to the organism: even a thin film of organic matter can severely restrict the action, if not abolish it. Optimum activity occurs at about 80–90% relative humidity although little difference is noticeable down to 50% relative humidity. Below this value, and above 90% relative humidity, the activity falls away rapidly. The temperature effect is difficult to determine because, under the conditions in which formaldehyde is commonly used, the effect of increasing the temperature is to increase the concentration of the gas; nevertheless, at 55–60°C the action is more rapid than at room temperature or at 37°C. It is inadvisable to attempt to sterilize at temperatures below 18°C.

Spores dried from suspensions in water succumb to a formaldehyde concentration of 1 mg/litre of air in 20–30 min, but for spores dried from peptone water the factor may be increased by up to tenfold, i.e., twice the concentration for 3–4 times the period of exposure.

For sterilizing bedding and medical surgical equipment a pre-vacuum treatment is essential to enable the formaldehyde vapour to penetrate. Alternatively, the items can be hung individually so that the vapour can circulate easily.

VI. FILTRATION

Filtration can be applied to the sterilization of both liquids and gases. Apart from the fact that both processes actually remove micro-organisms, there seems to be little in common between the mechanisms of the actions concerned. Both certainly can involve more than a simple mechanical sieving process.

A. Filtration of liquids

1. *Types of filters*

All types of liquids—aqueous solutions, oils and organic solutions—can be sterilized by filtration. The materials used include asbestos pad filters, sintered glass and metals, unglazed porcelain, diatomaceous earth discs or "candles" and cellulose membranes. All are available in a range of grades, only some of which are suitable for sterilization purposes.

The asbestos pad filters are made in various mixtures of washed asbestos fibre with cotton linter and other "filler" materials, the amount of asbestos present determining the efficiency of the filter. Some grades are suitable only for clarification purposes or for removing the larger organisms. They are manufactured in sheets a few millimetres thick and are cut into discs or squares ready for use. Being soft and pliable they are easily susceptible to damage, especially when wet, and so need careful handling, particularly when being sterilized. They will continue to function satisfactorily for several hours, but not for prolonged periods. This type of filter is strongly adsorbent, by virtue of the negative charge on the asbestos, and so is liable to remove active substances from the solution being filtered. Protein materials are particularly susceptible in this way and so are enzymes, viruses and similar complexes. The major losses occur in the first small volume passing through each pad, and decline progressively as the pad becomes "saturated". Further disadvantages are that the filters are liable to shed fibres and that sometimes a small amount of alkali is released during sterilization. Both can be dealt with by a preliminary washing through with water.

Sintered filters are not widely used for bacteriological purposes. Glass ones, by their nature, are only available in the smaller sizes, and the so-called "5-on-3" grade is the only one suitable.

The unglazed porcelain and the diatomaceous earth filters are available in a range of sizes as either discs or candles, the latter being in the form of cylinders closed at one end and with an orifice lead through the other end. The weakest point in all such filters, especially the metal-mounted ones, is at the cement jointing at the orifice end, which is likely to develop leaks with repeated usage. The flow direction with all candles must be inwards.

This type of filter cannot be used with safety for indefinite periods. It will remain efficient for several hours of use, but sooner or later, depending on the type of filter, the numbers of organisms involved and the potentially nutritive properties of the solution being filtered, organisms will begin to grow through and penetrate the filter barrier. The pores of the filter easily become clogged, but they can be cleaned by treatment with hypochlorite solution or by judicious scrubbing. They can be used repeatedly, but constant cleaning tends to wear down the wall thickness and so reduce filtration efficacy. For this reason they should be frequently checked by the bubble pressure test (see below).

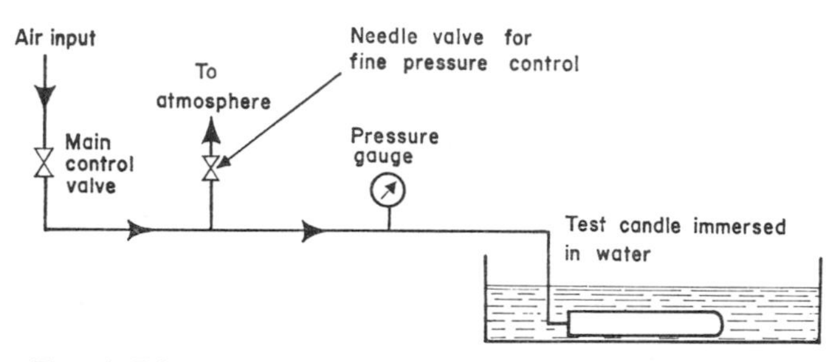

FIG. 6. Diagram of apparatus for the bubble pressure test.

These types of filter are graded according to their maximum porosity, the pores themselves being presumed to be cylindrical in shape, and this determines to some extent the efficiency of the filter. It is not all, however, because the maximum safe porosity for diatomaceous earth and certain porcelain filters is about 2·5 μm whereas for other materials it can be about 1 μm. The exact nature of the construction material plays some part.

Porosities are measured by the bubble pressure test, for which there is an established British Standard (British Standard Institution, 1963). Briefly, the method consists of saturating the candle with either water or carbon tetrachloride (according to the porosity of the material being tested) and then applying a slowly and gradully increasing air pressure to the inside of the candle and measuring the pressure at which bubbles just begin to break from the surface of the candle. A simple test apparatus for this is illustrated in Fig. 6.

The membrane filters, as other types, are available in a range of pore sizes. They are made with various types of cellulose and cellulose esters and, like the asbestos pad filters, can be used only once. They are paper-thin and so need careful handling, but they can be sterilized in the autoclave. Unlike asbestos filters, they do not absorb materials from the solutions being filtered.

2. *Mechanisms of filtration*

Although mechanical sieving must play some part in all filtration processes, other factors are also frequently involved. Of these, electrostatic and adsorption phenomena and the physical construction of the filter are the most important. Thus, membrane filters, because they are only thinly constructed, rely entirely on their pore size to hold back micro-organisms. A maximium pore size of 0·5 μm is desirable, therefore, to be certain of retaining all bacteria; for viruses it must be even less. The same probably applies to the sintered filters.

Asbestos filters, by the nature of their fibrous construction, have no porosity as such, and they rely for their efficacy on the adsorptive properties of the asbestos fibres themselves; filler materials seem to have little part, except for providing bulk and added thickness, and therefore mechanical interference. Evidence of this has been demonstrated photomicrographically by D. M. Wyllie (see Sykes, 1965).

Arguments have been raised from time to time on the relative merits of negative versus positive pressure filtrations and on the maximum pressure difference permissible with the various types of filter, but there is no conclusive evidence either way.

B. Filtration of air

The subject of the sterilization of air, and of other gases, is dealt with by Elsworth (this Volume p. 123), and so is discussed here only briefly.

With rare exceptions, micro organisms do not occur in the air in the naked state; they are found almost invariably attached to dust particles or fibres or else they are carried in dried saliva, mucus or similar particles. The size of the actual particle to be removed is, therefore, usually substantially larger than that of the organism itself.

Any granular or fibrous material—alumina granules, slag wools, coarse glass fibres and the like—will remove a proportion of airborne organisms, as has been shown by Sykes and Carter (1953, 1954). In practice, where large volumes of sterile air are required as in the fermentation industries, slag wool is commonly used, but for supplies to sterile rooms, etc., frame filters (the so-called " absolute" filters) made with finely woven spun glass fibre, and sometimes interleaved with various asbestos–cotton or –glass mixtures, are more usual. Like the filters for liquids they are supplied in a range of grades and sizes.

In passing, reference must also be made to the oldest known of our filter materials, cotton wool. On the small scale it cannot be beaten, provided it is kept dry, but at high air flow rates it tends to pack down and so become more resistant, with consequent build up of back pressure.

The efficacy of any air filter depends fundamentally on the number of opportunities the filter has of contacting and retaining the airborne particles, and this in turn depends on the size of the particles concerned, the diameter of the fibres or size of the granules of the filter material, its packing density and thickness, and the air flow rate. From mathematical considerations Stairmand (1950), Humphrey (1952) and Thomas (1952) have each shown that these chance contacts come about by (*i*) direct interception and trapping of the particles, (*ii*) inertial impaction of the particles, (*iii*) air flow turbulence and (*iv*) diffusion forces arising from Brownian movement of the particles: the strong influence of electrostatic forces is again evident. Dryness is essential for all types of air filter, and humidities over about 85% can begin to impair the performance of some of them.

Filter performance can be determined directly with airborne micro-organisms, including viruses where required, but this carries several difficulties and disadvantages, and so the tendency, especially in testing the frame-constructed filters, is to use particles of known size that can be more readily detected. Thus, the present British Standard (1957) specifies an aerosol of methylene blue of which the majority (99·8% by number and 90% by weight) of the particles measure 0·8 μm; in the United States, dimethyl phthalate, particle size about 3 μm, is widely used. Of recent years, the sodium chloride aerosol method, as devised by the Microbiological Research Establishment at Porton, has become increasingly in favour. The particles of this aerosol measure up to 1·5 μm but 85% of them are 0·3 μm or less. Penetration can be accurately determined by means of a flame photometer.

C. Electrostatic precipitation

This method of air treatment, although not strictly a filtration method, is mentioned here because of the considerable interest it has attracted. It is not an air sterilization method. It was first introduced for removing dusts and pollens from the atmosphere and is now used on the large scale in industrial establishments for providing clean, but not necessarily sterile, air.

The principle of the method is that when small particles are passed through a high voltage electric field they become charged, so that if they are then passed between closely spaced plates, alternate ones of which are raised to a high potential, they are attracted to and held by the plate of opposite potential. Potential differences of up to 20 kV have been used, but more generally the initial charge is developed from wires held at 14 kV with collecting plates held at 7 kV. The collecting efficiency of the system depends on the mass of the individual airborne particles and on the potential gradients developed.

Electrostatic precipitation as practised can remove some 98–99% of air-borne dusts and smokes, and about the same percentage of mould spores, but against the smaller bacterial cells they are rather less effective, in the range of 95–97%.

VII. CHEMICAL STERILIZATION

The number of chemicals that are reliable sterilizing agents, in contrast to disinfectants, is very limited and is virtually confined to chlorine, formaldehyde, 70% alcohol acidified with a mineral acid to pH 2 and ethylene oxide. Chlorine, usually as hypochlorite, is the only one widely used, and this is more commonly in a disinfecting, rather than a sterilizing, rôle.

Chlorine can be used in the form of hypochlorite solution or as one of the organic chlorine-releasing compounds such as chloramine-T, halazone, and the halogenated methylhydantoin and isocyanuric acid derivatives. Their properties as bactericides have been described in detail by Sykes (1965). They all depend for their activities on the amount of available chlorine present and sometimes on the pH value of the solution. Thus, hypochlorite solution containing 1000 parts of available chlorine/million will kill 99% of a suspension of *B. subtilis* spores at pH 11·3 in 70 min at normal temperature, but at pH 6·5 p.p.m. are effective in less than 15 min. The isocyanuric acid derivatives, on the other hand, are but little affected by pH value.

Opinions vary on the sporicidal activity of formaldehyde, but safe treatments are with a 1% solution at 37°C, or a 5% solution at normal temperatures. Both require up to 24 h. Acid alcohol will sterilize spores within 4 h, and ethylene oxide in 1% (w/v) solution is effective in about 2 h.

The chemical sterilizing agents have but little value in practice, and their applications are restricted to the treatment of apparatus and equipment where only surfaces are involved. They cannot be used for sterilizing solutions because they or their derivatives still remain in the solution. Ethylene oxide, in fact, has been suggested for sterilizing liquid culture media, but unfortunately it also impairs their nutrient properties, possibly because of the formation of ethylene chlorohydrin, as mentioned on p. 102.

And even in surface sterilization there are limitations. First there is the possibility of a reaction between the surface and the sterilant, then there is the question of absorption or retention of the agent in the surface and finally the problem of its removal at the end of the treatment without recontamination.

VIII. APPLICATIONS TO VARIOUS MATERIALS

Having discussed in some detail the different methods of sterilization available it is appropriate to conclude this Chapter by summarizing briefly their applicability to the sterilization of the various materials, culture media and other solutions, and apparatus and equipment used in the laboratory and elsewhere. In several instances more than one method can be used, and the choice is often an individual one. It is of some help, however, to know the pros and cons of each method in the context of the material to be sterilized.

A. Equipment

1. *Air supplies*

It is commonplace for microbiology laboratories to be served with a treated air supply in order to keep the airborne microbial population at a low level and so reduce the hazards of contamination. It is usual to have an extraction system as well as the input supply. In normal laboratories and in sterile areas, including closed screens, the supply should be maintained at a small positive pressure (to avoid external contaminants being sucked in), but in areas where infected animals are present or where pathogens are being handled a small negative pressure should be maintained (to prevent the escape of organisms).

In both situations the air should be introduced at ceiling level and the orifice fitted with a diffuser to prevent draughts; the extraction points should be near the floor. An upward directional air flow is undesirable because it tends to keep organisms airborne. A flow rate of about 8 changes of air/h should be maintained, but in sterile and other special areas up to twice this amount is recommended.

In enclosed screens for carrying out aseptic operations an even higher flow rate of up to 1 change of air/min is desirable. In such situations the general direction of the air flow should be to protect the essential site of operation within the screen from accidental contaminants introduced by the movement of the operator, etc.

In screens in which pathogens are handled the same air flow considerations are needed, but this time in the sense of keeping the screen surroundings free from contamination.

For most laboratory purposes, electrostatic precipitation is an adequate treatment for the incoming air, but additional safeguards are needed for sterile areas and for dealing with extracted air from pathogen handling areas. Both situations are met by fitting absolute or high efficiency filters at the input or extraction ports. Those intended for filtering infected, or potentially infected, air should be so fitted that they can be removed without undue disturbance and immediately covered and taken to the incinerator.

2. *Laboratory glassware and equipment*

The smaller units of laboratory glassware, such as flasks, tubes, bottles, pipettes and dishes can be sterilized by steam in the autoclave, by dry heat or with ethylene oxide.

Steam sterilization should be at 115°C for 30 min or 121°C for 20 min. During the heating stage all materials pick up a certain amount of moisture (from condensation on the cold surfaces) and so a final drying is necessary. This can be aided by releasing the steam pressure at the end of the process as quickly as possible and then applying a vacuum for a few minutes. Failing the latter, a hot air drying oven must be used. Sterilized materials in a damp condition are easily susceptible to recontamination, and sometimes even growth of the recontaminants takes place.

Dry heat, for a minimum of 1 h at 150°C (some authorities recommend 2 h at 180°C) is also commonly used. The method is not suitable, however, for containers plugged with cotton wool or covered or wrapped with paper: the former tends to char and the latter becomes brittle and discoloured. Nevertheless it is still common practice in many laboratories.

Syringes and other apparatus with closely fitting joints need careful attention because of their heat-change sensitivity. For this reason dry heat is unsuitable, and even in steam sterilization the rate of heating and of cooling must be controlled to avoid strain and breakage. The same applies to large pieces of glassware.

Radiation sterilization is unacceptable for glass because of the discoloration produced.

With rubber materials, such as gloves, tubing, plugs and closures, steam or ethylene oxide can be used. With the exception of silicone rubbers, dry heat is unsuitable. Repeated sterilization with steam, however, leads to mis-shaping and some denaturation, and with ethylene oxide an adequate degassing period must be allowed. The adverse effects of steam arise from the presence of small amounts of oxygen during the sterilization, and it can be eliminated by using the high pre-vacuum process. In terms of heat damage it is immaterial whether the treatment is for 30 min at 115°C or 3 min at 134°C.

Plastics are now used extensively for a range of apparatus and equipment, including "disposable" or "single use" syringes and Petri dishes, and they must all be sterilized by one of the cold processes (ethylene oxide or irradiation) because of their heat sensitivities. The disposable units as supplied commercially have all been cold sterilized. Repeated sterilization with ethylene oxide causes some types of plastics to craze and become opaque.

Preparation and storage. It is not easy to make specific rules concerning the way in which equipment should be assembled and wrapped before sterilization, and what is its subsequent safe period, because conditions vary so widely between different laboratories.

For steam sterilization the basic requirement for the wrapping or container is that air be easily removed and steam gain ready access. This means that metal canisters or drums must have properly sited portholes (which must be open during sterilization and closed immediately on removal from the autoclave) and that wrappings must not be completely sealed. To assist the steam treatment it is often convenient to assemble equipment wet or damp, but this can create subsequent drying problems. If the packages are not dried during the sterilization cycle by means of, for instance, a post-vacuum treatment, they should be transferred immediately to a hot-air drying oven, which itself should be maintained near-sterile in a clean atmosphere, taking precautions to avoid possible recontamination with dust, by handling or by contact with unclean surfaces.

These steaming and drying considerations do not apply, of course, to other methods of sterilization, but with ethylene oxide it is important to ensure that the gas can penetrate easily into the wrapping or container and that time is allowed after treatment for the material to be degassed.

No rule can be made regarding the optimum numbers of items, such as pipettes and Petri dishes, which should be sterilized in one package or container. It depends on several factors, the most important of which are the numbers of such items to be used at one sitting, the nature of the operations to be carried out—aseptic manipulations, for example, require higher standards than do other cultural procedures—and to some extent the conditions in which the work is to be done—a crowded, dusty area presents more hazards than does a clean, air conditioned laboratory. It is possible and permissible to re-seal a container after some of its contents have been used and to re-open it on a later occasion, but only on one such occasion. Aseptic handling of container and contents is a prerequisite for this. It is with these criteria in mind that each laboratory or individual decides for himself what is the most suitable pack size for the equipment he uses repeatedly and in large numbers. For pipettes, in particular, cardboard tube containers, each carrying one pipette, are the most suitable, and they can be sterilized repeatedly. Apart from these, containers should be made in metal or heat-resistant plastics and their closures should be such as will prevent access of microbial contaminants during storage.

Wrapping materials can include kraft paper, brown paper, close-woven cotton textiles, e.g., linen, and thin sheet plastics: paper and plastics should be free from pinholes. Only plastic packages can be completely sealed, but with such a seal they are unsuitable for heat sterilization: they are satisfactory for ethylene oxide sterilization provided the plastic is permeable to the gas and its absorbancy is low. Particular care needs to be taken against recontamination, not only during storage but also, and perhaps more importantly, during cooling after heat sterilization. This applies also to metal containers.

Sterilized materials of all types should always be stored in clean dry and draught-proof surroundings. The air in particular should be as free as possible from micro-organisms because temperature changes always cause packages to "breathe" and so bring about some interchange of the air within and without the package. If, therefore, airborne organisms are present there is a risk of some of them ultimately getting into the package. Under good conditions, sterile materials can be stored safely for several months: good housekeeping should ensure that they are not kept indefinitely.

These same principles of preparation, handling and storage apply to surgical materials.

3. *Surgical materials*

In this group are included surgical instruments, dressings, bandages, as well as special equipment, such as heart–lung machines and mechanical respirators. Syringes have been discussed in the preceding Section, and bedding is not included because it is not usually sterilized; it is only disinfected, and formaldehyde is the agent normally used.

The quickest and most reliable way to sterilize instruments and dressings is by the high pre-vacuum–high temperature steam cycle, but ethylene oxide or irradiation can also be used. A convenient way with ethylene oxide is to put the materials to be sterilized in a gas-impermeable bag, introduce the necessary amount of ethylene oxide, seal the bag immediately and leave it for 24 h. This avoids taking up sterilizer space for long periods, and it also avoids heat damage. Ethylene oxide is unsuitable for sterilizing adhesive dressings because it cannot penetrate the adhesive layer: irradiation is the only acceptable process.

Metal instruments, including knives and scissors, can also be reliably sterilized by boiling them in a 2% solution of sodium carbonate in water for 10 min (Anon, 1943). Such a treatment kills the most resistant spores, but along with other heat treatments it tends to blunt the cutting edges and lead to rusting of the less well plated instruments. This can be prevented by adding a small amount (0·1%) of sodium nitrite to the water. For obvious reasons instruments should never be left in the sterilizing solution, but removed and dried immediately. Dry heat sterilization can also be used.

Heart–lung machines and other complicated equipment are not usually sterilized in the proper sense. The treatment given is mainly to remove the hazard of respiratory cross infection. The only accepted way in which this can be done is with ethylene oxide. Again the whole unit can be placed in the sterilizer chamber or it can be enclosed in a gas-impermeable bag.

B. Culture media and other solutions

With certain exceptions, culture media, dilution fluids and the like can, and should, be sterilized by steam under pressure. Various times and temperatures are used but the recommended treatment is for 30 min at 115°C or 20 min at 121°C. The minimum acceptable is about 15 min at 115°C or 10 min at 121°C. Certain media, particularly those containing minced meat, may require a longer treatment; some years ago the author found that 45 min at 121°C was needed with meat from one particular source.

It should always be borne in mind that because of their relative complexities, culture media are always affected to a greater or less extent by any heat treatment. It is a maxim, therefore, that no medium should be subjected to more heating than is necessary. The actual duration of such a heating will depend on the size of the load to be sterilized and the size and nature of the individual containers (see p. 85), as well as the relative amount of steam available.

An alternative treatment is by tyndallization, but it is not recommended for solutions and media that can be sterilized in the autoclave. It is lengthy and unreliable for non-nutrient solutions, and the advantage to be gained in terms of heat damage to culture media is doubtful. It has a place, however, in the preparation and sterilization of solid egg or serum media, for which the usual inspissation treatment is followed by gentle steamings of 20 min duration on the two subsequent days.

Ideally, sugar media used for diagnostic purposes should not be sterilized in the autoclave because of the possible breakdown of the carbohydrate. Some workers recommend tyndallization, but the correct treatment is to steam sterilize the basal medium and then add to it the appropriate volume of a concentrated solution of the carbohydrate which has been sterilized either in the autoclave or by filtration—preferably the latter. The final mixture can be given a short steaming for, say, 10 min to kill off any accidental contamination. This is "a custom more honour'd in the breach than the observance" and in most situations it is probably of little significance; nevertheless it is a point to be borne in mind. It certainly does not apply to routine media, such as glucose agar or McConkey's media.

There is an increasing tendency to dispense with cotton wool plugged tubes and to use instead loosely fitting metal capped tubes or screw capped bijou or 1 oz bottles. Either reduces the rate of evaporation and loss of water from the medium, in comparison with cotton wool plugged tubes, and, provided they are stored cleanly, there is no risk of contamination, even in the metal capped tubes. They are generally more easy to handle and certainly less expensive and time consuming in preparation.

Nowadays several types of filling units, automatic or otherwise, are available by which large numbers of the same medium or solution can be dispensed. In some of these the essential parts of the unit can be detached and sterilized separately, thus allowing a pre-sterilized bulk of solution to be filled aseptically without the necessity of a possible re-sterilization in the filled tubes or other containers. It is used particularly with heat labile solutions, including some solutions for injection, but it is equally applicable to any liquid medium.

In terms of sterilization by filtration, attention must be drawn again to the possible loss by adsorption of small amounts of materials, especially from the first volumes of solution filtered. Asbestos pad filters are the worst offenders, but even with membrane filters there can be small losses. The adsorption is selective, but it includes enzymes, virus and phage particles, and some polysaccharides and antigens. It can be eliminated substantially by pre-treating the filter with a 0·1% gelatin or 0·05% sodium alginate solution (Hyslop, 1961). Sometimes asbestos pad filters also release alkali and trace amount of metals, notably iron. This can be met by prefiltering a weakly acid solution.

On the large scale, as in the antibiotics fermentation industry, media and other solutions are sometimes sterilized by means of heat exchange. This is a continuous-flow process, and the heat applied, usually from high pressure steam, is such that it continuously sterilizes the solution during its very short passage through the exchanger. The details of the process are beyond the scope of this book.

C. Contaminated materials

It should be axiomatic that all contaminated materials and apparatus are positively sterilized either in the autoclave or by burning without delay. It is not enough to immerse them for a period in a disinfectant and then simply wash or throw them away. Pipettes, slides, jars and other small individual pieces of glassware should be immersed immediately after use in a hypochlorite solution containing 1000–2000 p.p.m. of chlorine, and subsequently sterilized; apparatus such as Petri dishes, culture tubes, flasks, etc., should be placed in a lidded container and sterilized as soon as possible. They can then be washed safely. Combustible materials, such as cotton wool and paper, and even animal carcases, should be taken in closed containers to the incinerator.

Lysol and other phenolic disinfectants are not suitable substitutes for hypochlorite solution. They tend to leave oily residues on glass surfaces and, worse still, their volatility in steam causes them to contaminate the interior of the sterilizer and therefore subsequent sterilizer loads.

REFERENCES

Alder, V. G., Brown, Anne, M., and Gillespie, W. A. (1966). *J. clin. Path.*, **19**, 83.
Anderson, A. W., Nordan, H. L., Cain, R. F., Parrish, G., and Duggan, D. E. (1956). *Fd Technol., Champaign*, **10**, 575.
Anon. (1943). *Br. med. J.*, 633.
Bacq, A. N., and Alexander, P. (1961). "Fundamentals of Radiology", 2nd edn. Pergamon, Oxford.
Beeby, M. M., and Whitehouse, C. E. (1965). *J. appl. Bact.* **28**, 349.
Bowie, J. H. (1961). *In* "Sterilization of Surgical Materials: Report of a Symposium", p. 109. Pharmaceutical Press, London.
Brewer, C. M., and McLaughlin, C. B. (1954). *Science, N.Y.*, **120**, 501.
British Standards Institution (1957). "Specification for the Methylene Blue Test for High Efficiency Air Filters". B.S. 2831 : 1957.
British Standards Institution (1961). "Specification for Performance of Electrically Heated Sterilizing Ovens". B.S. 3421 : 1961.
British Standards Institution (1963). "Specification for Laboratory Sintered or Fritted Filters". B.S. 1752 : 1963.
British Standards Institution (1966). "Specification for Steam Sterilizers". B.S. 3970 : Parts 1 and 2 : 1966.
Coulthard, C. E. (1939). *Pharm. J.*, **142**, 79.
Darmady, E. M., and Brock, R. B. (1954). *J. clin. Path.*, **7**, 290.
Darmady, E. M., Hughes, K. E. A., and Tuke, W. (1957). *J. clin. Path.*, **10**, 291.
Elsworth, R., Morris, E. J., and East, D. (1961). *Chem. Engr*, No. 137, A47.
Ernst, R. R., and Shull, J. J. (1962). *Appl. Microbiol.*, **10**, 337.
Errera, M. (1954). *In* "Radiobiology Symposium", p. 93. Butterworths, London.
Fallon, R. J., and Pyne, J. R. (1963). *Lancet*, **i**, 1200.
Hannan, R. S. (1955). "Scientific and Technological Problems Involved in Using Ionizing Radiations for the Preservation of Food". H.M.S.O., London.
Hansen, N-H., and Riemann, H. (1963). *J. appl. Bact.*, **26**, 314.
Henfrey, K. M. (1961). *Hosp. Engr*, **15**, 260.
Henry, P. S. H. (1964). *J. appl. Bact.*, **27**, 413.
Humphrey, A. E. (1952). "Studies of the Mechanism of Bacterial Filtration from Air Streams". Special Rept. No. 2, Chemical Engineering Department, Columbia University. Columbia University, New York.
Hyslop, N. St. G. (1961). *Nature, Lond.*, **191**, 305.
Jordan, R. T., and Kempe, L. L. (1956). *Proc. Soc. exp. Biol. Med.*, **91**, 212.
Judge, L. F., and Pelczar, M. (1955). *Appl. Microbiol.*, **3**, 292.
Kaplan, C. (1960). *J. Hyg., Camb.*, **58**, 391.
Kelsey, J. C. (1959). *In* "The Operation of Sterilizing Autoclaves: Report of a Symposium", p. 22. Pharmaceutical Press, London.
Koh, W. Y., Morehouse, C. T., and Chandler, V. L. (1956). *Appl. Microbiol.*, **4**, 143.
Mayr. G. (1961). *In* "Sterilization of Surgical Materials: Report of a Symposium", p. 90. Pharmaceutical Press, London.
Medical Research Council (1959). "Report by Working Party on Pressure-Steam Sterilizers". *Lancet*, **i**, 425.
Medical Research Council (1961). "Report of Working Party on Pressure-Steam Sterilizers (2nd Communication)". *Lancet*, **ii**, 1243.
Nordgren, G. (1939). *Acta path. microbiol. scand.*, Suppl. 40.
Penikett, E. J. K., Rowe, T. W., and Robson, E. (1959). *J. appl. Bact.*, **21**, 282.

Phillips, C. R. (1961). *In* "Sterilization of Surgical Materials: Report of a Symposium", p. 59. Pharmaceutical Press, London.
Pratt, R., Durfendy, J., and Gardner, G. (1950). *J. Am. pharm. Ass., Scient. Edn,* **39**, 496.
Quesnel, L. B., Hayward, J. M., and Barnett, J. W. (1967). *J. appl. Bact.*, **30**, 578.
Rahn, D. (1945). *Bact. Rev.*, **9**, 1.
Report of the Committee on Formaldehyde Disinfection of the Public Health Laboratory Service. (1958). *J. Hyg., Camb.* **56**, 488.
Royce, A., and Bowler, C. (1959). *J. Pharm. Pharmac.*, **11**, 294T.
Royce, A., and Bowler, C. (1961) *J. Pharm. Pharmac.*, **13**, 87T.
Royce, A., and Sykes, G. (1955). *J. Pharm. Pharmac.*, **7**, 1046.
Scholes, G., and Weiss, J. (1954). *Biochem. J.*, **56**, 65.
Stairmand, C. J. (1950). *Trans. Inst. chem. Engrs.* **28**, 130.
Stark, W. H. and Pohler, G. M. (1951). *Ind. engng Chem.*, **42**, 1789.
Sykes, G. (1965). "Disinfection and Sterilization", 2nd edn. Spon, London.
Sykes, G., and Carter, D. V. (1953). *J. Pharm. Pharmac.*, **5**, 945.
Sykes, G., and Carter, D. V. (1954). *J. appl. Bact.*, **17**, 286.
Tallentire, A. (1958). *Nature, Lond.*, **182**, 1024.
Theil, C. C., Burton, H., and McClemont, J. (1952). *Proc. Soc. appl. Bact.*, **15**, 53.
Thomas, D. J. (1952). *J. Inst. Heat. Vent. Engrs*, **20**, 35.
Thornley, M. J. (1962). *J. appl. Bact.*, **25**, ii.
Thornley, M. J. (1963). *J. appl. Bact.*, **26**, 334.
Wagenaar, R. O., Dack, G. M., and Murrell, L. B. (1959). *Fd Res.*, **24**, 57.
Wilkinson, G. R., and Peacock, F. G. (1961). *J. Pharm, Pharmac.*, **13**, 72T.
Wilkinson, G. R., Peacock, F. G., and Robins, E. L. (1960). *J. Pharm, Pharmac.*, **12**, 197T.
Winge-Hedèn, K. (1963). *Acta path. microbiol. scand.*, **58**, 225.

CHAPTER IV

Treatment of Process Air for Deep Culture

R. ELSWORTH

Microbiological Research Establishment, Porton, Salisbury, Wilts., England

This Chapter deals briefly with the principles of filtration and heat sterilization of air and gives examples of suitable apparatus for sterilizing air supplies and for disinfecting effluent air from microbial cultures.

I. DISTRIBUTION AND NUMBERS OF ORGANISMS IN AIR

Although sterilization and disinfection of air may involve the removal of viruses and fungi as well as bacteria, the measurement of the removal of the latter is a practical assessment of the efficiency of removal of the other two. Thus a filter effective against bacteria will always remove larger fungal particles, and in many cases those particles that bear the smaller bacteriophages and viruses. Likewise, the effect of heat on all three classes of organism is similar, and a process that is effective against a bacterial spore will eradicate the other two.

As shown by Gregory (1961) the study of the bacterial flora of the atmosphere has been neglected. For the purpose of designing removal systems the following data represent almost all the available published facts. Colebrook and Cawston (1948) reported that air sampled in the open in London contained as few as or less than 2 organisms/cu. ft. The figure rose to 50/cu. ft in office conditions, and higher in more crowded situations. Outside air contained appreciable numbers of fungi as well as bacteria. Considering fermentation processes, in this laboratory we have slit-sampled air (Bourdillon *et al.*, 1941) in the vicinity of centrifuges used to separate suspensions of *Escherichia coli* and counted 110 particles/cu. ft. In other tests, effluent air from a culture vessel, producing the same organism, contained 40,000 bacteria-bearing particles/cu. ft. Telling *et al.* (1967) examined the effluent gases from deep cultures of Semliki Forest Virus (SFV) and Foot-and-Mouth Disease Virus (FMD) and reported a maximum of 50 plaque-forming units/litre, i.e., 1500 per cu. ft. They did not give any value for plating efficiency. In the subsequent calculations of the required removal efficiency of filters, a load of 2 bacteria/cu. ft will be assumed for sterilization of air supplies and 40,000/cu. ft for the treatment of effluent air.

II. ASSESSMENT OF DESIRED REMOVAL EFFICIENCY

Recent reviews on the practical aspects of filter design and selection include those of Cherry *et al.* (1963) and Richards (1967). Both draw on a method first put forward by Gaden and Humphrey (1956). In the following examples, which are intended to show the removal efficiency required in differing circumstances, the same procedure is employed.

Suppose an air rate (Q) of 200 cu. ft/min with a bacterial count (c) of 2 organisms/cu. ft is required for a bacterial batch culture where duration (t) is 24 h. The total organisms (N_1) to be removed will be given by—

$$N_1 = 200 \times 60 \times 2 \times 24 = 0{\cdot}58 \times 10^6$$

Assume that the allowable chance (p) of a contaminant penetrating the filter (or heater) is a thousand to one (10^{-3}), then the desired penetration, expressed as a percentage ($100\, N_2/N_1$), where N_2 is the number of organisms that pass the filter, is given by—

$$100\,\frac{N_2}{N_1} = \frac{100 \times 10^{-3}}{0{\cdot}58 \times 10^6} \approx 1{\cdot}7 \times 10^{-7}\%$$

Now consider the effect of reducing scale. If, as might be the case of a laboratory fermenter, only 0·1 cu. ft/min or about 3 litres/min of air is required but the other variables are the same as above we get—

$$100\,\frac{N_2}{N_1} = \frac{100\,p}{60\,Qct} = \frac{10^{-1}}{60 \times 0{\cdot}1 \times 2 \times 24} \approx 3{\cdot}5 \times 10^{-4}\%$$

so that a much less effective filter is required.

Next consider the treatment of the effluent air from the culture considered n the first example. Assume the same probability value of $p = 10^{-3}$. The bacterial count (c) is now 40,000 instead of 2, so that the desired penetration is now $0{\cdot}9 \times 10^{-11}\%$. Thus the removal process for effluent treatment must be much superior to that required for sterilizing the inlet air.

The maximum number of contaminant organisms that can be allowed to enter a growing culture before they manifest themselves is rarely known. The value of p is therefore a matter of judgement, which must include an allowance for the above uncertainty. However, an over-efficient rather than under-efficient process will always result by making calculations as above.

III. TESTING METHODS

A. Microbiological testing

A suitable method is as follows. An aqueous suspension of a micro-organism (preferably a *Bacillus* spore) is atomized in a stream of air, un-saturated with respect to water vapour, to produce a dried monodispersion of particles of approximately 1 μm in size. The bacteria-laden air is then passed through the filter or heat sterilizer under test, and is sampled simultaneously at the inlet and exit using a slit sampler or a microfilter (Elsworth, *et al.*, 1955). Bacterial counts are then carried out on the samples. For full details, see Cherry *et al.* (1963).

If the count on the inlet sample is N_1 and on the exit sample is N_2, then percentage penetration is $100\,N_2/N_1$. Earlier reports on air sterilization reflected the sensitivity of the method, which was poor, rather than revealing the performance of the apparatus under test. This was pointed out by Elsworth *et al.* (1961), who took larger samples of exit air and created test aerosols with higher bacterial concentrations. Thus they obtained bigger values for N_1. In common with the earlier workers, their estimate for N_2 either was, or approached, zero. Thus in examining similar equipment their estimate of penetration of $1{\cdot}75 \times 10^{-7}\%$ was understandably lower than those previously put forward.

B. Physical testing

Physical testing is a specialist operation, requiring expensive apparatus. Dorman *et al.* (1965) have described in detail an apparatus suitable for the routine testing of air filters passing up to 1000 cu. ft/min, which is the subject of British Standard 3928 : 1965, "Method of test for low penetration air filters." A particulate cloud of NaCl particles is generated by a battery of atomizers from a 2% NaCl solution. The cloud has an effective particle size in the range 0·02–1·0 μm. The exit air stream is sampled, and the salt content measured with a flame photometer. Penetrations from 100% NaCl to as low as 0·0003% can be determined in a few seconds.

Testing the same filter by the two methods, a value of 10^{-3} % by the sodium flame test corresponds to 5×10^{-6} % for a 1 μm monodispersion of *Bacillus subtilis* (H. M. Darlow, private communication). There is evidence that under natural conditions the organisms to be removed from an air stream exist either in clumps or attached to other materials. Because of this a practical conversion factor for design purposes is—

$$\% \text{ penetration of } 1\ \mu\text{m particles} = \% \text{ NaCl penetration} \times 10^{-2}$$

The ultimate sensitivity of the sodium flame test is thus 3×10^{-6} % in terms of 1 μm particles. This is similar to the sensitivity of the biological test previously cited.

IV. THE FACTORS INFLUENCING FILTER EFFICIENCY

The reliability coupled with the simplicity and the low operating cost of present-day filters rules out any case for using heat for sterilizing air supplies, though this is by no means the case for effluent air treatment. Although membranes are now being considered, even for the large scale (Aiba *et al.*, 1963), fibrous filters have received much study and are most commonly used. The examples that follow are restricted to the latter type. Dorman (1967) deals with the aerodynamics of small particles and the principles that a microbiologist should follow in choosing and using a filter. The following notes are based on Dorman's paper, which should be consulted for details.

There are three main factors influencing filter efficiency: air velocity, particle size and bed thickness.

A. Air velocity

Consider, as an example, the removal of 1 μm particles by a filter bed of glass-wool fibre which consists chiefly of 0·6–6·0 μm fibres packed to a density of 4–12 lb/cu. ft. Penetration increases with increased face velocity to a maximum in the region of 10–25 cm/sec, the actual value depending on the packing density. It then falls off as face velocity further increases. To get low penetration values, air velocities should therefore be chosen either much smaller than those at which the penetration is maximum, which means low pressure drop but a large-diameter filter; or a velocity larger than that at which maximum penetration occurs, which means a smaller-diameter filter but larger pressure drop.

B. Particle size

To remove particles smaller than 1 μm, low velocities are advisable. See, for example, Sadoff and Almof (1956), who filtered a virus of 0·08 μm dia. from an air stream and found that bed velocities must be less than 4 cm/sec. At such low velocities, penetration increases as particle size increases to 0·2–0·3 μm, passes through a maximum and then decreases as size further increases to 1 μm. At these velocities, the collection efficiency of a filter is as good for 0·05 μm as for 1 μm particles.

C. Bed thickness

For a cloud of one size of particle (monodisperse) penetration, as a function of bed thickness, is given by—

$$\frac{N_2}{N_1} = e^{-\gamma l}$$

where γ is the proportion removed by one layer of fibre and l is the number of layers in the filter. Although γ is small, in an efficient filter l is great, which results in a high removal efficiency. It follows that if two filters are connected in series the penetration of a monodispersion (i.e., a bacterial aerosol) through the pair is the product of the penetrations of the two individuals. Thus two filters each having a NaCl penetration of 10^{-3} % (which was stated previously as equivalent to 10^{-5} % for a bacterial aerosol) will have a penetration, when placed in series, of 10^{-12} % for bacterial particles. Note that the relation does not hold for NaCl penetration values because the NaCl cloud is not monodisperse.

V. STERILIZATION OF AIR SUPPLIES

A. Filter material

Porous ceramics, sintered glass or metals, porous membranes and beds of fibrous materials are all used in air filters. For laboratory purposes fibrous materials are best. High-efficiency material is available, and filters can be made up by the individual user. Above all, the filter bed can be designed for a particular duty, according to the principles already put forward.

Examples of fibrous material include non-absorbent cotton wool, vegetable fibre and glass-fibre papers, slag-wool mat and fibre-glass mat. Cotton wool is less efficient because of its large fibre diameter (17 μm). Fibre papers require special skills in mounting the elements to prevent leakage round the edges. They are not suitable for do-it-yourself manufacture. Further, they are prone to mechanical damage, which results in pin-holes. The loss in efficiency may not be significant and may go unnoticed at high gas velocities. At low gas velocities, or when used as a vent, a punctured fibre paper filter can be highly inefficient. These facts are well known to filter specialists. Slag wool or fibre glass are the best choices for making individual filters, which can be made up from standard laboratory items. The examples to be given will be restricted to the use of fibre glass.

From Sections II and III B it will be seen that for the sterilization of air supplies, a filter with a NaCl penetration of $<10^{-3}$ % will generally be adequate. In using a given fibrous material, there is a minimum packing density and minimum bed length for effective filtration. Whether a user

TABLE I

Operating conditions for slag wool and fibre glass

	Slag wool† <6 μm dia.	Fibre glass‡ 0·6–6·0 μm dia.	
Packing density			
lb/cu ft.	25	5	10
g/c.c.	0·40	0·08	0·16
Bed thickness, in.	3	$\frac{1}{4}$	$\frac{1}{2}$
Velocity, cm/sec	15	15	15
NaCl penetration, %	10^{-3}	$<10^{-3}$	$\ll 10^{-4}$

† Stilmed AST/13 filter medium, supplied 12·5 cm thick for compression to 7·5 cm: Stillite Products Ltd, London S.W.1.

‡ Grade AA glass fibre, Johns–Manville Co. Ltd, London S.E.1.

chooses greater values depends on whether he decides on a safety factor and how much pressure drop the system can afford. In regard to choice of air velocity it is better to use a value below rather than above the point of maximum penetration, even though this does mean a filter of larger cross-section. Recommended values for slag wool (Cherry *et al.*, 1963) and for a grade of fibre glass (R. G. Dorman, private communication) are given in Table I. Slag wool should be sterilized by dry heat at 160°–180°C for 2 h. Fibre glass may be steam-autoclaved or sterilized by in-line steaming at 5–20 p.s.i.g.

B. Autoclavable filters

For sterilizing the air to a small culture vessel the problem is to remove particles as small as 1 μm. The penetration should be $\not> 10^{-5}$ %. This is equivalent to 10^{-3} % by the NaCl test. Earlier (Section IV A, Table I) it was shown that a bed of glass-fibre mat (grade AA bonded fibre supplied by Johns–Mansville Co. Ltd, London S.E.1), packed to a minimum depth of ¼ in., at a density of 0·08 g/c.c., and operated at a face velocity of 10–15 cm/sec will perform this duty.

A 6 in. length of glass pipe line makes a suitable filter body. Bonded fibre-glass mat, as specified above, is cut into circular wads slightly larger than the body diameter. The bed is packed to a depth of 2½ in. This extra depth reduces the chance of edge leakage along the wall. Also the penetration is further reduced many orders below the acceptable maximum. The tubes are closed at each end with rubber bungs fitted with ¼ in. i.d. delivery

TABLE II

Packing details and operating rates for beds of grade AA glass fibre†

Body dia., in	Packing weight, g	Packing depth, in.	Air rate (litres/min) for removal of bacteria	phages & viruses
1	2·5	2·5	3·0–4·5	1·5
1·5	6·0	2·5	6·7–10·0	3·4
2	10·0	2·5	12·0–18·0	6·0

† 0·6–6·0 μm dia. (Johns–Manville Co. Ltd, London S.E.1).

tubes. The bed is retained under compression by filling the spaces on each side of it with stainless-steel wire-mesh pads (Knitmesh Ltd, South Croydon), or, less effectively, with ¼ in. i.d. ceramic insulating beads. The weight of packing and appropriate operating rates for 1, 1½ and 2 in. filters are given in Table II. At these air rates, the filters, packed as prescribed, have penetrations much less than 10^{-3} % NaCl. They are autoclaved each time they are used and re-packed if there is visual evidence that the bed is disintegrating.

C. Fibre-glass filters for in-line sterilization

In many applications, filters above 2 in. dia. are best made to engineering standards and to be suitable for in-line sterilization. The filter bed should be supported on each face by stiffened retaining grids that are flanged at the perimeter so that pressure is applied to the packing to prevent leakage round the edges. The filter should be located in the piping circuit with the faces vertical, and there should be drains to prevent condensate collecting in the bed during steaming or while in service. Richards (1967) has given design calculations for a filter that requires a bed depth of 18 in. and operates at an air velocity of 45 cm/sec. By contrast Dorman's experiments (Table I) show that for the material he tested, a bed thickness of $\frac{1}{2}$ in. will suffice at a packing density of 0·16 g/c.c. He also recommends the reduced velocity of 15 cm/sec, which implies a consequent increase in the area of the bed, relative to the example given by Richards.

D. Management of air supplies to filters

The air supply to a filter should not be saturated with water, to avoid the risk of water droplets forming in the filter bed with consequent loss of filtration efficiency. Laboratory operations are usually conducted at atmospheric pressure using a compressed air supply which is fitted with mist separators and water catchpots. At the point of delivery, while still under pressure, the air is generally water saturated. However, owing to expansion in volume after the control valve, accompanied by the taking in of heat from the surroundings, the air becomes unsaturated before it enters the filter. On a larger scale there may not be any significant reduction in pressure between the compressor after-cooler and the filter, so that the supply to the filter is virtually saturated. This can be avoided if the after-cooler can be run at a lower temperature than the filter body. In some installations in order to de-saturate the supply, a heater is included in the air stream before the filter (Rhodes and Fletcher, 1966). This also copes with any risk of fall in temperature if there is excessive pressure drop through the filter bed.

VI. TREATMENT OF EFFLUENT AIR

A. Reasons for treatment and methods available

The effluent air from most industrial fermentations requires no treatment apart from ensuring that the vent pipes are down-wind of the air intakes of the compressor used to supply process air. Disinfection of effluent air is limited to pathogenic cultures, such as those involved in bacterial and viral vaccine production. These are generally on a small scale, up to a limit of no more than 1000 litres working capacity. In the case of laboratory-scale deep cultures, even of non-pathogens, if it is not expedient to vent to the open air, it is advisable to process all effluent air to reduce the risk of infecting adjacent laboratory operations. For instance, consider effluent air containing 40,000 particles/cu. ft purged continuously at the rate of 3 litres/min. Then in a laboratory of 2000 cu. ft capacity, ventilated at 6 air changes/h, the average bacterial count in the room will be increased, on account of this source, by 20 particles/cu. ft. However, to get a true perspective on effluent air as a source of pollution, if the culture considered above lasts for 24 h, then the total emission of bacterial particles is 6×10^6.

This is equivalent to $\sim 10^{-3}$ ml of culture if the colony count is 10^{10}/ml. Thus to be consistent, the containment of the inevitable aerosols generated during sampling a culture requires similar, if not greater attention, than the treatment of effluent air.

The methods available for effluent-air treatment include scrubbing with disinfectants, such as oxidants or phenolics. This method does not bear serious consideration because regardless of the agent used, it is incapable of the removal efficiencies postulated in Section II. The effective methods are limited to heat treatment and filtration. The need for effluent air treatment emerged before filtration had reached its present stage of reliability and the earliest designs therefore involved heat treatment. Examples of both methods are given below.

B. Heat sterilization

1. *100 and 1700 litre/min sterilizers*

A design basis for heat sterilization was worked out by Elsworth *et al.* (1961). Figure 1 gives the temperature–time relationship for reducing incoming spores by a factor of 10^{-13} %. This shows that at 1 sec exposure time a temperature of about 350°C is required. At 200°C about 16 sec exposure is needed. Figure 2 shows the stainless-steel 100-litre/min unit from which the above data were obtained. A 1700-litre/min unit of similar efficiency was described by Elsworth *et al.* (1955). For other examples of heat sterilizers see the references in Elsworth *et al.* (1961).

2. *A 6 litre/min sterilizer for virus culture effluent gas*

The work described above on spores was extended to *Serratia marcescens* and T3 coliform phage. At a given temperature, the velocity constant for the destruction of each of these two organisms was about ten times that determined for spores. Thus if the conditions given in Fig. 1 are used for these two organisms, then the expected survival will be even smaller than 10^{-13} %. Deducing that virus particles will be affected in the same way, Telling *et al.* (1967) adapted a laboratory tube furnace, operating at about 300°C, to treat the effluent gas from 30-litre cultures of SFV and FMD. They measured a maximum of 50 plaque-forming units/litre in the untreated gas and detected none in the gas after passage through the sterilizer.

3. *A 5 litre/min electrically heated effluent filter*

Cameron *et al.* (1967) described a simple electrically heated effluent filter for air rates of 5 litres/min. It is used on 50-litre cultures of *Bacillus anthracis*, brucellas and salmonellas. It operates by trapping bacteria on a 10 cm dia. slag-wool filter bed maintained at 150°–200°C by an on–off

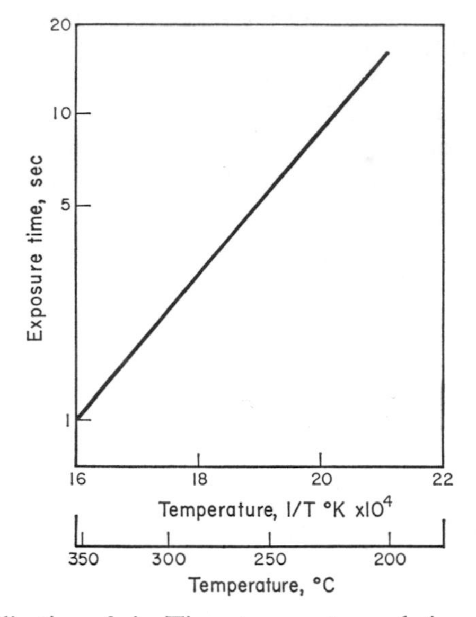

FIG. 1. Heat sterilization of air. Time–temperature relation, to give a survival of 10^{-13} %.

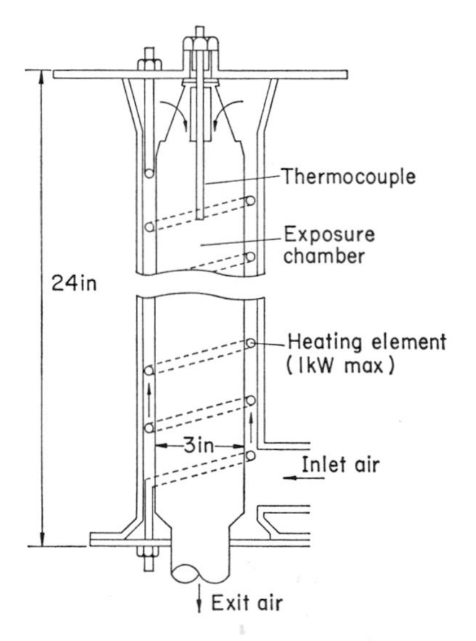

FIG. 2. Air sterilizer. Capacity, 100 litres/min (STP); working pressure, atmospheric.

controller. The bacteria are then killed by prolonged exposure at this temperature. The authors sampled inlet and exit air and, without indicating sample sizes, reported that the filtered air was sterile. The description of the filter material that they used and the bed velocity ($\sim$2 cm/sec at 200°C) is consistent with a NaCl penetration of appreciably less than 0·001%.

C. Effluent treatment by filtration

Besides the extra cost of electricity for heat sterilization, the equipment, whether for small or large scale, is complex and therefore costly. Filtration, if it is reliable, is an attractive alternative, and examples of its use are given below.

1. *Effluent filters for laboratory-scale cultures*

In most cultures the effluent air is saturated with water at a temperature above ambient. Therefore, if the air is passed without treatment into a filter at room temperature, condensation will occur at the filter bed, with consequent loss of filtration efficiency. On the small scale (3–15 litres/min) we have found that the insertion before the filter of a condenser cooled with mains water, consisting of an 18 in. length of 0·38 in. bore stainless-steel tube ($\frac{1}{8}$ in. B.S.P.), and a catchpot of 1 in. glass pipe line, 6 in. long, is sufficient to cool the air to less than ambient temperature and to collect condensate. This allows the filter to operate under unsaturated conditions. An additional precaution is to house the filter in a hot box in which a 60 W light bulb is fitted.

For bacterial and mould cultures, the exit air filter should be of the same size as the inlet air sterilizing filter (see Table II). When phage or virus is being propagated, the face velocity at the effluent air filter should not be greater than 5 cm/sec. The filter should be packed in the same way as a filter for inlet air. Table II gives appropriate velocities. Thus a phage culture operating at an air rate of 3 litres/min should be fitted with a filter of 1 in. dia. on the air supply and of $1\frac{1}{2}$ in. dia. on the effluent side. Again, owing to the safety factor that results from the increased bed depth, the penetration will be much less than the acceptable maximum.

2. *Effluent filters on pilot-scale cultures*

The precautions recommended above in the use of effluent filters do not mention the adverse effects of wetting a filter with culture foam. The results of such accidents are magnified as scale increases. Figure 3 illustrates a system in which the effluent filter is effectively isolated from the culture vessel. The filtration section is sterilized with formalin vapour, not steam. As operated, it treats 3 cu. ft/min of effluent air from spore cultures but its maximum capacity is 30–35 cu. ft/min. It consists of a fan, G, that exhausts laborabory air at a rate of 40 cu. ft/min through a 10 in. length of $2\frac{1}{2}$ in. i.d. fibre glass tube, D, connected to an expansion chamber, E, fitted with a drain cock at its base. The air then passes through a box filter, F. Its face dimensions are 8×8 in., the filter element is glass-fibre paper and the NaCl penetration is rated at 0·001 %. The purified air is exhausted through the

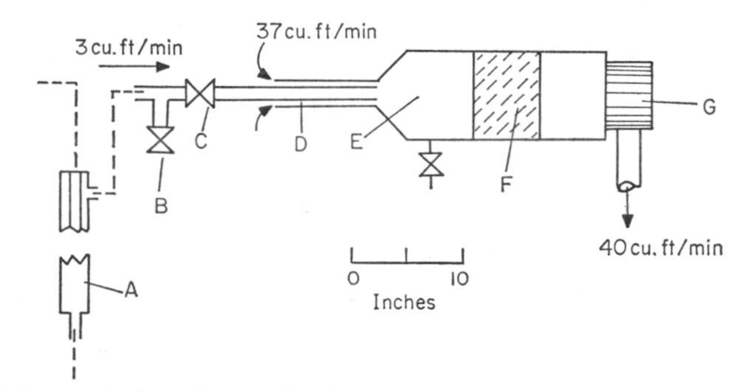

FIG. 3. Filter unit for effluent air. A, catchpot on culture vessel; B, C, valves on effluent air line; D, inlet tube for laboratory air; E, expansion chamber; F, filter element; G, exhaust fan.

fan and finally to the outside atmosphere. Air from the culture vessel at a rate of 100 litres/min ($\sim$3 cu. ft/min) is delivered through $\frac{1}{2}$ in. B.S. pipe via catchpot A into the expansion chamber. The dilution of the culture-vessel effluent air with laboratory air is sufficient to de-saturate the mixture that passes through the filter. Catchpot A separates any foam generated, and the expansion chamber, E, acts as a guard.

Before growing a culture, the system is steam-sterilized up to valve, C, and after seeding the fan is switched on. As air vents from the vessel, it is automatically exhausted through the filter. The filter unit need not be decontaminated at each culture, but can be treated periodically with formalin vapour.

The reduction achieved with a single filter is not to the standard suggested in Section II. Calculations based on the culture lasting 24 h with air loaded at 40,000 particles/cu. ft show that 4 bacteria can be expected to pass the filter during each batch. This is an acceptable value for a disinfection process.

Williams and Lidwell (1957) recommended that the air velocity through a port in a protective cabinet should be 100 ft/min in order to minimize the risk of contaminant organisms diffusing out of the cabinet and into the laboratory. At 40 cu. ft/min, the velocity of the air in the inlet annulus, D, is 1330 ft/min, which is much in excess of the above value.

Instead of relying on air velocity as the safeguard, a second unit is under development, in which there is a filter before the inlet tube. The air from the culture vessel is introduced directly into the expansion chamber. This removes all risk of back-diffusion of contaminant organisms into the laboratory, even in the event of fan failure, and this system of operation will then be suitable for treating the air from cultures of virulent organisms.

VII. FUTURE DEVELOPMENTS

In the past, air filters for use in microbial processes have had to contain packing material, often made originally for another purpose. However, there are signs that the makers have recognized present demands and are setting out to provide material specifically designed for filtration. It should be steam-sterilizable. It should be reproducible in quality, so that it can be installed with sufficient confidence to dispense with the necessity for conducting penetration tests before actual use. Materials likely to fall in this category in the future are steam-sterilizable resin-bonded glass fibre, such as is being developed by W. and R. Balston Ltd, Maidstone, Kent, or the membrane type of material reported by Aiba *et al.* (1963). It is also likely, as filtration methods are more properly understood and become even more reliable, that heat as a means of sterilizing air will lose its already marginal position and fall into complete disuse.

REFERENCES

Aiba, S., Nishikawa, S., and Ikeda, H. (1963). *J. gen. appl. Microbiol., Tokyo*, **9**, 267–278.

Bourdillon, R. B., Lidwell, O. M., and Thomas, J. C. (1941). *J. Hyg., Camb.*, **41**, 197–224.

Cameron, J., and Favelle, H. K. (1967). *J. appl. Bact.*, **30**, 216–263.

Cherry, G. B., Kemp, S. D., and Parker, A. (1963). *In* "Progress in Industrial Microbiology" (Ed. D. J. D. Hockenhull), Vol. 4, pp. 37–60. Heywood, London.

Colebrook, L., and Cawston, W. C. (1948). *In* "Studies in Air Hygiene", MRC Special Report, No. 262, p. 233. HMSO, London.

Dorman, R. G. (1967). *Chemy Ind.*, 1967, 1946–1949.

Dorman, R. G., Sergison, P. F., and Yeates, L. E. J. (1965). *J. Instn Heat. Vent. Engrs*, **33**, 390–396.

Elsworth, R., Telling, R. C., and Ford, J. W. S. (1955). *J. Hyg., Camb.*, **53**, 445–457.

Elsworth, R., Morris, E. J., and East, D. N. (1961). *Trans. Instn Chem. Engrs*, **39**, A47–A52.

Gaden, E. L. Jr., and Humphrey, A. E. (1956). *Ind. Engng Chem.*, **48**, 2172–2176.

Gregory, P. H. (1961). "The Microbiology of the Atmosphere", p. 113. Leonard Hill, London.

Rhodes, A., and Fletcher, D. L. (1966). "Principles of Industrial Microbiology" p. 54. Pergamon Press, London.

Richards, J. W. (1967). *Proc. Biochem.* (*inc. Biochem. Engng*, **2**, 21–25.
Sadoff, H. L., and Almof, J. W. (1956). *Ind. Engng Chem.*, **4**, 2199–2203.
Telling, R. C., Radlett, P. J., and Mowat, G. N. (1967). *Biotechnol. Bioengng*, **9**, 257–265.
Williams, R. E. O., and Lidwell, O. M. (1957). *J. clin. Pathol.*, **10**, 400-

CHAPTER V

Principles and Applications of Laminar-flow Devices

J. J. McDade

Biohazards Department, Pitman-Moore Division, The Dow Chemical Company, Zionsville, Indiana, U.S.A.

G. B. Phillips

Biological Safety and Control, Becton, Dickinson and Company, Cockeysville, Maryland, U.S.A.

H. D. Sivinski and W. J. Whitfield

Planetary Quarantine Department, Sandia Laboratories, Albuquerque, New Mexico, U.S.A.

I. CONTROL OF AIR-BORNE CONTAMINATION

Air has been considered an etiologic agent since the time of Hippocrates. In the 14th century, Fracastorius included air as one of the three factors he considered important in the transmission of disease. During the 18th century, the controversy over spontaneous generation involved noted scientists such as Needham, Spallanzani, Schulze, Schwann, Schroeder, and eventually Louis Pasteur. The resolution of this controversy, first by Schwann and later by Pasteur, came with the proof that air contains germs that can be the origin of ferments.

In England, John Tyndall's book, "Essay on the Floating Matter of the Air in Relation to Putrefaction and Infection", provided the final death blow to the spontaneous generation theory. Tyndall used an ingenious chamber to prove that micro-organisms present in the air were associated with air-borne dust particles. When the beam of light passing through the chamber did not scatter, and the air was free of dust, microbial growth did not result in open broth tubes exposed in the bottom of the chamber. Tyndall's work provided presumptive evidence in support of Lord Lister's theory that bacteria in the air were responsible for suppuration and putrefaction in some surgical wounds.

Speaking before the French Academy of Medicine on April 30, 1876, Louis Pasteur made a most succinct comment on the control of micro-organisms and the role of air-borne contamination in the transmission of disease. Pasteur stated:

> "... If I had the honour of being a surgeon, convinced as I am of the dangers caused by the germs of microbes scattered on the surface of every object, particularly in the hospitals, not only would I use absolutely clean instruments, but, after cleansing my hands with the greatest of care and putting them quickly through a flame (an easy thing to do with a little practice), I would only make use of charpie, bandages, and sponges which had previously been raised to a heat of 130°C to 150°C; I would only employ water which had been heated to a temperature of 110°C to 120°C. All that is easy in practice, and, in that way, I should still have to fear the germs suspended in the atmosphere surrounding the bed or the patients ..."

Thus, the need for control of microbial contamination in specific environments has been recognized for approximately 100 years. But, in the past 20 to 25 years, the parameters of interest and the degree of concern over the control of microbial contamination have expanded tremendously. This expansion has resulted in the need for greater and greater control over the many diverse processes involving or affected by micro-organisms.

More recently, the rising incidence of hospital-acquired infections has prompted initiation of detailed studies to determine the microbial profile of institutional environments (Bourdillon and Colebrook, 1946; Colebrook and Cawston, 1948; Duguid and Wallace, 1948; Blowers *et al.*, 1955;

Williams *et al.*, 1956; Engley and Bass, 1957; Fredette, 1958; Blowers and Crew, 1960; Blowers and Wallace, 1960; Greene *et al.*, 1962; Favero *et al.*, 1968; Shaffer and McDade, 1964a, b). These studies have indicated that large numbers of micro-organisms can, and do, exist on surfaces and in the air of intramural environments. Yet, when proper control measures are initiated and continuously employed, the degree of microbial contamination can be reduced and kept at a low level. Fortunately, then, microbiologists, chemists and engineers have developed many procedures, devices and sterilants that have provided increased safety and more reliable control over microbial contamination. For example, a variety of useful and efficient filters have been developed for removing micro-organisms and particulate matter from liquids and air (Decker *et al.*, 1962; Decker and Buchanan, 1965) Chemical agents such as ethylene oxide, peracetic acid, and beta-propiolactone have been incorporated into procedures for sterilizing thermolabile materials and for decontaminating environmentally controlled rooms and spaces (Hoffman and Warshowsky, 1958; Barret, 1959; Spiner and Hoffman, 1960; Phillips, G. B. *et al.*, 1962; Phillips, C. R. 1965). Food technology needs, and sterilization requirements for the exploration of space, have stimulated research into the kinetics of microbial destruction by both moist (steam) and dry heat (Pflug, 1960; Bruch *et al.*, 1963; Koesterer, 1964; Bruch, 1965; Schmidt, 1966). Sterilization of medical items by irradiation has also become a reality (Symposium, 1966). Improvements in the microbiological sampling of air and surfaces, as well as sterility testing, have allowed more precise identification of problems involving biological safety and the means for evaluating remedial procedures (Wolf *et al.*, 1959; Hall and Hartnett, 1964; Favero *et al.*, 1967). Finally, a variety of cubicles, hoods, cabinets, rooms, and enclosures have been designed to complement these and other advances in the control of microbial contamination (Wedum, 1953; Blickman and Lanahan, 1960; Jemski and Phillips, G. B., 1963; Wedum, 1964; Phillips, G. B., 1965; Dow Biohazards Department, 1967).

Concomitantly, industry has recognized a need for the control of inert particulate (fine dust, lint, fibres, etc.) contamination. The need for this control became obvious when it was discovered that fine particulate contamination caused seizure of small gears and bearings used in precision navigational instruments such as gyroscopes. Similarly, failures in electronic circuitry could often be traced to contaminating particulate matter. To eliminate the gross particulate matter associated with routine manufacturing processes, environmentally controlled work areas were developed. These environmentally controlled areas have been referred to as "white rooms" or "clean rooms". A clean room may be defined as an enclosed area employing control over the particulate matter in air with temperature, humidity, and pressure control as required (Anonymous, 1966).

II. CLEAN-ROOM CLASSIFICATION

Originally, clean rooms were classified into Class II, III or IV according to a U.S. Air Force Technical Order (Austin, 1963). More recently, this classification is being abandoned in favour of the Class 100, 10,000 or 100,000 system defined in Federal Standard 209a (Anonymous, 1966). Either system (Austin, 1963; Anonymous 1966) of classification is based on set limits of tolerable air-borne particulate matter of specific sizes. Conventionally, standard air cleanliness classes for air-borne particulate (nonviable) contamination are specified into three classes as indicated in Table I.

TABLE I

Classification of clean rooms (Anonymous, 1966)

Clean room class	Maximum number of particles per cubic foot (litre) 0·5 μm and larger	Maximum number of particles per cubic foot (litre) 5·0 μm and larger
100 (3·5)†	100 (3·5)	‡
10,000 (350)	10,000 (350)	65 (2·3)
100,000 (3500)	100,000 (3500)	700 (25)

† Numbers in parenthesis indicate metric system equivalent.
‡ Counts below 10 (0·35) per cubic foot (litre) are unreliable unless the number of samples taken is sufficient to give a statistically valid sample. Sample size is dependent upon the actual number of particles per cubic foot of room air. In general, the fewer particles per cubic foot of air, the larger the volume of air required to be sampled.

The classifications shown in Table I are based on a set number of air-borne particles of two specific sizes. The maximum allowable number of 0·5 μm and larger and of 5·0 μm and larger particles per cubic foot of air is indicated for each clean room class. Particle counts are to be taken during work activity periods and at a location that will yield the particle count of the air as it approaches the work location. Thus, in a Class 100 clean room, the particle count must not exceed a total of 100 one-half μm and larger particles per cubic foot (3·5 particles per litre). In a Class 10,000 clean room, the particle count must not exceed a total of 10,000 one-half μm and larger particles per cubic foot (350 particles per litre or 65 five μm and larger particles per cubic foot (2·3 particles per litre). Finally, in a Class 100,000 clean room, the particle count must not exceed 100,000 one-half μm and larger particles per cubic foot (3,500 per litre) and 700 five μm and larger particles per cubic foot (25 per litre).

Figure 1 depicts the three classes of clean rooms by statistical average particle size distribution curves (Anonymous, 1966). Special classifications may be used for particle count levels where special conditions dictate their use. Such classes may be defined by the intercept point on the 0·5 μm line in Fig. 1 with a curve parallel to the three established curves.

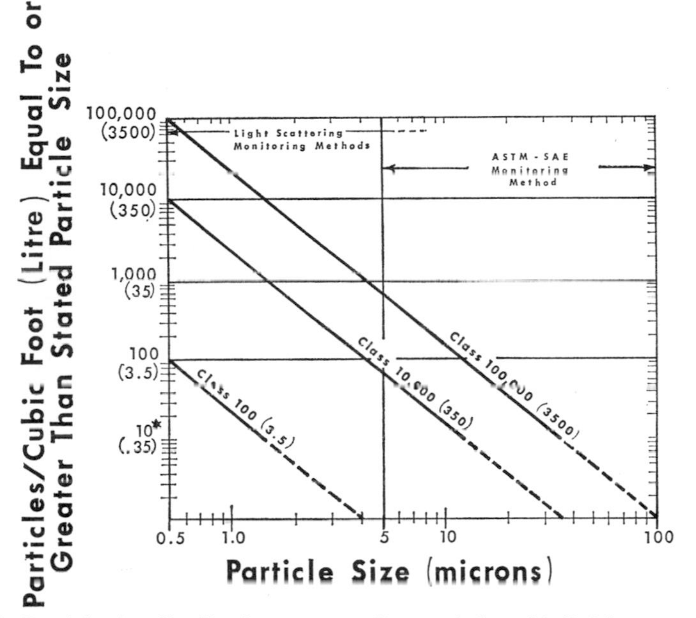

FIG. 1. Particle size distribution curves; Counts below 10 (0·35) per cubic foot (litre) are unreliable unless the number of samples is sufficient to give a statistically valid sample. Sample size is dependent upon the actual number of particles per cubic foot of room air. In general, the fewer particles per cubic foot of air, the larger the volume of air required to be sampled.

In addition to air-borne particulate matter, temperature, and humidity, air change rate or air flow, audio noise level, and vibrations within the clean room are also controlled. Depending upon the work being conducted, due consideration may also be given to environmental factors such as light level, microbial contamination level, electromagnetic radiation or fields, ionizing radiation, radioactive particles, gases, and vapours.

Throughout the discussion of the various types of clean rooms, benches and other contamination control facilities, the assumption is made that all filters are properly installed, sealed, and are free of leaks. Detailed information concerning filter installation, leak-testing and sealing is described in Federal Standard No. 209a (36). For continuity, a brief recapitulation of the filter leak-testing method described in Federal Standard 209a is included below as follows—

"... 50.1 Laminar flow rooms. Tests for laminar flow rooms should be as follows:

(a) An in-place filter test should be made to determine that the "HEPA" filter band contains no pinhole leaks. Tests should be made to determine leaks in (1) the filter media itself, (2) the bond between the filter media and the interior of the filter frame, and (3) the filter frame gasket and the filter bank supporting frames.
Leak tests should be made by introducing a high concentration of smoke or fog† into the plenum upstream of the "HEPA" filters (concentration should be of the order of 10^4 above the minimum sensitivity of the photometer used as the detector; this normally will be in the order of 80 to 100 micrograms per litre). The entire downstream surface of the "HEPA" filter installation is then scanned with an aerosol photometer probe at a sampling rate of 1 cubic foot per minute. The probe should be sized to provide approximately isokinetic sampling and should be held to 1 to 2 inches from the filter media and frame. (Thus, for measuring laminar flow equipment having air velocities of from 70 f.p.m. to 110 f.p.m., the probe should be sized from 1 to 1·5 inches squared.)
An aerosol photometer reading equivalent to 0·01 per cent of the upstream smoke concentration is considered a significant leak and should be sealed off.

(b) Airflow velocity should be measured through the cross section of the room and should conform to 40·2·12.

50·2 Nonlaminar flow rooms.

(a) Check for leaks (per laminar flow rooms, 50·1a) except that scanning is required to be done at the downstream surface of the "HEPA" filter, if it is readily accessible.

(b) Aspiration of contamination from leaks in the duct (between the filters and the room inlet) should be considered and tests made at the point of air inlet.

50·3 Laminar flow clean work station.

(a) All tests for laminar flow clean work stations should be (per laminar flow rooms, 50·1a), by introducing smoke into the intake ducts.

† Note: For example, cold generated DOP Fog. Ref: NRL 5929 "Studies of Portable Air-Operated Aerosol Generators", Echols and Young, Clearing House for Federal, Scientific and Technical Information, Springfield, Va., 22151.

(b) Airflow velocity out of the air exit of an unobstructed clean work station should conform to 40·2·12.

50·4 Nonlaminar flow clean work station. Check for leaks (per laminar flow rooms, 50·1a), except that scanning is required to be done at the downstream surface of the "HEPA" filter, if it is readily accessible . . ."

Item 40·2·12 referred to above is as follows—

"... 40·2·12 *Airflow velocity.* Airflow velocity through the cross section of the room normally is maintained with a uniformity within plus or minus 20 feet ($\pm$ 6·10 meters) per minute throughout the undisturbed room area . . ." (Anonymous, 1966).

III. NON-LAMINAR-FLOW CLEAN ROOMS

In a non-laminar flow clean room, filtered air is supplied to the room through ceiling ports or diffusers. The air filter in each supply duct is usually a high efficiency particulate air (HEPA) filter, rated 99·97 % efficient in removing particles 0·3 μm and larger from the air stream. Room air is exhausted through grilles or ports that may be located in the ceiling, in the floor, or in the wall(s) where the location may vary from the lower to the upper periphery. The normal volume of air moved is equal to about some 15 to 20 changes of the cubic room air capacity per hour, or a single change in three to four minutes. No effort is made to control the direction of air movement within the room. Cleanliness of the air within the room is attained by dilution of contaminated air with fresh, filtered supply air.

The non-laminar flow clean room supply diffuser-exhaust grille arrangements contribute to random air flow patterns (Fig. 2) that will vary according to the relative placement of air entrance and exit locations. Personnel movement and equipment relocation(s) may alter air flow patterns in the room, and these changes may further impede the flow of air and increase turbulence (eddy currents and static areas that develop as the supply air bounces off the clean room walls, floors, tables, equipment and personnel) within the room (Fig. 2). In fact, when turbulent air flow occurs near an operation that generates particles, it is often possible to have an increase in contamination within the clean room due to accumulation of air-borne particles in the recycling (turbulent) air flow. In general, a rather high static (positive) operating pressure is required to exclude exterior contamination from non-laminar flow clean rooms.

Ordinarily, if well controlled, the non-laminar flow clean room will operate at a Class 100,000 level of air cleanliness. However, the air cleanliness level of this type of facility is also dependent upon the ability of the janitorial staff to remove contamination brought into the area by clean room personnel, and to remove that generated within the room by both the personnel and the operations being performed.

Non-laminar flow clean rooms have very little self-cleaning capability, and attempts to operate them at air cleanliness levels below that of Class 100,000 requires extremely vigorous personnel controls and continuous

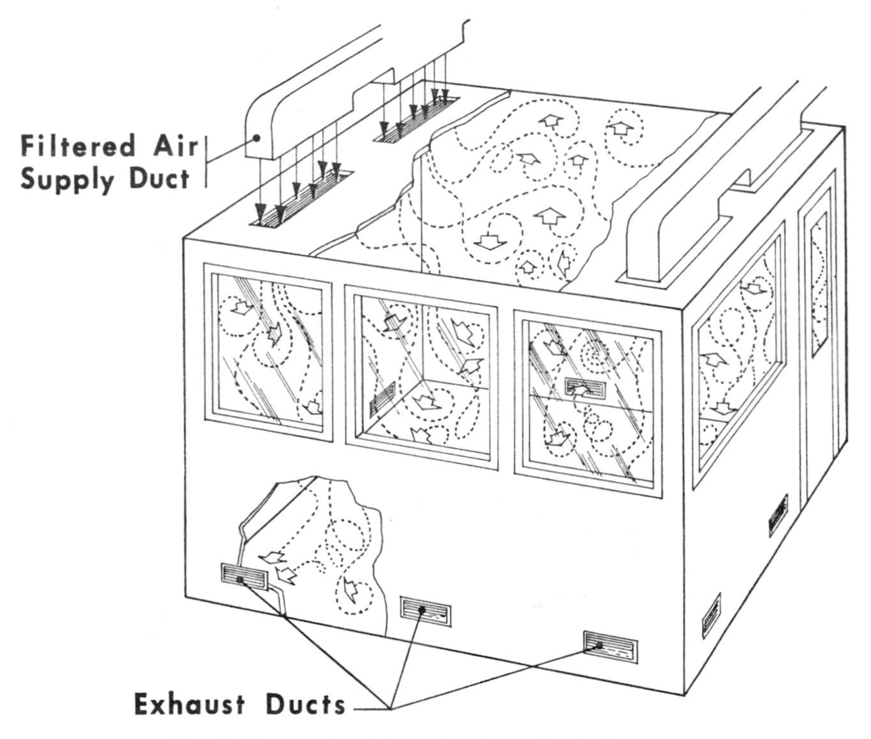

FIG. 2. Conventional (non-laminar flow) clean room.

janitorial cleaning, which can be impractical and uneconomical. Additional requirements include highly controlled clothing change facilities for both male and female workers, air showers, air-locked entrances and exits, and shoe cleaners or provision of clean room shoes for each worker.

It must be emphasized that most non-laminar flow clean rooms are operated for assembling specific products. For a number of reasons, including operating economy and personnel considerations, it may not be necessary to fulfill all the requirements imposed by a specific class for a particular non-laminar flow clean room. Product quality and reliability are overriding considerations.

IV. PRINCIPLE OF LAMINAR AIR-FLOW

During the past decade, the increasing need for cleaner environments for assembling aerospace hardware, miniature circuitry, and electronic equipment began to surpass the degree of air cleanliness attainable in a non-laminar flow clean room. A number of problem areas became apparent during operation of the typical, non-laminar clean room. For example, in this type of facility it was noted that under certain circumstances, particle contamination generated within the room would accumulate faster than it could be removed. A large portion of this room contamination settled onto the floor or onto horizontal surfaces, only to be rendered air-borne again by changes in air currents and by personnel activity in the room. Ultimately, this contamination had to be removed by manual cleaning. Also, air flow in this type of clean room was neither uniform nor directed to carry off particulate matter from the critical work area. During this period, Whitfield applied the principle of laminar air flow to existing clean room technology to develop the first laminar flow device (Whitfield, 1962, 1963, 1966).

Whitfield's work indicated that total control of clean room air was essential, not only in terms of cleanliness (particle-free), temperature, and humidity, but also in terms of the direction of air flow within the room. To accomplish total control of air flow, Whitfield developed the laminar flow clean room. In this facility, large volumes of air were introduced into the room through a bank of HEPA filters. The filter bank acted as a large air supply diffuser and reduced the velocity of the supply air considerably. The air tended to leave the filter bank with minimum turbulence and in a unidirectional manner. Then, this air was exhausted from the room through a series of grilles that were equal in area to that of the inlet diffuser. Air flow was either from one wall (inlet) to the floor (exhaust), from the ceiling (inlet) to the floor (exhaust), or from one inlet wall to an opposite exhaust wall (McDade *et al.*, 1967b). In such a system, filtered (HEPA) air made one uniform pass through the clean room in either a vertical (downflow) or in a horizontal (crossflow) pattern. Four advantages of this type of air flow control were obvious—

(1) it provided the clean room with a self-clean-down capability to remove contamination brought into and also generated within the room;
(2) it provided air flow patterns that carried air-borne contamination away from the work and the work area;
(3) it reduced personnel restrictions; and
(4) it reduced maintenance costs drastically.

The primary requirements for application of the laminar flow principle are: (1) that the space to be kept clean have walls or sides to maintain the laminar flow, and (2) that the air inlet and exit to the space each have an area equal to that of the cross section of the confined space. All of the laminar flow devices to be discussed later meet the above two requirements.

Fox (1967) has visualized the concept of minimum turbulence, unidirectional air flow (laminar flow) as follows:

> "... consider several sheets of glass laid one on top of the other and sliding down an inclined plate. While each individual sheet of glass may move down the plane with a somewhat different velocity relative to all of the other sheets of glass, each sheet of glass is confined to its own plane and does not enter the plane of any other sheet of glass. On the other hand, if the sheets were broken into small pieces and allowed to slide down the same inclined plane, the pieces of one sheet of glass would intermix with pieces from other sheets of glass in a random manner. This latter condition is called turbulent flow ..."

Experience with laminar flow equipment has shown that careful design, construction, and operation are necessary to obtain maximum benefits (Whitfield, 1962, 1963, 1966; McDade *et al.*, 1967b). For the remainder of this chapter, laminar flow is used as defined in Federal Standard No. 209a (Anonymous, 1966), "... airflow in which the entire body of air within a confined area moves with uniform velocity along parallel lines, with a minimum of eddies ..."

Thus, the design goals for laminar air flow devices are that—

(1) the air flow shall be uniform in velocity and direction throughout any given cross section of the device, and

(2) all air flow entering the device must be filtered through the inlet filter system.

Satisfactory application of the laminar flow principle to produce a clean environment depends upon two rather simple features—

(1) large quantities of air are introduced into the enclosure through a relatively large surface of HEPA filters; and

(2) the filtered air is used in such a way as to wash the entire volume of the enclosure by a single pass through it, thereby providing clean air to the work area and carrying away work-generated contamination.

These two features, especially the latter, represent the ideal situation, and one that should always be sought, at least before critical activities are introduced into the enclosure. The critical activity is then placed within the room in such a way as to minimize any adverse effects on the ideal situation. In concept, these two features appear simple, but, in practice, a number of subtle factors can influence the overall performance of a specific laminar-flow device. For example, the second feature described above, that of providing uniform, unidirectional flow of air, is easily demonstrated or monitored by observing smoke patterns in the air stream and by conducting air-velocity measurements. One of the chief problem areas encountered by designers of laminar flow equipment has been the failure to provide one entire surface of the enclosure as the filter bank and an opposite surface of similar area as the air exit. Care must be taken not to install the filter bank into a framework, thereby creating a solid surface around the periphery of the filter bank. When a filter bank is so enclosed, the offset of the filter from the adjoining side wall creates a dead air space along this side wall and generally results in turbulence or negative air flow. This condition is particularly important in clean bench application, as contaminated air outside the bench might be drawn inside along the side panels. Similarly, when an object is placed in a laminar flow air stream, "downstream" turbulence results, i.e., turbulent or non-laminar airflow conditions are produced behind or beneath the object placed in the laminar flow air stream. In general, past experience with laminar flow devices operating at velocities of 75 to 120 feet per minute indicates—

(1) when a laminar flow air stream passes along only one side of an item obstructing its flow, downstream laminarity is not re-established until the moving air reaches a distance of approximately six times the cross sectional diameter of the obstructing item, and

(2) when the laminar flow air stream passes along both sides of the obstructing item, downstream laminarity is re-established after the air stream reaches a distance approximately three times the cross sectional diameter of the obstructing item

Generally, these basic requirements are for an empty device. It has been shown by experience that if a laminar flow device will function correctly when empty, then it will function satisfactorily during use (Whitfield, 1962, 1963, 1966; McDade *et al.*, 1967b). The choice of type of laminar flow device is, of course, important to optimum air-borne contamination control for a given type of operation. A laminar flow clean room is a dynamic device, and cleanliness depends mainly on the performance of the device elements rather than on janitorial maintenance.

One final comment should be made before the individual types of laminar-flow devices are discussed: the laminar-flow principle has been successfully applied for the control of airborne particulate (both viable, i.e., microbial, and non-viable, i.e., lint dust, fibres, etc.) contamination and not for the direct control of surface or contact contamination. It should be apparent that laminar-flow devices will not prevent contact contamination from occurring. However, unless there is some control exerted over airborne contamination, exposed surfaces will be contaminated by sedimentation of the air-borne particles. This fact becomes more important as cleanliness levels are raised, especially when single particle control is necessary in very critical electrical and mechanical devices, and a single viable organism prevents sterility.

V. LAMINAR-FLOW DEVICES

The first laminar flow clean room (Whitfield, 1962), a wall air inlet to floor exhaust facility, is illustrated in Fig. 3. It proved the value of the laminar flow principle and it was the catalyst for adaptation of this concept in the devices described in the following paragraphs.

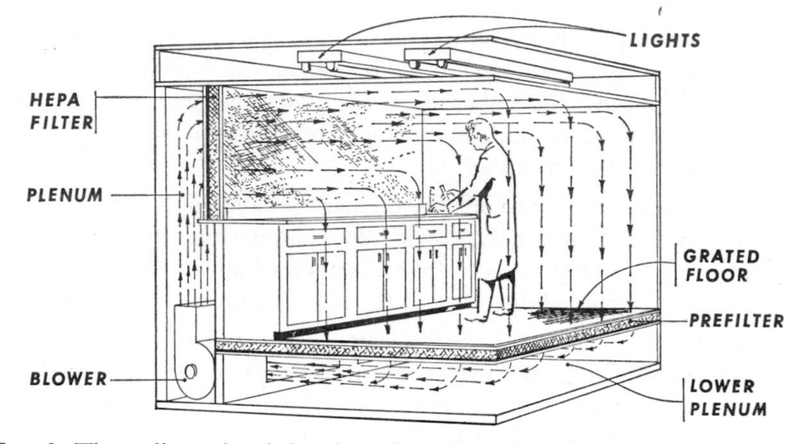

FIG. 3. Three dimensional drawing of a wall-to-floor laminar flow clean room.

A. Vertical laminar (downflow) flow clean room

The vertical laminar clean room (Class 100) employs the basic downflow principle demonstrated in Fig. 4. As shown here, supply air enters the room through a ceiling bank of HEPA filters and flows downward toward the floor in a laminar fashion. It is then exhausted from the room through a grated floor. After passing through the pre-filter (roughing filter), the return air is drawn up through plenums on either side of the room to be recirculated through the ceiling bank of HEPA filters. Thus, the room is ventilated with single-pass, unidirectional, minimum turbulence, clean air at a rate of 10 changes per minute or 600 per hour.

Normally, the vertical laminar flow clean room will operate within the Class 100 level of air cleanliness at air velocities up to 100 (usually 90 ± 20) feet per minute and will provide for very rapid removal of contamination generated within or introduced into the room. Under routine operating

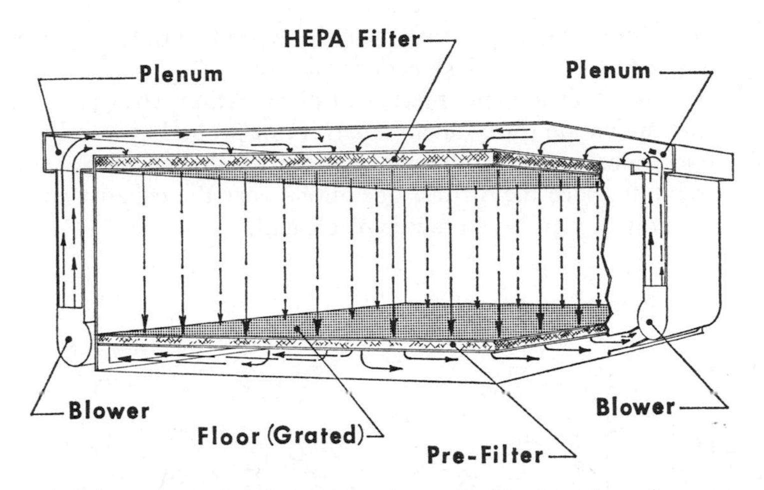

FIG. 4. Three dimensional drawing of a vertical laminar flow clean room.

conditions, the Class 100 clean room is under a slight positive pressure that is sufficient to prevent external contamination from entering when the facility door(s) is open.

In the Class 100 facility, contamination between adjacent operations is eliminated, janitorial attention is greatly reduced, and the effects of personnel activity on critical operations are minimized. Generally, air showers, air-locked doors, and shoe cleaners are not required during routine operation of this type of facility. However, some or all of these items may be used during an extremely critical operation, or possibly during assembly where microbial contamination must be considered.

B. Horizonal laminar (crossflow) flow clean room

The fundamental difference between this clean room and a vertical flow facility is the direction of air travel. Supply air enters the horizontal laminar flow clean room from a bank of HEPA filters in one wall, and is exhausted through the opposite wall, as illustrated in Figure 5. The plenums are located behind the HEPA filter bank and the grilles or louvers in the exhaust wall. Usually, the pre-filters are behind the exhaust grilles. The return air may travel over the ceiling, as a means of eliminating ducting, or through a side wall which has no openings, or under the floor via appropriate ducting.

A horizontal flow room of from 40 to 60 feet in length will normally operate within the Class 10,000 level of air cleanliness, although Class 100 air cleanliness conditions exist at the face of the HEPA supply filters. The air velocity within a horizontal laminar flow room should be not less than

100 feet per minute and, depending upon the type of work to be performed, it may be raised to as high as 140 feet per minute.

Care must be applied in the placement of operations that generate high levels of contamination (dirty operations). If not properly controlled, these "dirty" operations could contaminate adjacent or downstream work sites. It is desirable to place the dirtiest operations near the exhaust end of the room if the flow of work permits this placement.

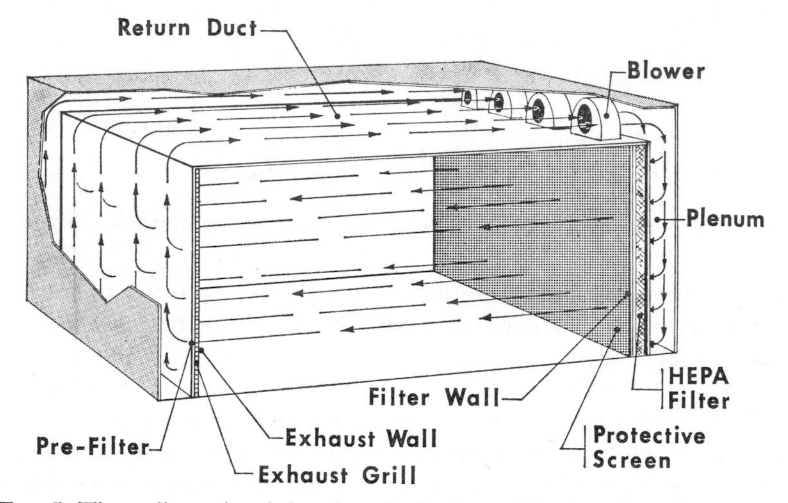

FIG. 5. Three dimensional drawing of a horizontal laminar flow clean room.

Horizontal laminar flow clean rooms possess the capacity to clean down very rapidly. This type of installation is lower in cost than a Class 100 room. The maintenance costs are also generally low, and highly restrictive personnel garmenting, air showers, air locks, etc. are not needed.

C. Laminar-flow tunnel

The laminar flow tunnel differs from the laminar horizontal flow room in that the tunnel is open at the end opposite the HEPA filter bank; the air is exhausted from the tunnel into the surrounding area, whereas in the horizontal room, at least a part of the exhausted air is recirculated; the tunnel depends upon air velocity to retard the infiltration of air-borne contamination and; temperature and humidity are not controlled—they will be the same as that of the area in which the tunnel is located.

The tunnel facility is frequently composed of a series of HEPA filter modules, fastened together and sealed to eliminate leaks between the modules. Figure 6 shows a cross section of this type of device. The walls and ceiling may be merely a sheet of transparent plastic film to permit utilization of the existing lights. The film must be fastened to the HEPA filter bank and at the floor to eliminate leaking of any air. It is normal to have an air velocity of from 100 to 130 feet per minute within this device and the air cleanliness level that may be expected will reach Class 10,000.

A laminar flow tunnel is likely to be the least expensive clean room attainable. It need not be a permanent installation, and can be moved from one location to another with a minimum of lost time and labour cost.

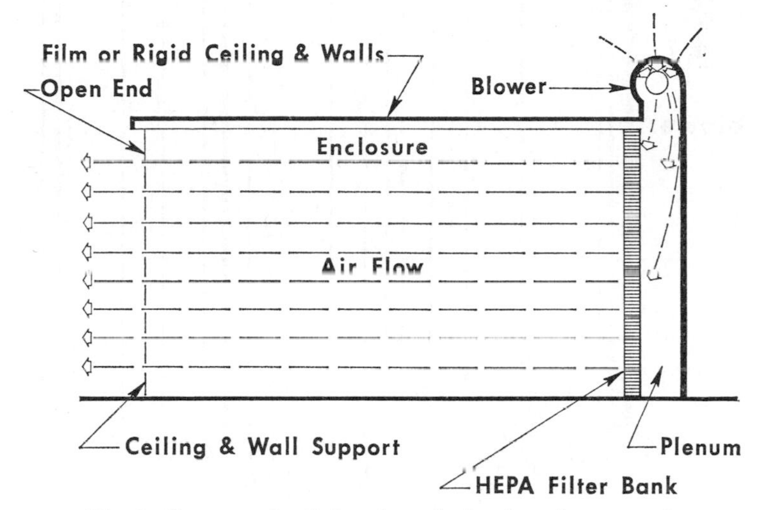

Fig. 6. Cross-sectional drawing of a laminar flow tunnel.

D. Vertical laminar-flow curtained unit

This configuration was developed to provide a "Portable Clean Room" with the air cleanliness of a Class 100 facility, yet remain mobile enough to be readily moved. Basically, the curtained unit is a vertical flow room having plastic curtains for side walls, blowers, a HEPA filter bank, support legs with casters for unrestricted movement to or from a structure, and is self-contained, except for electric power (Fig. 7). It is equipped with lights and uses conditioned (cooled or heated) air from the building in which it is located. It may be used in the extramural environment, even when a moderate breeze is blowing. The curtained unit will provide the same degree of air cleanliness (Class 100) as a vertical laminar flow unit when care is taken to prevent excess floor dirt from rising to the critical work area. This later consideration indicates that the air velocity should be determined by the environment in which the unit will be operating. Curtained units are not restricted to any particular size or height; although the larger they get, the greater the problem of moving and of storing.

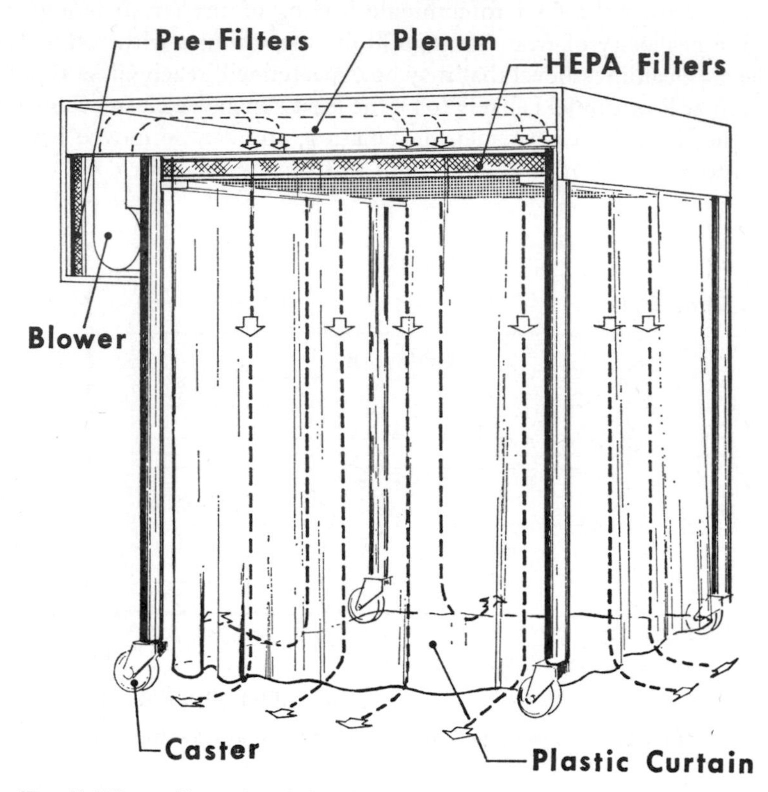

FIG. 7. Three dimensional drawing of a vertical laminar curtained unit.

E. Laminar-flow clean work stations

A clean work station is defined as a work bench or similar work enclosure characterized by having its own filtered air or inert gas supply (Anonymous, 1966). Thus, the clean work station is a device that provides a localized "clean zone" for bench-type operations. The units may be located in shop areas with high concentrations of particulate matter and, with a few restrictions, will function with full efficiency. A typical clean bench is shown in Fig. 8. Both horizontal and vertical configurations are available. Some have the air intake located above the plenum and work surface, instead of below it. The operational characteristics are similar to those of other laminar-flow devices. The laminar flow clean bench provides a uniform flow of filtered

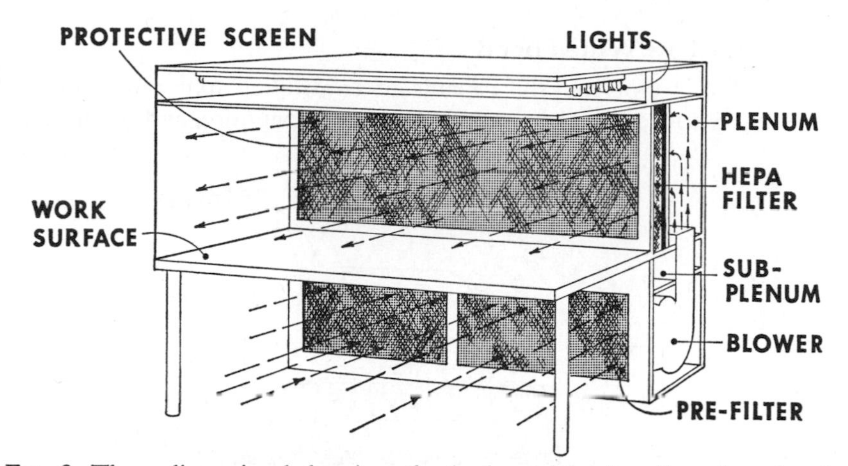

FIG. 8. Three dimensional drawing of a horizontal laminar flow clean bench.

air to remove air-borne dirt from the work area of the bench. Under operational conditions, it will normally provide Class 100 air cleanliness levels in the work area.

Features of this type of unit include rapid self-clean-down capability; ready access to the work area through the open front; reduction of personnel restrictions; and portability—it is easily moved from one location to another, requiring only electric power at the new location.

A Class 100 progressive assembly line can be set up by placing a series of laminar-flow clean benches end to end, with the end panels removed and the joints between benches sealed to prevent leakage at these points. The line may consist of as many benches as required and does not need to be installed within a clean room. It may be placed in a room that does not meet the air cleanliness class required for the operation to be conducted. The environment within the work area of the bench(es) constitutes the critical area. A side benefit from installing a laminar-flow clean bench or benches in any room is that proportionate to the room volume involved, the bench(es) will decrease the level of air-borne contamination within that specific room. This is due to the large volume of room air that passes through the HEPA filters (thereby cleaning it) of the bench(es) per unit time. For example, if 25% of the floor space in a room reasonably free of leaks is occupied by clean benches, it is entirely possible to have the air in this room cleaned to a Class 100,000 level—and to a Class 10,000 if personnel controls and janitorial procedures are rigidly followed and the benches are operated continuously (24 hours a day). Thus, the laminar flow clean work station may provide a method for upgrading the air cleanliness of a conventional (non-laminar flow) room.

F. Laminar-flow vented hood

Units of this type may be capable of controlling four environmental factors: air-borne particulate matter, fumes, temperature, and humidity.

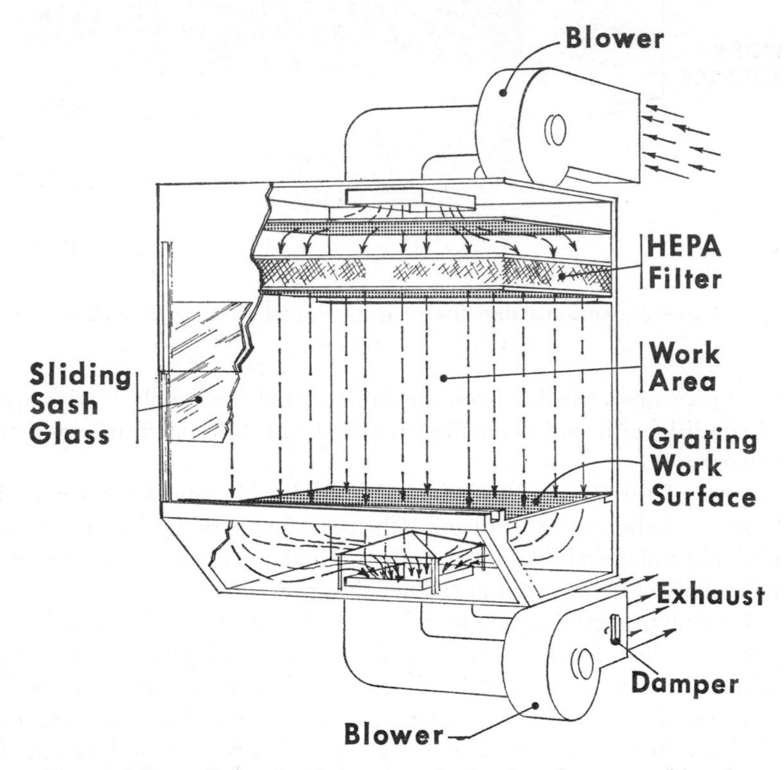

FIG. 9. Three dimensional drawing of a laminar flow vented hood.

As illustrated in Fig. 9, the vented hood provides a source of HEPA-filtered laminar air that moves down into the work area and then on through a grated work surface of the unit to be exhausted to the exterior of a given structure or building. Air movement within the cabinet is controlled by an exhaust blower with a variable flow rate. The exhaust blower is adjusted to remove air from the work area of the hood at the same rate that filtered supply air is supplied to the work area of the hood.

Basic design considerations for the development of a laminar flow vented hood should include a supply of HEPA-filtered air for the work area; a semi-confined work area with a vertical interior liner that is flush with the inside of the filter frame; a perforated (or grated) floor for the work area; and an exhaust blower with an adjustable air flow rate (under certain circumstances exhaust air may also be required to be filtered through HEPA filters before being exhausted to the outside).

G. Summary of laminar-flow devices

Under normal conditions, all laminar-flow devices will maintain Class 100 air cleanliness conditions for the first operational location downstream from the HEPA filter bank. The vertical laminar flow (downflow) room normally fulfills a Class 100 air cleanliness level throughout the entire room, and a horizontal-laminar-flow (crossflow) room of 50–60 feet in length will maintain a Class 10,000 air cleanliness level within the room.

Experience has shown that air velocities should be higher for crossflow rooms and hoods than for downflow units. For best control of contamination, crossflow rooms and hood devices should operate at air velocities from 100 to 130 feet per minute with the best control being obtained at the higher velocity.

The downflow rooms provide best contamination control at air velocities of 60 to 100 feet per minute, with best control being at 100 feet per minute. Below 50 feet per minute, air flow in the room is much more easily disturbed by personnel activity and the time required for removal of air-borne contamination is longer than at the higher air velocities. Velocities below 50 feet per minute have been used and usually have provided better control of contamination than conventional clean rooms; however, rigid control of contamination is not maintained as with the higher air velocities. The main reason for these differences is that air-borne contamination is carried from a critical area at almost the same rate as the air velocity. Therefore, the slower the air velocity, the slower the removal of air-borne contamination.

As a number of different laminar-flow devices exist, some confusion may occur concerning the type of equipment to be selected and the extent of the investment needed for a given requirement. An understanding should be gained of each type of device and its operating characteristics and capabilities before selections are made.

It is difficult to provide a current directory for commercial sources of laminar-flow equipment. This field is a relatively new one and in a dynamic state. New companies are being formed and older ones have gone out of business or merged to form a new organization. In any case, the 1968 issue of "Contamination Control Directory" is probably one of the most current listings of the manufacturers of laminar flow equipment in the United States (Anonymous, 1968).

VI. MONITORING A CLASS 100 (VERTICAL LAMINAR-FLOW) CLEAN ROOM

Due to its superior cleaning characteristics, the Class 100 clean room is replacing non-laminar flow clean rooms in many production and research applications. But, as enthusiastic as most new users are about Class 100 clean rooms, some do not yet appreciate the advantages of this type of room in terms of monitoring requirements.

The basic characteristic of the conventional, air conditioning type of clean room is fluctuation of the particle level because the room itself is an air-mixing chamber. Although air-borne particles in a conventional clean room are distributed fairly uniformly, monitoring must be continuous to measure fluctuations at the various work areas.

Monitoring a Class 100 clean room involves a different approach. Hour-to-hour and day-to-day air-borne particle counts are extremely low and particle-level fluctuations are negligible in a continuously operating room. Furthermore, since all the air in the room is flowing from the ceiling to the floor, air-borne particle counts at different locations in the room are also extremely low and variations are negligible. Monitoring a Class 100 clean room thus resembles the calibration of a reliable instrument; overall performance is monitored before it is put in use and then checked periodically, just as an instrument is calibrated initially and then recalibrated at regular intervals.

In conventional clean rooms, the general room conditions are approximately the same as the critical work area conditions. In the Class 100 room, however, particle counts within the general room area will be very low to zero. Direct monitoring of critical work areas in the Class 100 room may not be practical because contamination will be at an extremely low level unless the work station is seriously interfering with the air flow. Perhaps the most reasonable monitoring approach for a laminar flow room involves performance.

Monitoring clean room performance involves testing for air flow uniformity and velocity, and for HEPA filter and filter bank leaks. Generally the tests are made after installation and then on a periodic basis. Other tests such as temperature, humidity, etc., may or may not be a part of the clean room performance monitoring programme.

Leak testing of the HEPA filter system is necessary to maintain Class 100 conditions in all parts of the clean room. Probably the most practical method of leak testing is to test each filter in place to determine that there are no pinhole leaks. Tests should be made to determine leaks in the filter media, in the bonding between the filter media, and the interior of the filter frame, and the filter-frame gasket and the filter-bank supporting frames.

Uniformity of air flow and correct air velocities are necessary to maintain the required level of air cleanliness in the clean room. To assure that the required level of air cleanliness exists within the clean room, monitoring programmes must include air-flow and air-velocity tests. These tests are easily conducted and should be made upon completion of construction of the laminar-flow clean room, and after that on a periodic basis.

Careful attention should be given to air-flow patterns around equipment and other objects in the clean room. Monitoring of flow patterns may be done with air velocity measuring equipment or visually with the use of a small smoke generator.

Work location monitoring involves taking samples at the actual work location. However, in many instances, this is not practical because the sampling probe may interfere with the work in progress. If the sampling probe is located slightly to one side of the critical work area, it probably will not be exposed to actual work-location environment since the flow of air in a Class 100 clean room will tend to isolate one location from another. The need for work-location monitoring must be carefully evaluated before attempting to set up such a programme. Generally, this type of monitoring will be of greatest value during work location set-up for very critical work.

Clean-room performance monitoring should be scheduled. The schedule given below has proved satisfactory. Before use, clean room performance tests should be included in the clean room acceptance requirements. After 30 days of use, testing may be done on a spot-sample basis to check for equipment failure that did not show up in the initial tests. Items such as filter and filter-frame leaks due to vibration, settling of the clean room structure, loose drive belts, and motor-blower difficulties might be detected in this series of tests. If failures are found, a complete retest may be necessary before acceptable clean-room performance can be re-established.

At six-month intervals, complete air-flow tests should be conducted so that small changes in performance might be detected. At this time, other checks may also be made on a spot sample basis. After long shutdowns, performance tests should be made either on a spot or overall basis depending upon length of shutdown.

Performance monitoring tests are important when damage or malfunction occurs, especially when the HEPA filters, filter bank, or the air supply system are involved. The amount of performance monitoring done at this time will depend upon the extent and kind of malfunction. For example, for a small-area damage to the filter bank, local-area leak testing probably would be sufficient after repair, whereas if a large area of the filter bank were damaged by such things as fire or pressure-vessel failure, a complete retest might be necessary after repair.

When clean room equipment is changed or moved, air-flow patterns in critical areas could be affected. Under these conditions, patterns might be tested with such equipment as smoke generators, photometers, and particle counters. This test is intended generally to be a spot test or evaluation of a given location.

Finally, a warning is in order concerning work location monitoring. As mentioned, contamination levels generally are not constant or uniform in a laminar flow clean room. When analysing work-location contamination levels, enough data should be taken to establish confidence in such measurements; otherwise, the information should be treated as spot-test data.

VII. APPLICATIONS OF LAMINAR-FLOW DEVICES

A. Industrial

The electronic and aerospace industries have made increasing use of laminar-flow devices for the control of air-borne particulate matter. More recently, the National Aeronautics and Space Administration (NASA) has established a requirement for control of both viable (microbial) and non-viable (inert) particulate matter (Newell, 1967b). This requirement established the NASA bioclean rooms and is stated below:

> "... the application of technology related to the exploration of space, the Moon, and the planets of the solar system has resulted in increased sophistication for components and systems related to space flight hardware. This sophistication has brought with it many new requirements, one of which is the demand for new standards of cleanliness for the fabrication, testing, and launch environment of spacecraft. The policy for the United States space effort is directed toward preventing widespread or excessive biological contamination during exploration of the Moon and assure with a certainty of 0·999 that viable terrestrial organisms will not be transported to other planets until sufficient information has been obtained to insure that biological studies will not be jeopardized and that no hazard to Earth exists. Facilities providing the environmental control necessary for this degree of cleanliness are known as bioclean rooms, bioclean work stations, aerosol particulate controlled facilities, and microbially controlled facilities. Thus, a bioclean room or bioclean work station may be: a space in which airborne contamination, temperature, humidity, and microbial flora are controlled to a far greater degree than for conventional air conditioned areas ..." (Newell, 1967b.)

Consequently, several groups in the United States have conducted studies to evaluate the influence of clean-room operations and conditions on the levels of microbial contamination present within areas employing various degrees of environmental control (Portner *et al.*, 1965; McDade *et al*, 1965a, b, c; Favero *et al.*, 1966; Powers, 1967). In all of these studies, two main types of assays were performed. Volumetric air samples were collected with slit samplers and stainless-steel strips were used to measure the levels of air-borne microbial contamination accumulating on surfaces exposed within different environmentally controlled areas. Sample collection and processing was carried out according to the NASA standard procedures (Newell, 1967a). Tables II and III show comparative data collected during

TABLE II

Comparative summary of air-borne microbial particles within several environmentally controlled areas

Area Sampled	Average particle recovery per ft³ of air	
	Average	Range
Factory A	17·1	6.5 to 26·0
Factory B	13·5	6·4 to 19·7
Industrial clean rooms		
a. Class II, room A	5·1	0·5 to 18·0
b. Class II, room B	1·2	0·7 to 5·4
c. Class III, room C	0·2	0·0 to 0·6
d. Class III, room D	0·9	0·6 to 1·7
e. Class IV, room E	0·8	0·0 to 2·0
Laminar flow rooms		
a. Horizontal		
1. Filter wall	0·0	0·0
2. Exhaust wall	0·9	0·0 to 3·5
b. Vertical	<0·001	0·000 to 0·001

TABLE III

Comparative summary of microbial contamination on surfaces within several environmentally controlled areas

Area Sampled	Sampling interval	Total exposure of test surfaces	Range of aerobic mesophiles per ft² of surface
Factory A	3 weeks	25 weeks	3,000 to 29,500
Factory B	3 weeks	25 weeks	7,500 to 35,000
Industrial clean rooms			
a. Class II, room A	3 weeks	25 weeks	6,000 to 25,000
b. Class II, room B	3 weeks	25 weeks	5,000 to 26,200
c. Class III, room C	3 weeks	25 weeks	2,800 to 9,000
d. Class III, room D	3 weeks	25 weeks	100 to 400
e. Class IV, room E	3 weeks	25 weeks	100 to 500
Laminar flow rooms			
a. Horizontal			
1. Filter wall	1 week	7 weeks	None detected when filter intact
2. Exhaust wall	1 week	7 weeks	10,000 to 50,000
b. Vertical	Twice weekly	8 weeks	0 to 40

surface and air sampling studies in manufacturing areas, non-laminar flow clean rooms, and laminar-flow (horizontal and vertical) clean rooms (McDade *et al.*, 1967a). Although the results of these studies varied slightly, depending on the research group, the main consensus was that the degree of airborne and surface microbial contamination in a clean room was primarily dependent on the levels of personnel activity and density. These studies also indicated that as the environment of any given enclosure was controlled in a more positive manner with respect to reduction of nonviable-particulate contamination, the numbers of contaminating micro-organisms were also reduced. In addition, the lowest levels of microbial contamination were found in laminar-flow clean rooms. However, a comment is in order concerning the probable limited validity of air samples collected in laminar air flow with conventional microbiological volumetric air samplers. A microbiological air sampler operating at one cubic foot of air per minute and in a normally turbulent conventional (non-laminar) clean room will sample a representative amount of that environment. The same air sampler operating in a laminar air flow unit would sample only that streamlined column of air delivered from the filters to the orifice of the sampler. This column of air may not necessarily represent a statistical measure of other similar columns in the laminar air flow environment. It might appear, therefore, that results of air samples collected in laminar-flow devices be carefully analysed to avoid improper conclusions.

As an alternative, sometimes it may be advisable to evaluate performance of a laminar-flow device on the basis of the results obtained with procedures being conducted within the device. Thus, if sterility testing procedures conducted in laminar-flow units consistently prevented the occurrence of false positive tests, this may be a better basis for evaluating the performance of the device than sampling of the laminar air stream.

Laminar-flow systems also have been employed to eliminate background microbial contamination during the assay of samples containing extremely low levels of contamination (Favero and Berquist, 1968). For this purpose, laminar-flow benches have been used successfully. Experience has shown that it is virtually impossible to validly assay samples from a clean area in a conventional microbiology laboratory. This is especially true when extrapolations are made.

The pharmaceutical industries have constantly sought improved methods to protect quality and sterility. Terminal sterilization, when applicable, has played an important role in the manufacture of many pharmaceutical products, along with the use of quality ingredients, good manufacturing techniques, controlled environments and efficient testing procedures. However, since some items cannot be sterilized without physical or chemical deterioration, improved techniques for controlling the manufacturing environment to allow aseptic assembly assume added importance.

Modern techniques of ventilation engineering have resulted in the design of a number of non-laminar environmental-control systems for use in the production of pharmaceutical products. For example, dust hoods are often provided over work tables used for sterile filling operations. A typical unit delivers air through a HEPA filter and provides an air curtain with a flow rate of approximately 60 linear feet per minute over the working surface. These non-laminar flow devices use filtered, but turbulent, ventilation patterns to control contamination in the critical work area.

The first use of laminar flow equipment in the pharmaceutical industry, and probably the most widespread, is in sterility testing. Drug manufacturers usually operate under either the regulations set forth by the Food and Drug Administration or the Division of Biological Standards, National Institutes of Health. Each organization requires that samples of the finished product be monitored for sterility by testing in bacteriological culture media. An occasional organism in the air accidentally deposited onto a sample or introduced into sterility test media, might be registered as positive contamination and this could result in the loss of thousands of dollars worth of a product. Alternately, unnecessary retesting and resterilization might be required to demonstrate the presence or absence of contaminants. These problems are evidence that strict environmental control in the testing areas is essential.

To eliminate false positive tests, theoretically, sterility tests should be performed in a sterile environment. However, the difficulties and expense of working through absolute biological barriers negates utilization of such equipment for this purpose. Nonetheless, the Antibiotic Regulations of the Division of Biological Standards as well as the Ninth Edition of the U.S. Pharmacopoeia state that sterility tests should be performed in an area as free from microbial contamination as is possible to achieve. The use of non-laminar flow clean rooms employing turbulent air ventilation has definite limitations since viable particles shed by employees may be introduced into the work area. In this regard, laminar flow benches have provided an excellent environment for sterility testing.

Bowman (Bowman, 1968) recently reported her evaluation of three vertical laminar-flow benches used by the Food and Drug Administration for sterility testing antibiotics and transferring sterile enzyme preparations. Her tests were performed by sampling air in the units in several ways. One test included a procedure in which an aerosol of *Serratia marcescens* was generated in the room while air samples were taken within the bench. Her data indicated that the laminar flow benches maintained a clean, controlled environment. As a practical evaluation, Bowman reported that sterile enzyme solutions had been aseptically transferred in a laminar-flow bench to approximately 210,000 tubes of thioglycolate medium without a break in sterility.

Another use of laminar flow equipment has been in tissue culture procedures. Many vaccines are prepared in tissue cell cultures that must be kept free of contamination with bacteria and viruses. The use of laminar flow benches has contributed significantly to the prevention of tissue cell contamination. Generally, vertical flow benches are employed for manipulations of cell cultures and operations including trypsinization of cells and refeeding of cultures.

There has been some controversy with regard to the best type (crossflow or downflow) of laminar-flow bench to use for sterility testing. The relative merits and applications of each during agent containment and/or product protection tests have been described elsewhere (Favero and Berquist 1968; McDade *et al.*, 1968a, b).

Laminar flow rooms have also been employed in various other pharmaceutical operations. Bassett (Bassett, 1966) described the use of two laminar-flow tunnel rooms for bench operations requiring freedom from particulate contamination. Operations such as cleaning medicine droppers, putting cleaned plugs into jars and trays, and siliconing containers were carried out in these rooms.

Finally, laminar flow equipment has been applied to a number of diverse operations involved in the preparation of pharmaceuticals. Some of these applications include—

1. Subdividing and preparing samples into aliquots for various subsequent testing procedures. Frediani (Frediani, 1966) used laminar-flow benches in the preparation of standard sets of samples containing known amounts of penicillin contamination for testing purposes.
2. Draining and drying of glassware to be employed in antibiotic transfer operations where freedom from particulate matter is important.
3. Chemical assay of antibiotic products where minimization of product cross-contamination of the test sample is essential.
4. Elias and Vellutato (Elias and Vellutato, 1966) compared a vertical and a horizontal laminar-flow bench in filling operations for preventing penicillin contamination of drugs. They concluded that the vertical-flow hood was best under their test conditions, but they pointed out that the selection of equipment for the control of particulate matter must depend on the specific operation involved.
5. In aseptic fill operations of heat-labile materials. Podesta (Podesta, 1967) reported the use of horizontal laminar-flow hoods within a new room facility for aseptic filling of sterile materials.

B. Medical

In the medical field, the main application of laminar flow has been in two areas: patient isolation to minimize infection of susceptible patients and in surgical suites.

A number of patient isolator systems have been developed and used for minimizing infections following surgery (Levenson *et al.*, 1960; Burke, 1967), for experimental chemotherapy (Schwartz and Perry, 1966), and for treatment of patients who are highly susceptible to infection, such as badly burned cases (Shadomy *et al.*, 1965; Haynes and Hench, 1965). Primarily, these concepts utilized the germ-free isolator technique, modified as a plastic tent enclosing the patient's bed. The plastic tent provided a physical barrier between the air, the patient and the rest of the hospital. Nursing care was provided through attached gloves and sterile items were introduced through dunk tanks or through air locks.

Infection is a major cause of morbidity and mortality in cancer patients. This is especially true of acute leukemia, since 75% of these patients die of infectious diseases. The majority of infections occurring in patients with acute leukemia are caused by Gram-negative bacilli and fungi. Since the majority of patients who fail to achieve remission die of infection before receiving complete anti-leukemia therapy, it has been suggested that a significant reduction in risk of infection would allow more patients to survive long enough to receive adequate anti-leukemia therapy. One of the first experimental systems utilizing the laminar-flow principle to replace the plastic tent barrier has been studied at the University of Minnesota (Michaelsen *et al.*, 1967). This unit was developed to provide a protective environment for patients undergoing anti-leukemia chemotherapy. Shop window dummies sterilized with ethylene oxide and healthy individuals volunteering as "patients" were used to evaluate this module as a patient isolator system. During this evaluation, air sampling studies indicated that the levels of microbial contamination were at least one order of magnitude lower than that recorded for industrial conventional clean rooms and several orders of magnitude below that for critical areas of hospitals (Arnold, 1965; Fox, 1967). Also, during evaluation of this unit, it was found impossible to maintain sterile floors, walls, and other surfaces within the room. However, it was possible to reduce the number of micro-organisms on these surfaces. Items of furniture such as chairs and bedside tables, etc., were constantly the most contaminated items within the room. During simulated nursing tests with the sterilized shop window dummies, it was found that a very high percentage of sterile samples could be maintained when microbial contamination on personnel gloves was avoided and full barrier uniform was worn. If a partial uniform (less stringent barrier) was worn, there was a slight increase in the number of positive samples. However, when the gloves of nursing personnel were known to be contaminated, there was a marked increase in dummy contamination, regardless of the type of uniform worn. The transfer of contamination from personnel to a sterilized dummy simulating routine nursing care demonstrates the previously mentioned limitation of the laminar-flow concept, i.e., contact (surface) contamination is not prevented by this system. It must be re-emphasized that the laminar-flow principle was developed for the control of air-borne particulate matter and not to prevent contamination of items by contact with personnel or other items. Thus, in order to accomplish the goal of protecting susceptible patients from non-autogenous viable contamination, the laminar-flow principle must be supplemented by careful aseptic techniques, particularly when actual contact with the patients is unavoidable.

A second area of application of laminar flow has been in surgical suites. Two research installations currently exist, one at Bataan Memorial Hospital, Albuquerque, New Mexico (Whitcomb and Clapper, 1966), utilizing vertical laminar flow, and the other at the National Institutes of Health, employing a horizontal laminar-flow system (Fox, 1967).

Studies conducted in the downflow surgery have indicated that the level of air-borne microbial contamination is very low within the surgical theatre and extremely low in samples collected at the incision (McDade *et al.*, 1968c).

The crossflow surgery (Fox, 1967) was modified considerably in that there was no return wall opposite the supply wall, as in the conventional horizontal laminar-flow room. The exhaust opening was in the ceiling, at the downstream end of the room. Smoke testing, under varying air velocities and with personnel and several different equipment configurations indicated that a laminar flow of air tended to occur throughout approximately two-thirds of the room length. Then, the various layers of air began to bend upward toward the exhaust opening, until at about the last two feet of the room, the air travelled almost vertically to the ceiling. The data obtained during microbial aerosol and smoke tracer studies, during simulated patient trials, and during experimental surgery with dogs contributed to the existing literature confirming the efficacy of the laminar-flow principle in controlling air-borne microbial contamination.

C. Laboratory animals

At the present time, the application of laminar flow in the area of animal care and housing has been limited. There has been one reported instance of animals being housed in a laminar-flow room, although primarily for control of temperature, odours, and general health (Gerke-Manning, 1967). In addition, some commercial producers of laboratory animals have successfully employed the laminar-flow bench as a work station for manipulating virus-defined mice raised under filter-bonnet protection. Presently, however, there are no published results on the use of laminar flow as a barrier for the continued maintenance of animals.

Recently, it has been informally reported that a new laminar-flow animal-cage device has been designed and is being tested (Runkle, 1968). This unit resembles a horizontal laminar-flow bench on castors and provides adjustable shelving and baffles to create individual compartments for each cage. It has been suggested that this unit will make it possible to house and maintain pathogen free (SPF or virus defined) animals or animals whose defence mechanisms have been destroyed, without stringent physical barriers or time consuming techniques. However, this concept remains to be proven biologically as an effective microbiological contamination control barrier.

VIII. SUMMARY

An attempt has been made to describe the events that led to the development and application of the laminar-flow principle to the problem of air-borne particulate contamination and its control. The laminar-flow principle has been described in detail. Utilization of this principle in the development of a number of laminar-flow devices such as downflow and crossflow clean rooms, downflow and crossflow clean work benches, a laminar-flow tunnel, and a curtained (portable) laminar-downflow unit has been described. All available data indicate that laminar-flow devices are very effective in reducing and controlling both viable and nonviable particulate matter. The application of laminar-flow devices to various operations in industry and the medical and paramedical fields has been described. This application is constantly increasing and will, no doubt, extend considerably beyond the present scope of application.

REFERENCES

Anonymous (1966). "Clean room and work station requirements, controlled environment". Federal standard 209a. General Services Administration, Specifications Activity, Printed Materials Supply Division, Building 197, Naval Weapons Plant Washington, D.C.

Anonymous (1968). Contamination Control Directory. Blackwent Publishing Company, 1605 La Chauenga Avenue, Los Angeles, California 90028, U.S.A.

Arnold, V. E. (1965). Tech. Rep. SC-RR-65-47. Sandia Corporation, Albuquerque, New Mexico.

Austin, P. R. (1963). *In* "Conference on Clean Room Specifications", Air Force revised technical order 00-25-203. Tech. Rep. SCR-625. Sandia Corporation, Albuquerque, New Mexico.

Barret, J. P. (1959). *Proc. Anim. Care Panel,* **9**, 127–133.

Bassett, R. (1966). *Bull. parent. Drug Ass.*, **20**, 5–7.
Blickman, B. I., and Lanahan, T. B. (1960). *Saf. Maint.*, **120**, 34–36, 44–45.
Blowers, R., and Crew, B. (1960). *J. Hyg., Camb.*, **58**, 427–448.
Blowers, R., and Wallace, K. P. (1960). *Am. J. publ. Hlth*, **50**, 484–490.
Blowers, R., Mason, G. A., Wallace, K. R., and Walton, M. (1955). *Lancet*, **ii**, 786–794.
Bourdillon, R. B., and Colebrook, L. (1946). *Lancet*, **i**, 561–565.
Bowman, F. W. (1968). *Bull. parent. Drug. Ass.*, **22**, 57–65.
Bruch, C. W. (1965). Dry-heat sterilization for planetary-impacking spacecraft. Spacecraft Sterilization Technology, pp. 207–229. NASA SP-108. Office of Technology Utilization, National Aeronautics and Space Administration, Washington, D.C.
Bruch, C. W., Koesterer, M. G., and Bruch, M. K. (1963). *Devs. ind. Microbiol.*, **4**, 334–342.
Burke, J. F. (1967). *Hosp. Practice*, **2**, 23–29.
Colebrook, L., and Cawston, W. C. (1948). Microbic content of air on roof of city hospitals, at street level, and in wards. Studies in Air Hygiene. Med. Res. Council (Gr. Britain), Special Report Series 262, pp. 233–241.
Decker, H. M., and Buchanan, L. M. (1965). Spacecraft Sterilization Technology, NASA SP-108, 259–268. Office of Technology Utilization, National Aeronautics and Space Administration, Washington, D.C.
Decker, H. M., Buchanan, L. M., Hall, L. B., and Goddard, K. R. (1962). Pub. Hlth. Serv. Publn. 953. U.S. Government Printing Office, Washington, D.C.
Dow Biohazards Department (1967). Tech. Rep. BH67-01-014, The Dow Chemical Co., Pitman-Moore Division.
Duguid, J. P., and Wallace, A. T. (1948). *Lancet*, **ii**, 845–849.
Elias, W., and Vellutato, A. (1966). *Bull. parent. Drug Ass.*, **20**, 193–198.
Engley, F. B., and Bass, J. A. (1957). *Antibiotics A.* 1956–57, pp. 634–639.
Favero, M. S., and Berquist, K. R. (1968). *Appl. Microbiol.*, **16**, 182–183.
Favero, M. S., Puleo, J. R., Marshall, J. H., and Oxborrow, G. S. (1966). *Appl. Microbiol.*, **14**, 539–551.
Favero, M. S., McDade, J. J., Robertson, J. A., Hoffman, R. K., and Edwards, R. W. (1967). *J. appl. Bact.*, In Press.
Favero, M. S., Puleo, J. R., Marshall, J. H., and Oxborrow, G. S. (1968). *Appl. Microbiol.*, **16**, 480–486.
Fox, D. G. (1967). Thesis, University of Minnesota, Minneapolis, Minnesota.
Fredette, V. (1958). *Can. J. surg.*, **1**, 226–229.
Frediani, H. A. (1966). *Bull. parent. Drug. Ass.*, **20**, 189–192.
Gerke-Manning, J. E. (1967). *Contamination Control*, **7**, 21–25.
Greene, V. W., Vesley, D., Bond, R. G., and Michaelsen, G. S. (1962). *Appl. Microbiol.*, **10**, 561–566.
Hall, L. B., and Hartnett, M. J. (1964). *Publ. Hlth. Rep., Wash.*, **77**, 1021–1024.
Haynes, B. W., and Hench, M. E. (1965). *Ann. Surg.*, **162**, 641–649.
Hoffman, R. K., and Warshowsky, B. (1958). *Appl. Microbiol.*, **6**, 358–362.
Jemski, J. V., and Phillips, G. B. (1963). *Lab. Anim. Care*, **13**, 2–12.
Koesterer, M. G. (1964). *Devs. ind. Microbiol.*, **6**, 268–276.
Levenson, S. M., Trexler, P. C., Malon, O. J., Horowitz, R. E., and Moncrief, W. H. (1960). *Surg. Forum*, **11**, 306–308.
Levenson, S. M., Trexler, P. C., LaConte, M., and Pulaski, E. J. (1964). *Am. J. Surge.*, **107**, 710–722.

McDade, J. J., Favero, M. S., and Michaelsen, G. S. (1965a). Spacecraft Sterilization Technology, pp. 51–86. NASA SP-108. Office of Technology Utilization, National Aeronautics and Space Administration, Washington, D.C.
McDade, J. J., Irons, A. S., and Magistrale, V. I. (1965b). Space Program Summary 3735, **4**, 51–63. Jet Propulsion Laboratories, Pasadena, California.
McDade, J. J., Christensen, M. R., Drummond, D., and Magistrale, V. (1965c). Space Program Summary 3736, **4**, 27–34. Jet Propulsion Laboratory, Pasadena, California.
McDade, J. J., Favero, M. S., and Hall, L. B. (1967a). *J. Milk Fd. Technol.*, **30**, 179–185.
McDade, J. J., Whitfield, W. J., Trauth, C. A., Jr., and Sivinsky, H. D. (1967b). Techniques for the limitation of biological loading of spacecraft before sterilization. Presented at Committee on Space Research (COSPAR) Symposium on Sterilization Techniques for Instruments and Materials as applied to Space Research, July, 1967.
McDade, J. J., Sabel, F. L., Akers, R. L., and Herke, C. B. (1968a). Development of a laminar flow biological cabinet. Biohazards Department Tech. Rep. BH68-01-010. The Dow Chemical Co., Pitman-Moore Division.
McDade, J. J., Sabel, F. L., Akers, R. L., and Walker, R. J. (1968b). *Appl. Microbiol.*, **16**, 1086–1092.
McDade, J. J., Whitcomb, J. G., Rypke, E. W., Whitfield, W. J., and Franklin, C. M. (1968c). *J. Am. med. Ass.*, **203**, 125–130.
Michaelsen, G. S., Vesley, D., and Halbert, M. M. (1967). *Hospitals*, **41**, 91–106.
Newell, H. F. (1967a). Standard procedures for the microbiological examination of space hardware. NHB 5340.1. National Aeronautics and Space Administration, Washington, D.C.
Newell, H. F. (1967b). NASA standards for clean rooms and work stations for the microbially controlled environment. NHB5340.2. National Aeronautics and Space Administration, Washington, D.C.
Pflug, I. J. (1960). *Fd. Technol.*, **14**, 483–487.
Phillips, C. R. (1965). Spacecraft Sterilization Technology, 231–257. NASA SP-108. Office of Technology Utilization, National Aeronautics and Space Administration, Washington D.C.
Phillips, G. B. (1965). Spacecraft Sterilization Technology, 105–135. NASA SP-108. Office of Technology Utilization, National Aeronautics and Space Administration, Washington, D.C.
Phillips, G. B., Hanel, E., and Gremillion, G. G. (1962). Practical procedures for microbiol decontamination. Md. Tech. Manuscript 2. U.S. Army Biol. Labs., Ft. Detrick.
Podesta, J. W. (1967). *Bull. parent. Drug Ass.*, **21**, 63–70.
Portner, D. M., Hoffman, R. K., and Phillips, C. R. (1965). *Air Engng*, **7**, 46–49.
Powers, E. M. (1967). *Appl. Microbiol.*, **15**, 1045–1048.
Runkle, R. S. (1968). Development of a laminar flow rack for the maintenance of pathogen-free mice. Presented at the 12th Biological Safety Conference, Cincinati, Ohio.
Schmidt, C. F. (1966). *In* "Disinfection, Sterilization and Preservation". (Ed. C. A. Lawrence and S. S. Block), Ch. 32. Lea and Febiger.
Schwartz, S. A., and Perry, S. (1966). *J. Am. med. Ass.*, **197**, 105–109.
Shadomy, S., Ginsberg, M. K., LaConte, M., and Ziegler, E. (1965). *Archs envir. Hlth*, **11**, 183–200.

Shaffer, J. G., and McDade, J. J. (1964a). *J. Am. Hosp. Ass.*, **38**, 40–51.
Shaffer, J. G., and McDade, J. J. (1964b). *J. Am. Hosp. Ass.*, **38**, 69–74.
Spiner, D. R., and Hoffman, R. K. (1960). *Appl. Microbiol.*, **8**, 152–155.
Symposium (1961). Sterilization by ionizing radiations. *From* "Sterilization of Surgical Materials", pp. 7.56. The Pharmaceutical Press, London.
Wedum, A. G. (1953). *Am. J. Publ. Hlth.*, **43**, 1428–1437.
Wedum, A. G. (1964). *Publ. Hlth. Rep.*, **79**, 619–633.
Whitcomb, J. G., and Clapper, W. E. (1966). *Am. J. Surg*, **112**, 681–685.
Whitfield, W. J. (1962). Tech. Rep. SC-4673 (RR). Sandia Corporation, Albuquerque, New Mexico.
Whitfield, W. J. (1963). State of the art (contamination control) and laminar airflow concept. *From* "Conference on Clean Room Specifications". Tech. Rep. SCR-652. Sandia Corporation, Albuquerque, New Mexico.
Whitfield, W. J. (1966). Tech. Rep. SCR-66-956. Sandia Corporation, Albuquerque, New Mexico.
Williams, R. E. O., Lidwell, O. M., and Hirch, A. (1956). *J. Hyg., Camb.*, **54**, 512–523.
Wolf, H. W., Skaliy, P., Hall, L. B.,Harris, M. M., Decker, H. M., Buchanan, L. M., and Dahlgren, C. M. (1959). Sampling microbiological aerosols. Public Health Service Publication No. 686. U.S. Government Printing Office, Washington, D.C.

CHAPTER VI

Safety in the Microbiological Laboratory

H. M. DARLOW

Microbiological Research Establishment, Porton, Salisbury, Wilts., England

I. INTRODUCTION

The laboratory worker is more exposed to potentially lethal hazards than is the average citizen. His work can, and often does, involve toxic and explosive vapours, carcinogens, caustics, high voltages, radiation and many other injurious factors, to which in the field of microbiology is added infection.

It is to the mechanisms and prevention of this latter hazard that this Chapter is devoted. Infection differs from the majority of laboratory hazards in that its effects are not confined to the individual worker, nor to his co-workers in the same laboratory. Indeed the infected person may actually be regarded as a hazard. He can transmit infection in or on his person to remote situations both primarily and secondarily. His family, casual human contacts, domestic and experimental animals can all become involved, and it must not be forgotten that he can disseminate agents non-pathogenic to humans, but nevertheless capable of contaminating his cultures and reagents, or even of attacking plant life. This latter aspect, though less serious in its results, is germane to the present theme in that it shares many of the same causes and preventive principles. It is not intended to dwell upon it, but merely to stress the point that lack of care and awareness of the transmission of infection can produce results that extend far beyond the individual worker. Until relatively recently it was no cause for surprise, and indeed was almost traditional, that the solitary "backroom" pioneer should infect himself. With growing public enlightenment and the involvement of large research teams, however, the climate of opinion has changed, and the victim is no longer regarded as a benefactor, or as a martyr to medical research, but as a potential menace. Safety, therefore, if for no other reason, is of increasing importance, and its pursuit has, in fact, produced a very substantial dividend in the field of hygiene.

In nature, pathogenic micro-organisms invade the human body by routes which are dependent upon the natural history of the disease and the tissue preference of the organisms, these two factors being generally, but not always, obviously interdependent. For example, the causative organisms of typhoid fever, bacillary dysentery, cholera and the enteroviral infections are introduced to the alimentary tract in faecally contaminated food and water, or on feeding utensils and the fingers. They can survive digestive processes, and have a natural affinity for various tissues of the intestinal walls, or an ability to multiply rapidly within the lumen of the gut. Other than a temporary invasion beyond these sites is the exception rather than the rule. The infected organ becomes the ideal focus for transmission to other individuals.

Many organisms have the ability to invade and a preference for exposed mucosal surfaces, to which they gain access in a variety of ways. The importance of the naso-pharynx both as a portal of entry and a source of infection in a wide variety of diseases needs no emphasis. Here, cross infection occurs through the agency of exhaled droplets. In other infections, such as the venereal diseases, actual physical contact with diseased tissue, or surfaces contaminated with fresh exudate, is necessary, and whilst the disease may become widely disseminated in the host, natural transmission usually occurs via mucosae. The conjunctiva is not only susceptible to a number of specific diseases, such as trachoma, which may be transmitted by fingers, flies and dust, but can also serve as the portal of entry for other infections producing no local pathology. Even such unlikely infections as salmonellosis (Trillat and Kaneko, 1921) have been *inoculated* in this way.

Percutaneous infection occurs either through cuts, abrasions and punctures, as in tetanus, rabies and the more familiar forms of wound sepsis, or as the result of sub-cutaneous injection by arthropod mouth parts as in bubonic plague, typhus and the arbovirus infections. Invasion of intact, healthy skin is not known to occur with certainty, but local maceration, occluded glands and hair follicles and certain dermatological conditions predispose to the multiplication of and ultimate invasion by fungi, staphylococci and other organisms for example, generalized vaccinia in eczematous subjects; and micro-wounds must inevitably occur as the result of abrasive action between the skin and rough surfaces, or by the avulsion of hairs.

Infection of the lower respiratory tract usually occurs as the result of spread from another focus, either via the blood stream or lymphatics, or by extension downwards of an inflammatory process commencing in the upper respiratory mucosa. Once established, however, it readily gives rise to similar pathology in other individuals by means of exhaled droplets as in the classical example of pneumonic plague.

All these mechanisms can operate in the laboratory, but with a marked increase of emphasis on the respiratory route of infection due to the artificial conditions under which man and microbe meet. Three principal factors account for this: the ease with which small particles are produced by common laboratory techniques, the fact that many of these particles are sufficiently small to avoid capture by impaction in the upper respiratory tract, and the ability of the majority of human pathogens to invade the pulmonary lining, if given the opportunity. This does not necessarily imply actual pulmonary pathology as in pneumonic plague; indeed the symptomatology and course of the disease so contracted can be atypical. Furthermore, it has been shown that in many instances the number of organisms required to initiate infection by this route is much smaller than *per vias naturales*. In addition, whilst a worker is usually conscious of having run the risk of infecting himself by ingestion or wounding, and may be able to take timely preventive measures, he is generally quite unaware of the presence of an infective aerosol. Accidental infection by inhalation, therefore, assumes much greater importance in the laboratory than under natural conditions, and is, in fact, the commonest factor in microbiological laboratory morbidity.

Before proceeding to discuss the mechanics of laboratory infection, it is desirable to digress briefly and draw the reader's attention to some of the fundamental principles, albeit largely demonstrated in animals, which emphasize the contentions made in the above paragraph and add further stress to the importance of aerosols in the initiation of infective processes. As has already been pointed out, the lungs appear to act less frequently as a natural portal of entry of infection than the skin, mucosae and alimentary tract, at least as far as can be gathered from the occurrence of local pathological changes. Many of those diseases in which pathology does occur, such as pneumonic plague, pulmonary anthrax and tuberculosis, may be considered as camp followers in the march of civilization with its attendant overcrowding in ill-ventilated spaces. In this respect they closely resemble laboratory infection, and, indeed, all three examples have accounted for deaths in laboratory workers. On the other hand under experimental conditions the pathogens which have been shown to be infective to laboratory animals by the pulmonary route, regardless of their natural ecology and whether or not they produce primary pulmonary pathology, is now so extensive as almost to make it easier to count those diseases that have not been so induced. A similar range has accounted for accidental human laboratory infection, and, as only about 20% of such infections were preceded by a known accident, it has been concluded by Sulkin *et al.* (1963) that infection is probably airborne, though the number of instances in which exposure to aerosol can be demonstrated is relatively small. This also has been the experience of the present author (Darlow, 1960). No attempt can or should be made here to review the published records of human laboratory infection, since the number of references runs into many hundreds; but a few are quoted later to exemplify the hazards of particular techniques. It is relevant, however, to discuss particle size, lung retention and dose at this juncture.

The size of airborne particles has considerable influence on the degree of hazard in several respects. Clearly small particles will sediment slowly and remain inhalable for longer than larger ones which will tend to settle, and present contact and ingestion risks, or be redispersed as secondary aerosol from dusty surfaces. Moreover, Harper and Morton (1953) showed that particles of 1–4 microns escape capture by impaction in the nose, and are retained in the lung (Pattle, 1961). This may have a profound effect on the infectivity and resultant disease process. Infectivity itself depends on the size of the dose, individual susceptibility, and the virulence of the agent, which varies widely with organism and strain. It also depends on the site of invasion. Druett *et al.* (1953) showed that monodispersed aerosols of spores of *Bacillus anthracis* were far more infective for guinea pigs and monkeys than larger aggregates, and attributed this to a difference in the site of deposition in the respiratory tract. They also concluded that the number of spores required to induce infection increased with particle size. Druett *et al.* (1956b) demonstrated a similar phenomenon with *Brucella suis*, though they concluded in this case that the increased infectivity of monodispersates was in part due to subsequent rapid multiplication within the lung.

The site of deposition may also influence the clinical picture. For example, Druett *et al.* (1956a) showed that fine aerosols of *Pasteurella pestis* induced primary broncho-pneumonia in guinea pigs, whilst particles of about 12 microns caused septicaemia without a preliminary pneumonic phase. They also showed, incidentally, that cross infection to normal animals generally occurred when they were exposed to others infected by fine aerosols. It is worthy of mention here that pneumonic and septicaemic plague are far more lethal in man than the bubonic type in which the organism is introduced by the bite of a flea, and the infection tends to be localized.

A final example illustrates a somewhat different combination of the above mentioned factors. Darlow *et al.* (1961), using a strain of mice resistant to *Salmonella typhimurium*, failed to produce infection at an oral dose of 5×10^8 cells. They found, however, that a respiratory dose of only $4{\cdot}5 \times 10^3$ cells produced a rapidly fatal broncho-pneumonia of a somewhat different type from the secondary pulmonary changes induced by intra-peritoneal inoculation with the same dose. The alimentary canal remained uninfected. They suggested that salmonellosis might be transmitted in humans by the aerosol produced by flushing a water closet (Darlow and Bale, 1959). Tully *et al.* (1963) succeeded in infecting chimpanzees with *Salmonella typhosa* by the respiratory route, and rendered them immune from subsequent oral challenge.

It seems clear, therefore, that the principal hazard in the microbiological laboratory may be the aerosol, because it can reach a highly susceptible target undetected, where it can produce maximum effect in low dosage.

II. MECHANISM OF AEROSOL PRODUCTION

Microbial aerosols are generated by two main mechanisms, "atomization" of liquid suspensions and fine comminution of infected solid materials.

Outside the laboratory coughs and sneezes are proverbial in their ability to disseminate infection, but the bubbling and splashing of sewage, especially in the presence of detergent foam, the pounding of surf on the sea shore and the beating of rain on rotting vegetation have also been shown to produce measurable concentrations of viable organisms in the air, often at some distance from the source. Many laboratory techniques produce the same result by bubbling, splashing and frothing, and two additional mechanisms are involved, high frequency vibration and centrifugal force.

When a film of liquid bursts in air it collapses into spherical droplets, and this process is enhanced in bursting bubbles by the shearing action of the enclosed air as it escapes at the point of rupture, which in effect becomes a miniature scent spray. A falling drop of liquid tends to inflate like a parachute, producing the same effect (Mason, 1964). Centrifugal force, for example in the case of a leaking centrifuge tube, a dental drill, or any other high speed rotary apparatus, also produces a disruptive process. A drop of liquid falling upon a surface is shattered to a greater or lesser extent depending upon the texture of the surface, and gives rise to an explosive shower of satellite droplets on impact. If the surface itself happens to be liquid, it will contribute to the resultant aerosol. The depth of the liquid has some influence upon the magnitude of the aerosol, since in shallow films large numbers of droplets are produced by coronet formation, and in deeper pools by Rayleigh jet, which results in fewer and larger droplets (Hobbs and Osheroff, 1967). High frequency vibration, for example in ultrasonic apparatus, has a similar effect.

Any droplet remaining airborne for an appreciable time evaporates rapidly leaving a nucleus of suspended and dissolved matter. The size of this nucleus will depend on the concentration of solids present in the liquid. Nuclei produced from liquids of low solid content will remain airborne for much longer than those resulting from the atomization of more turbid liquids. Hence a dilute suspension of bacteria in water of low solute content can be far more hazardous than a thick slurry.

Aerosols are also produced by the disturbance of light-weight powders. Infected skin scales and comminuted hair, feathers, textiles, and house dust have all contributed to the spread of infection as vector particles. Dried secretions, pus, faeces, etc. have played their part both inside and outside the laboratory, where dried cultures, for example in the threads of screw caps, present an additional hazard. Freeze-dried, acetone dried and spray dried suspensions present a particularly dangerous problem. Another mechanism of dry aerosol production, often overlooked, is fungal spore shedding (Kruse, 1962), fungi being naturally propagated by the release, often explosive, of minute spores, and in some cases hyphal fragments. This also presents a health hazard both outside and inside the laboratory, when it can be a very troublesome source of incidental contamination.

III. HAZARDS OF COMMON LABORATORY TECHNIQUES

Having described the simple physical processes by which aerosols can be generated, and their importance in the transmission of infection, it is now necessary to review them in the setting of everyday laboratory practice by describing the hazards involved in each type of technique. Whilst the accent must still remain on aerosols, there is a residue of other hazards that it is convenient to discuss at the same time, particularly as some of these are secondary to aerosol deposition upon instruments and other surfaces from which organisms can be transferred to the skin, mouth and eyes.

A. Pipettes and pipetting

Pipettes are prepared for use in a variety of ways designed to preserve their sterility until needed. An almost universal practice, however, is the plugging of the upper end with a pledget of cotton wool to prevent the entry of dust. It is a common belief that the purpose of this is to protect the user. This is an erroneous and dangerous assumption. Not only is the plug a poor air filter, but it is also easily penetrated by organisms in liquid suspension. Furthermore, should it become wet, it obstructs the air flow and can readily be sucked out followed by an unimpeded gush of liquid. Whilst this plug probably serves its primary purpose adequately, it clearly adds to the hazards of pipetting by mouth. Oral pipetting should be avoided whenever infective or toxic materials are being handled, and preferably as a general rule, not only on the above grounds, but also because the upper end can become contaminated by the fingers and by deposition of airborne organisms. In addition Phillips and Bailey (1966) showed that aerosols could be generated within pipettes and suggested that these could result in infection in the absence of liquid aspiration.

Both accidental and intentional dropping of liquid from a pipette can produce an aerosol both during fall and on impact, particularly with hard or wet surfaces. Forceful ejection is even more hazardous, since it can result in splashing and frothing in the receiver, and spluttering and bubble formation at the pipette tip, especially when the last drop is discharged. High speed photography has shown that this can produce a very heavy aerosol of small droplets, mostly under 10 microns in diameter, and that larger droplets tend to bounce off agar surfaces onto the bench top (Johannson and Ferris, 1946). The situation is, of course, enhanced during the making of serial dilutions, not only because this may involve more operations, but also because, as the dilutions become higher, so the solid content of aerosol droplets and size of the resultant nuclei becomes less, thus reducing the sedimentation rate and increasing the probability of inhalation. Furthermore, efficient mixing of the dilutions involves agitation, in itself hazardous enough, but rendered more so if achieved by alternate sucking and blowing with the pipette, a procedure which is almost inevitably accompanied by bubbles.

It would appear, therefore, that all pipetting manipulations can be dangerous. Indeed Reitman and Phillips (1955) have shown that even the exercise of extreme care does not entirely eliminate the risk.

B. The platinum loop

The dissemination of organisms by sputtering on flaming contaminated inoculation loops, needles and spreaders is well enough known, though the hazard is not always appreciated. Whilst it is particularly liable to happen when the inoculum is semi-solid (e.g., spore slurries, mycobacterial colonies, or sputum), it can can also occur with relatively small liquid inocula (Anderson *et al.*, 1952), and give rise to measurable aerosols as well as to visible globules of unincinerated material, which can roll for some distance across the bench top. The same hazard arises when a hot loop is immersed in liquid culture, or makes contact with an agar surface.

A cold loop can be just as potent a source of trouble. Vibration or rapid movement in air can cause detachment of droplets, particularly if the contained film bursts. Splashing also occurs if the loop is used to mix cultures and suspensions, or during the preparation of slides for agglutination tests and microscopy. The simple withdrawal of a loop from a culture can draw out fine filaments of liquid, which break up into droplets, and this can also occur when a wet loop touches the wall of the container. Contamination of the mouth of a test tube or flask in this way leads to secondary contamination of the plug or stopper, which in turn can disperse airborne droplets or crusts, or contaminate the hands on subsequent removal. Even streaking out on a relatively dry agar surface is not without risk, especially if the surface is rough (Phillips and Reitman, 1956).

C. The agar culture

Apart from the above mentioned hazards attendant on inoculating agar plates, there are others which arise subsequently, particularly after incubation. A glass Petri dish culture dropped on the floor is a potent source of aerosol (Darlow, 1960), and the hazard is not entirely eliminated by using plastic plates. Whilst the inoculation of agar in bottles and tubes must carry less risk than in the case of plates, since resultant aerosols and splashes are to some extent confined, the breakage of such culture vessels must clearly be as dangerous as in the case of the dropped plate. When this kind of accident occurs, not only is an aerosol produced, but agar, condensate and potentially contaminated splinters of glass are splashed over a wide area of floor. The natural tendency is to stoop down and pick up the fragments. This brings the worker's nose close to the source of aerosol, risks contact contamination of his shoes and the lower fringes of his clothing, and invites punctured gloves and cut fingers.

When plates are removed from the incubator, it is not unusual for considerable condensation of moisture to occur under the lid. This liquid can readily become contaminated and can trickle out when the plates are handled, thus contaminating the bench top and the hands of the worker. Furthermore, it forms a film between the lid and the rim of the plate, which produces a visible sputter when the plate is opened. Even the opening of a dry plate is not without hazard if fungal colonies in the fruiting stage are present. Not only can this give rise to troublesome contamination, but it has resulted in human infection with several pathogenic fungi (Kruse, 1962).

A more bizarre hazard is the invasion of agar cultures by mites and other small arthropods. Their tracks, represented by chains of micro-colonies, can often be seen meandering across neglected cultures, and they can migrate from plate to plate, carrying organisms on their legs and mouth parts. The author was once called to a bench top literally swarming with spring-tails, which had emerged from a three day old culture of *Brucella suis*, when the lid was removed. Captured specimens were found to be heavily contaminated.

D. Plugs, caps and bungs

The removal of cotton wool plugs, screw caps, rubber bungs and other types of stopper from broth cultures, bottles of suspension, centrifuge tubes, etc., can create aerosols in a variety of ways. Firstly, the vessel may already contain aerosolized material as the result of shaking or stirring, which will be released on opening up. Under some circumstances, indeed, aerosol can escape when the closure is still *in situ*, as in the case of blenders (Anderson *et al.*, 1952) and other vessels subjected to violent agitation in shaking machines or ultrasonic vibrators.

Secondly it is not unusual for stoppers to become wetted with infective material either as a result of agitation of the contents of the vessel, or, inadvertently, when inoculating or sampling the contents. Not only may organisms survive in this moisture, but they may also multiply. Disturbance of the film, or residual dried crust, inevitably gives rise to a shower of particles when the vessel is opened. These may be of very small diameter, and Tomlinson (1957) showed that between 30 and 50% of those produced on removing a screw cap had diameters of 4 microns or less.

Opening vessels of dry materials can be equally hazardous, and in some cases probably more so. Reference has already been made to sporulating fungal colonies and the ease with which spores are dislodged, but freeze, acetone or spray dried material also presents an obvious risk, particularly if the container is sealed under a negative pressure, the sudden release of which causes an inrush of air, resulting in disturbance of the powdery contents and a subsequent back-surge of particle laden air.

In addition to the aerosol hazard the hands can become contaminated on contact with the wet or crusted stoppers, or vessel rims. The risk of infection is greatly increased, if the rim is sharp or chipped. Test tubes with everted rims are particularly liable to chipping, and hence to wounding of the fingers when the plug is grasped. Cutting open sealed glass ampoules is clearly doubly hazardous for this reason, and liable to the gravest consequences, if adequate protection of the fingers is neglected.

E. The hypodermic syringe

Beyond the inoculation of animals, hypodermic syringes are used in many other contexts all of which present hazards at almost every stage of the procedure. The experiments made by Hanel and Alg (1955) showed that the simple filling of a syringe is virtually free from aerosol risk, but microsplashes of aerosol dimensions can be produced, if the needle is caused to vibrate, as when withdrawn through a membrane, such as a vaccine bottle cap, or when the liquid being handled is viscous and liable to be drawn out into threads. They demonstrated the latter phenomenon with infected allantoic and amniotic fluids, and it is not unreasonable to expect that it would also occur when sampling animal body fluids, in which case the skin or visceral wall present the same hazard as a rubber diaphragm.

The filled syringe presents a fresh set of hazards. Liquid can escape from either end to drip on to the bench or hands, and the external surface of the plunger is frequently contaminated. The plunger itself may fall out, or the syringe can be dropped and its contents escape even if the barrel does not actually shatter. Wounding with the needle is one of the commonest laboratory accidents (Sulkin *et al.*, 1963), and by nature a potent cause of infection.

Before discharging the contents, it is customary to express excess liquid or air bubbles. This presents the same hazard as blowing out a pipette, especially as the needle must be directed upwards. Downward discharge, as when reconstituting lyophilized cultures, carries the same risks as pipetting, but enhanced by the fact that considerably more pressure is involved, and hence a much greater risk of splashing and frothing.

Almost all of the above mentioned hazards attend animal inoculation procedures and, indeed, are more liable to occur should the subject tend to be obstreperous. Apart from the threat of tooth and nail, the necessity for speed must detract from the exercise of due care, and a not infrequent result of this is the sudden explosive detachment of the needle, or even bursting of the syringe barrel, with consequent heavy contamination of the work area and surrounding air. The situation is described as "extremely dangerous" by Hanel and Alg (1955).

When the needle is withdrawn from skin or bottle cap the surface around the puncture must inevitably become contaminated to some extent and there may be subsequent leakage with resultant contact hazard.

F. Centrifugation

Centrifugation by its very nature is an effective method of generating an aerosol, which principle is put to practical use in fruit sprayers and other appliances. Apart from the actual disintegration of tubes, cups, trunnions, or rotor heads, all types of laboratory centrifuge present hazards at almost every stage of operation. This is particularly true of angle-head centrifuges (Wedum, 1964), which are probably more prone to accidental aerosol production than other types on account of their geometry.

When angled tubes are spun, the fluid level swings through 90° in response to centrifugal force. This inevitably results in overflow, if the tubes are too full, and, as is usually the case, lying at an angle of less than 45° from the axis of rotation. Even if the tubes are capped leakage past the seal, or even detachment of the cap, can occur, particularly when non-rigid components distort, either as a result of centrifugal force alone, or of the increasing pressure exerted by the confined liquid. These kinds of leakage generally occur during the running-up stage, before appreciable reduction of the concentration of suspended organisms in the upper layers of the supernate. In addition to this, should the lip or cap become wet during filling of the tube, the liquid trapped between the cap and the outer wall of the tube will be ejected (Whitwell *et al.*, 1957).

Droplets thrown off as a result of these faults are small enough to become widely dispersed in the laboratory and can readily be comminuted further by violent impaction on neighbouring surfaces. Their dispersion is enhanced by the cyclonic action of the air in contact with the revolving head. The use of sealed rotors does not entirely overcome these problems, since an air-tight seal, unless carefully designed, regularly inspected for deterioration of rubber O-rings, and kept well lubricated, is prone to liquid leakage, when spillage occurs from the enclosed tubes. Enclosure within a sealed evacuated bowl may not entirely eliminate such a hazard as particles may pass through the vacuum pump either as primary aerosol, or as secondary aerosol in oil mist. Furthermore, though the occurrence of a leak may cause obvious vibration due to imbalance, it is not usually detected until the tubes have been removed and examined. Unsuspected aerosol and surface contamination, therefore, may exist in the rotor and bowl, which will present a hazard on opening up the apparatus. It must be added that vibration itself, whether due to loss of contents of one or more tubes, or to primary failure to balance the tubes can generate aerosol.

Horizontal centrifuges are clearly less prone to the faults of the angle-head, since anything escaping from the top of the tube will tend to be thrown down into the cup. However, a fresh set of hazards is inherent in the swing-out rotors.

Cups can fail to swing completely leading to spillage, imbalance, or detachment, trunnions occasionally fracture, and both trunnions and slotted cups can jump centripetally at only moderate velocities resulting in detachment, or at least shattering of the contained tubes or bottles. The latter phenomenon is hard to explain, but the author has seen this occur on several occasions for no obvious reason. On two occasions the cups bounced back into their original positions on the rotor head, being found freely swinging at the end of the run, but with their contents shattered and with extensive evidence of abrasive contact with the inner wall of the centrifuge bowl.

Centrifuges of the Sharples, Laval, continuous and zonal types can and do present an aerosol hazard, the magnitude of which depends on whether or not effective sealing can be achieved. However, they all have to be opened up for the removal of their contents, and, owing to their relative complexity, are difficult to decontaminate.

G. Miscellaneous apparatus

In the preceding paragraphs the principal hazards of the commoner laboratory techniques are described. These hazards are shared in some measure by most other procedures, and, whilst the presentation of an exhaustive list would be tedious, some other instances are worthy of mention, if only to stress this point.

Shaking machines, fermentation and ultrasonic apparatus and other appliances used to agitate infective suspensions are all effective aerosol generators, and have contributed to laboratory morbidity, when inadequate precautions have been taken to avoid leakage during operation, or escape of material on opening up. Blenders used for tissue emulsification, and even the simple pestle and mortar have had a particularly bad reputation, perhaps because the material handled tends to be looked upon as more solid than liquid. Droplets derived from the resultant slurry are admittedly large, but none the less hazardous at close quarters.

Five other examples can be given to fill in the spectrum. Desiccators can implode and disseminate highly dangerous dried material. The Hughes press can explode, or emit jets of liquid or splinters of bacteria-laden ice; the author has witnessed an instance in which the operator was struck in the eye by such a fragment. Fraction collectors generate aerosols in the same way as pipettes, and water pumps can effectively aerosolize overflow from aspiration apparatus. Finally, even the autoclave itself is not without hazard, apart from mechanical mishaps and inefficient operation. It not infrequently happens that glass vessels burst, or boil over, during the pulling of the first vacuum. The liquid released passes down the drain line too rapidly for effective sterilization to be achieved in the steam ejector, and viable organisms can be recovered from the condense effluent line. Should this open into a tun dish, or other open receiver, an aerosol can be produced. This possibility appears to have been overlooked, but is simple enough to demonstrate even with heat labile organisms.

H. Laboratory animals

Animal handling involves not only those hazards inherent in human error and in microbiological techniques in general, but introduces a third factor, the perversity of the animal, which varies widely with species. Laboratory animals can bite, scratch, and shed organisms into the environment from body openings, hair, feathers and skin, and by disturbing droppings, exudates and contaminated bedding; they can carry infected ectoparasites, and can themselves escape when in a highly dangerous condition. They can harbour exotic and fatal infections communicable to man in addition to experimentally induced disease. Because they can be handled only with considerable inconvenience in hoods and other devices, contact with them is essentially more intimate, and their potential as agents of laboratory infections is proportionately increased.

IV. SAFE METHODS AND MATERIALS

A. Training

In the absence of overt accidents safe in the sense of delicate technique can and demonstrably does go a long way towards reducing the hazard of infection and laboratory contamination in general, but can only do so consistently if the worker is fully versed in the mechanisms of infection, and avoids haste and bad planning at all times. Even then in moments of distraction, or even pure zeal, mistakes can be made by the most careful of workers. Hence, it is always worthwhile to avoid risks and utilize available safety measures, no matter how time consuming and inconvenient. It must be stressed, however, that these can be overdone under certain circumstances, in that there may be a "last straw" which will break the patience or tolerance of some individuals, and lead to grave omissions in the absence of adequate supervision. Hence, physical and temperamental fitness, training and indoctrination, and sound planning of experimental work are all of prime and equal importance in maintaining safe conditions and, indeed, in producing reliable experimental results.

The teaching of safety techniques should be combined with the initial training of inexperienced workers. For example, the trainee learning how to handle a pipette and to drop out serial dilutions should be taught to do so with simulant organisms, and be made aware of the hazards and his inexpertize by taking swabs and air samples in the immediate neighbourhood. In the author's experience this invariably teaches a salutary lesson, and the same principle can be applied to other techniques. Nevertheless, though "Every man his own Safety Officer" is an excellent maxim, expert supervision at all levels coupled with a comprehensive set of safety regulations is very important, and it is desirable that one or more individuals should be given specialist training, provided with facilities for the investigation of hazards and accidents and their prevention, and be empowered to implement whatever precautions and regulations that may be deemed necessary.

Finally, it is very desirable to extend some degree of indoctrination in laboratory hazards beyond the limits of the staff immediately involved. The laboratory itself is fundamentally a "walk-in safety cabinet", and hence hazard awareness must be instilled into architects and maintenance engineers, responsible for design and services, and also into casual intruders such as firemen, caretakers and messengers.

B. Laboratory design

Whilst every effort must be made to prevent dissemination of infectious agents by the use of safe techniques described later, escape of organisms into the laboratory space is an ever present possibility. Ideally, the laboratory should be a self contained isolation unit designed as such (Wedum *et al.*, 1956; Phillips and Runkle, 1967). In practice, however, existing accommodation has to be adapted, which often results in a none too happy compromise. As far as possible the following basic principles should be applied.

The laboratory area should be divided into "clean" and "dirty" zones, separated by an intervening decontamination barrier, comprising at a minimum an air-lock-cum-changeroom and double ended autoclaves. Additional luxuries such as a window with speech diaphragm, or other "intercom" system, and a pass hatch, provided with means of sterilization by irradiation or fumigation, are undoubtedly desirable. The dirty zone should be sub-divided into cubicles, opening off a corridor to avoid through traffic, to ensure the isolation of individual agents and techniques, hence, reducing the number of persons at risk should an accident occur. Main valves, fuse boxes, ventilation fans and other machinery requiring the attention of the maintenance staff should as far as possible be situated outside the laboratory area, or at least in the clean zone.

Structural simplicity is essential to facilitate cleaning and decontamination, and to avoid situations in which dust can accumulate. Pipes and ducting should be kept to a minimum, and special attention should be given to the points at which these penetrate walls, floors and ceilings. Failing to seal such channels adequately predisposes to inefficient ventilation and fumigation, and to the migration of aerosols and vermin.

The water supply to the dirty area must be isolated from that to clean areas by installing separate cisterns. There must be no direct connection to mains supply. Elbow-operated taps are an advantage, and care should be taken to select taps designed to minimize dripping and splashing, which, whilst not dangerous under normal circumstances, can lead to the dissemination of troublesome contamination with water organisms that tend to colonize taps, sinks and drain traps.

All drains must be adequately trapped, and all traps must be kept topped up with water. Failure to do this leads to wide dissemination of aerosols and fumes, and can interfere with ventilation and fumigation. Pipes should be made readily identifiable, chosen to withstand corrosive chemical decontaminants, and be installed in such a way that, in the event of obstruction, upwelling in a lower part of the building cannot occur. Whilst it should be obligatory never to decant living cultures down drains, this cannot be avoided in some circumstances (e.g., animal room floor drains), and it is wisest to assume that all liquid effluent is contaminated. Though chemical treatment can be used on a small scale, decontaminants must be chosen with care and constant monitoring is necessary. Furthermore the release of excessive quantities of germicides can interfere with efficient sewage disposal. Heat treatment is undoubtedly more reliable, but demands considerable attention to detail in designing suitable equipment. The common problems that arise are corrosion, accumulation of sludge, sterilization of saturated displaced air, and overfilling of receiving tanks.

Though it has been attempted, continuous sterilization is rarely, if ever, called for, and batch treatment is the general rule. Various types of installation have been designed ranging from simple steam-jacketed boilers, in which effluent is both collected and sterilized, to more complicated systems comprising a receiving tank, sterilizing tank and delay tank, in which treated effluent can be retained until found sterile. Whichever type is chosen must be duplicated, and provision must be made for pumping effluent from one plant to another in the event of mechanical failure, and for steam sterilization of the entire plant, preferably right back to the drains. If sludge is allowed to collect it can lead to and conceal dangerous corrosion, and is always difficult to sterilize and remove. Its accumulation must be minimized by agitation with paddles or steam jets. Displaced air or steam can be vented efficiently through air filters, which, however, should preferably be heated to prevent obstruction by excessive condensation. Unless they perforate due to excessive pressure they constitute a barrier to contamination at all times. The type in current use on autoclaves, which includes a critical orifice to limit pressure surges, is satisfactory. Electrical vent heaters, such as that described by Elsworth *et al.* (1955), are highly efficient, but bulky, and ineffective in the event of a power failure. Furthermore, they can lead to explosion in the event of the decomposition of steam on red-hot metal parts, or the presence in the effluent of excessive quantities of volatile solvents.

Both visual and audible warning systems must be installed in duplicate to indicate when the maximum permissible fluid level has been reached in receiving and sterilizing tanks.

The ventilation system should be designed not only to dilute and eliminate aerosols, but also to ensure a flow of air from clean to dirty zones. There is a tendency on the part of designers to plan in terms of relative pressures, which, though it does in fact achieve this end, strains ingenuity and can render the system unnecessarily complex. The very human tendency to leave doors ajar makes a mockery of such efforts, and there is much to be said in concentrating upon extraction rather than attempting to provide a balanced supply and extract system with graded pressures across barriers. This does, however, raise two problems. Firstly, the room extract system must be integrated with that from safety cabinets, if these are not exhausted through filters back into the room (Darlow, 1967). Failure to do this can clearly result in a flow of air between two neighbouring dirty rooms and, hence, in the extension of accidental aerosols. Secondly, if air supply is by leakage through louvers in doors or other barriers (Chatigny, 1961), provision must be made for sealing these rapidly in the event of emergency or prior to fumigation.

Recirculation of part of the air is often recommended as an aid to temperature economy, where the ventilation rate is high. The air must of course be filtered and care is necessary to avoid build-up of humidity, gas fumes and so on. Small portable units are commercially available, ideal for use in cubicles in that they do not require adjustment of existing ventilation systems.

It has been suggested that, if all potentially hazardous work is confined to safety cabinets with adequately filtered extracts, room air can be expelled direct to atmosphere without treatment, provided that the outlet is sited well away from situations where an occasional escape of infective material might constitute a risk to the outside world, or where aerosols can re-enter the building through windows and ventilation shafts. This principle, however, cannot apply to the great majority of animal houses, and is difficult in built-up areas; added to which a cabinet system complex enough to cater for all hazardous operations is likely to tax the patience and the pocket of the user beyond reasonable limits.

Ideally, therefore, exhausted room air should be rendered safe. There are several methods of achieving this, but that which constitutes a barrier at all times and which is least expensive in energy is undoubtedly filtration (Darlow, 1961). Since accidental aerosols may consist of very small particles (Kenny and Sabel, 1968), ultra-high efficiency filters are recommended. Admittedly, they are prone to choking with house dust, etc., but their life can be greatly extended by good house-keeping, by ensuring that the air supply is reasonably clean, and, most important, by the fitting of disposable pre-filters. The filters should be installed in such a situation as to avoid contamination of lengths of duct-work and possible cross-contamination of rooms via a common manifold, in fact there is much to be said for mounting them within the dirty rooms. If sealed directly to the extracts near or in the ceiling they can be serviced within a dirty area by personnel versed in and protected from attendant hazards, thus reducing the number of individuals at risk and avoiding the necessity for complicated filter bank systems in otherwise clean service areas of the building. Suitable valves should be fitted in the duct-work so that the room can be isolated during fumigation.

Piped vacuum to laboratories is at present in vogue; it can easily constitute a convenient oubliette into which all manner of toxic substances can be discharged inadvertently, unless elaborate measures are taken to exclude infective aerosols and liquids at the service point. However, provided its use can be fully justified, and effective precautions are taken, there seems no reason to exclude it on the grounds of safety.

V. SAFETY PRECAUTIONS

In the preceding sections the hazards have been defined and the importance of indoctrination and laboratory design stressed. The stage is now set to proceed to the elaboration of preventive measures. These fall naturally into six categories—

(a) Careful technique.
(b) Methods of confining the agent.
(c) Physical protection of the worker.
(d) Intangible barriers (air disinfection).
(e) Disinfection in general.
(f) Immunological and kindred measures.

If it was always possible to perform all manipulations of infective material within ventilated hoods and prevent the escape of micro-organisms into the environment, most other precautions would be of secondary importance. Unfortunately this ideal is never achieved, not that it is physically impossible (except in the handling of large experimental animals for example), but because traditional open-bench methods die hard; indeed, little, if any, effort is made towards this goal in the majority of laboratories. Careful technique, therefore, is of prime importance.

A. Experimental techniques

When manipulating live micro-organisms at the open bench, the working area should be covered by a sheet of lint, or similar absorbent fabric, *moistened* with an appropriate disinfectant. This serves to absorb spills and prevent splashing. It also cushions the impact of dropped glassware. Adequate elbow room, a very appropriate term in this context, is essential, and overcrowding of apparatus and workers must be avoided. Movements must be slow and unhurried, and adequate time must be allowed for such procedures as discharging pipettes and flaming the necks of tubes and bottles.

Mouth pipetting should never be permitted under any circumstances. The dividing line between pathogens and non-pathogens is not clear cut, and cultures can be mixed in more than one sense; also it is unwise to employ two disciplines where one truly safe procedure will suffice. Many types of manual pipetting device exist, and the choice must be left to personal preference. It is, however, essential to choose one which can be disinfected either by heat or chemicals.

Great care must be exercised to avoid actions likely to give rise to frothing or splashing. Mark-to-mark pipettes are clearly preferable to other types in this respect as they do not require the expulsion of the last drop. Discharge should be as close as possible to the fluid or agar level, or the contents should be allowed to run down the wall of the tube or bottle wherever possible, not dropped from a height. Finally the pipette should be discarded vertically or at an angle into a pot (rubber for preference) containing sufficient disinfectant to flood the contaminated portion. It is unwise to rely on chemical disinfection alone, however, on account of air locks within the pipette, and the gradual neutralization of the disinfectant by organic residues. The pot and its contents, therefore, should ultimately be autoclaved before being washed up.

Loops should be made of platinum or soft non-vibrating alloy; the stem should be short, and the loop itself small. To allow adequate time for cooling, several loops should be at hand and used in rotation. Dissemination of organisms by sputtering in the bunsen flame can be eliminated by using hooded burners (Darlow, 1959; Winner and Quiney, 1953).

A wide range of plugs, caps and bungs is in current use, but there is much to be said for the retention of the traditional cotton plug as opposed to loosely fitting caps, which permit the immediate escape of liquid should a tube fall on its side. Whatever type is adopted, it should not be allowed to become wetted with culture. Before use the tops of tubes and bottles should be inspected for sharp edges, or irregularities likely to lead to injury, or inefficient fitting of the closure. Lipless tubes are much to be preferred in this respect.

Screw-capped glassware has two disadvantages in addition to the consequences of wetting the thread. If the caps are screwed on tight, the vessels tend to explode on cooling after autoclaving, whilst, if the caps are loosened beforehand, leakage can occur. The use of flat-bottomed disposable plastic vessels helps to eliminate both these hazards.

Hypodermic syringes are used in two principal contexts in the laboratory, animal inoculation and the removal of cultures or body fluids from rubber capped vessels or through animal tissues. The hazards have already been described and great care must be taken in avoiding them. If the manipulations cannot be performed in a safety cabinet, certain basic rules must be observed. Disposable syringes with built-in needles are to be preferred, but if glass syringes are used, the types provided with a locking mechanism to secure the needle should be chosen so that the needle cannot fly off unexpectedly. When drawing up infected materials through a bottle cap or skin, the needle should be withdrawn through a plug of cotton soaked in an appropriate disinfectant, into which air bubbles can subsequently be expressed. On completion of the operation, disinfectant should be drawn up into the empty syringe before the plunger is removed. Both should then be discarded in a disinfectant bath.

Freeze dried culture ampoules should always be opened in a safety cabinet using commercially available ampoule openers designed to avoid manual contact. When ideal equipment is not available, the ampoule must be wrapped in a disinfectant-soaked swab before breaking it open to minimize the risk of cutting the hands, and to a lesser extent of releasing aerosol of dried material. Wherever possible ampoules should be filled with dry nitrogen after freeze-drying, thus avoiding implosion that may occur during the sealing as well as opening of evacuated ampoules. The whole process of freeze-drying itself should be performed in a safety cabinet, and, though many organisms do not survive the rigours of the vacuum pump (some do, e.g., spores), filtration of the effluent air is desirable either up (preferably), or down stream of the pump.

By the same token precautions should be taken when evacuating desiccators and similar fragile vessels. If the procedure cannot be performed in a safety cabinet, at least adequate steps can be taken to prevent accidents from flying glass by boxing in the equipment. Precautions must again be taken to prevent the contamination of the vacuum pump, or preventing it from dispersing leaked material. This applies equally to vacuum pumps on high speed centrifuges.

The numerous hazards of centrifugation have been described above and the cure for some of the ills must be fully evident, yet, in spite of this, careless handling is all too frequent. The instruction manuals issued by the manufacturers provide much useful information on servicing, the correct filling of the tubes, metal fatigue, etc., and these instructions must be followed carefully, and the log books, generally provided, kept up to date; but the instructions unfortunately are not a child's guide to all possible eventualities.

High speed rotor heads are prone to metal fatigue, and where there is a chance that they may be used on more than one machine, each rotor should be accompanied by its own log book indicating the number of hours run at top or de-rated speeds. Failure to observe this precaution can result in dangerous and expensive disintegration. Frequent inspection, cleaning and drying are important to ensure absence of corrosion, or other traumata which may lead to creeping cracks. Rubber O-rings and tube closures must be examined for deterioration and be kept lubricated with the material recommended by the makers. Where tubes of different materials are provided (e.g., celluloid, polypropylene, stainless steel), care must be taken that the tube closures designed specifically for the type of tube in use is employed. These caps are often similar in appearance, but are prone to leakage, if applied to tubes of the wrong material. When properly designed tubes and rotors are well maintained and handled, leaking should never occur.

Simple as it sounds, the balancing of buckets is often mismanaged. Care must be taken to ensure that matched sets of trunnions, buckets and plastic inserts do not become mixed. If the components are not inscribed with their weights by the manufacturer, coloured stains can be applied to avoid confusion. When the tubes are balanced, the buckets, trunnions and inserts should be included in the procedure, and care must be taken to ensure that the centres of gravity of the tubes are equidistant from the axis of rotation. To illustrate the importance of this, two identical tubes containing 20 g of mercury and 20 g of water respectively will balance perfectly on the scales, but their performance in motion is totally different, leading to violent vibration with all its attendant hazards.

Cleaning and disinfection of tubes, rotors and other components requires considerable care. It is unfortunate that no single process is suitable for all items, and the various manufacturers' recommendations must be followed meticulously, if fatigue, distortion and corrosion are to be avoided. This is not the place to catalogue recommended methods, but one less well appreciated fact is worthy of mention. Celluloid (cellulose nitrate) centrifuge tubes are not only highly inflammable and prone to shrinkage with age and distortion on boiling, but can behave as high explosive in an autoclave (Silver, 1963).

Glass vessels in refrigerators and cold rooms maintained at a little above freezing point occasionally freeze, burst, thaw and void their contents. The resultant leaked material may contaminate a wide area; later it dries up and is less readily detected. It is wisest, therefore, to stand all such vessels in metal or plastic troughs of sufficient size to contain escaped material. The author recalls an extreme instance in which a glass bottle containing a suspension of spores and glass beads was placed too close to the refrigerator coils. On opening the door of the refrigerator the beads rolled all over the laboratory and out into the passage; a great deal of time and inconvenience was involved in cleaning up the consequences of this avoidable accident. In sub-zero refrigerators the risk is less, since the bottle contents remain solid until allowed to thaw after removal and careful inspection of the vessel walls. Nevertheless, the same routine should apply as a provision against failure of the refrigeration system. It would not seem out of place at this juncture to remind the reader that culture vessels in autoclaves also call for anti-spill precautions to prevent contamination of the drain line in the event of leakage during preliminary evacuation. Finally, refrigerators tend to become stores of forgotten material, and it is, therefore, important to label all vessels properly, giving full details of their contents. Without this precaution, which is all too frequently neglected, it is difficult to assess risk, and, hence, need for disinfection in the event of a spill.

B. Methods of confining the agent

The preceding paragraphs, though far from exhaustive, cover a wide spectrum of methodology in the absence of more elaborate means of protecting the worker. In many procedures the risks can be substantially reduced, if never entirely eliminated, by the exercise of care but there are many other procedures that are unavoidably dangerous, and demand more tangible barriers. Outstanding examples are the handling of lyophilized cultures, the opening of recently agitated culture vessels or bottles of suspension, the aeration of culture vessels, and the grinding of infected tissues especially with the use of a pestle and mortar. The logical solution to this kind of problem is to confine the agent by the use of some form of safety cabinet. Many types have been devised ranging from the simple chemical fume cupboard to hermetically sealed hoods, involving negative pressure, air locks and remote handling methods. Whilst some processes such as the continuous culture of pathogens can conveniently be carried out in the latter type, most conventional microbiological methods demand a simpler design combining the virtues of these two divergent systems without detracting from the safety factors.

In principle a safety cabinet consists of an enclosed work space fitted with an air exhaust system and openings for the hands of the operator. Air must be exhausted from the cabinet at a rate which will—

(1) Overcome any tendency for aerosol to diffuse out through the arm ports.

(2) Compensate for the air displacement of the operator's arms and the use of compressed air or gas within the cabinet.
(3) Ensure rapid dilution and removal of aerosol, gas fumes and heat.

The first two requirements can be met by ensuring a linear air flow across the arm ports of not less than 100 ft/min. Phillips (1965) quotes 50–100 linear ft/min, whilst a figure of 100–200 is laid down for fume cupboards in BS3202. Much will depend on the size of the openings. If these are small, much higher velocities will be developed when they are partially occluded by the operator's arms, whilst, if they are large, there is danger of partial reversal of flow due to turbulence in the immediate neighbourhood (Barbeito and Taylor, 1968). If a minimum velocity of 100 linear ft/min is achieved, standard arm ports of approximately $\frac{1}{3}$ ft^2 in area combine convenience with safety.

The fitting of gauntlet gloves undoubtedly improves the safety factor, but prevents effective air scouring of the cabinet unless provision is made for the admission of an adequate airflow through a valve, if necessary fitted with an inlet filter (Darlow, 1967a). Scouring and dilution are clearly important since they obviously reduce the hazard presented by residual aerosol in the event of ventilation failure or inadvertent opening of the cabinet before disinfection has been carried out. To ensure rapid dilution the total air through-put must be proportional to the volume of the cabinet. An exhaust flow of 100 c.f.m. through a cabinet of 15 ft^3 ensures an air change of 6·6/min. Using the calculations of Bourdillon and Lidwell (1948) it will be seen that an aerosol generated at time zero will be reduced by a factor of 735 in 60 seconds, whilst for a cabinet of twice the volume the factor would only be 27. Furthermore, not only do larger cabinets require higher rates of flow to ensure a reasonable margin of safety, but provision must also be made to ensure ventilation of stagnant pockets. Limitation in the size of this type of cabinet is, therefore, desirable.

Cabinets of more complicated designs to accommodate specialized procedures, particularly infected animal handling, have been reviewed by Wedum (1964), Chatigny (1961), Jemski and Phillips (1965) and many others. The principles involved, however, are materially similar.

Many methods have been used to sterilize the effluent air including gas-fired and electrical air burners, ultraviolet light, and electrostatic precipitation, but that which is generally agreed as being the most reliable and economic is ultra-high efficiency air filtration, which not only continues to constitute an effective barrier in the event of electrical or gas failure, but is free from the hazards of fire, high voltage, explosion, etc. This is not the place to discuss filter design, but certain essential criteria relevant to safety cabinets must be emphasized.

(1) The filter should have a methylene blue or sodium flame test penetration of not more than 0·003%. Filters of this efficiency are impenetrable to particles of the dimensions likely to be generated during most laboratory procedures as described by Kenny and Sabel (1968). They are also almost opaque to monodispersed virus particles, an unlikely challenge under normal working conditions.

(2) The filter should be mounted directly on the cabinet roof to avoid contamination of duct work, and care should be taken to ensure that the sealing and housing provide a tight fit and preclude the possibility of a by-pass (Darlow, 1967).

(3) The installation of a pre-filter within the cabinet, in which situation it can readily be decontaminated and changed with safety, is recommended to reduce premature obstruction of the master filter by cotton fluff, etc., and protect it from trauma and gross liquid splashes, which may result in a blow-out.

(4) A filter designed to handle a little more than the required volume should be chosen. This generally involves a pressure drop in the region of 1 in. w.g. and provision should be made for measuring this, and preferably, also, the total air flow in the effluent ducting, as a guide to the state of the filter and efficiency of the extract fan.

(5) Finally, suitable sampling points on the clean side of the filter should be installed to enable regular efficiency checks to be made with monodispersed spores or bacteriophage for the detection of leaks in the filter material, its peripheral sealing and the filter/cabinet seal.

Decontamination of the internal surfaces of the cabinet may be achieved by manual application or spraying of a suitable germicide, by ultraviolet irradiation, or by fumigation. Of these the latter is the most versatile, since it can also sterilize the filter and other inaccessible areas provided that the cabinet and exhaust duct can be rendered reasonably airtight.

Of the fumigants, ethylene oxide demands a high degree of air-tightness, and presents an explosive hazard, unless used under pressure as a mixture with CO_2 or a halogenated hydrocarbon. Furthermore, it is absorbed by rubber, and, as it is toxic to skin, may present a burns hazard, if absorbed by gauntlet gloves. Though it is sporicidal, it is somewhat slow to act even under optimum conditions of low relative humidity and raised temperature. β-Propiolactone, on the other hand, has much to recommend it. Its action is much more rapid than that of ethylene oxide, particularly under the conditions of humidity and temperature most usually encountered in safety cabinets. However, although it decomposes fairly rapidly, leaving no toxic residue, it also has vesicant properties, and is known to be carcinogenic at non-irritant concentrations (Roe and Glendenning, 1957). Formaldehyde and glutaraldehyde, though somewhat less active than

β-propiolactone, would appear to be the safest and most convenient fumigants. The vapour is generated by boiling aqueous solutions (Darlow, 1967), thus ensuring maximum concentration of monomer and optimum humidity simultaneously. There is no explosion hazard, and in the event of leakage, the odour is readily detected, if not intolerable, at a concentration well below the toxic level.

C. Laboratory clothing

In the more elaborate laboratory, designed for particularly hazardous operations, provision can readily be made for a major change of clothing. This situation is rarely achieved elsewhere, even where it might be considered desirable, except in animal houses. The average laboratory worker is content with something less than perfection. There are, however, certain criteria of conduct which can and must be accepted on logical and economic grounds, but which, nevertheless, are all too frequently overlooked.

The traditional laboratory coat tends to become a status symbol, but is ill-adapted for work at the microbiological bench. It opens down the front, which, in the event of a spill, predisposes to contamination of underlying street clothes and necessitates manipulation of contaminated buttons. Furthermore, it incorporates pockets and lapels in which forgotten sharp objects not infrequently find their way to the laundry. The more correct garb, surely, is something in the nature of a surgeon's gown, which can be inverted clean side out on removal. If not disposable, it must consist of a fabric capable of withstanding autoclaving or chemical decontamination before being sent to the laundry. Oliphant *et al.* (1949) record an instance in which six laundry staff members contracted Q fever as the result of neglect of this essential precaution. This overall garment must always be removed before leaving the laboratory for any reason whatsoever, and this precaution should apply equally to work with non-pathogens, since clothing is a well known vector of troublesome contamination.

As with pipetting it is wisest to adhere to one rule, and insist on the wearing of surgical gloves at the bench. Admittedly these can be punctured, but at least they minimize the risk of hand contamination, and are much more easily decontaminated than skin. A cap to protect the hair from splashes is important. Not only is hair difficult to decontaminate, but, if worn long, can easily swing into bunsen flames, open Petri dishes, etc. Facial protection can be provided in the form of commercially available plastic shields, or, partially by a surgeon's mask. It must be stressed that the latter is designed to prevent escape of droplets from the nose and mouth and is of little value in respiratory protection.

If hazards demand respiratory protection, then nothing but a full face respirator or ventilated hood will suffice. A half-mask of the paint sprayer's type is a half measure as it does not protect the eyes from *inoculation*, and the filters provided are often incapable of excluding small particles of the size encountered in laboratory work.

A change of footwear is desirable not only on the grounds of spread of contamination, but also on account of the destructive effect of autoclaving and disinfectants on leather shoes in the event of a spill. A wide choice is available ranging from light overshoes to gum boots, the latter being preferable in circumstances involving pools or troughs of disinfectant. When there is a risk of dropping sharp instruments or spilling infected liquids, blood, etc., the overall garment must extend below the top of the boots.

D. Intangible barriers

The infective potential of an aerosol accidentally released into the laboratory air may be reduced in three ways, by dilution, by physical removal from the air, and by the acceleration of natural biological decay by the use of germicidal agencies. As has already been suggested above, the principal purpose of ventilation is to limit the spread of aerosol rather than to dilute it, since to do so effectively would demand a ventilation rate in the order of that applicable to safety cabinets. This would clearly present more problems than it would solve, if the laboratory is large and complex. Much, however, can be achieved by adopting the second solution, which in effect implies filtered recirculation. Such a system can be more adaptable to changing circumstances, and it is obviously more economic to recirculate warm, relatively clean laboratory air, than to treat large volumes of cold, dust laden urban air. A high rate of recirculation is relatively simple to maintain in small areas of high risk. It must be stressed, however, that the purpose of recirculation is to boost rather than to replace an orthodox ventilation system.

Where no ventilation system exists, it has been suggested that opening windows wide after an accident could be of use. The value of this measure, however, is doubtful, since much will depend upon the force and direction of the wind and the size and distribution of the windows. Contamination could well be blown back into the building, and time will be wasted, thus increasing the chances of inhaling an infective dose. Immediate evacuation of the area is probably safer. However, something can be achieved under unventilated circumstances by the use of bacteriocidal vapours. This principle has been neglected of recent years, but the concept of a non-toxic, non-irritant chemical barrier between the worker and the potential aerosol has much to commend it, and, whilst the technique is still largely unexplored, a variety of substances has shown very definite potential, particularly in the case of freshly generated moist aerosols (Darlow *et al.*, 1958). In the present state of the art it would be unwise to attempt to formulate any firm policy, but it is considered that this field is worthy of further consideration.

Whilst direct sunlight and even diffuse daylight are slowly lethal to many micro-organisms, ultraviolet radiation at 220 to 280 nm, particularly around the 265 band, has been confirmed by many authors as rapidly lethal to a wide range of bacteria and viruses, and the application of UV as an air hygiene method is now fully accepted. In spite of this, however, it has many limitations and traps for the unwary, which are less well appreciated and which can lead to a false sense of security, if not actual primary hazard.

Low pressure mercury vapour germicidal lamps, emitting a high proportion of UVR at about 253·7 nm, are commercially available. They are of two principal types, hot and cold cathode. The hot cathode lamp is in essence a familiar "strip light" without a fluorescent coating. It requires the same circuitry as standard striplight fittings. Its life is mainly dependent on the life of the filaments, which tend to disintegrate if the lamp is turned on and off frequently. In addition the glass slowly becomes discoloured and opaque to UV, especially at low temperatures, at which the output in any case is lower than at room temperature, and there is in addition some reluctance to light up. The cold cathode lamp, on the other hand, starts instantaneously at low temperatures, and has a very much longer life. However, special high voltage circuitry is required. The choice must lie with the user.

Much has been published on the killing rates for numerous organisms at various intensities of illumination, for which the reader must refer to works on disinfection, but these cannot necessarily be translated into practical terms. Intensity meters are available commercially, but actual in-use tests with micro-organisms are desirable, and should be repeated at regular intervals. There is great variation in sensitivity, even from strain to strain of the same organism, and thin films of absorbent material can not only protect the organism, but may greatly reduce the output of the lamp, which itself can continue to glow attractively long after it has ceased to emit an effective intensity of the active waveband. Ensuring that the lamps are clean, therefore, is of prime importance, but equally important is to ensure that surrounding reflecting surfaces are clean and that they do not primarily absorb UV.

Germicidal wavelengths can produce damage to plastics and other materials either directly or through the agency of ozone. Perspex, for example, becomes visibly discoloured and crazed, but some rubbers and other plastics deteriorate insidiously, leading to unexpected failures. Oxidative processes in general are accelerated, and chemical changes, often bio-inhibitory, can take place in exposed materials (e.g., bacteriological media). Regular checking of materials is, therefore, necessary. By far the commonest effect, however, if also the most readily detected, is damage to exposed human skin and eyes. It is very easy to fail to notice a UV source in a well illuminated room. The resultant discomfort can be severe, if transitory, but permanent damage can be produced after repeated or chronic exposure. Means of avoiding this are discussed below.

UVR may be used to sterilize air, surfaces, and even thin films of liquid, but only if they are exposed to direct radiation or scatter from adjacent non-absorbing surfaces. It cannot, therefore, be regarded as an entirely satisfactory substitute for other methods of decontamination. Nevertheless, it has a valuable part to play in the field of laboratory safety in three principal contexts.

Firstly, it may be used for whole-room irradiation in walk-in incubators, refrigerators and similar spaces, where spills may occur, but where the ventilation rate is of necessity low or non-existent, or in situations in which the risk is particularly high, such as rooms housing shaking machines and apparatus used in aerosol studies. The skin and eyes of personnel entering such compartments must be adequately protected by gloves, fabric or plastic hoods, and transparent face shields, and warning devices must be exhibited. Switch gear may be integrated with the door of the room; but, though irradiation may be required only when the room is occupied, it may also be necessary to irradiate subsequently as a decontamination measure where fumigants cannot be employed. Reliance on manual operation is, therefore, more satisfactory. In the same context, UVR can be usefully employed in safety cabinets both as an emergency measure in the event of extract failure, and as a means of decontaminating exposed surfaces.

Secondly, it may be employed as a curtain to limit the spread of live aerosols along corridors, through air-locks, and between banks of animal cages and similar situations. Here the risk to human tissues is relatively slight and transient, but again warning devices are very necessary.

Thirdly, UVR may be employed in duct-work purely as an air sterilizing measure. In this situation it does not constitute a hazard, and is particularly useful in recirculating systems, where, owing to the low resistance to air flow (as opposed to that presented by filters), silent, low-powered fans can be installed. Here, however, inspection and cleaning are difficult and tend to be forgotten. Furthermore, dust accumulates rapidly on the surfaces of the lamps, and the cooling effect of the air stream reduces their output and life.

E. Disinfection

Much has already been said about disinfection, and this Chapter is not the place to discuss the finer details for which the reader must consult text books on the subject. However, it is relevant to discuss at least some of the practical aspects that have a direct bearing on safety.

The number of liquid germicides is legion, and, whilst it is widely known, it is not generally appreciated that many of these, especially the blander agents less likely to do damage to materials or human tissues, are only active against certain groups of micro-organisms and relatively or totally inactive against others. On the whole it can be stated that the hypochlorites and compounds with related activity (e.g., the chloroiso-cyanuric acid derivatives) have the widest germicidal spectrum, though there is still some controversy on their effectiveness against the mycobacteria. Being powerful oxidizing agents, they tend to be corrosive and somewhat unstable in solution, but are, nevertheless, the most valuable and active for general disinfecting purposes. The cresolic and phenolic compounds, which undoubtedly kill mycobacteria, are virtually non-sporicidal and only act slowly against many of the smaller viruses, even at concentrations that burn human skin. In the author's experience their germicidal spectrum is shared by the iodophores. Glutaraldehyde and formaldehyde solutions are generally effective, though too odorous for most purposes, and it must be remembered that though formaldehyde is an excellent fumigant, its solution, formalin, consists largely of hydrates and short-chain polymers which are relatively inactive. However, it has a synergistic action when mixed with the ampholytic detergent disinfectants (Perkins *et al.*, 1967) and other chemically compatible agents with discontinuous spectra. Where well authenticated information is not forthcoming, therefore, it is prudent, before choosing a routine disinfectant, to carry out trials with the micro-organisms to be handled, ensuring optimum conditions for survival and recovery, and not to accept a merely time honoured method for which there may be no basis in fact. To give an extreme example, it is all to frequently assumed that histological specimens and sections and other microscopical preparations are rendered sterile during staining and other techniques, but in many cases this is not so, particularly where bacterial spores are concerned. When the choice has been made, it may still be necessary to carry out periodic evaluations both of stock and ready-for-use solutions, when turn-over is not rapid, as activity may be reduced by decomposition, evaporation, or polymerization, or even, in the case of quaternary ammonium diluates, by bacterial colonization. The importance of maintaining adequate supplies of ready-for-use solutions in anticipation of an emergency cannot be overstressed. These should be freshly prepared at the start of each day's work as a routine.

Decontamination of rooms can be carried out by the application of liquid disinfectants, or fumigants. The former can only be applied in animal rooms and similar spaces where drainage is adequate, and risk of damage by moisture can be ignored. It has the advantage of combining disinfection with cleansing. A convenient technique is described by Perkins *et al.* (1967). Fumigation is most commonly and effectively carried out by vapourizing β-propiolactone or formalin. The lactone technique is well described together with attendant hazards by Spiner and Hoffman (1960). The vapour is generated by atomizing an aqueous solution by means of a spray device producing a large volume of aerosol in a short time; commercially available sprayers of the "spinning disc" type are most convenient. The method is clean, rapid and effective, but presents a toxic hazard. The formalin technique (Darlow, 1967b) involves boiling a mixture of formalin (40% w/v commercial grade) and tap water in an electrically heated boiler in a proportion of approximately 0·5 ml formalin to 1·0 ml water/cubic foot of space. The method is slower (an over-night holding time is recommended), but is equally effective. Care should be taken to avoid pools of water, dripping taps, etc., which rapidly absorb the vapour and reduce its concentration. The proportion of formalin required will vary with the area of absorbent surfaces and other conditions in the room, but it can be determined in advance for any given circumstance by carrying out preliminary trials with spore inoculated test pieces. After both procedures it is necessary to wear a gas mask fitted with a charcoal filter, when the room is entered for unsealing and airing, but residual formaldehyde can be eliminated rapidly by exposing ammonium hydroxide solution in a shallow pan, or absorbed on a piece of filter paper.

Routine cleaning of laboratories, as opposed to total decontamination, is not without its hazards and traps for the unwary. The principal object is the prevention of accumulation and dissemination of dust. This is most often achieved either by vacuum cleaning or by agglutination with waxes and oils. Both methods have their disadvantages. A vacuum cleaner, unless fitted with a highly efficient filter, only serves to disseminate small particles, whilst agglutination is a more laborious procedure and does not necessarily remove the dust. Both serve to smear contamination over bench surfaces and floor, and, unless provision is made for rapid decontamination of cleaning appliances, separate equipment will be required for each individual room, if cross-contamination is to be avoided. The author recalls an instance in which a rotary floor polisher, normally reserved for clean areas, was used to polish the bench tops of a laboratory where pathogenic spores were handled. Had this not been discovered in time, the sequelae might have been serious indeed. Surely, the simplest, if most frequently ignored, method is the mere swabbing down of dust collecting surfaces using appropriate disinfectant/detergent solutions, coupled with restriction of dust traps by good laboratory planning?

Germicidal paints must be mentioned, if only to dismiss them. Admirable as the concept of a self-sterilizing surface is, its effectiveness is limited by the activity of suitable germicides, the necessity for an intermediate moisture film twixt surface and germ, and the relatively small and ephemeral quantity of germicide which can be incorporated in paint, or adsorbed upon its surface. In moist conditions, however, frequent renewal of a germicidal residuum, such as in the case of hexachlorophene soaps and non-volatile disinfectants, is of generally accepted value in maintaining a low level of contamination, but should not be relied upon as a safety measure.

VI. LABORATORY INVESTIGATIONS

For several reasons safe working conditions cannot be maintained on the basis of a code of practice alone. Investigations are required to ensure an adequate standard of quality of materials, of performance of apparatus, and of human skills. New techniques and equipment must be put to the test, and the causes and consequences of accidents must be explored. Investigation should also be made an integral part of the teaching of trainees, particularly with a view to impressing them with the very existence of biological hazards, most of which pass undetected by human senses.

The importance of routine monitoring of air filters, ultraviolet lamps, germicides and other sterilizing agents has already been mentioned above, but the process can be extended in unexpected directions, which do not at first sight appear to have a direct bearing upon safety. To give an example—glassware, rubber liners, constituents of bacteriological media and other materials may often be contaminated with inhibitory but non-lethal impurities giving the impression of sterility, or that an inoculum was non-viable. This type of hidden hazard can only be revealed by curiosity and very little extra effort, amply repaid, not only in safety, but also in the saving of material and wasted experiments.

Routine sampling of air and dust, attractive and logical as it may sound, is generally not informative, unless safety precautions are inadequate. Even so, the hazard may have passed unnoticed by the time samples are taken and cultured, and labile organisms are likely to be entirely missed. In the best regulated circles routine sampling should not be necessary.

Tests with simulants, however, are of considerable value, the most widely used being *Bacillus globigii* spores, *Serratia marcescens* and the coliphages. These organisms are simple to produce in quantity, to disseminate as monodispersates in a Collison spray or similar device, and to collect with a Bourdillon slit sampler. They are also readily distinguishable from background contaminants.

Though numerous types of sampling device exist, many are not commercially available, but the above quoted air sampler is versatile and simple, and to be recommended for most laboratory investigations. It is to be preferred to settle plates, which, though commonly used, cannot be expected to collect particles of the size most readily inhaled into the lower respiratory tract, which constitute the principal hazard.

Much more could be said on the techniques of investigation, but it must suffice here to stress its value and the necessity to investigate before and during the planning stage of experiments, and not wait until an accident has occurred.

VII. MEDICAL ASPECTS

A high standard of personal hygiene is obviously important. The well known exhortation, "Now wash your hands" is even more applicable in the laboratory than the lavatory. Long finger nails are not only difficult to keep clean, but they interfere with delicate manual manipulations and traumatize rubber gloves, not to mention skin. Long hair and beards can readily become contaminated and are not easy to disinfect (Barbeito *et al.*, 1967); but beards carry a further risk in that they interfere seriously with the air-tight integrity of respirators.

Wounds, skin diseases, asthma, bronchitis, defective eyesight and many other disorders can render a worker more susceptible to laboratory infection than his fellows, and there is a strong case for the appointment of a permanent medical adviser, familiar with the activities of the laboratory and the diseases that may be contracted, empowered to control the activities of his potential patients where illness predisposes to infection, and on close terms with local public health authorities and family doctors. All sickness and accidents should be reported to him for assessment at the earliest possible moment. The family doctor must of course be appraised of the situation, and will not resent incursion on his preserves, if friendly liaison is maintained. If this is done and adequate measures taken, only a small proportion of laboratory accidents result in infection, and unusual illness resulting from undetected accidents, which may puzzle the family doctor, can be diagnosed and dealt with before progressive damage has been done. It is essential that the virulance and antibiotic sensitivities of organisms in use in the laboratory are determined in advance to expedite effective treatment, when the occasion arises. This information together with a brief description of symptomatology and references should be recorded for emergency use by any practitioner who may be called in to treat a condition with which he is unfamiliar.

It is therefore essential that the worker and his family doctor are impressed with suspicious symptomatology, and the importance of calling in the appointed consultant to advise on diagnosis and treatment. To this end it is a useful measure to provide the worker with an identity card to present to any doctor who may be called to see him, giving a brief warning on the possible significance of pulmonary or encephalitic signs, or of undiagnosed pyrexia, together with the name and address of the expert.

The value of protective inoculation needs no stressing, but it must not be assumed that it necessarily confers immunity. Immunity is relative to the size of the challenging dose, the portal of entry and the response of the individual, and in many cases is short lived. It cannot, therefore, be regarded as a substitute for other protective measures. If time can be spared, periodic checks on the state of immunity can be of value. Not only can these assist in assessing the potency of vaccines and provide confirmation of diagnosis in the event of sickness, but, by indicating sub-clinical infection, it can be of considerable use in detecting unsuspected accidents and careless technique.

VIII. SAFETY ADMINISTRATION

Rules by definition are rigid and, therefore, tend to be unpopular or even ignored, if they appear to be too inflexible to accommodate changes in techniques. Nevertheless, they are necessary, and can be rendered more palatable when garnished with explanatory detail. As experience and necessity add to their number, they tend to become tedious to the reader, but this can be overcome by division into indexed sections for ready reference (e.g., decontamination, isotope handling, protective clothing, etc.), and accompanied by appropriate summaries.

The task of formulating the rules is formidable. The legislator runs the risk of ridicule unless he draws up a practical and logical code. This demands experience of all facets of the laboratory activities, or the assistance of an advisory panel representative of a wide spectrum of disciplines. In all too many laboratories rules are rudimentary, if they exist at all, and this is often due to indecision as to how to start. However, once a foundation of "ten commandments" has been laid, the edifice will grow into a useful code of practice. A prerequisite is the appointment of a senior staff member, preferably medically qualified, to act as Safety Officer and empowered to maintain safety discipline and conduct investigations into hazards and accidents. Training naturally falls to his lot, and a plan for this is admirably summarized by Phillips (1962).

Ten basic rules are suggested below. Admittedly they read like the sayings of a Victorian moralist, but failure to observe each has been recorded as the cause of laboratory infection. Rule 10 sounds particularly prosaic, but does in fact emphasize the importance of training. The untrained may have no reason to doubt, and the partially trained may possess an exaggerated opinion of his ability; the doubter may be afraid to seek advice. The Safety

Officer, therefore, must maintain a co-operative and avuncular attitude to gain the confidence of his charges, and must himself entertain doubts until he has carried out the necessary investigations to ensure the safety of a novel technique or appliance.

(1) Always report accidents no matter how trivial at the earliest possible moment. Do so in detail and in writing. A lesson can always be learned from them, and the chances of repetition diminished.
(2) Always regard all micro-organisms as potentially pathogenic, even if only to your neighbours' tissue cultures, and label all infective material unequivocally to prevent confusion; code numbers are inadequate.
(3) Always report all sickness and injury as soon as possible. Delay in identification of active or potential laboratory infection may have grave consequences for others as well as yourself.
(4) Always use approved protective clothing and safety appliances.
(5) Always handle materials of unknown potential (e.g., samples of sputum, serum, faeces, unfixed pathological specimens, etc.) as if they were infective; they probably are.
(6) Always keep your laboratory clean, and ensure that all contaminated apparatus, cultures, lab. clothing, etc., are sterilized before they are washed up. Remove your lab. clothing before leaving the laboratory area to go to the toilet, canteen, library, etc.
(7) Always keep a ready-for-use supply of disinfectant at hand of sufficient volume to carry out emergency decontamination in the event of a spill. Ensure that the disinfectant is known to be active against the organism being handled.
(8) Never eat, drink or smoke at the lab. bench, or pipette with the mouth; and always remember that your hands, gloved or otherwise, may be contaminated and capable of transferring infection to body openings.
(9) Never perform any action hastily, particularly if it is liable to cause bubbling, frothing, spills and splashes, or contaminate surfaces which with care should remain clean (e.g., bottle caps).
(10) When in doubt, ask an expert.

IX. CONCLUSIONS

The epidemiology of laboratory infection has been described above, admittedly in somewhat skeletal form, but the ramifications of the subject are too extensive to allow a coherent condensation without omissions or resort to dogmatic platitudes. The literature on the subject is, in fact, very extensive, but many of the publications have appeared in unfamiliar journals and have consequently been overlooked. It is regrettable that the spilling of ink is so inaudible, and that positive action is only stimulated by a resounding spilling of pathogens.

One important point emerges. The handling of radioisotopes is standardized and controlled by statutory regulations, yet the study of pathogenic micro-organisms is not. Owing to their capacity for replication, they can be more deadly in far smaller dosage, and it is to be hoped that the future will bring procedural standardization, and raise safety from what at present is little more than an art to a respectable science.

REFERENCES

Anderson, R. E., Stein, L., Moss, M. L., and Gross, N. H. (1952). *J. Bact.*, **64**, 473–481.

Barbeito, M. S., and Taylor, L. A. (1968). *Appl. Microbiol.*, **16**, 1225–1229, BS. 3202. (1959) Laboratory furniture and fittings, p. 39, B.S.I.

Barbeito, M. S., Mathews, C. T., and Taylor, L. A. (1967). *Appl. Microbiol.*, **15**, 899–906.

Bourdillon, R. B., and Lidwell, O. M. (1948). *Spec. Rep. Ser. med. Res. Coun.*, No. 262, p. 347.

Chatigny, M. A. (1961). *Adv. appl. Microbiol.*, **3**, 131–192.

Darlow, H. M. (1959). *Lancet*, *ii*, 651.

Darlow, H. M. (1960). *Lab. Pract.* **9**, 777–779.

Darlow, H. M. *et al.* (1961). *Lab. Anim. Cent. coll. Pap.*, **10**, 65–69.

Darlow, H. M. (1967a). *Chemy Ind.*, 1914–1916.

Darlow, H. M. (1967b). *Lab. Anim.*, **1**, 35–42.

Darlow, H. M., and Bale, W. R. (1959). *Lancet*, *i*, 1196–1200.

Darlow, H. M., Powell, E. O., Bale, W. R., and Morris, E. J. (1958). *J. Hyg., Camb.*, **56**, 108–124.

Druett, H. A., Henderson, D. W., Packman, L., and Peacock, S. (1953). *J. Hyg., Camb.*, **51**, 359–371.

Druett, H. A., Robinson, J. M., Henderson, D. W., Packman, L., and Peacock, S. (1956a). *J. Hyg., Camb.*, **54**, 37–47.

Druett, H. A., Henderson, D. W., and Peacock, S. (1956b). *J. Hyg., Camb.*, **54**, 49–57.

Elsworth, R., Telling, R. C., and Ford, J. W. S. (1955). *J. Hyg., Camb.*, **53**, 446–457.

Hanel, E., and Alg, R. L. (1955). *Am. J. med. Technol.*, **21**, 343–346.

Harper, G. J., and Morton, J. D. (1953). *J. Hyg., Camb.*, **51**, 372–385.

Hobbs, P. V., and Osheroff, T. (1967). *Science, N.Y.*, **158**, 1184–1186.

Jemski, J. V., and Phillips, G. B. (1965). *In* "Methods of animal experimentation" (Ed. Gay, W. I.) Vol. 1, pp. 309–314. Academic Press, New York.

Johansson, K. R., and Ferris, D. H. (1946). *J. infect. Dis.*, **78**, 238–252.

Kenny, M. T., and Sabel, F. L. (1968). *Appl. Microbiol.*, **16**, 1146–1150.

Kruse, R. H. (1962). *Am. J. clin. Path.*, **37**, 150–158.

Mason, B. J. (1964). *Endeavour*, **23**, 136–141.

Oliphant, J. W., Gordon, D. A., Meis, A., and Parker, R. R. (1949). *Am. J. Hyg.*, **49**, 76–82.

Pattle, R. E. (1961). *In* "Inhaled particles and vapours" (Ed. Davies, C. N.) Vol. 1, pp. 302–309. Pergammon Press, London.
Perkins, F. T., Darlow, H. M., and Short, D. J. (1967). *J.I.A.T.*, **18**, 83–92.
Phillips, G. B. (1962). *Am. J. med. Technol.*, **28**, 291–295.
Phillips, G. B. (1965). *J. chem. Engng*, **42**(2), A117.
Phillips, G. B., and Reitman, M. (1956). *Am. J. med. Technol.*, **22**, 16–26.
Phillips, G. B., and Bailey, S. P. (1966). *Am. J. med. Technol.*, **32**, 127–129.
Phillips, G. B., and Runkle, R. S. (1967). *Appl. Microbiol.*, **15**, 378–389.
Reitman, M., and Phillips, G. B. (1955). *Am. J. med. Technol.*, **21**, 226–230.
Roe, F. J. C., and Glendenning, D. M. (1957). *Br. J. Cancer*, **10**, 327–362.
Silver, I. H. (1963). *Nature, Lond.*, **199**, 102.
Spinner, D. R., and Hoffman, R. K. (1968). *Appl. Microbiol.*, **8**, 152–155.
Sulkin, S. E., Long, E. R., Pike, R. M., Sigel, M. M., Smith, C. E., and Wedum, A. G. (1963). *In* "Diagnostic procedures and reagents" (4th ed.) pp. 89–104. American Public Health Association, New York.
Tomlinson, A. J. H. (1957). *Br. med. J.*, **2**, 15–17.
Trillat, A., and Kaneko, R. (1921). *C.r. hebd. Séanc. Acad. Sci., Paris*, **173**, 109–120.
Tully, J. G., Gaines, S., and Tigertt, W. D. (1963). *J. infect. Dis.*, **113**, 131–138.
Wedum, A. G. (1964). *Publ. Hlth. Rep., Wash.*, **79**, 619–633.
Wedum, A. G., Hanel, E., Phillips, G. B., and Miller, O. T. (1956). *Am. J. Publ. Hlth.*, **46**, 1102–1113.
Whitwell, F., Taylor, P. J., and Oliver, A. J. (1957). *J. clin. Path.*, **10**, 88–98.
Winner, H. I., and Quiney, A. (1953). *Lancet*, *i*, 376–377.

APPENDIX

List of manufacturers or distributors of some of the appliances referred to in Chapter

N.B. It is stressed that this list is included as a guide only, and does not imply that these are the only or best sources of supply.

Bourdillon Slit Sampler	C. F. Casella & Co. Ltd., Regent House, Britannia Walk, London, N1.
Air filters and portable recirculation units	Vokes Ltd., Henley Park, Guildford, Surrey.
	Microflow Ltd., 307, Lynchfield Road, Farnborough, Hants.
	Foramaflow Ltd., The Parade, Frimley, Surrey.
Safety Cabinets	Microflow Ltd.
	Foramaflow Ltd.

Rubber Pipette Discard Pot	P. B. Cow & Co. Ltd., Streatham Common, London, SW16.
Germicidal UV lamps	Philips Electrical Ltd., Century House, Shaftesbury Avenue, London, WC2. Engelhard Hanovia Ltd., Bath Road, Slough, Bucks.
Collison Spray	Aerosol Products Ltd., 26, Eccleston Square., London, SW1.
Ampoule Openers	Baird & Tatlock Ltd., Freshwater Road, Chadwell Heath, Essex.
Respirators, face shields, etc.	Siebe, Gorman & Co. Ltd., Neptune Works, Davis Road, Chessington, Surrey.

CHAPTER VII

Membrane Filter Techniques in Microbiology

J. G. Mulvany

Millipore (U.K.) Ltd., Wembley, Middlesex, England

I. CHARACTERISTICS OF MEMBRANE FILTERS

A. Introduction

Membrane filters are thin porous sheet structures composed of cellulose esters or similar polymeric materials. They act essentially as two-dimensional screens and as such, all particles, both biological and non-biological, which exceed the pore size are retained upon the surface of the filter from fluids which pass through (Fig. 1). In liquid filtration, a high percentage of particles somewhat smaller than the pore size are retained by secondary valence (van der Waals) forces, by random entrapment in the slightly tortuous pores, and by build up on previously retained particles. In gas filtration, because of the enormous specific surface and high resistivity of the filter, substantial electrostatic charges are generated and held by the filter. This charge prevents the passage of particles far smaller in dimension than the filter pore size. Thus, whilst the filter is capable of removing some particles smaller than the pore size the most important characteristic is that all particles LARGER than the pore size are positively retained on the filter surface, and lie in a plane where they can be readily examined and counted.

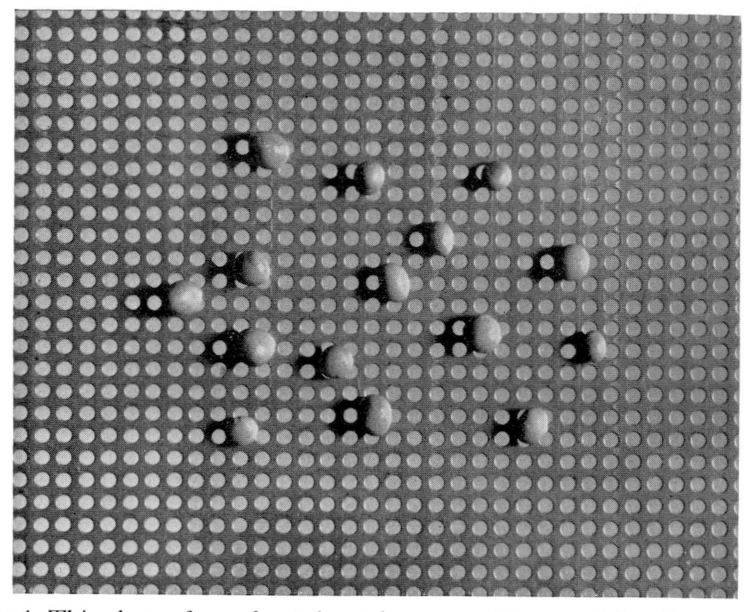

FIG. 1. This photo of a perforated metal screen and lumps of clay illustrates both the absolute surface retention and the integral structure of a Millipore filter.

Membrane filters differ generically from the class of filters known as "Depth Filters" among the most familiar of which are fibrous pads, papers, and diatomaceous earth. Depth filters consist essentially of a random orientation of matter in compacted mat or bed form. Fluid passes through the irregular channels defined by this orientation and particles are retained from the fluid by a combination of effects, but primarily by mechanical random entrapment and adsorption (especially in the sub-micron size range). The efficiency of depth filters depends on such factors as fibre or grain size, surface morphology, depth and degree of compaction, and other polar characteristics of the fibres or individual elements. Because of the random nature of the media, depth filters can never be assigned an "absolute" rating (unless concern is with virtual "boulders"). Even the most dense of useable depth media will have some passages well in excess of 1 μm.

B. Properties of membrane filters

1. *Introduction*

Membrane filters, as manufactured by the Millipore Corporation, are available in discrete pore sizes, ranging from 14 μm down to 0·025 μm. Membrane filters are also available from Gelman Instrument Company, U.S.A., and Sartorius-Membrane Filter, GmbH, Germany. The standard Millipore filters are composed of pure and biologically inert cellulose esters and are available in 12 discrete pore sizes as shown in Table 1. Flow rates for air and water are also shown and the conditions for these values is given below under "Flow Rate Determination".

TABLE I

Specification of membrane filters

Membrane filter Type	Pore size	Rates of flow	
		water†	air‡
SC	8·0 μm ± 1·4 μm	950	55
SM	5·0 μm ± 1·2 μm	560	35
SS	3·0 μm ± 0·9 μm	400	20
RA	1·2 μm ± 0·3 μm	300	14
AA	0·8 μm ± 0·05 μm	220	9·8
DA	0·65 μm ± 0·03 μm	175	8·0
HA	0·45 μm ± 0·02 μm	65	4·9
PH	0·30 μm ± 0·02 μm	40	3·7
GS	0·22 μm ± 0·02 μm	22	2·5
VC	100 nm ± 8 nm	3·0	1·0
VM	50 nm ± 3 nm	1·5	0·7
VF	25 nm ± 2 nm	0·5	0·3

† ml/min./sq. cm } Under 70 cm Hg
‡ litre/min./sq. cm } differential pressure.

2. *Porosity*

The pores in membrane filters are extraordinarily uniform in size. For example the total range of pore size distribution in a type HA, 0·45 μm, filter (Millipore) is ± 0·02 of a micro-meter. Each square centimetre of filter surface contains millions of capillary pores and the pore volume occupies approximately 80% of the total filter volume (Fig. 2). High porosity results in flow rates which are of the order of 40 times faster than flow rates through depth filters approaching the same particle size retention capability. For example, water will flow through a 0·45 μm, HA, filter at 38,400 litres/h/square metre, whilst a depth filter, with the same nominal retention efficiency, has a flow rate of only 750 litres/h/square metre of filter area (Osgood, 1967) under the same conditions.

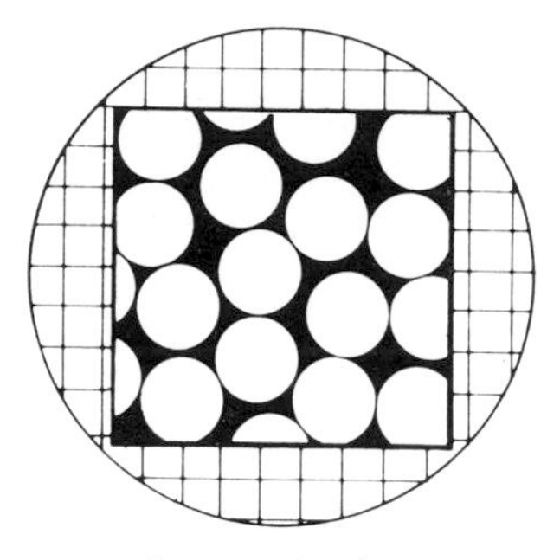

FIG. 2. This schematic approximates the large open volume (80%) of MF in proportion to solid material (20%).

3. *Chemical properties*

Millipore filters exhibit typical properties of cellulose esters with respect to chemical resistance. They are not attacked by water, dilute acids and alkalis, aliphatic or aromatic hydrocarbons, halogenated hydrocarbons or non-polar liquids; in fact, few fluids of interest to microbiologists will attack them. If a chemical compatibility problem exists, solvent resistant membrane filters are available in several pore sizes.

Since the membrane filter is free of biologically inhibitory factors it provides an optimum collection environment for micro-organisms. Any bacteriostatic agents present in the suspending fluid may be readily flushed free of the filter. There are no residual factors in membrane filters (Millipore) to block enzymatic systems or to interfere with proper organism identification by staining or culturing procedures.

4. *Thermal stability*

A membrane filter, made from cellulose esters, is stable in a dry state at temperatures up to 125°C in the presence of oxygen. Gradual decomposition takes place in air at temperatures above 125°C. They may be used in oxygen free atmospheres at temperatures up to 200°C. Membrane filters may be autoclaved satisfactorily at 121°C (15 p.s.i.) either individually for a period of 15 min. or if assembled in a filter holder, for a period which is specified for that particular holder.

5. *Colour*

White membrane filters (Millipore) are available in all twelve pore size types. Types AA (0·8 μm) and HA (0·45 μm), are also formulated in non-fluorescing black for examining contaminants under ultraviolet illumination and to provide a distinct optical contrast with materials whose colour and refractive index nearly match that of a white filter. White filters become transparent when their pores are filled with an immersion oil of matching refractive index. This property permits the examination of particles on the filter surface by common transmitted light microscopy methods with or without oil immersion objective lenses. The approximate refractive index for

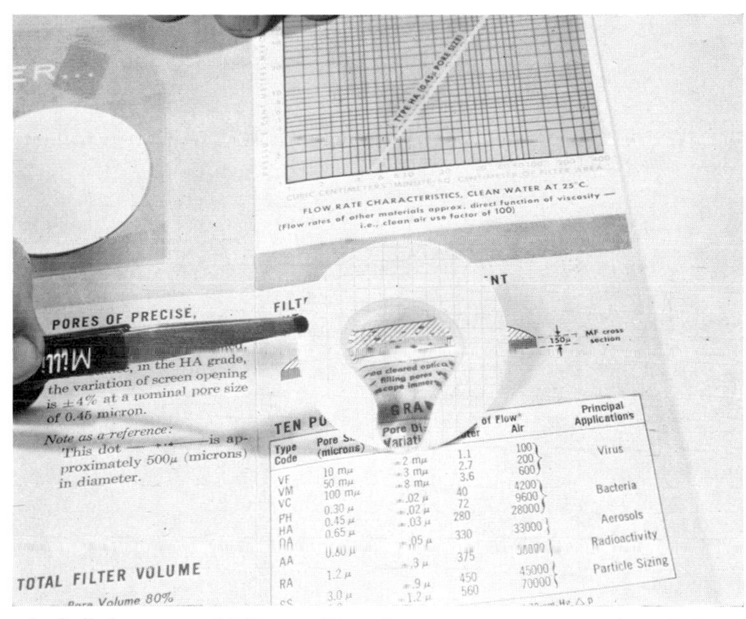

FIG. 3. Cellulose ester Millipore filters become transparent when their pores are filled with a fluid of matching refractive index.

the Millipore filter is 1·51 (see Fig. 3). Black membrane filters cannot be rendered transparent. Selected types of filters are available with imprinted grid marked surfaces thus facilitating statistical counts of particles or colonies on the filter suıface.

6. *Other properties*

The filter is an integral structure containing no fibres or particles which can work loose to contaminate a filtrate. It produces no ionic reaction with compatible fluids. If properly supported, Millipore filters will withstand at least 10,000 p.s.i. differential pressure without significant distortion of the pore structure. They require no treatment prior to use and are disposable after use.

C. Quality control procedures (by manufacturer)

1. *Pore size determination*

Skau-Ruska mercury intrusion measurements of pore radius are taken in which the formula—

$$pr = -2\gamma \cos \theta$$

relates pressure (p) and pore radius (r), to the (γ) coefficient of surface tension of mercury, and its contact angle (θ) with respect to the filter matrix. Pore size distribution curves are then plotted from these data and Fig. 4

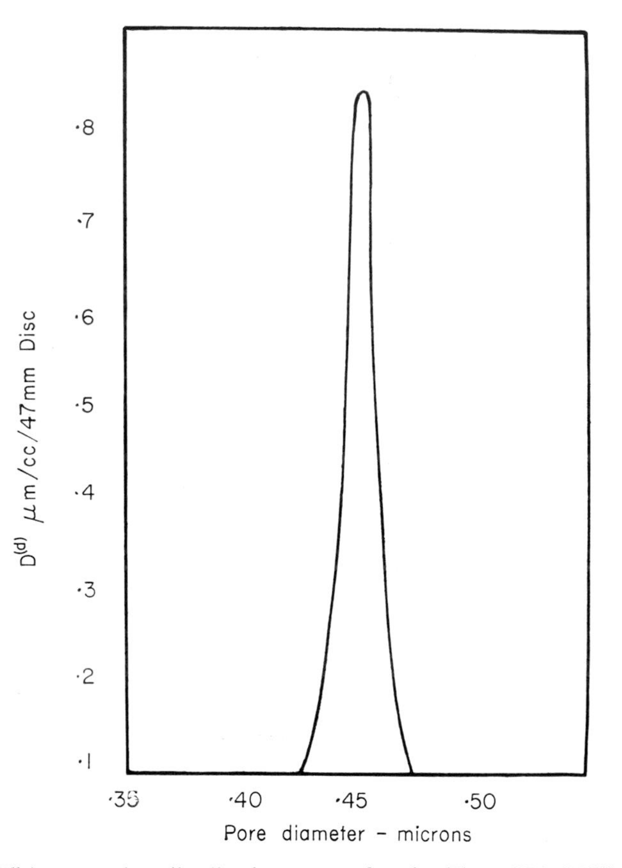

FIG. 4. This pore size distribution curve for the Type HA Millipore filter is derived from high-pressure mercury intrusion test data. Ordinate shows distribution of pores.

shows a typical result for the type HA (0·45 μm) filter. Pore sizes in the sample vary from 0·43 μm to 0·47 μm with the mean pore peaking at 0·45 μm.

Filtration tests are also run with contaminants of known type and size in which the filtrate is analysed to determine the largest particle size which can penetrate the filter.

2. *Bacterial growth determination*

For filters to be acceptable, cultures of organisms such as *Serratia marcescens* on the filter must compare favourably with agar plate control cultures for viable count, colony formation, and pigment production.

3. *Flow rate determination*

The time in seconds is recorded for 500 ml of "particle free" distilled water to flow through a filter disc at 25°C and 70 cm Hg differential pressure. Flow rates are expressed in terms of ml/min/sq cm of filter area. Air flow rates are measured as the flow in litres/min/sq cm of filter area with a pressure differential of 70 cms mercury. Table I shows values for the standard Millipore filters.

4. *Sterile filtration test*

A suspension of approximately 5×10^9 cells/ml of a known species and size of bacteria is passed successively through a test filter and a control filter. Passage of any cells into the filtrate from the test filter will be evident on the control filter upon subsequent incubation. To control the type GS, 0·22 μm, filter a suspension of *Pseudomonas* sp. is employed, and for the HA, 0·45 μm, a culture of *Serratia marcescens*.

II. TECHNIQUES FOR MICROBIOLOGICAL ANALYSIS (DETECTION AND ENUMERATION)

A. General techniques

1. *Introduction*

The use of membrane filters in microbiological analysis has many significant advantages over conventional methods. The two most important being—

(a) All micro-organisms present in large volumes of fluid can be concentrated on the filter surface.

(b) The suspending fluid, if inhibitory, can be thoroughly flushed from the filter.

These advantages are seen to their full extent in the sterility testing of antibiotics, where to overcome the inhibitory effects, the test was previously performed by very large dilutions of antibiotic in nutrient medium. Even with very large dilutions, micro-organisms still suffered from the inhibitory effects of the antibiotic and in many cases the results of such tests were meaningless. Another inherent advantage of the membrane filter technique is now immediately obvious, in that there is a large saving on nutrient medium as gross dilutions are not necessary.

Fortunately, new media and growth conditions are not required for culture on membrane filters. The filter, bearing organisms on its surface, may be placed directly on conventional solid media (Fig. 5) or placed on a pad

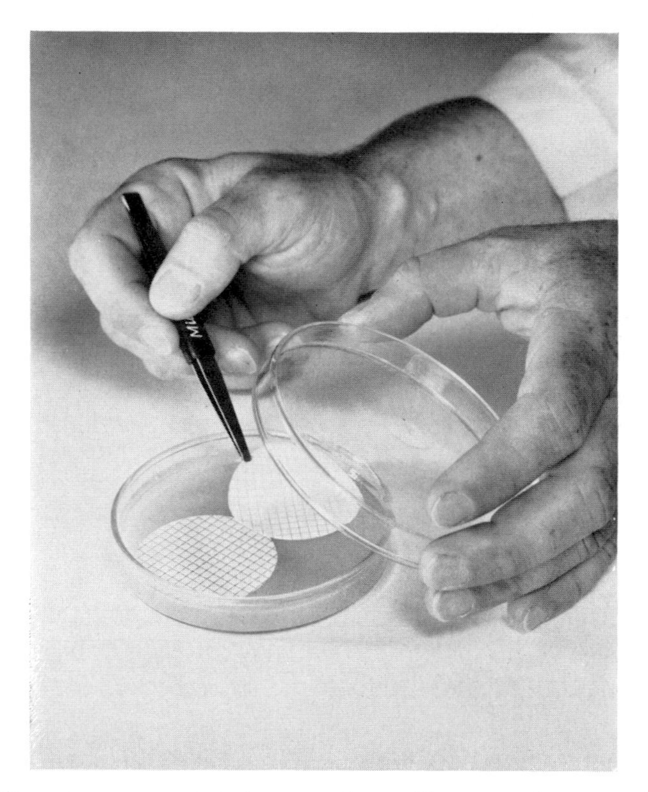

FIG. 5. Organisms concentrated on membrane filters may be grown directly on the filters placed on conventional agar plates.

saturated with broth medium (usually at double strength) Fig. 6; or plunged into a liquid growth medium, where detection rather than enumeration is desired, as in sterility testing. Colonies may be readily "picked" for further selective growth or staining, and both aerobic and anaerobic culturing techniques are in routine use.

The importance of staining in the microbiological field is well demonstrated. Staining techniques are particularly applicable when organisms—

either as individual cells or as developed colonies—have been isolated on the surface of a membrane filter. Such staining may be carried out to increase visual contrast between colonies and the filter; to differentiate specific organisms in a mixed culture; or stain individual cells for ease of study by transmitted light microscopy. In general, conventional staining techniques have been employed and are reported in the literature (Millipore Bibliography), some examples are detailed under Section 4 below.

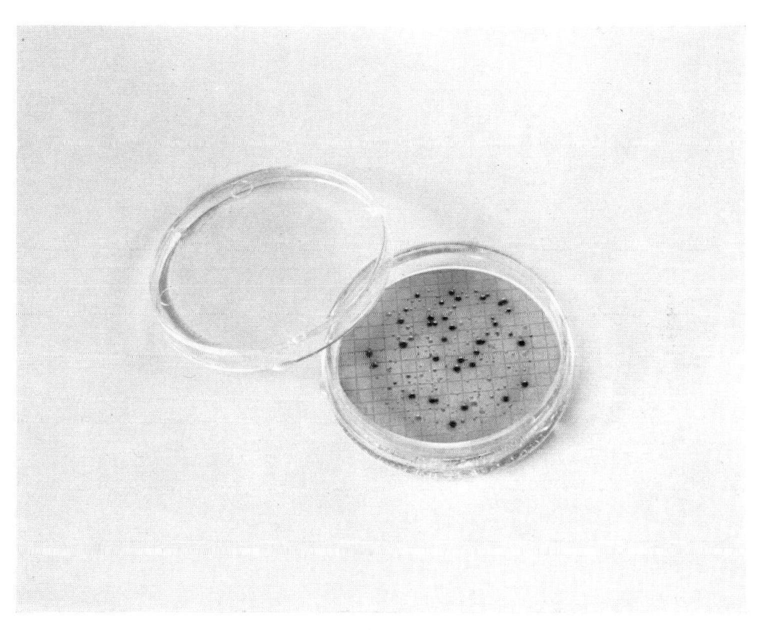

FIG. 6. Coliform colonies showing sheen after incubation on a pad saturated with MF-Endo medium. Colonies may be picked for sub-culture.

Finally, and of great importance, are the economic considerations of the membrane filter technique. Undoubtedly there are large savings to be made both on medium and glassware and, by virtue of simplicity, the techniques are less cumbersome than conventional methods.

2. *Apparatus*

A variety of filtration apparatus is available for microbiological analysis, however, the basic standard tool for many years has been the Pyrex Filter Holder (Millipore) as shown in Fig. 7. This holder employs a 47 mm diameter filter for liquid filtration in conjunction with a vacuum flask. Aseptic assembly of filter and filter holder after autoclaving is normal practice, and apparatus should be wrapped and autoclaved in the conventional manner.

A modification of this holder employs Teflon coating of the ground glass surfaces which clamp the filter, and this device may be sterilized by steam under pressure with a filter in place, providing the unit is thoroughly dry. The 47 mm diameter filter is placed between the funnel and the base and a special rubber cover (Fig. 8) is placed in the mouth of the funnel to equilibrate the pressure across the membrane filter during autoclaving. The entire assembly is then autoclaved for 15 minutes at 121°C.

FIG. 7. The sample is poured directly into the graduated funnel.

A new aseptic filtration system known as "Sterifil" has been recently introduced and is shown in Fig. 9. Sterifil is an autoclavable plastic filter holder and receiver flask combined in a closed system. Micro-organisms from liquid samples can be concentrated on the surface of a 47 mm filter placed in the holder without exposure to environmental microbiological contamination. Luer slip ports in the cover of the holder make it possible to introduce sample solutions, rinses, wetting agents and culture media with aseptic security. A small filter holder containing a microfibre glass filter disc is inserted in a cover port to vent the sterifil holder with sterile air (see Fig. 9). Ports not being used are covered with gum rubber caps. The Sterifil may be autoclaved with the filter in place; and following sample filtration may be filled with nutrient medium and incubated with the filter in place. This is a significant advantage in sterility testing as there is no possibility of

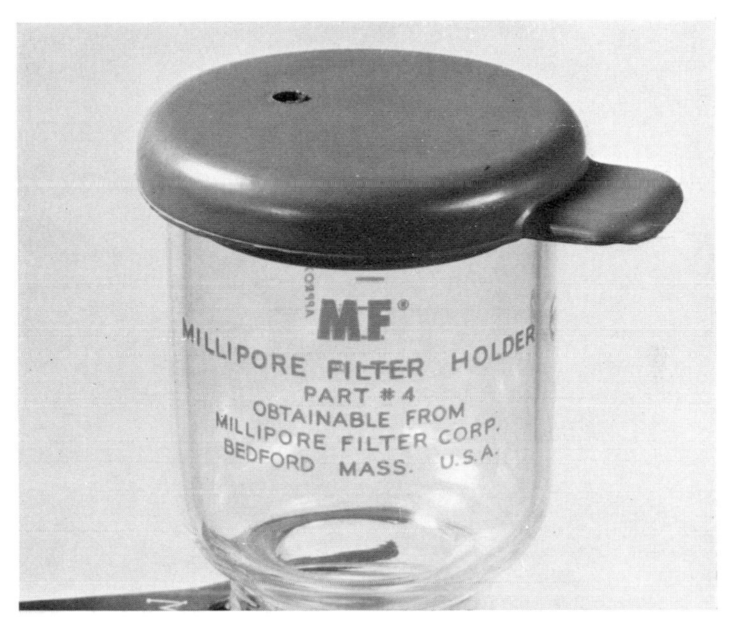

FIG. 8. A vented cover must be placed on the funnel when autoclaving an assembled filter holder and filter to avoid damage to the filter.

exposing the filter to environmental micro-organisms which could cause false positive results. Figs 10–13 show the versatility of the Sterifil filtration system in micro-biological analysis.

A special membrane filter (Millipore) has also been developed for use with the Sterifil system. The filter has a hydrophobic area at the periphery, approximately 3 mm wide, Fig. 14. This hydrophobic edge prevents the diffusion of solution beyond the filtration area where it could remain after flushing and, if inhibitory, interfere with the subsequent growth of collected organisms. The standard Pyrex Filter Holder suffers from the diffusion of solutions into the filter area covered by the ground glass surface, and it is impossible to adequately flush away this solution. There is an 80% reduction in absorption of solution with the edge-hydrophobic filter in the Sterifil, compared to the standard filter in the Pyrex Filter Holder. If the absorbed solution is strongly inhibitory to growth of organisms there will be an increase in the sensitivity of microbiological analysis when using the Sterifil system.

Some years ago "Field Monitors" were developed (Millipore) as a means of conducting microbiological analysis of water under field conditions. These devices have now been adopted in virtually every other microbiological application for which standard membrane filters and "laboratory-type"

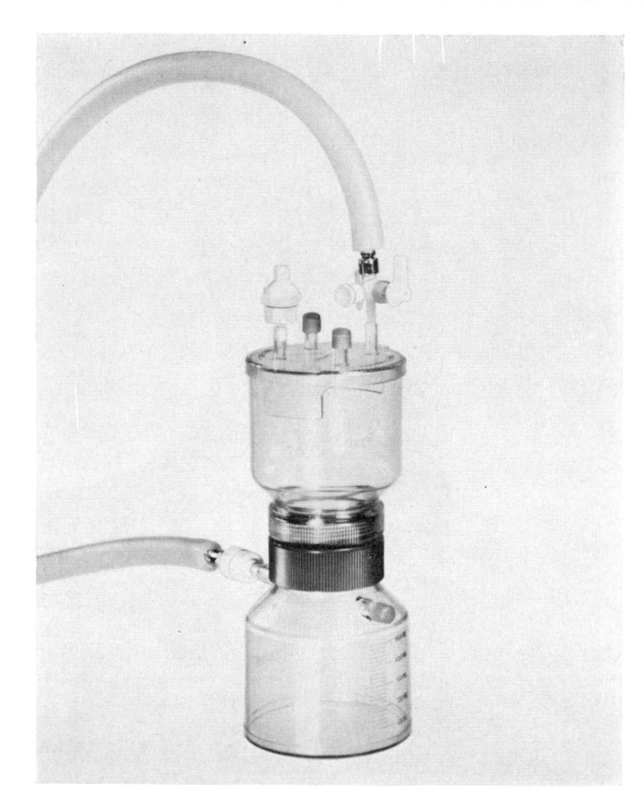

FIG. 9. Sterifil Aseptic Filtration System.

apparatus are used. Field Monitors (Fig. 15) are disposable plastic filter holders with a membrane filter tightly sealed between the top and bottom halves. A cellulose absorbent pad is placed beneath the filter to support it, and, if required, to act as a reservoir for nutrient media after filtration. Each half of the monitor is equipped with a port, covered (when not in use) by a plastic cap. Micro-organisms will be retained on the surface of the filter from fluids passing through and after filtration the monitor can be capped and transported to the laboratory for examination. Or, a nutrient solution may be introduced beneath the membrane filter by thoroughly

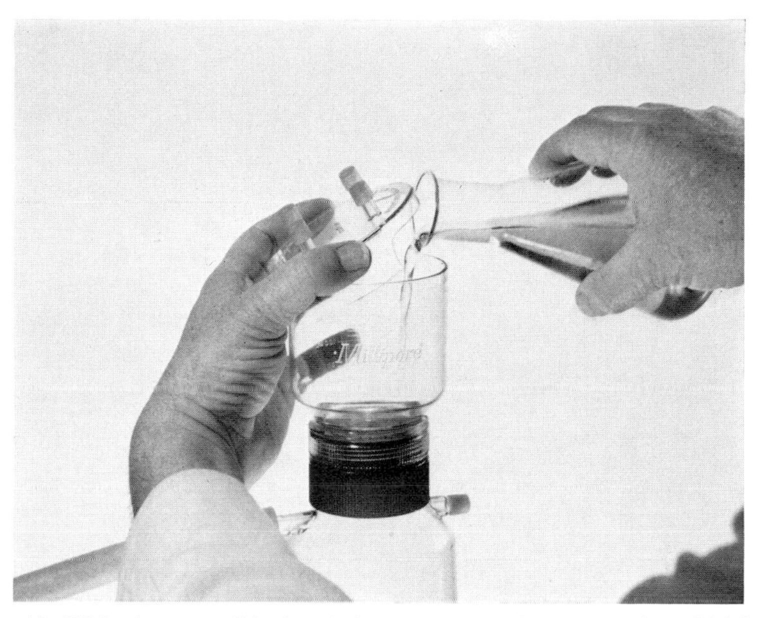

Fig. 10. With the cover lifted and the ports capped, test samples of biological solutions may be poured aseptically into the Sterifil funnel.

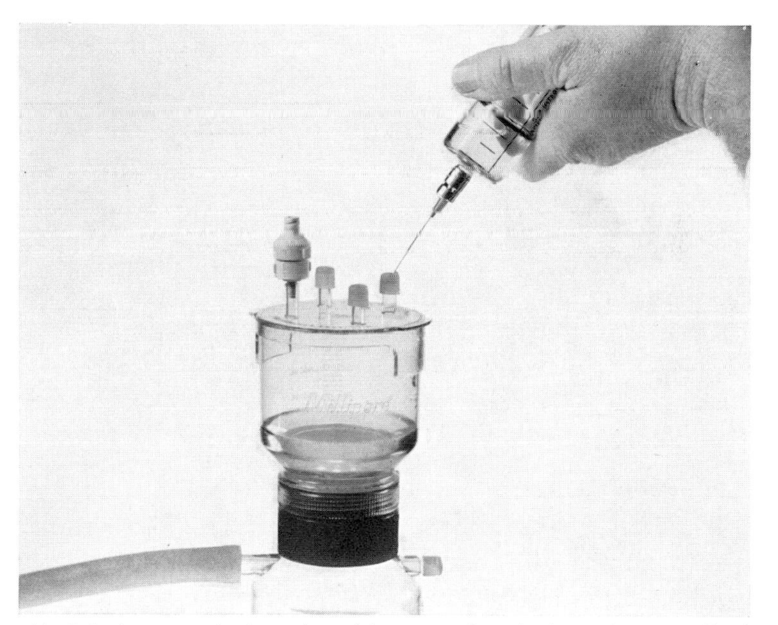

Fig. 11. Solutions may be introduced from a syringe by inserting a needle through a rubber cap covering one of the ports.

FIG. 12. After sample filtration and the introduction of broth media, the Sterifil Holder may be incubated with the filter in place. The Swinnex vent may be left in place if desired.

FIG. 13. The covered Receiver Flask, with both connections capped, may be used to store sterilized effluents following filtration. Side arm connections may be used as pouring spouts.

soaking the cellulose pad, and then the monitor may be incubated in the conventional manner to produce visible and identifiable colonies. The "Clinical Monitor" version of the "Field Monitor" incorporates a ring between its top and bottom halves to hold the membrane filter during

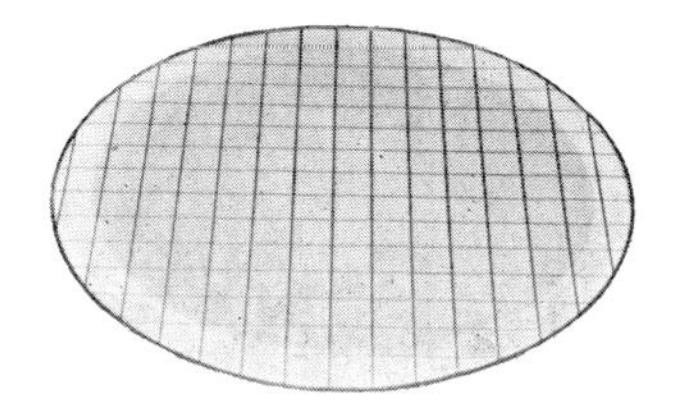

FIG. 14. Millipore filter wet with water to show hydrophobic edge area.

FIG. 15. Field Monitors are disposable plastic Millipore filter holders.

examination and to avoid accidental contamination. The monitor most often used in microbiological and clinical procedures is made of non-inhibitory polystyrene and is provided sterile, ready for immediate use.

Most microbiological and clinical analysis work employs either a white or black 0·45 μm (HA) membrane filter, generally with grid marked surfaces. A 0·65 μm filter is specified for milk analysis, however, to achieve higher sample throughputs.

3. *Culturing*

The membrane filter technique is readily adaptable to conventional methods of culturing. In sterility testing of antibiotics for example, the membrane filter is submerged in liquid thioglycollate broth for incubation of anaerobes and aerobes. A very common technique, as mentioned above, is to place an absorbent pad in a 47 mm plastic Petri dish and to saturate the pad with 1·8 ml of broth medium. The membrane filter can then be placed on top of this pad and the liquid medium will diffuse through the pores of the filter so providing organisms trapped on the surface with nutrient. After incubation, the top of the Petri dish is removed and the colonies are counted. If colonies are numerous and small, it is helpful to increase visual contrast by "shadow casting" with a low angle light, or to use a contrast stain as mentioned below. If cultured filters are to be transported or retained for future reference, they should be placed in a plastic Petri dish on a pad saturated with a preserving solution of 15% glycerine, 50% formaldehyde, 35% water. The dishes and filters may then be safety transported in this "wet" state, thus preserving colony morphology; or the filters may be dried after ten minutes for storage in cellophane envelopes.

4. *Staining*

Since most diagnostic staining procedures decrease viability, it is preferable to pick colonies for subculture prior to staining. When using the two contrast stains described below, however, colonies may be picked up to 15 min. after staining. Using an agar plate culture this technique is not possible.

In all staining methods it is important that solutions of dye, buffer, rinses, etc., to be passed through the sample filter, should be first filtered through an 0·45 μm filter. This step ensures that all particles larger than 0·45 μm are removed and thus will not obscure any organisms collected on the surface of the filter.

(a) *Staining filter to contrast with colonies.* A 0·01% w/v solution of Malachite Green Oxalate in distilled water is employed. After incubation and whilst still in a Petri dish, the surface of the filter is gently flooded with 2–4 ml of dye solution (Fig. 16). After 8–10 seconds the excess dye is gently poured off (Fig. 17), leaving colonies appearing white or yellow against a green filter background.

(b) *Staining colonies to contrast with filter.* 3 g of Methylene Blue dye are added to 300 ml of ethyl alcohol and the solution mixed with 1 litre of dilute potassium hydroxide (0·01% by weight). A fresh absorbent pad is placed in a clean Petri dish and soaked with 1·8 ml of dye solution. The filter with

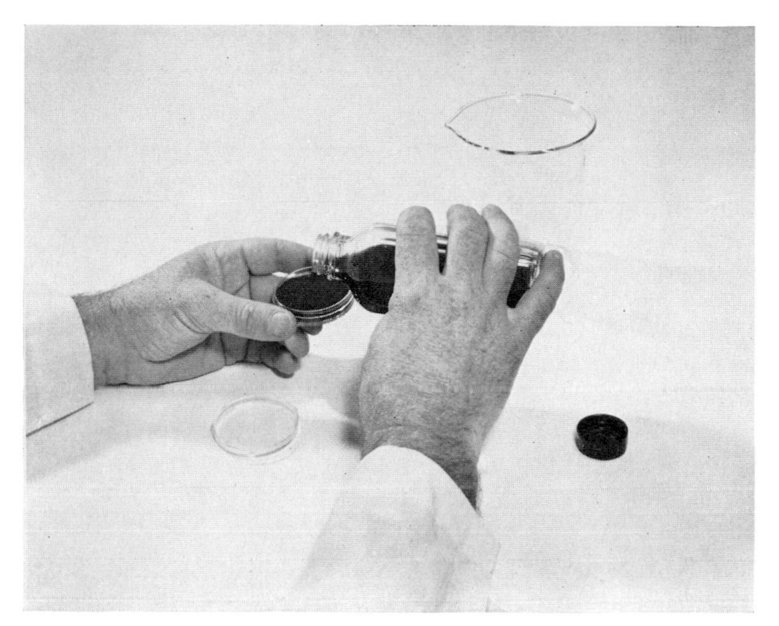

FIG. 16. Staining filter: Gently flood surface of filter with a solution of Malachite Green dye.

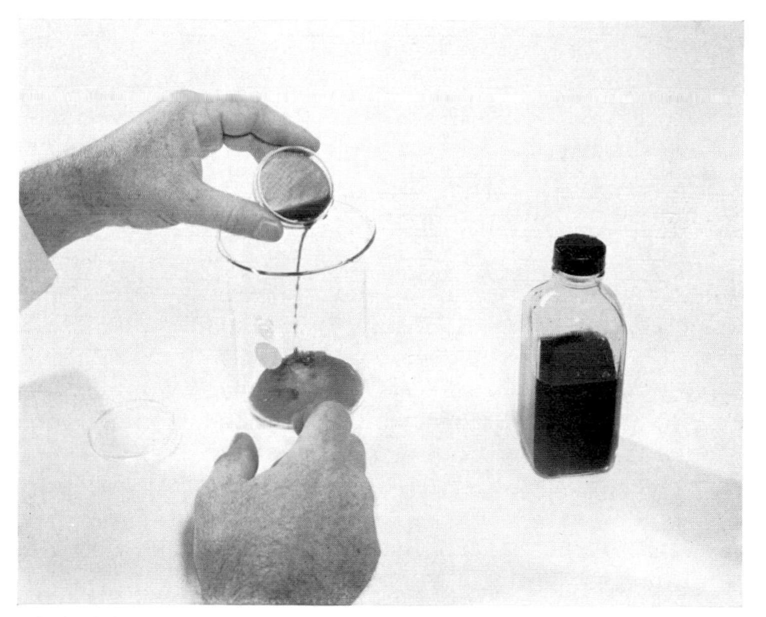

FIG. 17. Staining filter: Pour off excess dye. Colonies appear white or yellow against the green filter background.

developed colonies is transferred to the saturated pad and after a period of 15 min. the colonies will appear dark blue against a lighter coloured filter.

(c) A simple stain procedure for microscopy employs a 1% aqueous solution of Gentian Violet which can be used to rapidly identify bacteria and yeast collected on the surface of a 0·45 μm filter. The method employs the use of a Hydrosol Stainless Filter Holder as shown in Fig. 18, and all staining and

FIG. 18. Millipore hydrosol filter holder.

clearing is performed in the holder. After filtration the filter holder and filter are rinsed with 100 ml of filtered phosphate buffer (pH 5·2). 5 ml of filtered 1% Gentian Violet dye solution are then pipetted on to the filter and a contact time of 1 min. allowed before vacuum is applied to draw the dye through the filter. Successive rinsings with 100 ml phosphate buffer, 20 ml propanol, 20 ml propanol : xylol (1 : 1 ratio), 20 ml xylol are carried out. Finally, the filter is placed on a 2 × 3 in. glass slide, covered with immersion oil, and the organisms observed using an oil immersion objective. This method is very rapid but does not give such good resolution as the following method, which is more lengthy.

The technique employs a dye made by adding 60 ml 5% (w/v) aqueous phenol, 5 ml ethanol, 4 ml glycerol, 125 mg basic fuchsin (95% (w/v) dye content), and 120 mg methylene blue to 1 litre of distilled water. The dye solution is stirred, whilst being gently heated for ten minutes and then allowed to stand for 12 h in a refrigerator. Before use, the dye solution is filtered through a coarse grade asbestos pad followed by a 0·45 μm (HA) filter.

After sample filtration, the filter is left in the holder and 10 ml of dye solution is added to the funnel and a contact time with the filter of ten minutes is allowed. Then vacuum is applied to draw the dye through the filter which is subsequently flushed repeatedly with filtered distilled water until the filtrate is colourless. Finally, the filter is cleared with immersion oil for microscopy in the same manner as described for the above Gentian Violet stain procedure.

(d) As mentioned above, the staining techniques employed with filter collected organisms are, in principle, the same as conventional methods. For example the staining of acid-fast bacilli collected on the filter surface employs a modified Ziehl–Neelsen method. The stains are prepared as follows—

Fuchsin stain—3 g of basic fuchsin are added to 100 ml ethanol and stirred. 10 ml of the fuchsin solution are made-up to 100 ml with a 5% (w/v) phenol solution in distilled water. (The basic fuchsin concentration used is only one-third of that used in the standard method.)

Counter stain—0·3 g Methylene Blue (80% (w/v) dye content) is added to 30 ml aqueous ethanol (80% (v/v)) and 100 ml distilled water (the standard Ziehl–Neelsen method employs an alkaline counter-stain by using 0·01% potassium hydroxide solution instead of distilled water).

The decolourizing solution is made by adding 30 ml concentrated hydrochloric acid to 970 ml ethanol.

After the sample has been filtered, the filter is flushed with 10 ml filtered physiological saline. 10 ml of fuchsin stain is then added to the funnel and left in contact with the filter for 30 minutes, when a vacuum is applied and the stain pulled through the filter. (In the conventional method the slide is flooded with fuchsin stain, heated, and left in contact for five minutes. The filter method employs a 30 minute contact time at ambient temperature.) The next step is to add 50 ml of decolourizing solution, leave in contact for 2 min, and filter before flushing with decolourizing solution until the filter appears white. (In the conventional method 20% sulphuric acid is used for decolourizing; this treatment would damage the filter matrix.) The filter is then flushed with 20 ml of physiological saline and finally 20 ml of counterstain is added to the funnel and immediately filtered. The filter is removed from the holder and air dried before placing on the surface of 2 ml of immersion oil in a Petri dish. The filter is finally placed on a 2×3 in. slide for oil immersion microscopy.

(e) A method of staining in suspension before membrane filtration has been reported as an improved procedure for staining *Bacillus subtilis* var. *niger* and vegative cells of *S. marcescens* (Shanahan *et al.*, 1956).

Four drops of 2% Crystal Violet dye are added to a given concentration of spores contained in 10 ml of diluent; three drops of Triton-20 (10% solution) (Rohm and Haas Product) are also added. The mixture is brought to a boil, the heat source is removed, and the mixture allowed to stand for 8 min., after which it is cooled and filtered through a 0·45 μm (HA) filter. The filter is then washed with 5 ml of equal parts of 50% ethanol and distilled water. This clears the membrane of dye. The filter is then air-dried and mounted on a slide with oil of 1·508 refractive index for microscope examination. The spores of *B. subtilis* var. *niger* are observed to be a deep violet and may be easily counted.

Staining of dye resistant *S. marcescens* suspended in tryptose-saline is accomplished as follows—

Three drops of 2% (w/v) alcoholic fuchsin and 2 drops of 5% (w/v) phenol solution are added to 10 ml of cell suspension and the mixture heated gently for five minutes. The heated and then cooled suspension is filtered through a 0·45 μm (HA) filter, which is then washed twice with distilled water, and prepared for examination as described above. *S. marcescens* stains red while the filter remains colourless.

These procedures are advantageous in certain instances since the heat applied results in a more intense staining of the micro-organisms.

B. Clinical microbiology

1. *Introduction*

No area of membrane filter technology has afforded greater opportunity, or greater challenge, than the application to clinical fluids where the unique characteristics of the filter are so clearly useful. The filter technique is especially valuable to the pathologist in those instances where—

(a) The organisms of specific interest are of low concentration in large volumes of fluid.

(b) The suspending fluid is bacteriostatic, as in the case of spinal fluids and urine from antibiotic-treated patients. These agents may be easily flushed away through the filter, leaving the organisms on the filter surface for culturing free of inhibitory effects.

(c) The filter method allows virtually all the suspending liquid being sampled to be washed free from the organisms present in that liquid. In nearly every clinical test, the liquid being sampled will have some inhibitory action on the growth of organisms. For these reasons more rapid growth rates are possible with the membrane filter method, leading to earlier diagnosis of pathological conditions. (Winn *et al.*, 1966).

(d) The use of membrane filters, as in total viable counts of urine, avoids time consuming dilutions.

The membrane filter is easily clogged by proteinaceous and colloidal material which can quickly form a layer over the surface and prevent filtration. Thus, the direct filtration of body fluids often calls for some ingenuity to avoid filter clogging from suspended or non-fluid components so prevalent in these materials. The acceptable method of fluid pre-treatment, where large volumes must be filtered, will often be dictated by the organisms to be recovered and by the fluid itself, but in general five techniques have been suggested.

2. *Pretreatment*

Table II shows typical body fluids which have been examined by the membrane filter technique, and also suggests the method of pretreatment to allow rapid filtration of the sample.

TABLE II

Treatment for various clinical fluids

Typical Fluid	Typical volume	Treat-ment
Sputum	5–15 ml	A
Nasopharangeal washings	10–50 ml	A
Gastric washings	20–30 ml	B
Pleural fluids	20–30 ml	A
Ascitic fluids	15–25 ml	A
Bursal fluids	10–20 ml	A
Urine—24 hr sample	1–3 litres	B
Urine—Fresh sample	100 ml	C
Spinal fluids	2–10 ml	D
Whole blood	5 ml	E

(a) *Treatment A* (*Mucoprotein Digestion*). The sample, for example, 10 ml Sputum, is collected in a sterile tube and 30 ml of N-Acetylcysteine (NAC)—sodium carbonate solution is added. (This digestant is prepared by adding 5 g of NAC to 1 litre of a 4% solution of sodium carbonate in distilled water). The tube is capped and agitated on a vortex mixer for 30 seconds, then allowed to stand at room temperature until mucolysis is complete; (with sputum this will require 15–30 mins). 10 ml of sterile 1% Triton X–100 solution (Rohm and Haas Product) is added to the tube and mixed by inversion. The mixture is then ready for filtration through a 0·45 μm membrane filter placed in a Sterifil Holder.

(b) *Treatment B.* (*Large Volume, Heavy Detritus, e.g., Gastric Washings*). The sample is allowed to settle for 12 h in a refrigerator and then filtered through a sterile cheese cloth to remove the heavy debris. A 20–30 ml representative aliquot is removed and treated as in method A (or C for urine).

(c) *Treatment C.* (*Normal Urine Samples*). A 100 ml sample of urine is added to 400 ml of an autoclaved 0·1% solution of Triton X-100 held in a water bath at 45°C. The entire 500 ml volume is then immediately filtered through a sterile assembly containing a 0·45 μm filter, for culturing or staining.

(d) *Treatment D* (*Small Volume, Clear Samples, e.g., Spinal Fluids*). No pretreatment is required and the sample may be immediately filtered as drawn and before clotting initiates. The 0·45 μm filter should be flushed with 20 ml of sterile physiological saline.

(e) *Treatment E* (*Whole Blood*). Recent attempts to establish more effective techniques for filtering blood using membrane filtration have been thwarted by rapid clogging of the pores of any bacteria-retentive filter. This is hardly surprising when it is recalled that the diameter of a red blood cell is approximately 7 μm, while the effective pore size of a filter that will retain bacteria is 0·45 μm. The procedure given here is still in an experimental stage and awaits further clinical confirmation before it can be used with complete confidence in all situations. Five ml of the whole blood sample, with a small amount of anticoagulant, is lysed in a sterile, closed Sterifil Filter Holder with a mixture of 50 ml 0·05% Triton X-100 solution, and 50 ml of 0·8% (w/v) sodium carbonate solution. This destroys the cell structure and the liquid becomes completely clear enabling filtration to be readily performed through a 0·45 μm membrane filter, and leaving on the filter surface only the bacteria contained in the blood sample. After filtration, the filter is flushed with normal saline to remove all sample traces that might inhibit subsequent bacterial growth on the membrane filter. The filter used is the special edge-hydrophobic type, mentioned above, having a water-repellent rim.

3. *Culturing for identification and enumeration*

After filtering the sample through a sterile 0·45 μm membrane filter, the filter is transferred aseptically to the surface of nutrient agar in a Petri dish, or to a plastic Petri dish containing a sterile absorbent pad saturated with 1·8 ml of double strength broth. The dish is then inverted and incubated in the conventional manner (usually for 24 h at 37°C). One of the contrast stains described earlier will facilitate enumeration at a low magnification. Colonies may also be picked for sub-culturing.

4. *Specific applications of the membrane filter technique.*

(a) *Rapid diagnosis of bacteraemia.* To fully utilize the advances in antibiotic therapy, early identification of the offending organisms in blood is essential. To this end, the membrane filter technique is an invaluable supplement to present methods of analysis. Growth of colonies is normally well developed within 24 h and in negative tests no growth has been observed after 72 h incubation. One of the significant disadvantages of the current standard method, that of inhibition of bacterial growth in broth culture due to the normal bacteriocidal properties of blood, is overcome. After filtration of 5 ml of the blood sample as detailed in Treatment E (above) the flushed filter is aseptically transferred to Casman's blood agar and incubated in an atmosphere of 5% carbon dioxide, so providing optimum conditions for both aerobic and anaerobic types. Casman's blood agar is prepared as follows—

"Add 8·6 g of Casman's agar base (2% tryptose, 0·3% beef extract, 0·5% sodium chloride, 1·5% agar and 0·3% dextrose) to 200 ml of distilled water, mix thoroughly, heat to boiling with frequent agitation and boil for 1 min. Sterilize by autoclaving at 121°C for 15 min. Prepare a blood solution by lysing 0·2 ml of sterile blood with 0·6 ml of sterile distilled water. Remove the Casman base medium from the autoclave and cool to 45°–50°C. Add 10 ml of sterile whole blood and 0·3 ml of the blood solution. Mix gently, and pour into 60 mm Petri dishes. The quantity prepared will be sufficient for 10 to 12 plates".

Research work carried out on an earlier method of bacteraemia identification using a membrane filter technique (Winn *et al.*, 1966) showed that after 24 h of incubation, extra cellular bacteraemia could almost always be ruled out if the filter was negative. This was not the case in parallel broth cultures where growth was often delayed. Typically, the availability of filter colonies permitted probable identification of *Staphylococcus aureus* isolates after a mean period of 17 h compared with 42 h for the parallel broth cultures. Fig. 19 shows a typical colony development after 24 h incubation at 35°C.

(b) *Detection of* Mycobacterium tuberculosis *in sputum.* Probably one of the most difficult tasks facing a clinical laboratory is that of isolating *M. tuberculosis* from sputum. With the recent advance in Mucoprotein digestion as detailed in treatment A., the task of isolating *M. tuberculosis* from sputum (10 ml) is now comparatively simple. After digestion as in treatment A. (above) the mixture is filtered through a 0·8 μm membrane

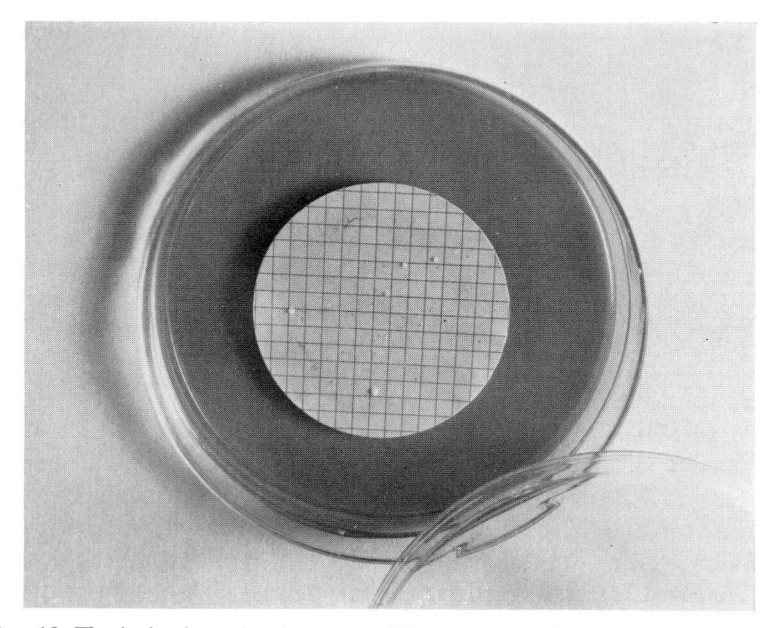

FIG. 19. Typical colony development of *S. aureus* after 24 h incubation at 35°C.

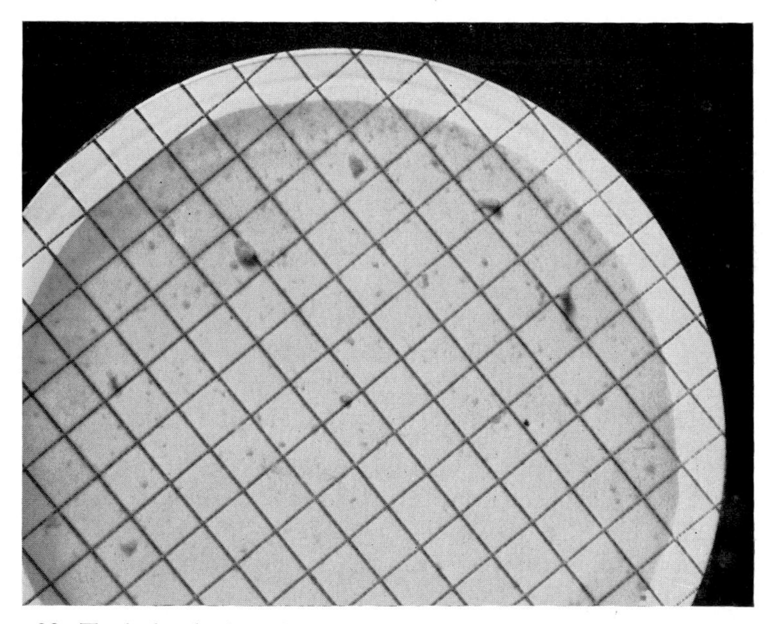

FIG. 20. Typical colonies of *M. tuberculosis* from sputum, after isolation and culturing on MF-Millipore filter.

filter which is then flushed with 30 ml of sterile normal saline. All bacteria will be retained on the surface of the filter which is aseptically transferred from the Sterifil holder to a sterile Petri dish containing Lowenstein Jensen medium with 1·5% Tween-80 (Baltimore Biological Laboratory, Baltimore, Maryland, U.S.A.). The Petri dish is covered, inverted, and incubated in an atmosphere of 10% carbon dioxide at 35°C. Identifiable colonies are often observable as early as the fifth day, and are illustrated in Fig. 20. Haley and Arch (Haley and Arch, 1957) showed that *M. tuberculosis* could be positively identified using the membrane filter technique in an average of five days; in contrast to an average of approximately 27 days when material is conventionally cultured on Lowenstein–Jensen medium, and 18 days when it is conventionally cultured on Tarshis blood agar (Tarshis and Frisch, 1951) (see Table III).

(c) *Viable count of uncatheterized urine*. Total viable counts in urine by conventional plating procedures require extensive dilutions to detect a pathological condition of 10,000–100,000 organisms/ml. A membrane filter technique employing a Clinical Monitor (Millipore) is more direct and simple. With a flamed bacteriological loop, 0·01 ml of urine is placed in a sterile tube containing 10 ml of sterile Brain Heart Infusion Broth and the

TABLE III

Number of days required for growth of tubercle bacilli from sputum

Specimen number	Membrane filter	Lowenstein Jensen	Tarshis' Blood agar
1	6	sterile	sterile
2	4	23	19
3	4	24	14
4	5	24	19
5	4	23	14
6	3	sterile	15
7	6	28	22
8	4	13	13
9	6	17	17
10	5	34	26
11	6	27	21
12	6	23	13
13	7	34	20
14	6	27	19
15	5	35	14
16	5	40	19
Average	5	26·6	17·7

whole is well shaken. The plugs are removed from the Clinical Monitor and a syringe and valve assembly is attached to the outlet of the monitor. 2 ml of the inoculated medium is drawn into a sterile volumetric pipette which is inserted into the inlet hole of the monitor. The entire 2 ml of the medium is filtered through the monitor by means of the syringe (Fig. 21)

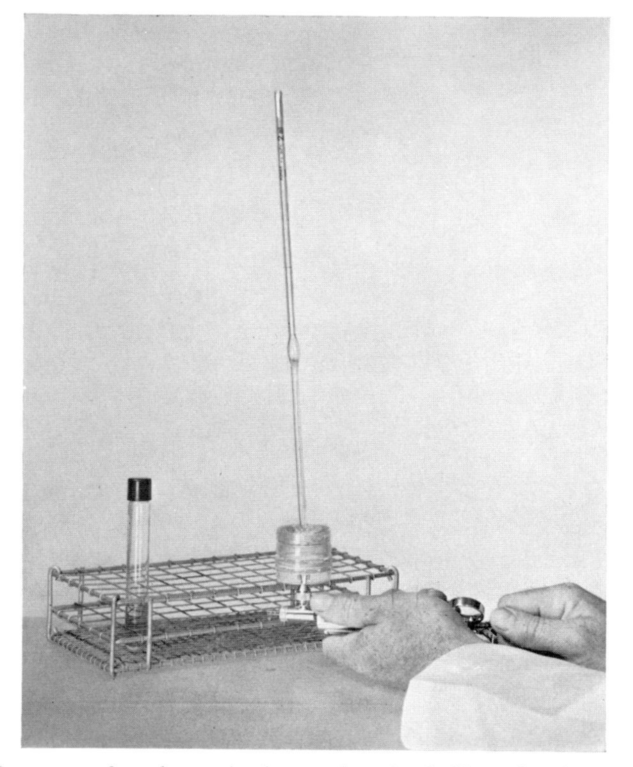

FIG. 21. By means of a volumetric pipette, draw 2 ml of inoculated medium through the Monitor with the syringe.

and filtration is stopped just as the last drop of fluid disappears from the filter surface, so that sufficient medium will remain in the underlying absorbent pad to support organism growth. The monitor is then recapped, inverted and incubated at 37°C for 24 h. A contrast stain may facilitate colony counting, and the number should be multiplied by 500 to obtain results for 1 ml of urine sample.

(d) *Other pathogens*. The literature contains many additional techniques of clinical significance (Millipore Bibliography). Methods for the isolation of *Bacillus anthracis*, *Brucella*, *Coccidioides immitis*, *Histoplasma capsulatum*,

Neisseria gonorrhoea, *Pasteurella*, *Pneumococcus*, *Salmonella*, *Shigella*, *Vibrio*, and many other organisms are referenced (Millipore Bibliography). In general, any medium in common clinical use may be used as a nutrient substrate for the membrane filter techniques, providing a degree of precision and sensitivity not obtainable with conventional methods—due largely to the ability to concentrate even a few organisms from large fluid volumes, and to culture these organisms free from the effects of bacteriostatic agents.

5. *Airborne organisms*

Spores from fungi and certain spore forming bacteria may be collected without significant loss of viability by direct filtration impingement on a 0·8 μm membrane filter. However, many airborne micro-organisms—those which do not exhibit a protective capsule or cannot revert to a protective spore stage—desiccate rapidly when impinged on a solid surface and cannot be collected viably on a dry filter surface. Thus, sampling for all forms is best done by first impinging the organisms into a liquid medium

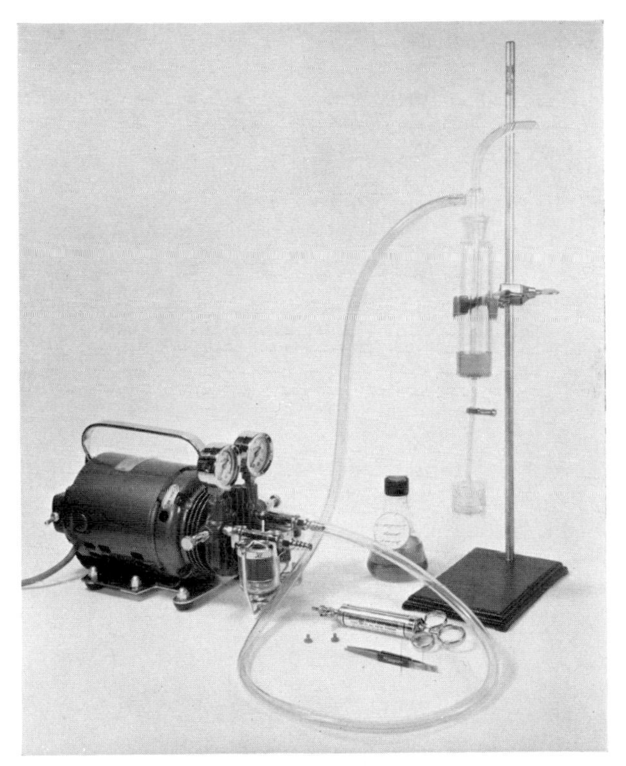

FIG. 22. Impinger, with 30 ml of impingement fluid, is set up in sampling area with clamped-off Monitor on Sampling Tube.

and then filtering the fluid, by techniques already described, to concentrate the micro-organisms for staining or culturing. Sampling requires the use of an all-glass impinger containing 30 ml of sterile impingement fluid, which is buffered gelatin broth. A Clinical Monitor is attached to the bottom of the impinger, Fig. 22, by means of a sampling tube which is clamped during sampling.

The standard procedure is to draw approximately 12·5 litres of air per minute through the impingement fluid and, in a period of 23 min this will yield a sample volume of approximately 10 cubic feet. After sampling,

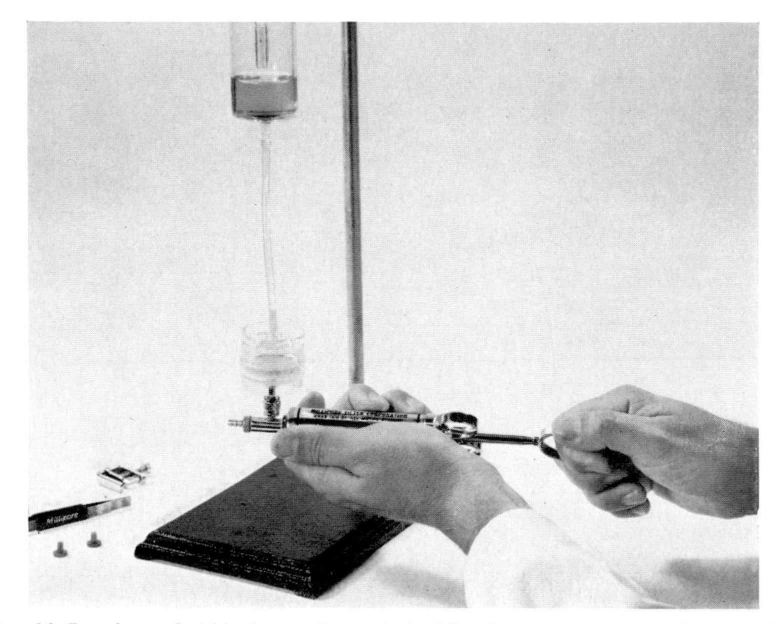

FIG. 23. Impinger fluid is drawn through the Monitor to concentrate the organisms.

a syringe and valve is placed in the base of the monitor and the clamp on the sample tube released. All but 1 or 2 ml of the impingement fluid is drawn through the filter, Fig. 23, and then the impinger is washed down with 10 ml of sterile broth medium. The remaining fluid is then drawn through the monitor and filtration is stopped just as the last few drops of medium disappear from the filter surface. The monitor is then capped and incubated.

C. Water microbiology

1. *Introduction*

The coliform group of bacteria originate in the intestinal tract of man and animal. Their presence in drinking water is indicative of a potential public health hazard because of the possible presence also of pathogenic enteric organisms responsible for such diseases as typhoid fever, dysentery and cholera. Both coliform and water-borne enteric pathogens exist under like conditions, although millions of coliforms usually occur for every pathogen. Presently known methods for the isolation and identification of specific pathogenic organisms are much too complex and time-consuming for routine use.

The coliform group of bacteria, on the other hand, is easily isolated and identified. Accordingly, standards for drinking water quality have been based on the relative number of coliform organisms present in a water sample, and maximum allowable concentrations of coliforms have been established to assure proper and safe sanitary control.

Until membrane filters became available the "Multiple-Tube Fermentation Method" was used to estimate the number of coliform organisms present in a water sample. This older method, also known as the MPN (most probable number) Test, relies for interpretation on the production of gas bubbles in small Durham tubes by the action of coliform bacteria in fermenting a lactose broth medium. From 48 to 96 h are required to obtain results.

By contrast, the new method provides a direct enumeration of coliform organisms in larger, more representative water samples, and does so in approximately 18 h. In many instances, the filter procedure will detect the presence of bacteria where the MPN method fails.

Many countries have now established drinking water standards on the basis of the membrane filter technique. Although the growth media may differ slightly between procedures, the basic advantages of the membrane-filter technique for bacteriological examination of water are inherent in all the procedures and may be summarized as follows—

(a) Shorter incubation; 18–24 h in most cases.
(b) Use of a larger sample; the ability to concentrate on a small surface all the micro-organisms contained in a large volume of sample.
(c) No interaction between micro organisms which are separated from each other on the filter surface, and cultured by diffusion of growth medium through the pores.
(d) The colonies are easily counted.
(e) The elimination of bacteriostatic or bacteriocidal agents likely to inhibit the growth of micro-organisms.
(f) Easier identification of various groups.

The basic simplicity of the membrane filter technique has, in few cases, lead to difficulties in application. While the technique represents an improvement in simplicity over earlier methods it is by no means foolproof and like most procedures in microbiology, it has limitations which must be thoroughly understood if the procedure is to be used to best advantage (Rose, 1966). Most microbiological analysis work employs a 0·45 μm pore size filter and except for *Vibrio comma*, the spirochetes, and some species of *Pseudomonas* found in sea water, organisms of sanitary significance are generally larger and are retained quantitatively on the filter surface. These include *Salmonella*, *Shigella*, *Proteus*, *Staphylococcus*, faecal streptococci, and coliforms. Water samples should be selected, if possible, to produce

FIG. 24. Basic equipment and materials required for routine coliform water analysis in the laboratory.

no more than 20–80 coliform colonies on a single filter so as to facilitate counting. Potable water samples should preferably be 100 ml. If this is not possible, at least 50 ml should be examined. Waters with few bacteria and little particulate contamination permit running 200 ml or more. For turbid water a minimal sample may be divided between two filters; for raw waters, an appropriately smaller sample size is used.

2. *Standard procedures*

The basic equipment and materials required for routine coliform water analysis in the laboratory are shown in Fig. 24. The procedure is very simple and is detailed as follows—

(a) Place a sterile 0·45 μm white filter—grid side up—aseptically in a sterile filter holder (or use an autoclave sterilized Sterifil Filter Holder assembly).

(b) Pour the water sample into the funnel and apply the vacuum to start filtration.

(c) Rinse the funnel walls with 20 ml of sterile phosphate buffer. This rinse is sufficient between samples from potable sources. Funnels should be resterilized when switching from raw to potable sources.

(d) Remove the funnel and transfer the filter aseptically to a plastic Petri dish containing a sterile absorbent pad saturated with 1·8 to 2·0 ml of M-Endo Broth.

(e) Invert the dish and incubate at 35°C for 18–24 h.

(f) Remove the filter from the dish and dry it for one hour on an absorbent paper at room temperature.

(g) Place the dry filter between two 2 × 3 in. glass microscope slides (hinged with tape) and count the number of "sheen" colonies with the aid of 5–10 × magnification. The illumination (incandescent) should be placed as directly above the filter as possible.

The M-Endo Broth is the membrane filter medium recommended for the recovery of coliform organisms from water. It is made by dissolving the following ingredients in 1 litre of distilled water to which 20 ml of 95% (v/v) ethanol has been added—

M-Endo Broth

Tryptone or polypeptone	10·0 g
Thiopeptone or thiotone	5·0 g
Casitone or trypiticase	5·0 g
Yeast extract	1·5 g
Lactose	12·5 g
Sodium chloride	5·0 g
Dipotassium hydrogen phospate	4·38 g
Potassium dihydrogen phosphate	1·38 g
Sodium lauryl phosphate	0·05 g
Sodium desoxycholate	0·1 g
Sodium sulphite	2·1 g
Basic fuchsin	1·05 g
Distilled water (with 20 ml 95% ethanol)	1 litre

pH 7·1–7·3

The medium is brought to boiling point and then heating is stopped; the pH of the final medium should be between 7·1 and 7·3. The appearance of visible green or golden metallic sheen on a colony constitutes a positive test for coliform organisms.

Colonies not giving this reaction cannot be considered significant. The broth also suppresses the growth of non-coliform colonies, thus aiding in the differentiation between these and the coliform types which exhibit the characteristic sheen. If any doubt exists colonies should be transferred to a "Brilliant Green Bile Salt Lactose Broth" for 24–48 h at 35°C and observed for gas production, which is presumptive evidence of coliforms. This broth is made by dissolving 10 g peptone and 10 g lactose in not more than 500 ml distilled water. A 10% (w/v) solution of oxgall in distilled water is added and the whole is made up to 975 ml with distilled water. The pH is adjusted to 7·4 and then 13·3 ml of 0·1% solution of brilliant green in distilled water is added. Distilled water is then added to make one litre.

Although the presence of coliform bacteria in a potable water gives an indication that the water may be polluted, there is still no assurance that the water has been contaminated from a faecal source. Recently, a new method has been developed for the recovery of faecal coliforms utilizing the membrane filter technique (Geldreich *et al.*, 1965). This method has two distinct advantages over the tube technique—

(a) The organisms to be tested do not have to be in pure culture or grown in non-selective media.

(b) The results for both total and faecal coliform tests can be obtained simultaneously for a much more meaningful report. A new medium has been developed, MFC Medium, for use with the membrane filter procedure to recover the faecal coliform strains.

MFC Medium

Tryptose	10 g
Proteose Peptone	5 g
Yeast Extract	3 g
Sodium Chloride	5 g
Lactose	12·5 g
Bile Salts No. 3 (Difco)	1·5 g
Aniline Blue	1 g

Dissolve of 1 grosolic acid in 100 ml of 0·2 N sodium hydroxide. Add the ingredients to distilled water containing 0·01% rosolic acid sodium salt prepared from this solution to make 1 litre of medium. Heat to boiling point.

This broth utilizes high temperature incubation (44·5°C water bath) for selectivity and an indicator system of aniline blue and the sodium salt of rosolic acid. The sample is filtered as above for the standard coliform procedure and the membrane filter is placed aseptically in a plastic Petri dish containing a sterile absorbent pad saturated with 2 ml of the MFC Medium. The Petri dish is inverted, wrapped in a waterproof plastic bag and incubated, submerged, for 24 hours in a 44·5°C water bath. After incubation the filter surface is examined and all blue colonies are counted as faecal coliforms. Non-faecal coliform colonies appear grey to cream coloured.

FIG. 25. Taking the sample from a river.

FIG. 26. Attaching a sterile sample tube.

3. *Field procedures*

When laboratory facilities are not available, microbiological examination can be conducted in the field, using field monitors (described earlier). Figs. 25–30 describe pictorially the sampling, filtration and introduction of nutrient medium stages in a typical field test. Use is made of a sterile field monitor, syringe and valve, and a sterile sampling tube (Fig. 26) to draw the water from the sampling cup. Sterile ampouled culture medium is

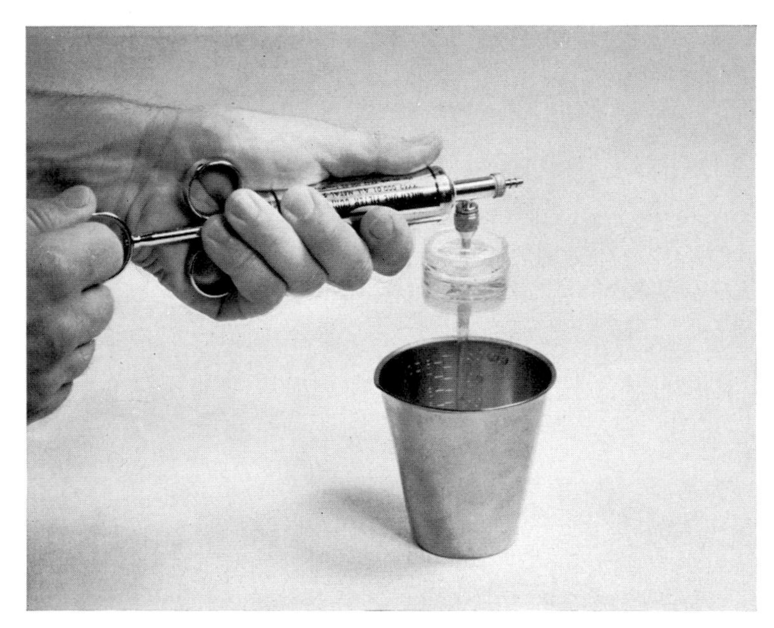

FIG. 27. Drawing the sample through the filter.

introduced into the base of the monitor to fully saturate the absorbent pad (Fig. 29). The monitor is then plugged (Fig. 30) and placed in a portable incubator (Fig. 31) which may be operated by a car battery, for example.

To summarize, the membrane-filter procedure is an excellent technique for the microbiological examination of water and is capable of yielding a more precise and rapid answer than the MPN test (Geldreich *et al.*, 1967). However, proper care must be paid to all aspects of the procedure.

D. Industrial microbiology

1. *Introduction*

The membrane-filter method is also an important standard technique in the microbiological examination of industrial fluids. Typical of this

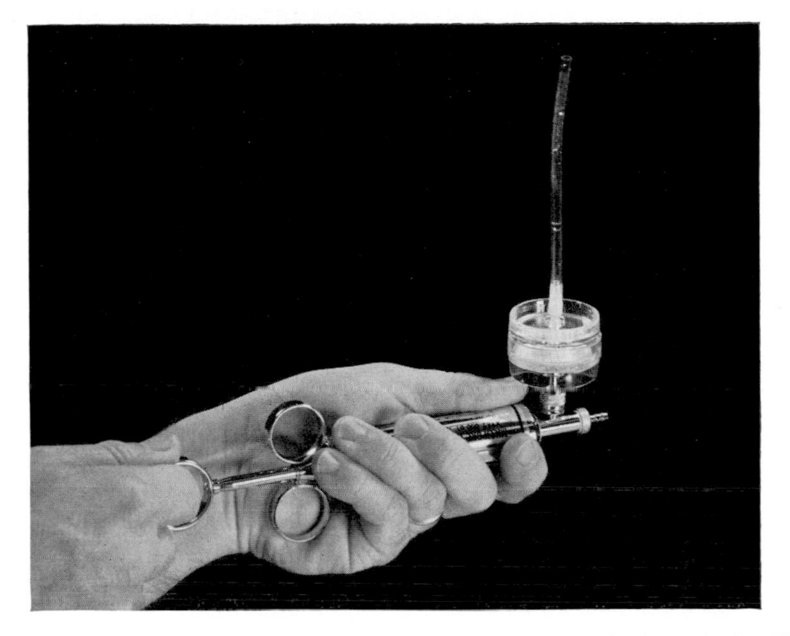

FIG. 28. Invert the Monitor to draw the last of the sample through the filter.

FIG. 29. Add the nutrient media to the underside of the monitor.

FIG. 30. Cap the monitor ready for incubation.

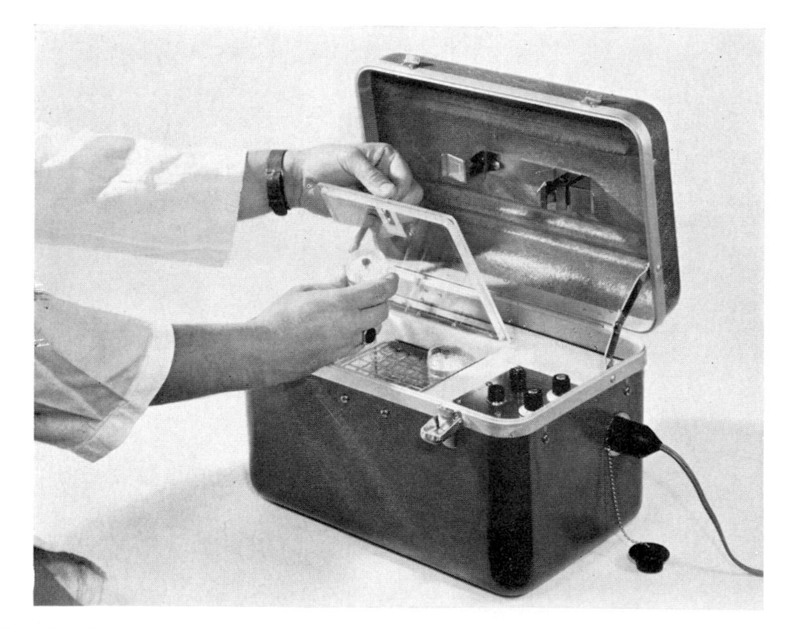

FIG. 31. Place the capped monitor into a portable incubator.

importance is the sterility testing of antibiotics where earlier procedures utilized huge volumes of medium in an attempt to overcome the inhibitory effects of the product. Even with this precautionary dilution the conventional method was a very insensitive test. The membrane filter procedure has also been very satisfactorily applied to the microbiological examination of beer, wine, soft drinks, syrups and sugar products. The procedure for collecting and culturing airborne organisms has proved very useful for the routine analysis of air in sterile chambers, such as a sterile beer bottling room.

2. *Sterility testing of antibiotics*

For the reasons mentioned above the membrane-filter technique has been adopted as the standard for sterility testing of antibiotics by nearly all countries in the world. Lightbown (Lightbown, 1963) reported that seven-year-old batches of dihydrostreptomycin sulphate tested by filtration procedures revealed the presence of contaminating organisms which had survived in the dry powder. The contamination had not been revealed by earlier conventional dilution methods. Bowman (Bowman, 1966) reported that 40 vials, each containing 10 ml of a suspension of procaine penicillin G in dihydrostreptomycin sulphate solution were tested by both the membrane-filtration method and a direct method. The membrane-filtration method recovered a Gram-positive bacillus contaminant whilst the direct method showed no growth in the 40 tubes of thioglycollate medium after incubation for seven days at 32°C.

As sterility testing of antibiotics normally requires the examination of several samples, a six place filtration manifold, Fig. 32, has been designed to allow samples to be run simultaneously. A flushing manifold is provided so that sterile distilled water can be used to rinse away residual antibiotic after filtration. The Therapeutic Substances General Regulation, 1963 (U.K.) requires—"that a test for sterility shall be applied to a sample from each batch of the substance; such that the number of containers taken for test from every batch shall be 2% of the containers in the batch or 20 containers, whichever is less, taken at random, and an additional 2 containers from each thousand or part of a thousand after the first".

"The sample is dissolved and passed through a membrane filter with an average pore diameter not greater than 0·75 μm (in practice a 0·45 μm filter is always employed). The filter shall be suitably washed to remove any bacteriostatic activity before submerging in a fluid nutrient medium."

The Sterifil Filter Holder is an ideal aseptic filtration system for sterility testing of antibiotics, and the hydrophobic edge filter should always be used, to prevent the diffusion of antibiotic solution beyond the filtration area, where it could remain after flushing and interfere with subsequent growth of collected organisms. After incubation, the tubes of fluid medium are observed for visual growth, and if growth is indicated this should be confirmed by microscopic examination.

3. *Microbiological examination of oil and oil based solutions*

Oils and oil based solutions, such as fuel, hydraulic fluids and cutting oils, as well as oils used in pharmaceutical preparations, cause a membrane filter to become hydrophobic when they are filtered. The residual oil must be flushed away, therefore, so that the aqueous base nutrient used for

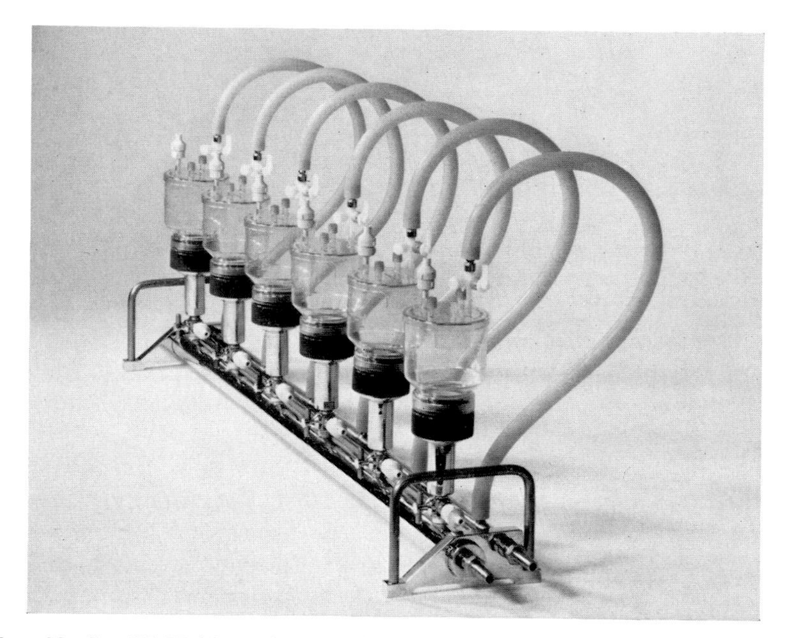

FIG. 32. Sterifil Holders, fitted with neoprene stoppers may be used with the Millipore Sterility Test Manifold when large numbers of samples must be run in a short time.

analytical culturing can diffuse through the membrane filter to support micro-organism growth. The procedure entails the filtration of the oil sample through a sterile 0·45 μm grid marked membrane filter. After sample filtration, the filter funnel and filter are rinsed with at least 100 ml of sterile 0·1% Triton X-100 solution and this is followed with a 20–30 ml rinse of sterile normal saline. The filter is then transferred to a plastic Petri dish containing a pad saturated with 1·8 ml of a suitable total count or selective broth medium. The dish is inverted and incubated for the specified time and temperature.

4. *Microbiological examination of beer, wine and malt beverages*

The membrane filter technique for microbiological analysis is especially useful for determining the number of micro-organisms in samples of beer, wine, rinse waters and containers where counts are normally low and, therefore, the sample volumes must be relatively large. The sanitary quality of empty bottles can be tested by adding 50 ml of sterile normal saline to the bottle, gently swirling so that the saline comes into contact with all the internal surfaces and finally pouring the solution through a sterile 0·45 μm filter. The filter can then be placed on a nutrient medium and incubated following standard procedures. When filtering a bottle or can of beer it is advantageous to add a few drops of octanol to the filter flask to prevent excess beer foaming and carrying over to the vacuum pump. Octanol can also be added to the beer sample in the filter funnel to reduce the foam. After filtration, the 0·45 μm membrane filter should be placed on WL Nutrient Medium and incubated at 33°C for 48 h. WL Nutrient Medium permits the growth of yeast and in those instances where the yeast count is low certain bacteria can be cultured. WL Differential Medium may be used to determine a bacterial count when the yeast count is relatively high, since the actidione in the medium will inhibit the growth of yeast without interfering in any way with the growth of bacteria found in beer. Like WL Nutrient, WL Differential Medium reflects only those organisms capable of growing at pH 5·5 or below. The differential medium should be incubated aerobically for culturing *Acetobacter* and anaerobically for culturing *Pediococcus* and *Lactobacillus*. WL Nutrient Medium and WL Differential Medium are prepared according to Gray (Wallerstein Lab. Comm., 1950, 1951).

It is also possible to carry out a direct staining procedure for microscopic examination of the micro-organisms collected from beer and wine. After filtration the membrane filter is submerged in Loeffler's Methylene Blue dye for 3 min. Excess stain is removed by dipping gently in clean water and the filter is then submerged in n-propanol for 5 min followed by submerging in xylol for 7 min when the filter becomes transparent. The filter can then be observed by oil immersion microscopy. Yeast cells are seen as lightly stained with a darker nucleus and may be oval, oval with buds, or elongated. Bacteria are normally darkly stained rods (long or short) and spheres. Protein precipitates are normally seen as darkly stained irregular aglommerates.

E. Radiomicrobiological techniques

The membrane filter provides an excellent medium for the recovery and subsequent detection of radioactively labelled material from liquid suspension. Due to the absolute retention characteristics of the filter, all labelled material will be retained in a single plane on the surface. There will be no depth penetration and, provided the sample size is adjusted to prevent overcrowding, there will be an even distribution of label. This means that absorption of radiation in the filter matrix or in clumped material on the filter surface, is avoided. By this technique, very sensitive, quantitative analytical systems can be achieved.

Methods employing labelled bacterial cells, virus particles and nucleic acids have been reported (Weissman, 1965; McLeod *et al.*, 1966) and some specific examples are mentioned below.

The uptake of lactose-1-^{14}C was used to measure the relative amounts of galactoside permease in suspensions of *Staph. aureus* of concentration 250 mg/ml (McLatchy and Rosenblum, 1963). After induction the cells were incubated at 37°C in the presence of labelled lactose (200 μmoles/ml of activity 0·04 μc) and samples were removed at 10 and 20 min for counting. The induction medium was prepared by dissolving 10 g Proteose peptone, 1 g Yeast Extract, 2 g dipotassium hydrogen phosphate and an appropriate inducer in one litre of distilled water. The inducers employed were Lactose, Galactose and β-D-Thiogalactosides. The samples (1 ml) were filtered through an HA (0·45 μm) filter and then the filter was washed, dried and counted in a thin window gas flow counter (Nuclear Instrument Co.) for not less than 2000 counts. Allowances were made for background counts, and the counts per minute per milligram (dry weight) of cells were calculated.

In another application (Weissman, 1965) the amount of ^{14}C-labelled single strand RNA combined with a heat denatured non-labelled (natural) replicative form of MS2 RNA was estimated by annealing the two and determining the amount of label in double strand form.

As well as direct radioactivity measurement, it is possible to place the filter in contact with photosensitive surfaces for autoradiographic studies (Berlin and Rylander, 1964).

F. Fluorescent antibody techniques

The use of specific antibodies which have been labelled with a fluorescent dye is becoming of increasing importance in diagnostic parasitology and bacteriology. The great advantage of the technique lies in its ability to demonstrate, by fluorescence, those cells in a heterogeneous suspension to which the antibody specifically binds. Individual species or serologically distinct strains may be rapidly and easily identified within groups of organisms.

It has been recently shown (Danielsson, 1965; Danielsson and Lurell, 1965), "that bacteria suspended in tap water or cultured in nutrient broth and then collected on the surface of a non-fluorescent membrane filter, could be identified within one hour by means of the fluorescent antibody technique." The black membrane filter had to be used due to the strong autofluorescence of white membrane filters. Briefly, the method used is to concentrate the organisms present in the solution on the surface of a black membrane filter. The organisms are then stained by allowing the cells to combine with a fluorescent antibody for a period of about one hour. The filter is then washed and inspected under a fluorescent microscope using incident illumination.

III. STERILE FILTRATION TECHNIQUES

A. Introduction

With varying success, filtration has been used for many years as a means of sterilizing and clarifying biological solutions. Not until the introduction of membrane filters, however, has filtration become a truly reliable and economic method for sterilizing not only heat labile materials but also solutions that are normally processed by autoclaving. The uniform pore size, absolute retention and high porosity characteristics of membrane filters combine to make them the ideal device for routine laboratory and production use. There are several technological and economic advantages of using membrane filters for the sterilization of biological fluids and these will be considered below.

B. Advantages of membrane filters

1. *Sterility*

A 0·22 μm membrane filter should normally be used in sterile filtration operations, as all reported bacteria are larger than 0·22 μm and therefore will be physically retained on the surface of the filter under all conditions. Thus, provided that the membrane filter holder assembly is autoclaved according to instructions, and all other operating parameters are carefully observed, there is no chance of a product being non-sterile after filtration through a 0·22 μm filter.

2. *Bacterial grow through*

To date, no reports of bacterial grow through when using the (Millipore) membrane have been received. Once the bacteria are trapped on the filter surface they will not grow through the filter and therefore cause non-sterile filtrates during lengthy filtration runs.

3. *pH change*

The pure composition and homogeneous structure of membrane filters contributes neither ionic or particulate contamination to the filtrate. There is absolutely no pH change.

4. *Bubble testing*

This is a non-destructive method for checking the filtration assembly prior to operation. The procedure is especially advantageous since sterility is not broken and no product is lost. The bubble point for each membrane filter type, with any specific liquid, is defined as the pressure required to force air through a liquid saturated filter. The air pressure has to overcome the surface tension of the fluid in the pores before breaking through. A filter assembly free of all leaks, therefore, will hold air pressure indefinitely under these conditions—failure to hold the pressure specified for the filter type and liquid, is evidence of a passage larger than the filter pore size. The bubble point with water for a 0·22 μm filter is 55 p.s.i., and for a 0·45 μm filter is 32 p.s.i. In general, the bubble point will be slightly higher with oils and slightly lower with alcohol solutions. The filtration assembly used for sterile filtration of biological fluids should be bubble point tested after autoclaving and just before the filtration is commenced. A satisfactory bubble point guarantees that fluid passing to the downstream side of the filter is in a sterile condition.

5. *Flow rate*

The high porosity and pore configuration results in flow rates nearly 40 times faster than through a conventional filter approaching the same particle size retention capability. Membrane filters may be used at high inlet pressures without fear of breakthrough of bacteria, as the screen configuration of the filter will not be modified even under pressures of several thousand p.s.i.

C. Filtration systems

At present both the 0·45 μm and the 0·22 μm filters find use in sterile filtration applications, although the 0·22 μm has become the more widely applied of the two, because the pore size is smaller than any known bacteria and thus, sterility is assured. The 0·45 μm filter, having larger pores, yields flow rates approximately three times those of a 0·22 μm filter but should only be used where prior experience indicates that organisms smaller than 0·45 μm are not present in the solution to be filtered. The 0·22 μm filter should always be used with solutions containing Sera, Plasma or Trypsin, where species of *Pseudomonas* or other bacteria of a small size are known to occur.

Selection of the proper filter holder for each application depends chiefly on the volume of liquid to be processed, however, other factors must often also be considered. Some liquids, because of their high viscosity, particulate density, or molecular composition, require relatively large filtration areas

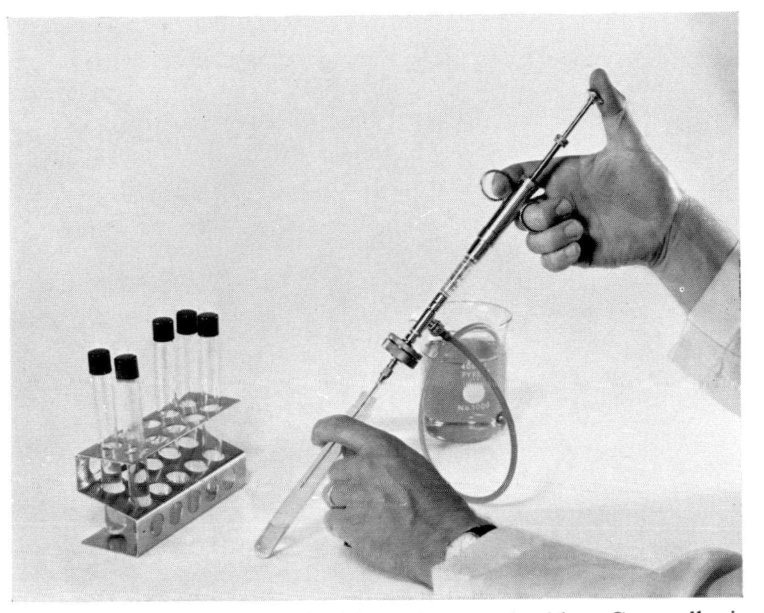

FIG. 33. Micro-Syringe Filter Holder, when used with a Cornwall pipetting syringe (Becton-Dickinson), makes an ideal device for dispensing identical quantities of sterilized tissue culture medium.

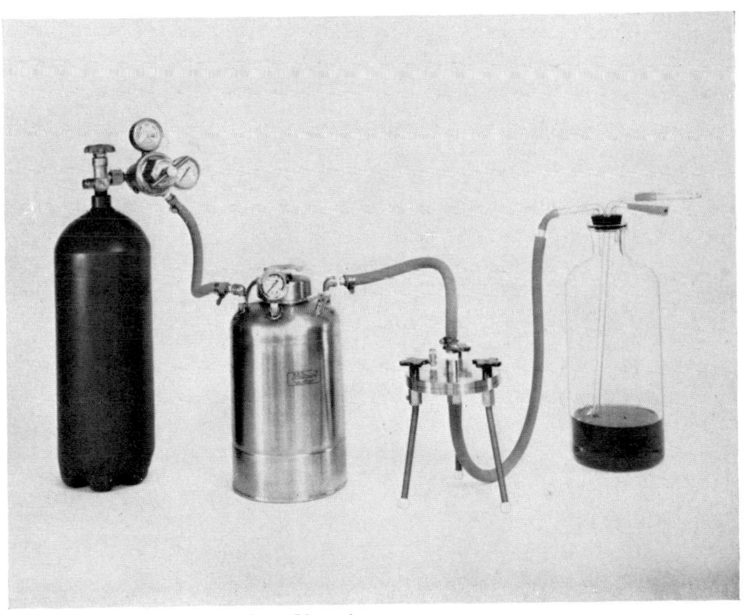

FIG. 34. A typical sterilization filtration set-up.

to achieve practical throughputs and flow rates. These materials, serum and plasma, for example, also typically require some preliminary processing such as centrifugation or prefiltration prior to sterilization by filtration.

In designing a sterilizing filtration system it is very important that positive pressure is used to force the liquid through the filter. A negative pressure filtration system suffers from : (i) the possibility of leaks drawing contaminated air into the sterile products; (ii) a maximum pressure differential of 1 atmosphere, and (iii) the necessity to transfer the sterile liquid from the vacuum flask to the point of use. When filtering protein solutions it is imperative to use positive pressure to reduce foaming which could result in denaturation. Providing the membrane filter is correctly supported, there is no possibility of forcing bacteria larger than the pore size through the filter. The above disadvantages of negative pressure filtration systems

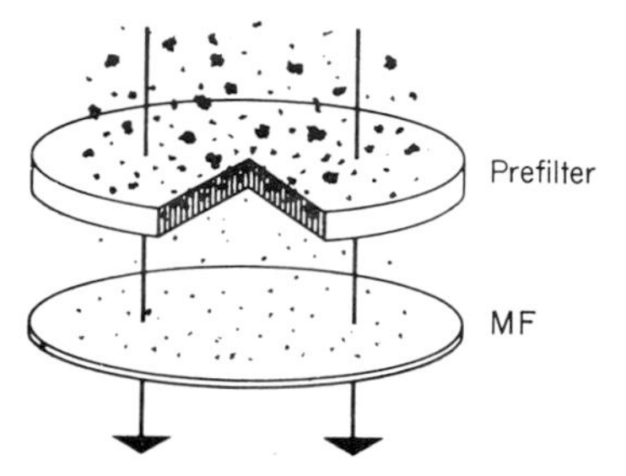

FIG. 35. A Microfibre Glass prefilter directly before the MF helps to extend filter life.

are overcome by employing positive pressure, and also higher flow rates are obtainable from a relatively small area of filter by using high pressures. Fig. 33 shows positive pressure filtration on a small scale using a luer-lok syringe, and Fig. 34 shows a larger scale, using a pressurized vessel and a 142 mm filter holder.

If the liquid to be sterilized has a high burden of suspended particulate matter which will rapidly clog the membrane filter then it is advantageous to use a fibre glass prefilter directly before the membrane filter. The prefilter is a depth type filter, as described in an earlier section, and is capable of removing a high percentage of particulate matter and so extending the life of the membrane filter (Fig. 35). Fibre glass prefilters are extremely inert, will not contribute particulate contamination to the filtrate and may be sterilized by steam under pressure in place with the membrane filter.

The assembled Filter Holder System and connecting hoses should be sterilized together (Fig. 36) and a final aseptic connection to a sterile receiving vessel made. If the autoclaving schedule for each assembled filter holder is carefully followed, complete and positive reliability may be expected.

Normally the filtration system will be autoclaved at 121°C for a period between 10 min and 45 min, depending on the size of the filter holder. Assembly of separately sterilized filters and filter holders is sometimes practised under very carefully controlled aseptic conditions, but is not generally recommended.

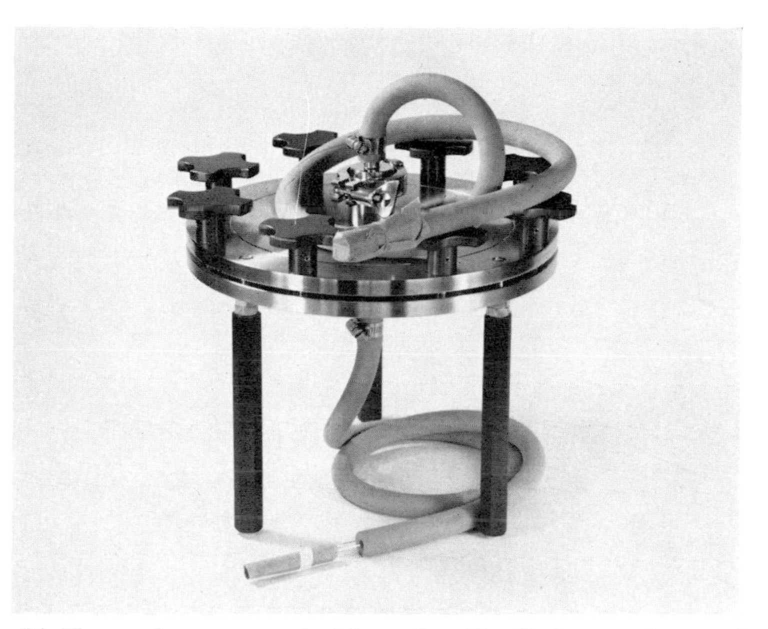

FIG. 36. Hose ends are wrapped with good quality Kraft paper for autoclaving. Hoses must not be pinched off.

D. Bubble point testing

As mentioned above the bubble point test should be performed on the filter system after autoclaving and after aseptic connections have been made. Fig. 37 shows a typical set-up for bubble point testing where the liquid to be filtered is contained in a stainless pressure vessel and filtration is achieved by using pressurized air or nitrogen. After filtration the liquid will pass into a sterile receiving vessel. The procedure for bubble point testing is as follows—

(a) With valves B and D closed, open valves A and C and allow sufficient liquid to flow through the filter to assure thorough wetting. The vent valve may be used to bleed off any entrained air.
(b) Close valves A and C and check handwheels for tightness.
(c) Aseptically attach tube F to a sterile receiving bottle. Open valves B and D.

(d) Pressurize upstream of filter holder to 20 p.s.i. through valve B. Residual liquid trapped in the filter holder will pass into the bottle. A few bubbles may be seen at this time due to air trapped in the downstream side of the filter. Only continuous bubbling constitutes a bubble point.
(e) Increase the pressure on the upstream side until continuous bubbling is seen and note the pressure.

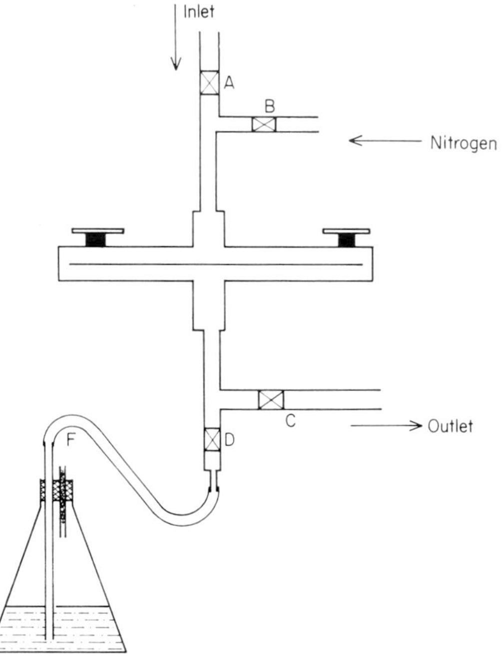

FIG. 37. Set up for Bubble Point Testing a filtration system.

(f) If the bubble point is satisfactory disconnect the nitrogen line from B and vent excess pressure. Close valves B and D.
(g) Open valve A and allow liquid into the filter holder and use vent valve to bleed off any entrained air.
(h) Open valve C and continue with filtration.

The bubble point test is a most critical method of determining the integrity of a filtration system just prior to the commencement of filtration.

E. Sterilizing filtration of specific liquids

1. *Introduction*

It would be impractical to present data for every product now being routinely sterilized by membrane filtration, so a few commonly sterilized liquids are discussed below with approximate filtration specifications.

2. *Liquids not requiring pretreatment*

(a) *Distilled water.* The 0·45 μm pore size membrane filter is frequently used for filtration of distilled water as normally it is used for make-up purposes where some terminal sterilization will be performed. A fibre glass prefilter is placed on top of the membrane filter in a large system involving tanks and extensive piping as these often contain filter clogging algae, rust particles and scale. Glass systems, and those of laboratory size, rarely require a prefilter in addition to the 0·45 μm filter. Filtration should always be carried out at the point of use and only in the volume immediately required. Approximate performance, would be 50 litres/min through a 293 mm, 0·45 μm, filter at 20 p.s.i. for up to 10,000 litres total throughput.

(b) *Tissue culture media (without serum).* Hanks Medium will flow at 3 to 4 litres/min. through a 293 mm 0·22 μm filter and a fibre glass pre-filter system, at 12 p.s.i., for a total throughput of approximately 500 litres. Bacteriological broth media (synthetic) will give essentially the same performance.

3. *Liquids requiring some pretreatment*

(a) *Tissue culture media.* 2% Serum Media will flow at 1·5–2·5 litres/min. through a 293 mm, 0·22 μm filter and fibre glass prefilter, at 12 p.s.i., for approximately 240 litres total throughput. Media containing 5% and higher concentrations of serum will require separate processing of the serum before addition to the medium. Process the serum by serial filtration as is indicated below for frozen serum. After adding the filtered serum, sterilize the final medium through a 0·22 μm filter and fibre glass prefilter as for the above 2% media. Prefiltration is especially important for media containing serum, lactalbumin and trypsin.

(b) *Bacteriological broth media (non-synthetic).* Protein content contributes to membrane filtration clogging, but the performance will essentially be the same as that for the above tissue culture media.

(c) *Phage and virus suspension.* Performance and method of handling depends upon particle size and suspension condition. Pure suspensions (no cells) may be filtered for sizing directly through the appropriate membrane filter with no significant loss in titre. Propagated cultures containing cells and cellular debris, even following centrifugation, cannot be filtered for clarifying without some loss of titre due to adsorbtion and entrapment of the virus and phage by cellular matter on the filter surface. A 0·45 μm membrane filter is used to advantage for most clarifying work on propagated cultures.

4. *Liquids requiring extensive pretreatment*

Products in this class are predominantly protein solutions and require separate and individual processing through a series of successively finer filters before they may be passed through a 0·22 μm filter for sterilization. This process is referred to as "Serial Filtration". Only the final 0·22 μm filter assembly needs to be sterile, although normal precautions against contamination are advised for prior steps as well. Serial filtration may be performed by employing a separate filter holder for each pore size of filter to be used. Recently, a technique has been introduced allowing two or three membrane filters to be stacked on top of one another in the same filter holder, thus giving serial filtration conditions. This stacking procedure is possible by placing a monofilament fibre mesh made from a non-toxic sterilizable material such as dacron (terylene) between each filter in the stack.

Protein solutions should be filtered as fresh as possible for maximum throughputs. Old, heavily contaminated materials will filter in smaller volumes due to their higher densities of filter-clogging contaminants. Although protein materials are characteristically chilled for handling and storage, experience indicates that warming these materials, especially prothrombin and serum to a maximum of 37° C results in a 4–7 times increase in the amount which can be sterilized through any given 0·22 μm filter disc. No apparent harm to the product comes from this warming procedure.

Positive pressure is preferred for protein filtration since vacuum causes products to foam which may result in denaturation. Pressures should be limited to a maximum of 50 p.s.i. to avoid exceeding the bubble point of the 0·22 μm filter. Reports indicate no change or loss of activity from membrane filtration (Millipore) of fibrinogen, antihaemophilic factor (AHF Factor VIII), prothrombin level and accelerator globulin (AcG Factor V).

Serum. It would be impossible to discuss all the various techniques that have been used for filtration of serum which by its nature will vary grossly depending primarily on the source and history. A typical example of a system for filtering frozen serum is to first centrifuge at 2,500 r.p.m. for 25 minutes and follow this by serial filtration through a fibre glass prefilter, 1·2 μm (RA) membrane filter, 0·45 μm (HA) membrane filter and finally through a sterilizing 0·22 μm membrane filter. Batches of 20–30 litres total throughput before clogging are possible with 293 mm filters when using this system.

IV. SUMMARY

In this chapter it has been possible only to scratch the surface of the many biological applications for membrane filters. The characteristic which makes a membrane filter such a valuable tool is that all particles both biological and non-biological which are larger than the pore size will be positively retained on the filter surface. Membrane filters were originally used strictly as an analytical tool in the microbiological analysis of potable waters. However, due to their "absolute" nature, the membrane filter has gradually become an important device for the process sterilization of fluids.

REFERENCES

Baltimore Biological Laboratory, Baltimore, Maryland, U.S.A.
Berlin, M., and Rylander, R. (1964). *J. Hyg., Camb.*, **61**, 307–315.
Bowman, F. W. (1966). *J. Pharm. Sci.*, **55**, 818–821.
Danielsson, D. (1965). *Acta. path. microbiol. Scand.*, **63**, 597 603.
Danielsson, D., and Lurell, G. (1965). *Acta. path. microbiol. Scand.*, **63**, 604–608.
Difco Laboratories, Michigan, U.S.A.
Geldreich, E. E., Clark, H. F., Huff, B. H., and Best, L. C. (1965). *J. Am. Wat. Wks Ass.*, **57**, 208–213.
Geldreich, E. E., Jeter, H. L., and Winter, J. A. (1967). *H.L.S.*, **4**, 113–125.
Haley, L. D., and Arch, R. (1957). *Am. J. clin. Path.*, **27**, 117–121.
Lightbrown, J. W. (1963). Proc. Round Table Conference on Sterility Testing, London.
McClatchy, J. K., and Rosenblum, E. D. (1963). *J. Bact.*, **86**, 1211–1215.
McLeod, R. A., Light, M., White, L. A., and Currie, J. F. (1966). *Appl. Microbiol.*, **14**, 979–984.
Millipore Bibliography, Millipore (U.K.) Ltd., Heron House, 109 Wembley Hill Rd., Wembley, Middlesex.
Osgood, G. (1967). "Sheet Filtration and Filter Sheets." Paper presented to the Filtration Society, 4th April, 1967 (London).
Rohm and Haas Product. Lennig Chemicals, 26 Bedford Row, London, W.C.1.
Rose, R. E. (1966). *Wat. Sewage Wks.*, **113**, 1–4.
Shanahan, A. J., Fornwald, R. E., Chapman, D. M., and Bleachey, A. N. (1956). *J. Bact.*, **71**, 499–500.
Tarshis, M. S., and Frisch, A. W. (1951). *Am. J. clin. Path.*, **21**, 101–113.
Wallerstein Lab. Comm. (1950). **13**, 357.
Wallerstein Lab. Comm. (1951). **14**, 169.
Weissman, C. (1965). *Proc. natn. Acad. Sci. U.S.A.*, **54**, 202–207.
Winn, R. W., White, M. L., Carter, W. T., Miller, A. B., and Finegold, S. M. (1966). *J. Am. med. Ass.*, **197**, 539–548.

CHAPTER VIII

The Culture of Micro-organisms in Liquid Medium

C. T. CALAM

Imperial Chemical Industries Ltd, Pharmaceuticals Division, Alderley Park, Macclesfield, Cheshire, England

I. INTRODUCTION AND SURVEY OF CULTURE METHODS

The object of the present chapter is to give an account of the methods used for the growth of micro-organisms in liquid culture. This process is fundamental to both academic and industrial work and many lines of development have arisen, with a correspondingly large number of procedures. As detailed accounts of the more advanced and specialized applications will be given in other Chapters, the present Chapter will deal only with batch culture and with apparatus using simple types of instrumentation. After a brief survey of the different types of apparatus used for culture work it is proposed to discuss first theoretical aspects of the subject, such as agitation-aeration and medium development, followed by an outline of the biochemical principles involved. This will be followed by a description of apparatus used for experimental work. The Chapter will be concluded by short accounts of the development of typical fermentation processes, which are intended to illustrate the way the methods and apparatus described are put into practical use. The Chapter will also deal with subjects such as the prevention of infection, handling of cultures and special fermentation procedures such as the use of hydrocarbons as substrates. A glance through a copy of a journal like "Biotechnology and Bioengineering" is sufficient to show the enormous field that is covered by the title of the present Chapter. It is only possible here to give a brief account of the field based on a choice of those subjects which seem most important, and it is regretted that so much must be omitted.

A. Submerged culture

Submerged culture methods have been used for many years for the culture of bacteria and yeasts. More recently they have come to predominate in work with fungi and actinomycetes. Originally stirring and aeration were not used, though cultures were sometimes aerated by a stream of air. This proved inadequate for fungi which produce a thick growth and require mechanical stirring. Until the thirties workers with fungi were usually satisfied with surface cultures, but in 1933 Kluyver and Perquin described the use of shake cultures to provide a more uniform type of growth. The drive towards stirred and aerated cultures came in the early forties with the development of the penicillin fermentation, and many difficult problems of engineering design were rapidly solved, particularly by groups at The Northern Regional Research Laboratory, Peoria, Illinois, U.S.A., The University of Wisconsin, U.S.A., and E. Merck and Company, Rahway, New Jersey, U.S.A. At that time most of the apparatus used was home-made, but during the fifties, commercially made fermentation apparatus became available and the subject became specialized into the fields of continuous culture, bacterial cultures, culture selection, screening and biochemical engineering. As a result different types of apparatus were developed to suit the various types of work. The Istituto Superiore di Sanita in Rome provided many of the ideas incorporated in modern apparatus. In the meantime techniques of culture maintenance, inoculum production and prevention of infection were improved so that fermentations can now be run for several weeks if required.

Submerged culture techniques have now become well established and the construction and installation of apparatus is common. Nonetheless care is still required to obtain the most suitable equipment and use it to the best advantage. The mention of specialization leads conveniently into the next section in which the different types of apparatus will be reviewed and their suitability for different types of work discussed.

B. Stirred culture apparatus

Apparatus for submerged culture varies widely. Bacteria and yeasts will grow in submerged culture by simply inoculating them into a quantity of stationary medium. Growth is anaerobic or microaerophilic. For aerobic cultures, oxygen must be supplied at least by shaking or by bubbling air through the liquid. Simple apparatus of this type has been used for important work, for instance the experiments of Monod (1942). Culture times are short and light growth is adequate. The short culture cycle reduces the risk of infection. With mycelial cultures such as moulds and actinomycetes, the culture soon becomes thick and structured, so that bubbling of air no longer circulates the liquid; instead, the bubbles pass through the stationary culture providing very little oxygen, and the organism changes its metabolism to a system more suited to low oxygen conditions. Bubbling of air can maintain cultures up to 5–10 g/litre of dry cells and may be used for production of inoculum though more vigorous inocula are prepared in shaken or stirred culture.

Shaken cultures represent an advance over the bubbling of air as a means of increasing cultural effectiveness. Flasks may be shaken on rotary-reciprocating or wrist-action shakers. Flasks from 25 ml to 5 litres in size may be used with mild or vigorous agitation. While single or a few flasks may be used, the great advantage of the system is the provision of identical conditions in hundreds of flasks, so that it is particularly suitable for screening, optimising action of media or similar work.

Apparatus for stirred culture is made in various sizes and styles. In the laboratory, fermenters 6 in., 9 in. and 12 in. diameter are usually used, following the sizes of industrial glass tubing, and holding 3–20 litres of culture, with various speeds of stirring and degrees of elaboration in instrumentation and ancillary equipment. For many types of work such as the production of experimental batches of metabolites, very simple apparatus can be used with a minimum of ancillary equipment. Simple apparatus can also be used for teaching and for radioactive tracer work, though in the latter case precautions must be taken to prevent the escape of radioactive substances. Simple apparatus is also suitable for process development work especially of the type used to follow up the results of shake flask screening. In this case numerous fermenters are needed and elaboration would only be time wasting. Elaborate fermentation equipment that is required for continuous cultures is described by Evans *et al.* (this Series, Vol. 2). Complex control systems are also needed for many types of metabolic study, but in this case the critical study periods are often relatively short and manual supervision is possible so that complicated experiments can often be performed in relatively simple apparatus.

Simple apparatus can often be used for chemical engineering studies, since the object is usually the simulation of industrial processes. Effort is devoted more to the ancillary equipment used to measure factors such as stirring patterns, mixing, bubble distribution, power input, foaming, viscosity and heat exchange. A rather large fermenter is usually desirable so as to allow the removal of large samples and the insertion of probes, which would be impossible with small apparatus.

Pilot plants are usually intended to provide basic technical and economic data for plant process development. Frequently extraction and purification of products is more important than fermentation. Pilot plants usually operate on a works basis with shift cover. The atmosphere is quite different from laboratory work, the inventor's personal skill and knowledge being replaced by the workaday attitude of the process operators. The process must be carefully defined and free from ambiguities. Equipment must be strong and reliable. Fermenters of 100–500 gallons are common, but pilot fermenters of 5 litres to 10,000 gallon capacity are available. A very common type of fermenter contains about ten litres of culture. The ratio height/diameter is 2·5 : 1. Such fermenters are used in large numbers, mounted permanently in groups and temperature is controlled by water-jackets. Safety is of great importance, both from accidents and from any dangers arising from organism and medium. If solvent extraction methods are used, the fire hazard requires careful attention. In industrial work the transmissibility of data from one type of apparatus to another is important, and nomenclature used in laboratory, pilot plant and the main plant should be similar.

In choosing fermenters the question of sampling must be considered. It is undesirable to reduce the medium volume too much. Thus with a culture of three litres (actual) it is inadvisable to remove more than 500 ml in samples. For practical reasons, these are likely to require 25–50 ml each, so that not more than 10–15 can be taken, or 2–3 per day for 5 days. This and other factors must be considered when determining fermenter size.

This short review is intended to stress the different types of apparatus used for different purposes. Apparatus is expensive and can take up a great deal of space. Economy must be the object, though to obtain cheap and unreliable apparatus is wasteful. As an illustration, for testing mutants or investigating media it is essential to use 6–12 fermenters at the same time to obtain controlled experiments. To keep up a steady stream of work more fermenters are needed so that experiments can be overlapped, and up to 50–60 or more fermenters may be provided. This type of work will require a special laboratory and a staff of workers, the system being geared to the production of experimental data. At the other end of the scale many workers find one or two fermenters quite adequate. They prefer a compact self-contained unit which occupies a small space in the laboratory and which is used occasionally as circumstances require. Yet other workers will find a small shaker adequate or even a single flask held in a wrist-action shaker for a few hours incubation. When apparatus is selected, the emphasis must be on matching it to the job and not on appearance, or degree of sophistication. In the research laboratory versatility may be of great importance and several commercially available small-scale fermenters are designed to satisfy a wide range of cultural conditions.

C. Sources of fermentation apparatus

Apparatus suitable for these purposes may be made by the worker in the workshop attached to the laboratory, or purchased from a manufacturer of scientific apparatus. The choice of the best way to obtain apparatus is not easy as it depends so much on circumstances. Purchased apparatus or specially designed and made apparatus is usually expensive. Only apparatus of very good quality should be used, and it is usually necessary for quality to be built in at the start. Frequently experiments will last for 2–14 days and during this time it is necessary for all conditions to be held constant. Apparatus, especially shaking machines, must be capable of virtually continuous operation for weeks or even years, and this can only be achieved by careful design and construction. The most successful pioneers such as M. J. Johnson at Wisconsin University, used rugged but effective apparatus. This facilitated successful experimentation that would not have been possible with weaker but more elegant equipment. Much apparatus made for chemical or biochemical experiments is only intended to run for a few hours and is not rugged enough for fermentation work. It also does not give adequate protection against contamination and difficulties arising from foaming; attempts to improve it by minor modifications will probably be unsuccessful. The best approach is to visit a laboratory where experimental work is in progress and see different types of apparatus in use. The advice offered may not be accepted, but such visits give the best idea of the problems involved and their solution.

Later in the Chapter examples will be given of commercially available apparatus which includes not only fermenters but also ancilliary apparatus such as feed-pumps, pH-control systems and the like. The makers referred to are typical of the best available. The reader is advised to make enquiries, compare prices and delivery-dates and choose appropriately. A great diversity of apparatus is used successfully, each type having its particular advantages, which have to be matched with the user's special requirements.

D. Surface culture

Surface cultures may be grown on stationary liquid media in any kind of vessel. Conical flasks are frequently used as well as special flasks giving a large surface area, for instance Roux bottles. Liquid layers 1–2 cm deep are suitable. The most economical depth (yield/ml × volume) is often about 2–3 cm. Inoculation is usually important, since if this is inadequate growth fails to cover the surface or may develop in a semi-submerged condition. When spores or cell-suspensions are used for inoculation, they tend to go to the edges or form islands, from which outgrowths develop. Much of the inoculum sinks and is lost. This problem is most serious when large numbers of large flasks are being used and the supply of inoculum is limited. The culture period may be 5–15 days depending on the organism. With fermentation temperatures of 25° care must be taken to avoid overheating in large stacks of culture flasks. Growth starts with the formation of a thin layer of cells. With moulds and actinomycetes this is followed by the growth of aerial hyphae and sporulation.

Kluyver and Perquin (1933) proposed the use of shaken submerged cultures instead of surface cultures because growth was more uniform and it avoided using cultures in different stages of growth. This argument still applies to a considerable degree, though ageing is sometimes a serious factor even in submerged cultures. The main reason for the adoption of submerged culture is convenience, economy and reduced requirements of space, together with greater reliability in inoculation and growth patterns. In any case submerged culture is essential for large scale work and it is a waste of time to initiate work in surface culture which will be useless for further development. For these reasons surface culture is used only occasionally or for those exceptional cases in which submerged culture cannot be used.

E. Pathogenic organisms

The handling and culture of pathogenic organisms is a highly specialized operation of which the writer has no personal experience. Laboratories known to specialize in this field are the Karolinska Institute, Stockholm, the Microbiological Research Establishment, Porton, Salisbury, England, Wellcome Research Laboratories, Beckenham, England, and at several places in the United States. Heden (1958, and later papers) has described apparatus and methods used at Stockholm; Elsworth and Stockwell (1968) have described apparatus used at Porton and drawn attention to features important for the culture of pathogens; no doubt advice could be obtained from this source. A detailed article by Chatigny (1961) describes American methods to control the risks involved in the use of dangerous cultures. Some firms, mentioned later, make apparatus specially designed for work with pathogens. Dangers arise both from obvious spills and leaks and also from invisible sprays and aerosols given off by the air-stream and through vents. Care must be taken with all bacteria unless they are known to be perfectly safe. Some fungi, e.g., *Aspergillus fumigatus*, are also dangerous.

II. THEORETICAL ASPECTS OF THE CULTURE OF MICRO-ORGANISMS

A. Physical factors

The physical factors which will be discussed here are agitation, aeration and temperature. Together they govern all aspects of the growth of the cell. Thickness and viscosity of the culture are regarded as part of the problem of agitation, as is also foaming which is dealt with elsewhere in this Series (Volumes 1 and 2).

1. *Agitation-aeration*

When cells grow in stationary liquid medium the nutrients around them are rapidly depleted. The cells may sink or float giving rise to layers of growth in which the medium becomes exhausted and the cells die. Apart from nutrient depletion, the occurrence of a stationary layer immediately surrounding the cells provides a diffusion barrier which restricts the rate at which nutrients can enter the cells. It is the object of agitation to overcome such obstacles to nutrient supply. Circulation is required to overcome layering and the formation of regions of starvation, and turbulence to reduce the diffusion barrier in the immediate vicinity of the cells. Agitation is also needed to facilitate transfer of oxygen to the liquid. The weight of oxygen required is roughly equal to that of the other nutrients. The fact that the medium is not saturated with oxygen facilitates solution, but if the concentration falls too low, metabolism will be seriously disturbed.

With very small single-celled cultures such as bacteria and yeasts the diffusion barrier is relatively small, and the cells tend to remain in suspension. Even an unagitated culture will give a fair amount of growth, but Monod (1942) found in his classical work on growth that aeration by bubbling air gave improved results. With aerobic bacteria the rate at which oxygen will dissolve in the medium largely controls the level of cell concentration that can be achieved. Sargeant (1968b) reports high yields of cells of *Escherichia coli* (10–12 g/litre) with a well aerated stirred fermenter, while shaken flasks gave only 1–2 g/litre.

Mycelial cultures are more difficult to deal with since the hyphae are larger and clumping soon takes place together with layering. Growth is improved by bubbling air but is limited to a few g/litre. The provision of a stirrer giving moderate agitation (5–8 cm dia., 500 r.p.m. in a laboratory fermenter 15–25 cm dia.) produces a further improvement and is adequate with cultures of 10–15 g/litre of dry cells. A larger agitator will be required for up to about 25 g/litre, in laboratory fermenters. Cultures differ very much in these matters and the figures are only intended as rough indications.

Many types of work are carried out with cultures of this density and, with media containing 30–50 g/litre of sugar as nutrient. Specific activities of cells (activity/gram) are high and many growth studies have been made at quite low cell densities. With bacteria, cell densities are usually quite low, often one g/litre or less. In all these cases relatively moderate agitation is therefore adequate for mixing and turbulence, but perhaps not so adequate for oxygen transfer because power input may be too low especially with fast growing bacteria. In such cases, agitation is an oxygen-transfer problem (c.f. Sargeant, 1968b).

With industrial experiments using moulds or actinomycetes media are usually more concentrated and mycelial weights higher, 50 g/litre or more, with apparent viscosities up to 1500 centipoises. The types of agitation described above would be inadequate as the turbulence provided would be unable to deal with the diffusion barriers provided by clumps of mycelium, while it can also be difficult to provide adequate circulation. With such cultures, adequate transfer of dissolved oxygen and nutrients to the cells is probably more difficult to realise than is transfer of oxygen to the liquid.

The problem of agitation will appear different to workers using different processes. For many types of work it will present no difficulty. The problem only becomes difficult when it is desired to work with concentrated media, or when for some reason only very poor agitation is possible.

The mechanical requirements for designing agitation equipment are discussed by Oldshue (1966). Common types of agitator are shown in Fig. 1. Oxygen transfer and circulation should be adequate with an agitator tip

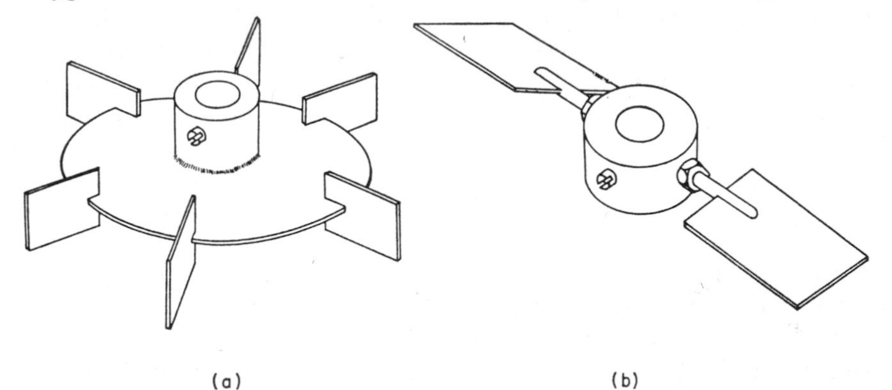

FIG. 1. Common types of fermenter agitators, (a) bladed disc agitator, (b) pitched-paddle agitator.

speed of about 500 cm/sec. An agitator 10 cm in diameter would give this rate at 1000 r.p.m. In fact good results can be obtained with agitators of this size at lower speeds than this, for instance 500–700 r.p.m. The agitator would preferably be of the pitched paddle type, throwing the liquid downwards to improve circulation. Circulation is also helped by reducing the baffling to the minimum required for adequate turbulence. Bladed disc agitators are often used, but they are not particularly suitable for the small fermenters being discussed here (up to about 20-litres capacity) and their advantage would seem to lie in their geometrical similarity to those used in large fermenters. In large fermenters the possibility of using much larger agitators greatly eases the problem of getting good agitation and turbulence.

The other factor of importance is turbulence, required to help the passage of nutrients through the diffusion layers around the cells. Turbulence is thought to be related to Reynold's number, R_e, which equals—

$$\frac{\text{Revs/h} \times \text{agitator diameter}^2 \times \text{density}}{\text{Viscosity}}$$

Assuming full baffling, a Reynold's number of 10^5 gives good turbulence. A 10 cm diameter agitator would give this in cultures of low viscosity, but with very thick mycelial cultures (apparent viscosity, e.g., 1500 centipoises) Reynold's number would fall to as low as 5×10^2. To bring it to a reasonable value a very high speed of stirring is needed, or the agitator must be several times larger, or only low-viscosity cultures can be dealt with. This leads to the view that very thick cultures are best handled in large fermenters (c.f. Nixon and Calam, 1968), a view confirmed by experience in a number of cases. As regards baffling, full baffling is provided by a single baffle 10% of the vessel diameter (in small fermenters, two in larger); the top should preferably be below the surface of the liquid, so as to avoid suppression of circulation at the surface, which increases the tendency to foaming.

Aeration-rates are usually half to twice the culture volume/minute. Much more oxygen is available in the air than is required by the organism.

Summarizing the position, with bacteria and yeast (low viscosity) it is relatively easy to obtain adequate agitation and aeration, especially if the cell concentration is low. With mycelial organisms the problem is more difficult, but the difficulties are not great with moderate cell concentrations (10–15 g dry cells/litre). With such cultures a moderate degree of stirring will give sufficient mixing but oxygen transfer to the medium can be a limitation. With high mycelial concentrations (30–50 g/litre) it is more difficult to achieve adequate agitation especially turbulence. In this case transfer of oxygen to the medium is less likely to be limiting than the passage of nutrients across diffusion barriers in the thick mass of hyphal tufts. Adequate circulation can also be difficult to achieve, especially in conjunction with adequate turbulence.

To exemplify degrees of agitation-aeration in laboratory fermenters the following rates of stirring and agitator sizes are given—

Slight: air bubbled through liquid,

Moderate: agitator 3 in. diameter (8 cm) at 500 r.p.m., with baffling, in a 6 in. diameter fermenter.

Intense: agitator 6–9 in. diameter 1000–1500 r.p.m., or a smaller agitator at a higher speed of stirring.

In discussing agitation, attention is concentrated on the size of the agitator rather than the diameter of the fermenter. Agitator diameters are usually about half the fermenter diameter, ranging from one-third to three-quarters. Thus especially as regards turbulence, fermenter size, which limits agitator size, puts a limitation on the density and viscosity of culture which can be handled successfully.

The above views on agitation are presented with considerable reserve. There are great differences in the viscosity of different cultures and general growth requirements, and it is impossible to lay down rules to cover in advance every situation that can arise.

Agitation in shaken flasks is not well understood. As a rule, media are more dilute than with stirred fermenters, but cell concentrations are higher than may be expected by analogy with similar media in stirred fermenters and yields of antibiotics and other products are unexpectedly good. With rotary shakers, the culture usually spreads over the walls of the flasks and direct transfer of oxygen to the culture may occur. Suggestions are sometimes made that shaken cultures are almost anaerobic, but experiments by the writer (Calam, 1965) showed that with *Penicillium* cultures, although the level of dissolved oxygen was lower than in stirred culture, it never showed signs of reaching zero, and there were no signs that cotton-wool plugs offered a serious barrier to diffusion of oxygen into the flasks. Foaming may occur in submerged cultures; problems arising from this and methods of controlling it are described by Bryant (this Series, Vol. 2).

2. *Temperature*

Cultures are grown at various temperatures from 20°–45°C, moulds usually at 25°C, yeasts, bacteria and actinomycetes usually at higher temperatures (30°–37°C). The effect of temperature is to increase growth-rate with consequent changes in other processes. With moulds, death often occurs at 32°–35°C, setting a rather sharp limit to the use of higher temperatures. With other organisms, higher temperatures are better withstood and in the choice of temperature the fear of killing the culture is less prominent.

The working temperature is of considerable practical importance. At lower temperatures, infecting bacteria have less advantage as their growth is slower. At higher temperatures evaporation becomes greater both from water-baths used to control the temperature of fermenters and from the medium itself. Water vapour may condense in outlet lines and loss of volume must be made up by addition of water or in other ways. On the other hand, at higher temperatures cooling to counteract heat generated by the growing culture is easier because the differential between culture and cooling water is greater and overheating is less likely to occur. The environmental temperature can affect fermentation apparatus in various ways: electric motors and bearings may overheat; the characteristics of some instruments may be affected; at low temperatures oils thicken and solidify making mixing and addition difficult.

When apparatus is situated in a laboratory at 18°–22°C, the usual size of heater may be inadequate for high temperature working; this effect can become more serious if space-heating is reduced at weekends. These problems are not serious in themselves but can be inconvenient if an apparatus usually used at one temperature has to be used at a very different one.

Similar effects occur in surface culture. With mass cultures, overheating due to poor heat transfer from stacks of bottles can be a serious problem and adequate ventilation is essential. With large stacks of bottles, temperatures 5° above the ambient may occur.

B. Media for the culture of micro-organisms

Representative media, used for different purposes, are given in Table I. (1) Czapek-Dox solution has been used for mould culture for many years. It can also be used for actinomycetes. It is a very simple synthetic medium capable of supporting a moderate amount of growth. The nitrogen source is nitrate and therefore the reaction of the culture solution approaches neutrality as the anion is utilized. The presence of 50 g/litre of glucose allows the production of 10–15 g/litre dry weight, but the quantity of nitrogen is limiting. (2) The bacterial medium contains only peptides and amino-acids which favour the growth of those bacteria that have complex growth requirements. (3) For yeasts, an ammonia plus molasses medium is used (Pyke, 1958). To obtain maximum growth of cells, which is very rapid, the molasses is added at an exponentially increasing rate, just avoiding exhaustion of the available oxygen supply. The low oxygen concentration increases the rate of solution of oxygen in the medium, however the oxygen concentration must not be allowed to fall to zero or alcohol production starts at the expense

TABLE 1
Representative media for the culture of micro-organisms

Medium	Czapek-Dox	Nutrient Broth	Molasses + Ammonium sulphate	Corn-steep	Glucose + Urea (glutamic acid production)
Type of organism	Moulds	Bacteria	Bakers' yeast	Moulds and Actinomycetes	*Brevibacterium divaricatum*
Ingredients g/litre	Glucose 50	Peptone 10	$(NH_4)_2SO_4$ 2·7	Corn-steep solids 20–40	Glucose 100
	$NaNO_3$ 2	Meat extract 10	$Na_2HPO_4.12H_2O$ 2·3	Carbohydrate 60	Urea 10
	KH_2PO_4 1				Wheat bran extract (5% of bran extracted with water) 40 ml
		NaCl 5	Beet Molasses (55% sugar) 55·6	Chalk 5	
	KCl 0·5				
	$MgSO_4.7H_2O$ 0·5	pH adjusted 7·4–7·5		pH adjusted to 5·5–6·5	KH_2PO_4 1
	$FeSO_4.7H_2O$ 0·01				$MgSO_4.7H_2O$ 0·5
	pH 4·5				pH adjusted to 7·3 Several additions were made of urea and ammonium tartarate.

1. Corn-steep solids: 100 g contains 7–8% nitrogen, mostly as amino-acids, 5–15% of lactic acid and inorganic matter, mainly Mg, K and phosphate. Obtainable from Corn Products Ltd. (in U.S.A. and other countries); Brown & Polson, Manchester; Manbre and Garton, London.

2. Molasses: cane molasses contains sucrose 32, other sugars 30, nitrogenous substances 3, organic substances 7 and ash 15%: beet molasses sucrose 50, organic non-sugars 20, ash 9·5 and water 20%.

of growth. Molasses is mainly a source of sugar but also contains amino-acids which act as a supplementary source of nitrogen. (4) Corn-steep media are typical of those used industrially. The amino-acids in the corn-steep give rapid growth, carbohydrate (e.g., starch or lactose) maintains respiration in the cells while product formation takes place. The corn-steep may be replaced or supplemented by other cheap sources of organic nitrogen, such as soya flour. Vegetable oils may also be added both as anti-foam agents or nutrients. Media of this type tend to work at neutral pH values, chalk is also often added to neutralize acids. Industrial workers refer to media of this type as "robust", meaning that they give good results without the need for careful balancing of the ingredients which is often necessary with synthetic media. (5) This medium was used for production of glutamic acid by *Brevibacterium divaricatum* (Ajinomoto, 1961). Media of this type, usually with additions of biotin, give yields of 30–50 g. glutamic acid/litre. The nitrogen level is controlled to give low growth and maximum conversion of sugar to glutamic acid. The glutamic acid fermentation is further discussed at the end of the chapter.

In most recipes all the substrates are added at the start. It is increasingly the practice, even in batch processes, to add materials in the form of feeds started at the appropriate stage of the growth-cycle. This is particularly so with industrial developments of media of the corn-steep type. The precise methods for doing this are worked out empirically.

Hosler and Johnson (1953) showed that although penicillin was usually made with media containing complex nitrogen sources, it was possible to obtain comparable results by supplying glucose and ammonia alone, provided that they were correctly balanced and pH was controlled at the right level. Their experiments draw attention to the main principle behind development which is to provide optimal conditions for the organism to grow and then hold it in a productive state for as long as possible. Whether this is done by starting with a medium containing all the components, or by adding them gradually, is a matter of convenience and economics. Media of the synthetic type may become more important in future but to obtain really good results elaborate instrumental control systems will be necessary.

The recipes given in Table I make little reference to the mineral require-ments of the organisms, a subject that is covered in the article by Bridson and Brecker (this Series, Vol. 3A).

The utility of simple synthetic media is limited for two reasons. Firstly pH control is difficult since buffering power is poor. Secondly sugar is the only source of carbon for growth of cells, respiration and biosynthesis. As the first two predominate it is likely that little sugar will be left for bio-synthetic purposes. These defects are reduced in corn-steep or other "natural" media, in which growth is based on carbon from the corn-steep liquor. As the corn-steep liquor is broken down, a variety of acids and ammonia are released and sugar is used more slowly. Under these conditions the pH tends to remain nearer neutrality and there is a longer period for biosynthesis and more materials are available for it. Thus Moyer and Coghill (1946) found that the addition of corn-steep liquor increased the yield of penicillin from 2–6 u/ml to 160–220 u/ml.

Hydrocarbons are now being extensively investigated as nutrients for micro-organisms. Organisms capable of utilizing hydrocarbons have enzymes which enable them to oxidize terminal carbon atoms, thus allowing the parent carbon chains to serve in the same way as fatty acids. These substrates present practical difficulties but are not essentially different from the nutrients which are employed in other fermentations though oxygen demands are higher than with more conventional substrates.

It can be seen from the foregoing that there is a general pattern in media and organisms, which relates the recipe of the medium to the amount of growth produced and the metabolic activities of the organism. It is now necessary to consider the quantitative aspects of this a little further. It should be mentioned in passing that there is probably too great a tendency to put organisms into classes. Thus bacteria are usually regarded as organisms grown on weak complex media and producing relatively few metabolic products. They are also regarded as being rich in enzymes and capable of bringing about many transformations: an expression sometimes used is "bag of enzymes". Fungi and actinomycetes on the other hand are regarded as vigorous producers of metabolic products such as antibiotics and organic acids. In fact not only do both types of organism have a wide range of activities but many fungi produce only cells and carbon dioxide.

The products of micro-organisms fall into several groups, (1) cells, consisting of cell-walls, proteins, storage products (starch, fats) and nucleic acids, (2) primary metabolic products such as citric acid and amino-acids which occur naturally in the organism's growth pattern, and (3) secondary metabolic products such as antibiotics that are formed for no apparent purpose. The first group is obviously the main product derived from the nutrient materials consumed during the growth of cells. The second group occurs normally in tiny amounts but accumulates under special conditions. The third group likewise only accumulates when conditions are favourable. There is some resemblance between the last two since both require the diversion of material which is normally used for cell growth. The other product that takes up the bulk of the organism's food supply is (4) carbon dioxide. While growth is often regarded as the process that consumes most of the materials participating in metabolism, large amounts of sugar are required for maintenance energy (Pirt, 1965). It has been suggested (Righelato, 1967) that, since the production of secondary metabolites is often independent of the rate of growth, these substances must arise from the maintenance side of the activities of the cells.

The data in Table II give an idea of how the materials supplied in medium are related to the material produced by the growth of the culture. Mould fermentations have been used for this purpose, the first, using *Aspergillus terreus* being due to Calam *et al.* (1939), the second from experimental records.

TABLE II

Examples of material utilization in fermentations

1. *Aspergillus terreus*, in stationary culture.

	Carbon conversion g	Carbon conversion % of total
(i) Carbon supplied (glucose)	21	100
(ii) Substances produced—		
Mycelium	6·1	29
Itaconic acid	1·6	8
Carbon dioxide	13·3	63
(iii) Nitrogen in mycelium, 2·5%.		

2. *Penicillium chrysogenum*, penicillin production in deep culture.

	Carbon conversion g	Carbon conversion % of total
(i) Carbon supplied (sugar, corn-steep, oil)	57	100
(ii) Substances produced—		
Mycelium	18	31
Unused, in filtrate	5·6	10
Penicillin	2·4	4
Carbon dioxide	31	55
(iii) Nitrogen in mycelium, 6·2%.		

These results stress the extensive conversion of carbon to carbon dioxide. This is especially the case with the *A. terreus* fermentation. The amount of carbon appearing in the product is low. The mycelium is at first relatively rich in nitrogen, but when uptake slows down, mycelium weight continues to increase as storage materials are laid down. In the case of itaconic acid the product contains only carbon, hydrogen and oxygen. With penicillin, of 1·56 g/litre of nitrogen removed from the medium, 0·33 or 21% was returned as penicillin. This is an appreciable amount of the nitrogen being metabolized and could reach serious proportions in a high yielding fermentation. Uptake of sulphur for penicillin formation can also reach an appreciable level.

These brief examples serve to show some of the principles involved in media for fermentation work. They stress in particular the wasteful nature of the average fermentation. With the itaconic acid fermentation the tendency towards the acid side is not harmful, but with penicillin production the culture must be held neutral, not only to increase productivity but also to avoid destruction of penicillin. Penicillin is a typical secondary metabolite and production starts when growth begins to slow down after the initial rapid growth stage. The problem is to balance the fermentation to keep production going as long as possible.

The metabolism of bacteria and yeasts differs from that of moulds and actinomycetes. The difference is largely quantitative, as similar biochemical mechanisms are involved. Since the cells are particulate and multiply by division, growth is strongly exponential and then ceases when the nutrients are exhausted. The cells remain active but no longer multiply, and the peculiar effects due to regrowth of fungal mycelium are not observed. The stationary cells may possess powerful enzymic action, for instance proteolysis, toxin production or in cases like the glutamic acid fermentation, conversion of sugar to new products may be observed.

A variety of materials is used in preparing media, apart from pure substances such as glucose, sucrose, starch and organic and inorganic sources of nitrogen. These are mainly cheap sources of sugar such as molasses, or protein hydrolysates, corn-steep liquor and meals made from soya beans, oil-seed pressings and the like as sources of nitrogen. Vegetable oils are also frequently used as anti-foam agents and as supplementary sources of carbon. Although the use of natural materials seems unscientific, it often leads to increased productivity. This is because such materials break down relatively slowly, thus providing a slow supply of nutrient. Thus, as has been mentioned above, a medium containing lactose, corn-steep liquor and chalk, in the penicillin fermentation, gives results as good as or better than can be obtained with glucose and ammonia with carefully calculated supplies of nutrient and with pH-control; it thus provides ease of working and development with excellent efficiency as regards production. For a further discussion of media, see Calam (1967), and other chapters in this series.

C. Metabolism, growth and productivity, process development

The present section discusses the general patterns of metabolism which occur in microbial cells, and the way in which they are related to media, agitation and product formation. It is based in the first instance on a study of a graph (Fig. 2) showing in generalized form the pattern of metabolism shown by microbial cultures, involving consumption of nutrients and formation of cells, and passing through several phases of growth and production.

After considering the effects of aeration and agitation, there will be a discussion of methods used in process development. Process development is not meant here to imply particularly industrial work, but the sort of work that is nearly always necessary to provide optimal conditions in laboratory studies. Process development is usually empirical, and depends largely on skill and experience, based on a minimum of data correlated as well as intuition and experience will allow. The collection of sufficient data for a genuinely scientific study usually takes too long for the conclusions reached to be available by the time the process has to be brought into practical use. If the present empirical methods are to be replaced it will be necessary to devise theoretical and practical systems of working which will give results much faster than is possible at the present time. In conclusion another model of cell-metabolism will be discussed, which is now becoming more commonly applied to these questions.

Micro-organisms and fermentations vary so much in their rates of reaction and quantities of cells involved, that only a general account can be given here. The reader will have to adjust the figures quoted to suit the timing and material usages involved in his own work.

1. *General pattern of metabolism*

The fermentation, which is represented by fig. 2, can be divided into four main phases. These are indicated in the figure.

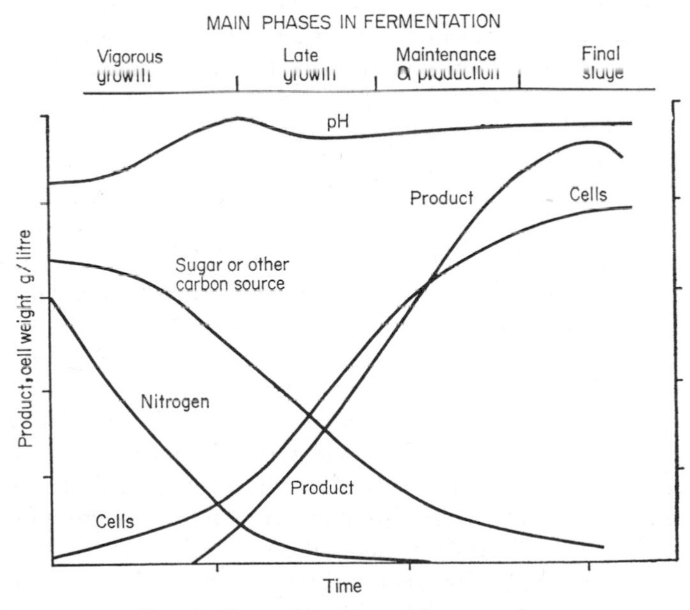

FIG. 2. General pattern of fermentation.

(a) *Growth*. Cell growth, shown here as starting from a small inoculum, proceeds rapidly with utilization of carbon and nitrogen. The growth curve shown is more typical of moulds than bacteria which show a lag-phase and then a logarithmic increase in growth until growth ceases. The differing behaviour of moulds is due to the nature of the mycelium, which shows variation in its composition during the fermentation. During this phase nutrients and oxygen are freely available and growth is therefore only restricted by the characteristics of the cells. The doubling time of bacteria and yeasts is about half an hour or less, of mycelial organisms three to six hours. If growth of cells is the only object, the experiment would be nearly complete at the end of this phase.

(b) *Late growth*. Growth continues during this period but restrictions have begun to appear. These are usually due to the approaching exhaustion of nitrogen or to the limits of the efficacy of the agitation-aeration system being reached or to unfavourable pH changes; less commonly, inhibitors may be formed. If specific growth rate (weight increase in g/g cells/h) is measured, it will now be found to be decreasing, in fact true growth involving synthesis of proteins is about to be replaced by other types of metabolism. These involve either formation of reserves or secondary metabolism; at the same time maintenance energy must be supplied (Pirt, 1965). It seems likely that the enzymes involved in these processes are constitutive, since the organism changes to these activities without an inducer. However some adaptation can occur for example if several nitrogen sources are available (amino-acids, ammonia, nitrate), the steady flow of metabolism may be held up while adaptation to these take place. Product formation commences as a result of all these changes. Righelato (1967) suggests that the formation of product-synthesizing enzymes occurs at this stage, because the specific rate of product formation (g product/g cells/h) increases for a limited period of time. The same effect could be produced assuming the enzyme was present but only became operative gradually as competing growth reactions died out. The specific product formation rate is governed by the culture used, and improved mutants owe their efficiency to this factor.

At this stage, as cell weight (g/litre) reaches a considerable value, agitation efficiency may fall. With bacteria and yeasts, if the tip-speed of the agitator is inadequate, oxygen uptake becomes limiting; with mycelial organisms reduced turbulence will cause a second type of limitation. In either case the result will be a fall in the level of dissolved oxygen. Since the rate of solution of oxygen is proportional to concentration in air/degree of unsaturation of liquid, this will facilitate dissolution of oxygen in the medium. If, however, growth goes too far, dissolved oxygen may fall to a very low level, for instance with bacteria 1–2%,with moulds 5–15% of saturation, with even lower effective levels in the cells. In some cases anaerobiosis or near anaerobiosis may be achieved.

Low oxygen concentrations can affect the metabolism of the cells. Thus Harrison and Pirt (1967) investigating *Klebsiella aerogenes*, found that below 15 mm Hg partial pressure of oxygen, production of butanediol, alcohol and volatile acids replaced growth and carbon dioxide production. With yeasts, growth changes to alcohol production (Pasteur effect) (Pyke, 1954). Quite often a low oxygen tension in the liquid continues for a long time, and if sugar is fed, a reduction in feed-rate may not restore the level of oxygen to the higher level. Oxygen electrodes are of great value in studying this stage and are discussed by Brown (this Series, Vol. 2).

With mycelial organisms another phenomenon often occurs at this stage, that is fragmentation of the mycelium. This is probably because the ratio cell-walls/protein tends to become too high, the protein concentrates at the ends and in branches, leading to a break-up of the threads. This process can be observed under the microscope and is discussed further by the writer in another Chapter of this book (this Volume, p. 567). It may occur now or at a later stage in the fermentation. As a result of fragmentation the viscosity of the culture falls sharply and agitation is improved. During fragmentation, the culture may become very thin, but builds up again as some kind of re-growth occurs. If growth has been stopped by inadequate agitation it will restart and more nitrogen will be consumed.

(c) *Maintenance and production.* Following the end of growth, marked by a cessation of uptake of nitrogen, a maintenance period occurs when the culture metabolizes sugar, or possibly fat, and product formation continues at the original rate (g/litre/h). Some increase in cell-weight may occur due to the formation of storage products, but the previous high viscosity never returns. Nitrogen in the medium remains at a low or constant level, but some nitrogen turnover appears to continue as a little ammonia is present (0·02–0·05 g/litre). Nitrogen metabolism is probably necessary to maintain some biochemical reactions involved in biosynthesis, e.g., methylation. It is interesting that many changes occur in the mycelium of fungi during the fermentation, but production continues steadily, indicating the relative independence and persistence of the biosynthetic enzymes.

During the late-growth and maintenance phases production continues while a certain balance is maintained between supplies of nitrogen and carbon source (usually sugar). This period is prolonged by certain medium constituents such as corn-steep liquor or crude proteins, or by the use of carefully adjusted feeds. A feature of this stage is usually a slow increase in mycelial weight which possibly has a stabilizing effect. As the wet mycelial weight is usually two to four times the dry weight, a considerable amount of water is transferred from the medium to the mycelium. Consequently the concentration of substances in the medium may appear higher than expected. Other aspects of these subjects will be discussed later.

(d) *Final stage.* After a time, which may be from 70 to 400 hours, the mycelium appears to become exhausted and the fermentation comes to an end. It seems likely that this is due to the steady decay of enzymes and proteins. The breakdown of the cells which occurs at this time is associated with foaming.

In concluding this section it should be stressed that with different organisms the phases vary considerably in relative length, and that the phases often overlap. The foregoing account assumes that the appropriate temperature was used; this has to be determined for each organism. For a further discussion of temperature and other physical factors, see Aiba *et al.* (1965) and Patching and Rose (this Series, Vol. 2).

2. *Effect of aeration and agitation on metabolism*

The effects of agitation and aeration on metabolism are very difficult to describe because they are so subtle and because different types of fermentation respond so differently to them. A few points can however be made which should serve as a general guide. In the account already given of agitation, three aspects have been mentioned, rate of solution of oxygen, general mixing (circulation) and turbulence (transfer of nutrients and oxygen to cells). The levels of agitation used and their suitability for different types of culture were also indicated.

During the early growth phase, general mixing is the most important factor, since during this period of free growth, limitations have not yet arisen. When restrictions begin to appear in late growth they may be in oxygen transfer or in general nutrient transfer (turbulence). As has already been explained, the subsequent behaviour of the culture will depend on whether the restriction in growth occurs due to agitation or to exhaustion of the medium. The operator has a choice as to which it will be; very likely a balance will be chosen restricting but not preventing growth. With cultures of bacteria and yeasts of low viscosity, solution of oxygen will probably be the limiting factor (c.f. Sargeant, 1968b) and the decision as to what to do is practical and economic.

With viscous mycelial cultures a limit to turbulence is bound to be reached, though very high mycelial weights can be handled, e.g., 50 g dry mycelium/litre or more, especially after fragmentation has occurred. Hockenhull and MacKenzie (1968) have referred to "biological space" as a factor in fermentation work, that is the quantity of mycelium a fermenter can handle, which can be used in calculating the behaviour of cultures. Although the term has disadvantages, it is useful in enabling the properties of a fermentation system to be visualized, and was used by Hockenhull and MacKenzie for calculating feeds for the growth of *Penicillium chrysogenum*. The "biological space" available in a fermenter could be found by filling it with a salts solution (phosphate, magnesium etc.) and fitting a pH-control system using ammonia, inoculating and then feeding with sugar at an exponentially increasing rate so as to give a high growth-rate (0·1 g/g/h) until an oxygen electrode showed dissolved oxygen to fall to a low level (25%) of saturation. The cell concentration would then indicate the biological space available, and if necessary the agitation system could be adjusted to increase it.

After a mycelial culture has entered the maintenance and production phase, probably with some degree of fragmentation, a lower intensity of agitation would probably suffice. At this point an oxygen electrode usually shows a return to a high oxygen concentration, showing that the demand is being easily met. When variable speed agitation is available, the operator often considers the possibility of increasing the rate of stirring at the time when growth is thick and the oxygen electrode is showing a low concentration of oxygen. The writer has never been able to decide whether this is advisable or not. The effect would be to temporarily increase oxygen and nutrient supply and carry growth forward to a higher level. This might bring on the late-growth and maintenance phases, with a reduction of oxygen demand, or it might bring on a worse condition if the faster agitation rate itself became overloaded. The conditions at the end of the growth phase would in any case be altered, and mycelial weight would increase above the level originally planned. Only experimental work could show the best way to deal with these problems.

The agitation-aeration system thus characterizes the fermenter, be it large or small, and the metabolism of the culture has to be adjusted to suit it. There is however great flexibility in the metabolic system of micro-organisms, as is well shown by shaken-cultures, in which sulphite-oxidation tests show very poor oxygen transfer and yet quite high cell-concentrations can be maintained. Fig. 3 compares the general metabolism of *P. chrysogenum* in plant fermenters and shaken flasks, and shows results which are remarkably similar in spite of very different rates of respiration. This figure was taken from a paper by Nixon and Calam (1968) in which scale-up and agitation problems were discussed in some detail

Finally, agitation should be carefully adjusted to provide good growth conditions with a minimum of splashing and foaming. This involves the careful positioning of the baffles, and the correct choice of volume of medium, speed of stirring and size and shape of the agitator. Practical experience is required for this, but the articles on fermenter design, previously referred to describe systems which are suitable.

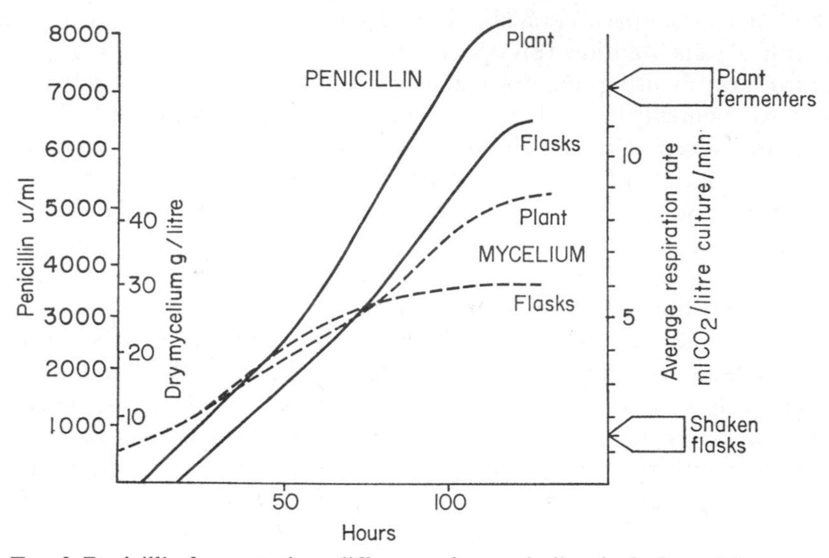

FIG. 3. Penicillin fermentation: differences in metabolism in flasks and fermenters.

3. *Principles of process development*

The expression process development is used here to mean the *ad hoc* development or optimization of a process which is required when work is being started on a new fermentation process, or with an organism and medium which has been used for some time under conditions which are felt to be sub-optimal. The resulting process could then be used as a basis for a more extensive and scientifically planned programme.

The introductory phase involves three steps, (1) an examination of the organism along the lines suggested under the heading "culture" in the section on the conduct of fermentations, so as to choose the best available clone, (2) a survey of any relevant literature, and (3) the testing of the culture on a variety of media (say 10–20), both synthetic and complex, in order to find a starting point for further development. In considering results it is necessary to take into account the errors which are inherent in biological work. Having found a medium which gives a reasonable amount of growth and of a product, a further set of media should be tested in which the compositions are varied over a wide range. Shaken flask culture is convenient at this stage. At this time it is also worth while to carry out tests at different temperatures. If a complex medium shows promise, it is advisible to try several samples of the materials concerned, as often one source is better than another.

After a certain amount of work on the lines suggested some idea of the possibilities of the process should be obtained, along with any technical difficulties that might be involved. It is useful to prepare graphs on the lines of Fig. 1, even if only very rough data are available, and to determine what is limiting the yield, what is the fate of the carbohydrate supplied, and some data about the pH-pattern and the optimum pH. If pH-control is difficult, the addition of chalk is often helpful.

Work lasting a few weeks or months should give a degree of improvement. At this stage it is as well to consider carefully what it is best to do, since it will now be possible to forecast what the future rate of progress is likely to be. If further progress is likely to be slow it may be best to start strain selection work, including mutation, as this often gives a marked improvement in half to one year. Alternately it may be better to turn to pilot-plant production so as to obtain supplies of product without delay. Suggestions about this are made elsewhere in the chapter. It is inadvisable to continue experimental work at random, as progress is unlikely to be made unless a definite lead is available.

4. *A mechanistic model of the metabolism of the microbial cell.*

Graphs such as Fig. 2 have been used for many years to represent microbial metabolism. Recently a different approach has been introduced, for instance by Bu'Lock (1965) to show the mechanism of secondary metabolism. This approach is based on a quantitative analysis of material balances during fermentation referred back to a mechanistic model. This model takes into account the metabolic routes which are known to exist in the cell (c.f. Nicholson, 1967). Some of them are summarized in Fig. 4. Four main metabolic routes are (1) breakdown of substrates to intermediates, (2) formation of proteins, cell-walls and storage materials, (3) respiration and hydrogen-transfer systems and (4) product formation. The balance of these processes, directed by the nucleus, gives the fermentation its characteristic features. This balance is affected by the availability of substrates, rates of oxygen supply and other factors, which are governed by the agitation system employed.

As an example let us consider an imaginary but typical fermentation involving a fungus or actinomycete, in which was consumed 150 g/litre of glucose and 30 g/litre of corn-steep solids and produced 30 g/litre of mycelium and 10 g/litre of a secondary metabolite made from six acetyl-CoA units and with a molecular weight of 300. It is assumed that 65% of the carbon in the starting material was converted into carbon dioxide. This leads to the following carbon balance sheet (weights in g/litre)—

Carbon added		*Carbon used*	
Glucose 150, 42% C,	71	Carbon dioxide, 65% of the C,	55
Corn-steep 30, 45% C,	13·5	Mycelium 30, 40% C	12
		Product 10, 80% C	8
			79
		Unaccounted for	9·5
	84·5	Total	84·5

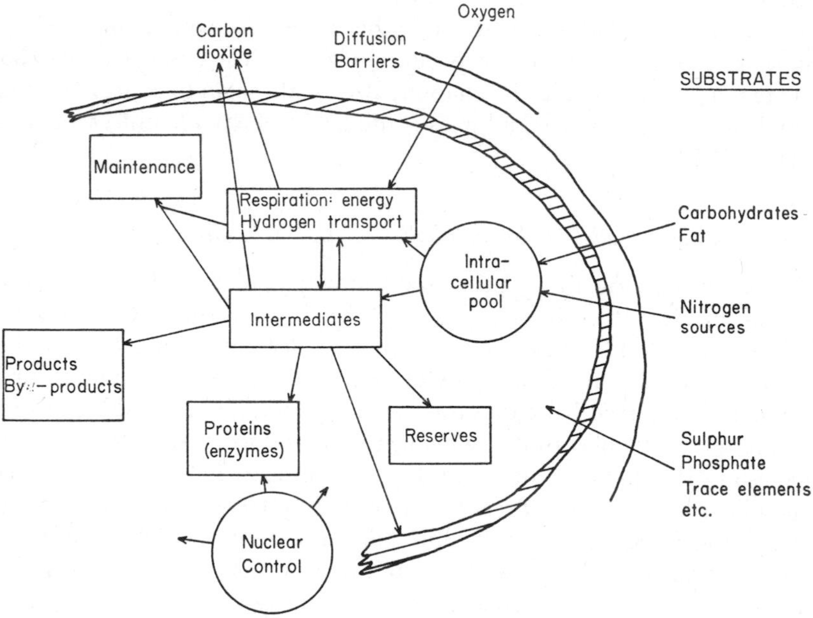

FIG. 4. Sketch of cell and cellular metabolism. The cell wall is cross-hatched.

These figures may not be precisely representative of actuality, but they bring out several points of importance, particularly the very closely balanced quantities of carbon. Of the 9·5 grams unaccounted for, only part would be available as some is always left unused in the filtered metabolism solution. An increase in mycelial weight to 40 g/litre would remove half of the apparently unused carbon, while circumstances causing a 10% increase in the respiration rate would have a like effect.

Another consideration is that the biosynthesis of the product at 10 g/litre is equivalent to 20 g/litre of mycelium in terms of carbon used, while the potential competition for carbon would be increased if more nitrogen was made available to increase the weight of cells produced. Restriction of the growth of the cells would release more carbon for product formation. Cells often contain a pool of amino-acids which may disappear during vigorous growth or starvation (Calam and Davies, 1961; Bent, 1966), thus indicating the competition that occurs for nitrogenous materials, or differences in the rates of their biosynthesis and utilization.

Passage of material into the cells is resisted by the normal diffusion barrier around them. Oxygen supply is resisted by a second barrier raised by the necessity for it to dissolve in the medium before it can become available. The effect of inadequate agitation is to increase the amount of already dissolved food available, relative to the amount of oxygen. If there is only a minimum of oxygen the cells will remain active or grow only slowly, at the same time decomposing nutrients via pathways which require less oxygen, such as the formation of carbohydrates, alcohols or aldehydes (c.f. Harrison and Pirt, 1967). This will change the balance between the other reaction systems in the cells.

A number of biosynthetic routes may operate competitively as parts of a general system. A possibility that such a system suggests is that resonance might arise leading to rhythmical changes in the rate of growth. Duckworth and Harris (1951), (see also Calam, this Volume, page 367) described rhythmical changes in the life-cycle of *P. chrysogenum* that might have arisen in this way. Resonance might be started at inoculation, early vigorous growth being followed by overshoot and a subsequent retardation while the biochemical system readjusted itself. Consider the position when a culture reaches 15 g/litre, inoculated with a small inoculum (0·3 g/litre) or a large inoculum (5 g/litre) such as might be used industrially, consisting of cells that had been growing rapidly in a medium which had just begun to run out of sugar. The small inoculum would grow smoothly in a new rich medium through six generations up to the 15 g/litre, while the large inoculum which had just begun to be starved, would resume growth in the rich medium and reach 15 g/litre after only 1½ doublings in weight requiring much less time. Whatever results were observed the two inoculation systems might be expected to initiate different metabolic patterns in the mass of cells that developed, which would lead to different patterns of resonance, and resonance could be exaggerated by control systems applied to the growing culture.

The mathematical simulation of biological systems is a difficult subject of growing importance. Several kinds of approach to this problem have been made. Pirt and his colleagues (cf. Pirt, 1965; Righelato *et al.*, 1968) resolve the metabolic activities of the organism into growth-rate dependent and growth-rate independent components. The growth-rate independent maintenance energy is particularly important in determining secondary metabolism, while growth-rate affects enzyme stability. Mathematical models are being devised in many laboratories. The following references may be of interest: Koga *et al.* (1967), Moraine and Rogovin (1966), Maxon and Chen (1966) and Ramkrishnan *et al.* (1967). Yamashita and Hoshi (1969) have described the computer control of glutamic acid production, in which a mathematical model is used to continuously predict the progress of the fermentation, which is monitored continuously, thus allowing production to be optimized. It must be emphasized that this Section, though largely of a theoretical nature, is not an attempt to summarize fermentation theory, but rather to suggest some ideas which the practical worker can consider for use in the course of development work.

III. PRACTICAL METHODS FOR CELL GROWTH

A. Apparatus for stirred and aerated culture

Fig. 5 gives a diagrammatic representation of a typical fermenter. It is not intended to represent any particular design but to indicate the important features that are present in all fermenters. Discussion of the various parts of the fermenter and different types of fermenter is followed by descriptions of ancillary apparatus and special fermentation problems. These Sections do not describe in detail how fermenters should be made; references will be given to such descriptions and to sources of commercially available apparatus. It is intended instead to discuss practical problems in using fermenters and the questions which are important in setting up and carrying out experiments with micro-organisms. Table III lists papers describing typical laboratory fermenters such as are commonly used.

Table IV lists a number of firms that manufacture fermenters. It is impossible in an article like this to review them all in detail, but it can be mentioned that the New Brunswick Scientific Co. has been making fermentation equipment for many years, and its products are widely used in the U.S.A. All the manufacturers supply ancillary apparatus, such as temperature control units, pH control units, feed pumps and the like. Commercial apparatus has been extensively reviewed by Solomons (1967). It is recommended that enquiries be made for catalogues from manufacturers or their agents and opportunities be taken to see apparatus in exhibitions or at work in laboratories. Improvements are constantly being brought out, resulting in new and improved designs.

Solomons (1967) has discussed the cost of different makes of fermentation apparatus. Prices are usually fairly high, the cost of an individual small

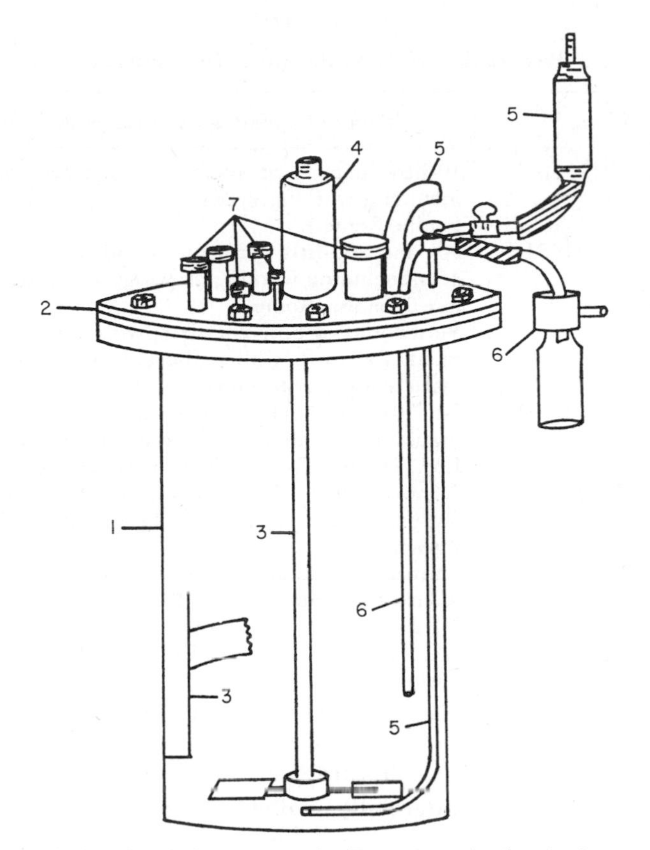

FIG. 5. Typical stirred fermenter. 1. Fermenter body, 2. fermenter cover, 3. agitator and baffles, 4. bearings, gland and drive, 5. air supply, inlet and outlet, 6. sample line, 7. entry points to fermenter.

fermenter is about £400. The question is sometimes raised as to whether it is better to buy fermentation apparatus or to have it made to one's own design. Where only one or two fermenters are required it is probably better to buy, but when a large number are needed a specially designed installation may be preferred as it can be more economically fitted into the space available. Construction to one's own design also enables special features to be incorporated. However, design and construction is a long and expensive business, and although the apparatus may cost less, other factors such as time spent on design and delay in construction should also be taken into account, especially the possibility of mistakes arising from inexperience.

Fig. 5 lists the main parts of the fermenter and its associated equipment, and these will now be discussed in turn.

TABLE III

Papers describing laboratory fermenters

Reference	Types of apparatus and Laboratory of origin
Chain *et al.* (1954)	10-litre metal fermenters and ancillary apparatus, including working drawings. (Istituto Superiore di Sanita, Rome.)
Paladino *et al.* (1954)	90- and 300-litre fermenters and ancillary apparatus, including working drawings. (Instituto Superiore di Sanita, Rome.)
Hedén (1958)	Fermenters of different sizes, with ancillary equipment, for work with pathogenic bacteria. (Karolinska Institute, Stockholm.)
McCann *et al.* (1961)	5-, 30- and 600-litre fermenters used in an industrial laboratory. (Imperial Chemical Industries Ltd., Pharmaceutical Division, Manchester.)
Feichter (1965)	A novel type of stainless steel fermenter, with bottom entry agitator. (Technical University, Zurich.)
Hepple (1965)	Fermenters for the culture of pathogenic bacteria. (Glaxo Ltd., Stoke Poges, England.)
Elsworth and Stockwell (1968)	3-, 20-, 100- and 400-litre fermenters for general use. (Microbiological Research Establishment, Porton.)

TABLE IV

Some commercially available fermenters

Manufacturer	Apparatus
Biotec, A.B., Stockholm.	3-, 10-, 100- and 6000-litre fermenters. Bottom entry agitators or magnetic drives available. Rubber membrane sampling points.
New Brunswick Scientific Co., New Brunswick, N.J., U.S.A.	A wide range of fermentation equipment in all sizes and for all purposes.
Chemap, A.G., Männendorf, Switzerland.	Fermenters from 3 to 100,000 litres capacity, mostly with bottom entry.
Marubishi Laboratory Equipment Co., Tokyo.	Fermenters of 50 to 2000 litres capacity.
Taylor Rustless Fittings, Leeds, England.	Fermenters made to customers designs; a number of standard designs are available.
L. H. Engineering Co., Stoke Poges, Bucks., England.	Self-contained 10-litre glass fermenter units.

B. Parts of the fermenter

1. *Fermenter body*

The whole construction and design of the fermenter depends on the choice of the fermenter body, its size, shape and material of construction.

Popular sizes are 15 cm diameter, 30 cm deep holding 3 litres; 22·5 cm diameter, 45 cm deep holding 10 litres; and 30 cm diameter and 45–60 cm deep holding 15–25 litres. Fermenters of these sizes can all be sterilized in laboratory autoclaves. The largest size mentioned weighs about 40 kg when filled with medium, which is as much as can be lifted without undue strain and lifting by means of a pulley is strongly recommended. All three sizes are useful for different purposes, the larger ones in particular for work involving the extraction of products. Evaporation, absorption of water in the mycelium and sampling usually involves the loss of 0·5–1 litre of medium This is much more serious with the smaller fermenters and makes it difficult to obtain an accurate quantitative estimate of yield. The 15 cm size is also about the smallest in which really good agitation can be provided (but see Evans *et al.*, this Series, Vol. 2); for work involving very thick cultures or experiments on agitator design the larger sizes are recommended. In considering the working capacity of the fermenters a clearance between the medium and the lid of 10–15 cm is allowed; this is desirable to allow for foaming.

For many purposes, stirred fermenters of quite small sizes are used (100–500 ml). These are usually made of glass, and agitated with magnetic stirrers. Temperature is controlled by working in a constant temperature room or by a water-jacket or bath. Such fermenters are specialized in application to situations when a low level of agitation is enough and foaming is slight. A particularly elegant system of small conical fermenters is offered by Biotec, Stockholm. No worker should be put off from using very small fermenters or adopting unorthodox shapes.

Fermenters like that shown in Fig. 5 and discussed in the first paragraph have a height/diameter ratio of about 1·5. For the medium itself this ratio is about unity. For laboratory work this ratio is not very important; it allows a considerable degree of agitation to be applied. A shallower layer of medium is not so satisfactory in this respect. Height/diameter ratios up to 1·5 or even up to 2·0 can be stirred adequately. In the plant this ratio is often 2–2·5 : 1 and calls for two or three agitators on the stirrer shaft. Some workers use rather tall laboratory fermenters with two agitators, but it is more convenient as a rule to use a ratio of 1·0–1·5 and a single agitator even if this gets away from a geometrical relationship to plant fermenters. This type of relationship probably has little significance at the fermenter sizes being considered.

The fermenter body may be made of stainless steel or glass. Sections of industrial glass tube closed at one end may be used, attached to the cover by the normal type of flange. A range of glass jars is made of Pyrex glass (Jobling, Dundee, Scotland) and these can be fitted to the normal type of cover, using a supporting cage. Interchangeable glass and stainless steel bodies are thus possible, but as the glass is thicker than steel, it may be found that the internal fittings are too long or too wide for the glass, and this should be provided for in the original design. For regular use stainless steel is to be preferred owing to its resistance to rough handling.

The fermenter body and cover form a closed vessel which must be able to resist any pressure likely to be applied to it. With portable fermenters attached temporarily to an air supply line this is not a serious matter, but when the fermenter is permanently attached it becomes a pressure vessel and must be strong enough to meet all the safety regulations which are legally required.

2. *The fermenter cover*

The fermenter in Fig. 5 is shown with all its main points of entry through the cover, this being a very common arrangement. Important features here are—

(1) the cover should be rather thick (at least 8–10 mm) so that it can be modified and extra entry points added without distortion in welding;

(2) a strong flange must be provided on the body to hold the cover firmly or tie bars with a baseplate;

(3) the underside must be machined flat to engage closely with the flange on the body using a rubber sealing ring;

(4) great care must be taken in arranging the various entry points and other fittings to the best advantage; when many fittings are to be used it may be worth making a model of the cover as it is easy to run out of space for later use.

There is an increasing tendency to put all the entry points through the upper part of the sides of the fermenters and the agitator through the bottom, leaving the fermenter cover quite plain. With suitable designs any risk of the fermenter contents flowing out via the entry points can be avoided. The gland for the agitator shaft must be perfect. The system has however considerable advantage especially for fixed systems, the plain cover is light and easy to remove, which makes agitator changing easy and cleaning is also greatly simplified.

3. *Stirrer and baffles*

Fig. 5 shows a plain pitched paddle agitator of a type very suitable for small fermenters, with a diameter two thirds or three quarters of that of the vessel (see Fig. 1). For 15 cm fermenters agitators of 7·5–11 cm are suggested, of the pitched paddle type set to throw downwards. These should have small blades 3 cm long, 2 cm wide at 15–45° to the horizontal. For 30 cm fermenters two or four similar blades, or rather larger may be used, giving an agitator 15–20 cm in diameter (Hosler and Johnson, 1953). Stirring should be in the range 400 r.p.m. (mild agitation) to 1000–1200 r.p.m. or more depending on the angle of pitch of the blades. Although a pitched paddle is recommended, many workers prefer the bladed-disc type of agitator. These are best with fermenters of 30 cm diameter and upwards. Some use a small agitator rotating at about 1500 r.p.m. with no baffle and with air passing over the surface. Air is drawn down the vortex and passes into the liquid. Agitation of this type is only suitable for lower population cultures.

Various types of agitators are described in the literature, but the two mentioned here meet most requirements. With small fermenters the type and size of agitator can be important, especially with thick cultures, but experiment is desirable to obtain the best results. A single agitator is preferable to two unless the medium depth exceeds 20–30 cm, when two may be used in narrow fermenters. It is advisable to allow $\frac{1}{10}$–$\frac{1}{4}$ horse power drive motors per fermenter when designing apparatus, although the full power of these is unlikely to be necessary. A single small baffle should be provided, mounted on a removable ring, one tenth of the vessel diameter. With a 30 cm vessel, two baffles may be used. The baffles should be well immersed or circulation at the surface may be reduced which encourages foaming and reduces dispersion of antifoam.

With larger fermenters, bladed-disc agitators are usually used, $\frac{1}{3}$–$\frac{1}{2}$ the vessel diameter, with four baffles.

4. *Gland, bearings and power input*

Great care must be taken to ensure a thoroughly sound design for the gland and bearings of small fermenters. Circumstances are not ideal, the shaft may not be quite straight and considerable strain may develop. If the bearings are unsuitable, trouble may arise resulting in excessive maintenance. A working life of one year between major stoppages for maintenance should be expected, but regular maintenance is essential.

For shafts 30 cm long a diameter of $\frac{3}{8}$ in. (9 mm) has been found adequate but with larger fermenters $\frac{1}{2}$ in. (12·5 mm) is better. Some workers use the sampling and air inlet lines to support a bottom bearing, but this is unnecessary with shafts up to 50 cm long. The upper part of the shaft should be mounted in two ball bearing races as far apart as convenient together with gland or seal. A suitable arrangement includes two self lubricating sterilizable ball races (Fafnir Bearing Co., Wolverhampton, Staffs) protected by oil seals ("Gaco", G. Angus and Co., Wallsend on Tyne, Northumberland).

Other fermenters have used two Oilite bushes with an asbestos-graphite stuffing box between. The former works well with a $\frac{1}{2}$ in. shaft 45 cm long up to 1500 r.p.m., the latter with a smaller shaft up to 800 r.p.m.

Elsworth and Stockwell (1968) describe the use of mechanical seals lubricated with sterile water for the glands of their fermenters (these seals are made by Crane Packing Co., Slough, Bucks, England.). This system is recommended as giving absolute protection against ingress and egress of organisms, which is particularly important when using pathogens.

Top-entry agitators may be driven from a descending vertical shaft coupled to an electric motor, via a universal joint, or a pulley system can be used. Flexible cable drives, preferably attached to a vertically mounted motor provide a convenient method. They are very reliable and quiet and there are no exposed moving parts; the top of the fermenter can be reached from all sides without difficulty. Several fermenters have been seen with open drives or pulleys: frankly, they are a menace to any assistants called upon to use them.

It is desirable to be able to use a variety of speeds in the range 200–2000 r.p.m., according to the type of agitator chosen for different purposes. Variable speed is best provided by an electronically controlled D.C. electric motor such as the "Stardrive" (Lancashire Dynamo Electronic Products, Rugeley, Staffs., England) which gives an accurate setting of speed over a ten-fold range by merely setting a knob. It is advisable to provide ample power, e.g., $\frac{1}{4}$ h.p. for each fermenter (up to 30 litres), though some workers consider $\frac{1}{10}$ h.p. adequate for most purposes. Speed may also be varied by pulley-changing (with belt drives, "timing" belts are recommended) or by gears such as the Kop gear box (Allspeeds Ltd., Accrington, Lancs). If gear systems are used, care must be taken to avoid excessive noise which can be very troublesome.

Commercial fermenters are supplied with both top and bottom entry agitators. Feichter (1965) has given details of the construction of a small fermenter with a bottom-entry agitator. Top entry drives are more convenient for portable fermenters, since with a bottom drive special arrangements would be necessary to protect the shaft when the fermenter was detached from its mounting.

5. *Air supply, inlet and outlet*

Fig. 5 shows a very simple aeration system. Air enters the fermenter via a small air filter (10 cm long, 2–5 cm in diameter depending on fermenter size, packed firmly with cotton-wool) (for air sterilization, see Elsworth, this Volume, p. 123), through a tube (5 mm bore) extending below the agitator. The filter is attached to the inlet line by a rubber tube, secured with a screw clip, and a small cock which can be shut during autoclaving to avoid medium rising into the filter is provided. Air (pressure 5–10 p.s.i.) is metered to the fermenter via a Rotameter gauge. The outlet is a tube (12 mm bore) rising vertically for 5–7 cm and then bending over to an angle of 45°. During sterilization in an autoclave it is plugged with cotton wool and wrapped with paper. During the fermentation the plug may remain, but it is usually better to remove it aseptically and slip on a long, wide, sterile rubber tube descending to a bucket. This will receive any foam (see also treatment of air effluents by Elsworth, this Volume, p. 123). This system works very well with moulds and actinomycetes. The pressure in the fermenter is that of the atmosphere. Fermenters discussed here, based on fig. 5, should be capable of withstanding pressures of 15–30 p.s.i., or more, without leaking. The small cocks fitted in the air and sample lines may be of the type used for gas which are convenient but often must be replaced frequently to avoid leakage. They are not suitable for work with pathogens. Saunders stainless-steel diaphragm valves, in the $\frac{1}{8}$-inch size are better (The Saunders Valve Co., Wolverhampton).

(a) *Evaporation*. With the system described and an air-flow of 1·5 litres/min., at 25°C, a 3-litre culture loses about 150 ml of water by evaporation in the course of 4–5 days. For most purposes this is not a serious matter; at higher temperatures evaporation is more serious but even so can usually be disregarded or can be compensated by an addition of an equivalent volume of water. Some workers (Borrow *et al.*, 1961) correct for evaporation by humidifying the inflowing air by passing it through a tank of water at working temperature. This can lead to deposition of water in air lines and filter that can only be avoided by heating, or preferably by using heaters for air sterilization. It is questionable whether this system is worthwhile for routine use.

(b) *Working under pressure*. By fitting a valve in the outlet the fermenter can be operated under pressure. Preferably an automatically controlled valve should be used so that a steady value will be obtained. The main problem is that of foaming or blockage of the outlet. Preferably a receiver should be provided before the valve so that trouble of this kind can be avoided. The effect of pressure in the fermenter is to force fermenter contents into the sample line. If the cock is not a good one leakage can occur followed by infection. The line may become blocked with mycelium. Fermenter contents may leak at other points and gravity feeds may be reversed.

(c) *Escape of dangerous organisms or gas*. If bacteria are used, especially pathogens, the air outlet is a potential source of danger. There is also a danger of proteinaceous sprays being liberated. These risks must be carefully assessed. An obvious precaution is to allow the rubber tube from the outlet to pass into a large flask via a long cotton-wool plug through which the air escapes. This will catch any foam and at the same time prevent the escape of organisms or spray. If, however, pathogenic organisms are involved the air must be passed through a special filter. To avoid condensation of water this should be heated. Alternatively a heater may be used to sterilize the air (see Elsworth, this Volume, p. 123). The apparatus should be ventilated or placed in a special cubicle for safety. A pilot-plant for growing pathogens is described by Hepple (1965), and Biotec (Stockholm) supply apparatus specially designed for this purpose. Certain cultures produce dangerous volatile metabolites, others may give off radioactive isotopes or inflammable gases. Special precautions must be taken in such cases and adequate ventilation provided.

(a) *Air flow-rate: diffusers.* With stirred fermenters an air flow of 0·5–1·0 vol/vol culture/min. is usually sufficient. Larger volumes rarely do better and can cause foaming. In some cases a porous diffuser is used to break up the air stream. With vigorous agitation this is unnecessary, and with moulds and actinomycetes, growth over the diffuser or clogging of the pores soon makes it ineffective. When no stirring is used a diffuser may be advantageous. Norris (personal communication) has found the use of a very fine diffuser in the form of a stainless steel filter of porosity grade 4 (Gallenkamp, Ltd.) to be advantageous as an aeration device for unstirred cultures. The filter forms the entire bottom of a small (1 litre) fermenter and the air takes the form of minute bubbles in a vigorous state of agitation. It is necessary to have a culture medium of relatively low surface tension for correct bubble formation to occur. Thus a peptone-containing medium is adequately aerated by this means whilst a mineral-salts solution may not be. With a suitably formulated medium aeration (as judged for instance by growth and sporulation of *Bacillus thuringiensis*) is fully comparable to that achieved in a vigorously stirred fermenter at similar rates of air flow.

6. *Sample line*

Fig. 5 shows an orthodox system in which samples are withdrawn via a sample tube into a small bottle (1-ounce wide mouthed MacCartney bottle with a screw neck). The bottle is held in a small attachment (c.f. McCann *et al.*, 1961). The attachment may contain an air-filter pad but it is better to have an air filter in the line attached to the side arm of the attachment. To take a sample, the cock is opened and a bottle full of material removed and discarded. A further bottleful of material is then taken and retained. Other devices of a similar kind are used; the sampling device described by Elsworth and Stockwell (1968) reduces unnecessary loss of material to a minimum. Samples can be taken through an entry point using a pipette, but this is less satisfactory as regards the risk of infection. Another sampling system involves removing the sample with a hypodermic needle through a rubber diaphragm. This method, which is advantageous when working with pathogenic organisms, will be referred to again in the next section. The sample points mentioned above satisfy the principle that when a fermenter has been inoculated it should never be opened to the atmosphere. Biotec (Stockholm) supply $\frac{1}{2}$ in. and 1 in. connectors for making attachments to fermenters without risk of infection.

7. *Entry points*

Entry points are necessary for making additions to the fermenter or for inserting electrodes and probes of various kinds. Fig. 1 shows the usual kind, consisting of a short length of tube closed by a screw cap. It is necessary to avoid any risks due to material being spilt on the top of the fermenter. The use of a cap makes it possible to preserve the sterility of the top of the entry point. If entry points are frequently opened to make additions, there is a serious risk of infection; regular additions should be made by an enclosed transfer system. The opening of an entry point would present a serious risk if a pathogenic organism was in use.

Convenient internal diameters for entry points are 0·5, 2 or 3 cm. Narrow entry points are used for inserting feed tubes and the larger sizes are used for fitting electrodes or for adding material to the fermenter. Separate entry points should be provided for all the items considered likely to be used, with at least one spare (an electrode may fail and it may be advisable to fit another to check its performance). Two or three narrow tubes may be inserted via a wider tube, using a special cap carrying narrow metal tubes through which the feed-tubes can pass. All the entries and caps must be made so as not to leak under pressure, otherwise if foaming occurs, weeping and infection may be experienced.

Electrodes or feed tubes may be inserted by passing them through a well-fitting rubber bung which is inserted in the tube and then held in place by using a piece of adhesive tape of the type which resists autoclaving (e.g., Lassovic P.V.C. tape No. 92, T. J. Smith and Nephew, Hull, England). Rubber bungs made of red rubber (B.S. 2775) should be wrapped with P.T.F.E. tape to prevent them sticking to the metal. Alternatively a screw-on holder can be used with a rubber O-ring. Narrow tubing can be attached using a Luer-Lok type connector. Feed tubes should extend down to the surface of the liquid or drops of feed or antifoam may not penetrate surface foam for a considerable time, or may slowly flow down the cover and walls of the fermenter.

Great care must be taken in locating the entry points. The possibility of more being required later exists, and the ones selected should be placed to leave room, if possible, for more. They should also be arranged conveniently for any electric leads, feed-tubes and other equipment so as to avoid a tangle around the fermenter.

A useful type of entry system involves a rubber disc through which material is added or withdrawn through a hypodermic needle. A metal tube and a perforated cap can be made to accommodate a circle of butyl rubber about 2·5 mm thick of a type that seals after piercing. To make an addition, the solution is drawn aseptically into a sterile syringe fitted with a suitable needle and injected via the rubber disc, which has previously been wiped with alcohol. Additions can be made by an unskilled operator and the rubber disc can be used a dozen or more times. Swedish workers have developed this system to a high degree and Biotec (Stockholm) provide rubber diaphragms held between metal grids which allow a large number of withdrawals or additions each through a new area of rubber. Sampling points of this type can be placed on the side of the fermenter allowing direct access to the culture for sampling.

8. *Small fermenters*

The fermenters so far described have been of a fairly large size, usually holding 3–15 litres. Very similar fermenters are used with working capacities down to one litre. Below this size the production of fermenters of orthodox shape requires specialized equipment (see Evans *et al.*, this Series, Vol. 2), splashing and foaming cause more trouble than with larger fermenters, and it is harder if not impossible to get a vigorous agitation pattern.

Several workers have described small fermenters, for instance Parker (1966), and Puziss and Hedén (1965). The latter fermenter holds 100 to 1000 ml and is conical in form. Recent work with it has been described by Brookes and Hedén (1967). It is now available commercially from Biotec (Stockholm). It may also be used for continuous culture work. See also Evans *et al.*, this Series, Vol. 2.

9. *Improvised fermenters*

The need for an improvised fermenter may arise quite suddenly and the possibility of temporary installations naturally comes to mind, especially as chemical glassware or stainless steel vessels may be available. The main principles are, that the apparatus should be sufficiently strong and that the limitations of improvised equipment should be realized and no attempt should be made to go beyond them. These limitations usually are (1) the degree of stirring possible is usually moderate or low, (2) apparatus usually tends to wear out quickly and becomes difficult to keep in good order, (3) risks usually have to be taken over infection, and (4) the apparatus may be unsafe and liable to give rise to injury. Perhaps the most unfortunate possibility is that an attempt will gradually be made to convert an unsuitable vessel to a permanent fermenter, with the result that a great deal of money is wasted and permanent difficulties established. This can be particularly so on the hundred gallon scale when the cost of temporarily converting a chemical vessel is high and the result likely to be an expensive liability. The easiest way to provide a stirred fermenter is to use a cylindrical thin-walled Pyrex bottle with a magnetic stirrer. This can be sterilized and placed on the magnetic drive, which should not produce overheating. The apparatus is placed in an incubation room. Aeration is by the vortex; by injecting a current of sterile air, oxygen transfer can be increased. This is suitable for bacteria and for other cultures of fairly low cell densities. Glass chemical reactors are another possibility, but they are often not strong enough, and the stirring equipment is unlikely to run long enough or resist infection or autoclaving. Another possibility is in using a stainless steel storage vessel with a detachable lid. The lid and joint will, however, be probably too lightly constructed to give satisfaction. The small variable-speed electric motors often used in chemistry are not generally continuously rated and have poor torque characteristics. Air-motors made by Desoutter (Hendon London, Type MA–1) are much better in this respect.

The author obtained useful results with fermenters made from wide necked, spherical bolthead flasks of 2 litres capacity fitted with two extra necks. One was short and fitted with a bung and a curved air-inlet tube reaching the bottom of the flask. The other was 20 cm long, 2·5 cm bore and plugged with cotton wool and served as an inoculation and sampling point. A brass bearing inserted in a large rubber bung was fitted to the main neck. Through this passed a 6 mm stirrer shaft with a single blade at the bottom 35 × 20 mm (flag stirrer). The neck, bung, bearing and part of the shaft were wrapped in lint and tied to prevent infection. The fermenter with medium (1·2 litres) was sterilized in an autoclave. Stirring was from a bicycle hub bearing with a pulley on top with a belt to a $\frac{1}{4}$ h.p. motor, mounted vertically, and a rubber tube connection to the shaft giving 500–600 r.p.m. The shaft bearing was held in a screw clamp and the flasks stood on a cork ring in a constant temperature room. A long series of experiments on the penicillin fermentation was made with fermenters of this type (Calam *et al.*, 1951); the limitations of the apparatus are obvious.

Lengths of industrial glass piping (Q.V.F., Ltd., Stoke-on-Trent; Jas. Jobling, Wear Glass Works, Sunderland.), with diameters 15 or 22·5 cm, length 30 or 45 cm with flanged ends can be used with mild steel plates (3–6 mm thick) bolted on using the standard joint rings and rubber gaskets. If desired, contact with the steel is prevented by interposing a thin P.T.F.E. or stainless steel sheet. In the centre of the lid a short, threaded vertical tube is welded to form a stuffing box. An "Oilite" bush (Manganese Bronze and Brass Co. Ipswich) is placed at the bottom to support the shaft and the stuffing box filled with one of the proprietary types of prepared graphite-asbestos packing. Holes are bored in the cover to take rubber bungs through which pass the air inlet and other inserts, and a larger hole is provided for inoculation, sampling and making additions, e.g., antifoam. The sample point is best provided with a short vertical tube (20–25 mm bore) with a screw-on cap or bung. The tube could be screwed in and brazed, to avoid distortion by welding. Fermenters of this type which are very strong were used very successfully for preparing small batches of cultures of antibiotics. Industrial tubing is available in short lengths, closed at one end, but these do not provide a flat base and make the fermenters awkward to handle.

C. Installation and ancillary equipment

1. *Mounting and installation of fermenters*

Fermenters are of three types, (1) small portable fermenters which are filled with medium and sterilized in an autoclave, (2) portable fermenters which are sterilized in an autoclave and filled aseptically with sterile medium, and (3) fixed fermenters which are sterilized *in situ*. The first type, usually with capacities of 3–15 litres, are often used in waterbaths, sometimes in small numbers but often in groups of a dozen or more. The second type is often used for research, usually in ones or twos, with independent temperature control. As with the first group, sterilization in an autoclave is easy, as the whole fermenter is sterilized inside and out. Sterilization of fermenters *in situ* is more difficult as cool spots may occur where sterilization is imperfect. Fixed fermenters are usually made in larger sizes (50–litres capacity or more). They are extensively used in industrial work. 50–Litre fermenters (approximately 25 cm diameter, 100 cm high) are used in industrial pilot-plants, often in groups of 20–30. Either the cover or the body is fixed, one or other being removed for cleaning. The fermenters are sterilized empty, pet-cocks being provided so that steam can fill every part of the fermenter and side-branches. Medium is made in a stirred vessel of about 250 gallons capacity, and transferred aseptically. This is also helpful when starchy or proteinaceous media are used. The quantity of liquid in the fermenter is observed via a sight-glass or by an electric probe. Fermenters of this type, with complete ancillary equipment, are made by the New Brunswick Scientific Co. ("Fermacell" and "50-litre industrial").

Each type of fermenter has its special advantages, which should be dispassionately considered when fermenters are ordered. If a fermenter is fixed permanently, it becomes a pressure vessel for insurance purposes, and requires a heavier standard of construction for itself and the appliances attached to it. With fixed fermenters, designs with bottom agitators and a plain easily openable cover should be considered because of greater ease of cleaning: adequate space must be provided to give easy access for servicing.

Some workers prefer to place fermenters on the bench when and where required without any fixed mounting. This has considerable attraction, since quite often fermenter units require an appreciable amount of space and an apparatus that can be brought in and out will have an advantage. A fermenter of the type shown in Fig. 5, incorporating temperature control and with a suitably attached motor could readily be used in this way. Portable fermenters are commonly small in size and stirred magnetically, e.g., models made by Biotec, Stockholm. If a fermenter is used regularly and especially if it has a lot of ancillary equipment, it will occupy bench-space permanently and a special mount is desirable.

The way a fermenter is mounted is a matter of convenience. The installation should include basically—

(i) Temperature control and recorder, relays and the like.
(ii) Air supply and control, air filters.
(iii) Stirrer drive.
(iv) Antifoam system.
(v) Points for compressed air, water, vacuum and electricity.
(vi) Space for flasks etc. containing materials for additions or feeds, feed-pumps and the like.
(vii) Sampling apparatus.
(viii) Switches, indicator lights etc.
(ix) Special control equipment for pH etc.
(x) Motor speed-control units with dials and possibly recorders.
(xi) A valuable addition is a log-timer (e.g., Londex, London) consisting of a digital electric clock which can be started at the time of inoculation and gives directly the age of the fermentation.

This ancillary apparatus, with the associated pipes and cables, can require a great deal of space, especially if several fermenters are in use. Ideally around each fermenter there should be at least 18 in. (45 cm) of clear space on which things can be placed, together with racks to hold electric meters, automatic valves and the like. If top-entry drives are used with flexible drives, there is no need to fix the motors to the frame holding the fermenters, they could well be hung from the ceiling in the same relative position, so that space is available at the back as well as at the front of the fermenters. When planning the installation, a rough model could be well worth while, and any instruments should be looked at critically, and those chosen which require least space and which can be fitted up with maximum convenience. Instruments and dials are best situated remote, but visible from the fermenters. Lighting should be good. Indicator lights should be provided to show whether the drives are working or not. Great care should also be taken to insist on quietness in operation, engineers do not normally take this into account. Individual variable speed motors which are electronically controlled, are quieter than geared systems, while belt drives give off dust. The apparatus must be easy to clean, and unaffected by spills and foam-outs which are apt to find their way into inconvenient places. The fermenters should be easily fitted into place and secured.

2. *Sterilization*

The sterilization of portable fermenters presents no special problems, as they can be put into an autoclave and submitted to steam pressure in the usual way (see Sykes, this Volume p. 77). The fermenter, sample line and sample vessel, and the air filter, with the cocks in the lines closed, should be sterilized after assembly. Some autoclaves have an automatic pressure release: this is inconvienient as the sudden fall in pressure may lead to a boil-out. With the larger sizes of fermenter, trouble may arise if very thick media (e.g., containing seed-meals) are used, since the meals swell and circulation is prevented. It may be necessary to cook the medium beforehand to make it more fluid or to increase the sterilization time. Experiments showed that 5-litre fermenters containing 3 litres of medium lost 120 g in weight through evaporation of the medium during sterilization for 30 min. at 15 p.s.i. Cooling took 45 minutes.

Sterilization of fixed fermenters presents a variety of problems as it is difficult to apply the steam pressure fully to all parts of the lines and attachments of the fermenter, especially if the fermenter contains medium which could be pushed out into the lines and air-filter. This is not a problem with larger fermenters as steam leads can be fitted to the pipework allowing steam to pass both ways into the fermenter and out through the lines, under pressure. Steam can be bled off at the sample point and in other places to sterilize all points. If a Crane seal is used for the agitator this can also be sterilized with steam. If glands are used they should be located below the lid so that they are thoroughly heated during sterilization, and the same goes for bottom-entry drives. A method used for small fixed fermenters is to sterilize them empty and then add the sterile medium aseptically. To sterilize the fermenters, formaldehyde solution is put in, and live steam blown in at atmospheric pressure, which passes with the formaldehyde to all parts of the fermenter. After steaming for 30 minutes with steam emerging from all entry points, the formaldehyde has disappeared and the fermenter is ready. If air filters are to be sterilized by the direct action of steam a packing such as glass fibre should be used as cotton wool and some other packings are destroyed by the passage of steam. Direct steam sterilization of medium leads to condensation and a volume increase.

Infection troubles rarely arise from poor sterilization, except occasionally when a new medium or technique is being used and normal conditions are inadequate. These are cured by a longer sterilization time. Infection usually arises from the use of non-sterile pipettes etc. used for inoculation, additions of materials that were not properly sterilized, from leaks or from mistakes in technique.

3. *Temperature control*

The most accurate control is obtained when the sensing element is inserted in the fermenter and controls the temperature through a small direct heater (Elsworth and Stockwell, 1968). Cooling must also be provided either by a jacket or a small coil or "finger". The insertion of large "fingers" to control temperature should be avoided as they may interfere with circulation and reduce agitation. Care should be taken to ensure that the system has "fail safe" characteristics, as sometimes minor faults lead to unexpected overheating or cooling.

Fermentation temperatures may be controlled by circulating water (at controlled temperature) through water jackets or by placing the fermenters in a bath. In both cases care must be taken to ensure excellent circulation so that all the fermenters are at the same temperature; the sensor must be placed so that it measures the temperature near the fermenters rather than in a water storage tank, where the temperature may be different. The storage tank should be large so that if the circulating water drains back it will not cause a flood. Owing to the large volume of water involved, systems of this type give very steady temperature readings.

Whichever system is used, the control apparatus should respond to changes of 0·25° or less, preferably 0·1° or less. For methods of temperature control, see Patching and Rose, this Series, Vol. 2.

When temperature is controlled by water baths or water jackets held at constant temperature, the temperature inside the fermenter usually differs from this by 1°–2°. This may be allowed for, but the difference may vary throughout the fermentation, depending on the extent of heat production. A temperature difference of this sort is not important when comparisons are being made using similar processes, so that this type of temperature control is convenient when large numbers of fermenters are being used to test mutants or media, or producing samples of new antibiotics. For fundamental research, however, this can be a disadvantage and individual control is to be preferred.

Large changes in the temperature of the fermentation laboratory should be avoided (e.g., if the heating is cut off at the weekend, as it may affect the temperatures of the fermenters).

TABLE V

Pumps for supplying feeds to fermenters

Maker	Types of pumps
Technicon Ltd., Chertsey, Surrey, England.	Multi-stream fixed and variable speed pumps intended for analytical work.
Watson-Marlow Ltd., Marlow, Bucks., England.	The MHRE/72 micro-metering pump gives varying rates and is suitable for electronic control. Larger models are available.
	"Delta" pumps intended for "on-off" work.
Quickfit and Quartz, Stone, Staffs, England.	Multichannel peristaltic pump, variable speed.
Sigmamotor, Middleport, N.Y., USA.	Kinetic clamp peristaltic pumps giving up to 4–6 streams. Speeds variable using a "Zeromax" gearbox (Zeromax, Minneapolis, USA.)

All these pumps can be set to deliver from 1–2 ml/h upwards to 100 to 200 ml or more, with an accuracy of about 1%.

4. *Pumping and feed systems*

It is often necessary to apply feeds of sugar or other substances to fermenters. This is best done by pumping. Gravity feeds are sometimes used but are unreliable owing to back pressures and other factors. A serious problem is that feed rates of only 1–5 ml/h are often the most suitable, while the system must be capable of working under sterile conditions. For this purpose peristaltic pumps are the most suitable.

Table V lists a number of firms that supply pumps that have been found satisfactory in practice. The Sigamamotor pumps are cheapest (about £80), the others cost about £120 to £180 though the Technicon pumps are more expensive than this. Multichannel pumps are relatively cheaper as they can supply 6–12 channels, and these may give different rates of flow if tubing of different size is used. Usually some experience is needed with any particular pump before good results are obtained, but once this is gained, feeds can be maintained with an accuracy of 1–2%. Very low feed rates can be obtained by inserting a timer into the system, so that feeding occurs only at intervals. Timers may also be used for antifoam additions etc. when feed is only added at intervals. In the author's laboratory "on-off" additions are made using multichannel peristaltic pumps which work continuously; a T-piece is inserted in the inlet tube which is opened and closed by a magnetic valve which admits a leak of air thus starting and stopping the feed. The Watson-Marlow MHRE/72 type includes a transistorized speed controls system, giving zero to full rate on applying 0–24 volts and is suitable for proportional control.

Pump tubes made of "Tygon" or silicone rubber are commonly used. Accurately calibrated tubes are supplied by Technicon Ltd. Quickfit and Quartz make tubing for their own pumps. The pump is connected to the fermenter by fine-bore polythene tubing ("Portex" tube, bore 0·5–2 mm; Portland Plastics, Hythe, Kent). To ensure good mixing, the tube leading into the fermenter should continue, inside a narrow supporting tube, so that it ends near or below the liquid level. The tubes may be sterilized by pumping 10% formaldehyde solution through them for 5–15 minutes, and then water, taking care that all parts that will come into contact with the medium will be sterilized.

Some workers use the hydrolytic gas method to displace feed into the fermenter, using a hydrolysis cell taking a current of 10–20 mA. This system is satisfactory if temperature changes in the laboratory are reasonably small.

5. *Systems for the control of foam*

Although foaming is sometimes troublesome, this is not usually the case, for instance Elsworth and Stockwell (1968) regard it as a minor problem. However, it can be serious especially when high mycelial weights of moulds or actinomycetes are grown in rich media. Cultures with synthetic media are easiest to control, and healthy vigorously growing cultures are easier than older ones where fragmentation and autolysis has occurred. It is important to deal with foaming systematically and decisively, and to add antifoam before the trouble becomes serious, otherwise it may get out of control. Glass fermenters are convenient as the culture can be seen, but if growth occurs on the walls this advantage is soon lost. Details of foam control are given by Bryant (this Series, Vol. 2).

Various types of antifoam agents are used. These are—

(1) Vegetable oils, e.g., arachis oil, maize oil or cotton-seed oil, possibly with addition of 1% octadecanol. Using these oils also provides another nutrient for the organism.

(2) Mobilpar-S (Mobil Oil, Caxton House, London S.W.1), a suspension preparation of paraffin wax.
This may be mixed with technical white oil (light).

(3) CLRS, a series of antifoam agents made by Hunko Products, National Dairy Products Corporation, Memphis, Tennessee, U.S.A. They consist of a specially treated mixture of vegetable oils and fats.

(4) Polypropylene glycol (M.W. 2000; Shell Chemical Co.).

A variety of methods is used for adding antifoam agents. Metering pumps of the types already mentioned can be used to add antifoam continuously or intermittently, using a timing device. Foam may be controlled by an addition system operated by an insulated probe dipping into the fermenter and driving a relay sensitive to changes in resistance (250–1000 ohms) when the foam rises in the fermenter. Such a system was used by Hosler and Johnson (1953). They are readily usable in fermenters of 20–30 litres capacity or larger, but in small fermenters the space available for the probe is usually limiting. A short probe soon becomes overgrown with mycelium and ceases to respond to the foam since it loses its resistance. McCann *et al.* (1961) gave details of a pneumatic system which is very effective for foam control. The antifoam agent is put in small wash-bottles and forced by air pressure (5–10 p.s.i. for 20 seconds) which is applied every few hours ($\frac{1}{2}$–3 hours), through a No. 2 Record needle which allows about 0·1 ml to pass on each occasion. This system is especially effective when a number of fermenters require treatment, but there is a danger of the needles being blocked by particles of rubber in the antifoam. It is also necessary to control the temperature of the laboratory to some extent in case the antifoam thickens. A cruder method is to attach a test-tube of antifoam to the fermenter through a wide flexible tube, a small quantity being poured from it at intervals: this is naturally very apt to result in overdosing.

For occasional or routine additions of antifoam agent it is convenient to keep handy some 1 ml Tuberculin syringes with Record No. 19 1 in. needles, already filled with 0·5 ml quantities of polypropylene glycol 2000. These should be sterilized and stored in long test-tubes with metal caps. When required, the antifoam can be injected into the fermenters through a rubber diaphragm fitted in one of the inlet points. This is readily done by an unskilled operator.

Mechanical systems for foam breaking are also used and discussed by Bryant (this Series, Vol. 2).

It is sometimes alleged that additions of antifoam agents interfere seriously with oxygen transfer. This may be true in some cases but our experience with the antifoam agents listed above is that the additions produce only very small changes if any, in the reading of an oxygen electrode monitoring the fermentation (see also Brown, this Series, Vol. 2; and Bryant, this Series, Vol. 2). Destruction of foam can indeed be advantageous as it improves the agitation in the culture. Excessive additions must be avoided at all costs, and if trouble is being experienced it is inadvisable to make repeated additions: some other steps must be taken to reduce foaming, such as a temporary reduction in rate of aeration or an alteration to the speed of stirring. Sometimes a change in anti-foam agent is effective. It should be remembered that many antifoam agents also act as nutrients, and excessive additions may seriously distort the metabolic pattern of the micro-organism.

Only experience will show the best method of controlling foam, assuming of course that the problem is a serious one which cannot be dealt with by a few occasional additions. With cultures of 3–15 litres a small addition (1–2 ml) is usually desirable at the start, which will probably be effective for the first 24–36 hours. After this, a small quantity should be added every few hours (0·1–0·25 ml). Occasionally more frequent dosing is necessary, and some of the Tuberculin syringes of polypropylene glycol described above, are a useful stand by. It is not advisable to add more than 2–5 ml of antifoam/litre of culture during the course of a fermentation. Many fermentations benefit from the vegetable oil antifoam agents that are frequently used, while in shaken flasks a mixture of white oil with vegetable oil increases penicillin production. With shaken flasks, if vegetable oil is not being used, a drop of polypropylene glycol is usually sufficient to control foaming. It has to be emphasized that each type of fermentation is different and the anti-foam system must be adjusted to meet requirements. Polypropylene glycol is at present the best general antifoam agent, but there are occasions when it fails and when one of the others gives better results.

If foaming becomes impossible to control it usually means that the culture is infected, or that the medium and agitation system are unsuitable and it can only be dealt with by making basic changes to the fermentation system. Severe foaming is sometimes associated with the strain of organism used, and a selection programme may be necessary to find a more suitable culture.

6. *Inoculation*

Inocula for fermenters are usually prepared in shaken flasks or in smaller fermenters. Although some organisms will grow from a very small inoculum it is usually best to use 1–10% of a vigorously growing culture. With very small inocula or suspensions made from slopes, growth is often uncertain or delayed, while with mycelial organisms it may lead to pellety growth. The choice of inoculum conditions is a matter of experience; it is nearly always well worthwhile to make trials to find the best system.

Inocula grown in fermenters is sometimes superior to that grown in shaken flasks, though bacteria usually grow well in flasks. Care is needed with the inoculum itself to provide an adequate seed, which may well be a smaller shaken culture. Fermenters are useful because it is easy to follow growth to the right point, as determined by a rapid cell weight determination, pH, sugar or nitrogen exhaustion. Several lots can then be drawn off in bottles to give identical inocula for a number of fermenters.

Fermenters are inoculated by pouring the inoculum into one of the entry points, or better by attaching the bottle to the sample point and blowing in with a brief application of compressed air, or by some other enclosed transfer technique. The reliable inoculation of fermenters is very important to successful and regular working. The labour of operating fermenters is considerable and it is a pity to jeopardise it by using miscellaneous slopes and erratic inocula to start the cultures. Plant workers naturally insist on a proper culture system so as to obtain maximum reliability and economy. It is as well to take the same stand in the laboratory.

7. *The fermentation laboratory*

Many workers use only one or two fermenters which are placed in an ordinary laboratory, there being no special requirements. When larger numbers of fermenters are used special laboratory requirements become apparent. While all kinds of laboratories can be used, the following points may be emphasised.

(i) Ample space round the fermenters is desirable, together with wall space for attaching meters and other equipment. The floor should be well made and easily cleaned after spillage. There should be ample services (water, vacuum, compressed air, gas and possibly steam), and plenty of electricity points (two or three per fermenter). Space should be available for trolleys of equipment. Large cupboards should be provided near the laboratory so that every unnecessary item can be cleared away. The lights should be located to illuminate the fermenters in a convenient manner.

(ii) Benches will be required for several purposes such as cleaning and preparing fermenters, examination of samples, writing notes, weighing etc., and a small workbench with a vice is convenient. A large and easily accessible sink with a large drain is needed for washing up: for 30-litre fermenters a low sink, preferably lead-lined is convenient. With large fermenters a heated pan for medium making is convenient.

(iii) Attention should be given to ventilation and to reduction of noise.

(iv) Any hazards arising from the fermentation or the equipment should be considered. These may arise from pathogenic bacteria or fungi. Dangerously dermatitic substances are given off by some organisms (the author has encountered two causing serious trouble), and danger could arise from inflammable metabolic products or medium constituents. If dangerous organisms are to be used the whole laboratory should be laid out and organized to reduce the risks involved. It is important not to allow dangerous gases or organisms to escape into nearby rooms via holes provided by service lines or the ventilation system.

(v) It is important to see that good provision is made for cleaning the laboratory and keeping it tidy. Fermentations often run through the week-end and the cleaning system may be disturbed, increasing the risk of infection and accidents. Some supervision of fermenters during the night is valuable, if only by unskilled personnel who can at least spot breakdowns and make minor adjustments and records.

(vi) A reliable electricity supply is essential, as power failures lasting more than a few minutes can seriously disturb a fermentation. Equipment should be designed to re-start if the interruption is a short one. In some areas falls in voltage occur, during peak periods, resulting in a reduction of speed of stirring apparatus. Steps should be taken to prevent this. To reduce danger, portable apparatus, magnetic valves and the like should be operated at 25 or 50 volts instead of at the full level of the power supply.

8. *Control of infection*

Infection arises from three main causes, (i) poor design and construction of apparatus, (ii) failure in sterilization, (iii) unsystematic working and general untidyness. To ensure that the apparatus is in good order is mainly a matter of forethought and design; at the present time all the technical problems involved with laboratory fermenters have been solved. Sterilization of equipment is usually straightforward in principle, but it may be necessary to extend the heating time and regulate formalin treatments and the like. If this does not produce success it usually means that there is some weakness in the general system that needs correction. Once the apparatus has been brought into correct use and a sound system of sterilization evolved, it is unlikely that any trouble will be experienced.

Most trouble arises from unsystematic working and general untidyness which leads to mistakes, such as accidentally adding a non-sterile solution or using the wrong pipettes. This is curable by the usual laboratory good housekeeping system, but in fermentation work a series of complex operations is often involved, frequently at an inconvenient hour of the day or night, and it is perhaps not surprising that mistakes are occasionally made. They are usually brought under control quickly enough.

Rapid severe infections are usually caused by mistakes, slight infections which develop gradually arise through inadequate sterilization. Infection is usually detected by plating a drop of culture after diluting, or by inoculating into liquid medium or by adding a little solution to an agar slope. Each type of organism requires a suitable method. Mycelial cultures may be filtered through a sterile filter-paper cone held in the upper part of a culture tube by a wire ring to give a clear solution that falls directly into the medium. Microscopy is useful, but a culture has to be heavily infected before the infecting organisms can be seen.

If the laboratory is dogged by frequent and unexplained infections, the whole system of working should be investigated and the apparatus given a detailed and critical inspection. In this way all possible weaknesses are eliminated. Sometimes methods and apparatus are adequate for simple experiments, but not good enough for more elaborate kinds of experimental work.

D. Growth of cultures in shaken flasks

The first proposals to use shaken cultures were made by Kluyver and Perquin in 1933 but a variety of difficulties delayed the general acceptance of the method. The development of antibiotic production in the forties and fifties led to the need for shake-flask screening systems and established shaken culture as a valuable working tool. The main problem at that time was to produce shakers of sufficient reliability, and even today it is still necessary to make sure that any shaker to be used for fermentation work is thoroughly reliable.

Although the culture conditions in shaken flasks seem rather unpromising, the yields that can be obtained are usually comparable with those in stirred apparatus, though the metabolic pattern may be slightly different. Shaken cultures are essential when any kind of screening work is involved. For general microbiological work shaken culture and stirred-aerated culture are complementary. One or two fermenters take up less space than a shaker capable of dealing with the same volume of medium, and have other advantages. A small shaker is useful for carrying out subsidiary tests and for preparing inoculum cultures.

In dealing with the subject it is proposed to consider four separate aspects, (1) the flasks used for culture, (2) the shaker table, (3) the drive for the shaker and the mechanical supports and (4) the incubation chamber. These are of course all interlocked but each involves its own particular set of important considerations. Although shakers are commonly seen operating at high speeds for antibiotic production, there are many occasions when quite slow speeds are appropriate, and the shake-culture system is more flexible than may appear at first sight.

Although most readers will probably wish to buy their shakers, a considerable number of laboratories use shakers of their own designs. A design for a rotary-shaker that has given good service is that of Chain *et al.* (1954). The following firms offer rotary shakers for sale: New Brunswick Scientific Co., N.J., U.S.A.; L.H.E. Engineering Co., Stoke Poges, Bucks; Gump shakers, Blaw-Knox Inc., New York, U.S.A. New Brunswick supply shakers in incubation cabinets, as do Gallenkamp, London (Orbital Shaker). New Brunswick supply both rotary and reciprocating shakers. Firms supplying chemical apparatus offer silent reciprocating shakers which are useful for some purposes, but which tend to run too fast for shaking flasks. The speed of shaking is usually 220–250 cycles/min., describing a circle 2·5–5 cm diameter.

Although rotary shakers are probably more convenient when using large numbers of flasks, there are occasions when reciprocating shakers can give a more vigorous type of mixing. On the other hand reciprocating shakers probably require a more robust type of construction as the strains are greater.

It is very difficult to say how to choose a shaking machine, probably the best approach is to discuss the matter with other workers who have experience in the field. Points to bear in mind are (*a*) the need for reliability and ease of maintenance, (b) general soundness of construction, and (c) ease of attaching and removing flasks, with uniform conditions for all the flasks involved. A shaker should be able to operate continuously for 6–12 months without attention and require only a minimum of maintenance over the years. All the parts should be readily replaceable so that if a failure occurs there is no unnecessary delay. The moving parts should be easily available for inspection. A shaker has to withstand considerable strains, especially if slightly out of adjustment and any faults in design will eventually become apparent, though it may be quite a long time before stress-cracking appears. These points should be the subject of enquiry when considering a shaker. They will be discussed in more detail below. As with fermenters a reliable electricity supply is necessary, and delay switches should be provided to prevent the shakers being stopped by momentary power failures, and arrangements made for them to be restarted after a breakdown. It is also important to arrange the switch gear so that the shakers cannot be accidentally re-started while loading or unloading is in progress.

Large scale shaken-flask work may call for the storage, washing-up and sterilization of large numbers of flasks, as well as their transport to and from the shakers. Facilities for aseptic methods of inoculation are also needed. In any planning, ample allowance should be made for these requirements. The general methods for operating shaken cultures are similar to those used for stirred-aerated cultures and need not be described in detail.

1. *Flasks for shaken cultures*

On rotary shakers narrow-necked conical flasks are most usually used. Common sizes for fermentation work are 500 ml (holding 50–100 ml of medium) and 2000 ml (holding 300–600 ml). Smaller flasks are sometimes used but usually hold only a small quantity of solution, giving inconveniently small samples for testing. Small shake flask cultures have many possibilities however, and a great many can be put in a small space. For quantitative screening work it is best to work with a single size of flask so as to obtain standardization. The 500 ml size gives a good degree of agitation, and will hold enough medium (50 ml) to provide sample material for quantitative analysis. For many analyses very little material is needed, but usually 10–15 ml of filtrate is convenient. With a thick culture less than 50 ml may be insufficient.

Much consideration has been given to the most economical size of flask. If it is desired to produce large volumes of culture the volume per unit area of shaker is rather independent of the flask size, so that a small number of large flasks is not more economical than a greater number of small ones. There is a variety of occasions when larger flasks are preferable, for instance when growing inoculum and it is desired to avoid transfers.

Some workers, using rotary shakers, prefer spherical flasks with longish necks. These have been seen in antibiotic research laboratories in U.S.S.R. and Czechslovakia. Whatever their advantages they do not stand up by themselves as will the conical form. Spherical flasks are also used for inoculum and spore production for industrial plants, as spherical flasks better withstand vacuum and pressure during transfers. Such flasks are often provided with a short sidearm to which a tube is attached for transfer of the culture.

Test-tubes (2·4 cm dia) may be used on rotary shakers, but the circulation pattern is not suitable for mycelial cultures, reciprocating shakers can be used successfully however for many purposes. The roller-tube system used for tissue cultures does not seem to have become common for work with micro-organisms. The best rotary shaker for test-tubes, seen by the writer, was in Prof. I. S. Alikhanian's laboratory at the Kurchatov Institute in Moscow, where the test tubes (approx. 2·5 cm × 25 cm) were held in two plastic sheets with suitable holes, attached to a shaker table sloping at 15° to the vertical and shaken by a rotary drive. These tubes held 15 ml of medium and gave good results in antibiotic mutant screening. The shakers held 500–700 tubes. Test tubes are suitable for shake cultures using reciprocating shakers.

As yet plastic disposable flasks do not seem to have come into use. Their advantages and disadvantages would be mainly dependent on the economics of washing up, as in the case of plastic Petri dishes. However, with flasks, the addition of medium would be more difficult.

Attempts have been made to improve agitation in shaken flasks, by heating and indenting the walls, by inserting a coil of stainless steel wire around the bottom or by fitting a baffle. Agitation is also improved by reducing the volume of culture medium. Rates of transfer of oxygen achieved in 500 ml flasks are summarized in Table VI with the results from a 30-litre fermenter for comparison.

These results show the considerable effect of baffling of different kinds. Oxygen uptake by cultures was slower than by sulphite solution. Matilova

TABLE VI

Oxygen transfer in shaken flasks

Flask conditions	Flask contents	Rate of oxygen transfer, mM/litre/h	Reference
Plain	culture	4	Calam (1965)
Wire spiral	culture	9–13	Jensen and Schultz (1966)
Plain	sulphite solution	12	Gaden (1962)
Baffled	sulphite solution	108	Gaden (1962)
Plain	sulphite solution	36	Smith and Johnson (1954)
Plain	sulphite solution	18	Matilova and Brecka (1967)
Indented	sulphite solution	108	Matilova and Brecka (1967)
30-litre fermenter	culture	21	Author's records

Results in 500 ml flasks with 50 ml of solution or culture, on rotary shakers at 250 r.p.m., 2·5 cm circles.

and Brecka (1967) found that when flasks were modified to increase the rate of oxygen uptake by sulphite solution, these modifications also increased the rate of production of bacitracin by *Bacillus licheniformis*. Modifications increasing oxygen uptake from 16 mM/litre/h to 43 mM/litre/h raised bacitracin production from 115 u/ml to 343 u/ml, but further improvements in oxygen transfer had relatively little further effect. Calam (1965) found that in penicillin cultures dissolved oxygen did not become exhausted, though it fell to a low level. Schultz (1964) also considered the transfer of oxygen into the flasks, but did not find this to be limiting. Although indentation or other forms of baffling can increase the productivity of shaken cultures its success depends on the type of circulation set up. Under unfavourable conditions the mycelium separates from the medium and is deposited on the walls, alternatively foaming may develop. For large scale use, plain flasks are most suitable and ensure similarity of conditions. Where only a few flasks are in use, it is easier to experiment. As will be mentioned in the next section, flasks (e.g., 500 ml) vary slightly in different countries and this can cause trouble if the clips which hold them do not fit well.

With reciprocating shakers the situation is rather different. Reciprocating shakers may operate at fast or slow speeds. At fast speeds splashing is a serious problem unless special flasks are used. With slow speeds conical flasks are satisfactory. Thus for penicillinase production, following the method of Pollock (1951), *Bacillus subtilis* was grown in 5-litre conical flasks, each containing 1 litre of medium, on a shaker giving 100 strokes (5 cm) per minute. For making glutamic acid with *Brevibacterium* or *Micrococcus glutamicus* the Japanese workers (c.f. Ajinomoto, 1961) recommend that an oxygen absorption coefficient be maintained of $5–6 \times 10^{-6}$ gram mol. of oxygen/ml/min. (240 mM/litre/h). It was found when repeating this work that oxygen absorption at this rate was achieved by using "Revenue flasks" of 750 ml capacity, which have a cylindrical body about 10 cm diameter and 10 cm high, on a reciprocating shaker giving 200 strokes/min., 5 cm long. The normal neck was replaced by a tube 25 cm long, 2·5 cm diameter, as there was considerable splashing. With reciprocating shakers the deposition of culture on the walls and neck can be more troublesome than with rotary shakers.

Test tubes can be shaken very conveniently on reciprocating shakers with a short throw (2·5 cm at 240 cycles/min.). Rapid growth of bacteria can be obtained and this is a useful method for rapid tests for infection in mycelial cultures. Thick growth can be obtained in 8–12 h. On the whole rotary shakers are more generally useful than reciprocating shakers, but the latter have special advantages for a number of purposes.

2. *The shaker table and attachment of flasks*

With rotary shakers the table supporting the flasks is conveniently made of a duraluminium sheet (8 mm thick) to which the flasks are attached by clips. The drive to the sheet and points of attachment should be made via a light steel reinforcement with well fitting bolts. The plate will normally be driven at the centre. Considerable strains can arise at this point and if the attachment is unsound stress-fractures develop and the plate will become broken. Some makes use a steel frame with detachable trays to hold the flasks. These are very convenient but they are often clumsy and heavy. In a small incubation room it may not be easy to slide the trays off and on to the shaker. It is a good rule to keep the shaker table as light as possible, as every increase in weight increases greatly the strain on the bearings. Wood (multi-ply) can be used for shaker tables but is less satisfactory than duraluminium with more than about 36 flasks.

A shaker table 90 cm square will hold 100–500-ml flasks. The number of flasks can be increased by providing two decks of flasks. This is done in several types of shaker, the flasks on the lower deck being mounted on removable trays. It is often more convenient to have two shakers (each for 100 flasks) mounted one above the other, rather than a single shaker. A large single shaker used by several workers is subject to stoppages from all the experiments in progress, and this can be inconvenient. Separate shakers are more easy to load and unload quickly. The shakers should have a space 45–60 cm high between them or the flasks on the lower shaker cannot be reached conveniently. Lack of space is especially disliked by girls whose hair is upset.

Flasks were originally held on the shakers by strips of wood or metal with holes for the necks of the flasks. The modern practice of using clips is better. Clips may be short, gripping the bottom of the flasks, or longer coming well up the sides. In choosing a shaker the clips should be inspected very critically to make sure the flasks are held firmly, but also that the fit is comfortable and the flasks are easily inserted or removed. We have found that British flasks are slightly larger than American, and some American clips are in consequence dangerously tight. Other clips are too loose and the flasks rotate and are cut. When flasks are inserted or removed they very occasionally collapse, and light leather gloves should be worn with the palm and roots of the fingers reinforced with a strong piece of leather stuck on with rubber latex.

It may be convenient to have some of the clips fixed to thin duralumin plates attached to the shaker table by nuts and bolts. These can be replaced by other plates having clips of a different size, e.g., for 2000 ml flasks. Pieces of plywood can be obtained with suitably placed holes, which can be seated over the necks of some of the flasks and held down by springs or rubber bands to hooks on the shaker table. This can be used to support small flasks and to grow inoculum for a further generation of cultures. The shaker table and flasks must be carefully balanced, this is discussed in the next section.

With reciprocating shakers the shaker table is usually oblong but is otherwise similar to those used with rotary shakers. Flasks may be attached with clips, and special types may be used for round or cylindrical flasks. A frame with Terry clips may also be used. Reciprocating shakers often run relatively slowly and it is possible to attach flasks with adhesive tape sticky on both sides, however with large flasks the adhesion may be too strong and the bottoms of the flasks may be broken when they are removed from the shaker.

3. *Drive and supports for the shaker*

Rotary shakers are usually driven at the centre by an eccentric mounted on a short vertical shaft, with a belt-drive from a vertically mounted electric motor. The weight of the shaker-table may be supported by the eccentric, in which case thrust bearings are used, or by eccentrics or cranks at the corners. If the table is supported at the centre it must be steadied, and this can be done by corner supports or by vertical legs with spherical ball-races at the ends. Some shakers are driven by gears instead of a belt drive. Whatever the system it is necessary to make sure that it is very well made with easily replaceable ball-bearings of standard sizes, and that there is no play. Any weakness of mis-alignment will be shown up when the shaker is in use and a faulty design can be most expensive to keep in order. It is important also that moving parts should be easily accessible so that trouble can be located without stopping the shaker and a faulty part replaced during a stop of only 10–20 minutes. Shakers with concealed mechanisms are at a disadvantage in most respects. The drive should be via a continuous toothed timing belt (J. H. Fenner Ltd., Hull, England) with special pulleys to avoid slipping. These are readily available. An ample motor is desirable, e.g., $\frac{1}{2}$ or 1 horsepower for 100–200 flasks, $\frac{1}{4}$ h.p. for smaller sizes. A split pulley on the vertical shaft of the shaker is desirable to facilitate changing the belt. It is not usually necessary to vary the speed of rotary shakers.

A counter-weight should be provided to balance the shaker table; this should be arranged so that the centre of gravity is placed at about half the radius of the shaker table. The shaker table should be loaded with flasks and medium and the shaker tilted so that the shaker table is vertical. The balance weight is then adjusted so that the shaker rotates freely and is balanced in all positions. A well balanced shaker will work for a long time without attention, but the driving belt should be changed every 9–12 months so as to avoid a sudden collapse. Wire net guards should be fitted round the moving parts of the shaker and the edges of the shaker table should also be guarded. The motor should not be mounted under the shaker table, but to one side, so as to avoid a local hot spot. Easy access for lubricating the shaker should be provided. Wear is detectable because knocking starts, and the worn bearing should be replaced as soon as possible (within 48 h) as damage may spread widely into other areas. Shakers are usually fixed down either by screws in the floor or by special adhesive pads.

The best type of reciprocating shakers are those in which the shaker table is, in effect, hung from cranks thus allowing a natural return action as the flasks move to and fro. As with rotary shakers a good quality of manufacture is essential. Speed changes are usually made by altering pulley sizes Very good chemical shakers such as Griffin and Tatlock's "Silent-bloc" shaker are often used in chemical laboratories. These allow only a rather short to and fro motion which is, however, adequate for most purposes. The wooden box often provided should be replaced by a flat board to which flasks may be attached with clips.

An alternative to the reciprocating shaker is the wrist-action shaker in which individual clips are held in retort clamps which are vibrated to give a vigorous wrist-like shaking action. The chemical type of wrist-action shaker is not intended for very long runs, and cannot be relied on to run unattended for more than a few hours. A small reciprocating shaker is probably better, or if a single flask is involved, it may be stirred by a magnetic stirrer.

4. *Incubation chambers for shaken flasks*

Shaken cultures are carried out on shakers which are placed in suitable incubation chambers. These may be cupboards surrounding the shaker, preferably with the motor outside, or the shaker may be placed in a large incubation room.

An incubation room should be provided with a fan system to circulate the air together with heating and cooling to maintain the required temperature. As a rule, when working at about 25°C, the main problem is cooling and suitable refrigeration must be provided. Careful tests must be made to ensure evenness of temperature, especially when several shakers are in use and there is shelving. In our experience local warm spots have been occasionally encountered causing unexpected experimental results. If there is any doubt, flasks should be randomized on the shaker table. In incubation rooms space is often short, and it may be desirable to set up a rough model before installing a shaker as often a move of a few centimeters provides much greater convenience, while it is also necessary to allow for the entry of trolleys for transporting flasks.

Several makers provide elegant cabinets for shakers, fitted with heating and cooling and all other necessary conveniences. These certainly are very attractive and useful. Some of these take rather a small number of flasks. In our experience 36 flasks are enough to cover many types of experiment. Good accessibility of the shaker drive is also necessary. Alternatively a wooden cabinet may be used or a laboratory cupboard employed. These work well at slightly elevated temperatures (28°–30°C) but cooling is a problem. If a temperature of 22°–25°C is required the shaker may often be used in an ordinary laboratory with air conditioning, for instance tucked under a table, but this is only suitable for occasional use.

E. Special types of fermentation

1. *Growth of cultures in aerated medium*

In a previous section details have been given of the methods used for stirred and aerated cultures. The methods used for aerated cultures are the same except for the absence of the stirrer, and it is unnecessary to describe them in detail. In order to make better use of the current of ascending air it is usual to use a taller column of medium; the height/diameter ratio being 3–5/1. It is also possible to obtain some advantage by passing the air through a diffuser. For this purpose a sintered glass disc (Grade 2) diffuser is suitable. Diffusers are convenient for yeasts and bacteria, but with mycelial organisms lasting more than a day or two the mycelium frequently grows into the diffuser causing blockage of the pores. In the early days of penicillin research the Distillers Company workers used aspirator bottles which were aerated by jets of air so arranged as to improve the circulation of the medium, but it is probably easier in the long run to use an impeller. The rate of air flow used should be of the order 1–2 vols/vol medium/min. More air can be used but it may easily cause foaming, which can be troublesome in the absence of circulation.

An obvious development, especially to save air, is to use a tall narrow fermenter or tubular fermenter. This method is often used for bacteria and yeasts. The industrial production of baker's yeast is usually carried out in large aerated but unagitated fermenters, and a variety of small yeast fermenters is seen in different laboratories, the simplest being a measuring cylinder. The tower fermenters developed by A.P.V. Ltd., London, (Royston, 1966) exploit very well the special properties of the yeast fermentation. The use of a narrow tube, however, can greatly restrict circulation, especially with mycelial organisms. For these reasons tower fermenters are not generally used with fungi. In the submerged culture process for citric acid production very restricted growth is required, and in this case tower fermenters have given good results (cf. Miles Laboratories, 1968).

A development of the aerated fermenter is the air-lift fermenter. This consists of a fairly narrow vertical tube with an expansion tank at the top from which a return tube leads back to the bottom. Air is admitted at the bottom and rises up the vertical tube causing circulation of the medium and some degree of mixing. Temperature control can be achieved by electric heating tape wrapped round the tube. Air-lift fermenters have been used by Blakeborough *et al.* (1967) but their application is rather limited.

Aerated fermenters provide the easiest approach to improvisation, as a bottle with an air inlet tube and a plug in the neck is the simplest way to produce a fermenter. Owing to the poor degree of agitation only cultures of low density can be supported, but this method has been used for inoculum production with some fungi. Growth should be restricted by using a dilute medium.

2. *Anaerobic fermentations*

The methods currently used for the growth of anaerobic bacteria have been reviewed by Sargeant (1968) and Willis, Hungate, Hobson and Barnes, this Series, Vol. 3B). These organisms may be grown by inoculating a batch of fresh oxygen-free medium in a bottle or flask with cells from a previous culture. Growth can be vigorous until some inhibiting factor or exhaustion of nutrient arises. However high yields of cells are seldom obtained in this way and better results are produced by using orthodox stirred fermenters in which conditions are carefully controlled.

Examples of recent work at the Microbiological Research Establishment, Porton, given by Sargeant (1968) were the growth of *Thiobacillus denitrificans* and *Clostridium pasteurianum*; the object being to obtain high yields of cells. *T. denitrificans* in bottles, gave only 30 mg/litre dry weight of cells, but in a 140 litre stirred fermenter with an atmosphere of nitrogen and the pH controlled at 6·9 with sodium bicarbonate, yields as high as 700 mg/litre were obtained. With *Cl. pasteurianum* yields of cells as high as 5 g/litre could be obtained by adjusting the medium and controlling the pH at the optimum value, in this case 5·6–5·9. In fermentations of this kind growth seems to be limited by the concentration of inorganic ions, which tends to become excessive when the pH has to be controlled in the presence of the organic acids which accumulate as spent energy sources. Other fermentations on similar lines gave increased production of enzymes and toxins. Sargeant also referred to the possible use of biphasic and dialysis cultures for the growth of anaerobes. The use of stirred fermenters for the commercial production of bacterial toxins by anaerobic bacteria has also been described (Hepple, 1965, 1968).

3. *Hydrocarbon fermentations*

Considerable interest has been aroused recently by the possibility of using hydrocarbons as a substrate for microbial growth. The production of fodder yeast in this way appears to be a possible means of alleviating protein shortages in many countries. A variety of organisms, moulds, actinomycetes and bacteria can grow on hydrocarbons, but certain yeasts of the genus *Torula* seem particularly suitable.

Information is available in a series of patent specifications prepared by B.P. Ltd and other firms. In an early patent Filosa (1963) described the growth of *Candida lipolytica* on crude paraffins in shaken flasks or aerated bottle cultures. In later patents, 5-litre fermenters were used with high speed vortex aeration (e.g., Filosa, 1967); in others, fermenters of 60 litres working capacity were used, with vigorous stirring and aeration. In the B.P. work heavy gas oil was used, the yeast utilizing only the long chain normal paraffin fraction. Thus Vernet and Filosa (1967) inoculated 40 litres of medium with 20 litres of *Candida* inoculum. After 25 hours the cell density was 18 g dry cells/litre. In this case one litre of heavy gas oil was added at the start and the pH was controlled with 10 N ammonia and the working temperature was 30°. When 20 ml of ammonia had been added, hourly additions of gas oil were made, totalling 17 litres, i.e., 250 g/litre; normal paraffins in the oil fell from 10% to 1·8%. The medium contained ammonium phosphate 2 g/litre together with small quantities of NaCl, Mg, Zn, Mn, Fe and yeast extract. In other patents the process was improved by adding the oil as an emulsion, thus reducing the doubling-time of the cells (Laine *et al.*, 1967). It was necessary to adapt cultures by growing the inoculum on paraffin-containing medium.

Substantial yields of glutamic acid were obtained when *Arthrobacter simplex* and other bacteria were grown on decane, using an aqueous medium containing acetate (Kyowa, 1967); cultures were in flasks shaken at 220 strokes/min., on a reciprocating shaker. A paper by Hamer *et al.* (1967) describes shaken and stirred culture experiments and discusses safe methods for working with inflammable methane gas.

4. *Brewing and effluent disposal*

These two subjects embrace important aspects of the fermentation field, in which a large number of specialized techniques are used. The range of activities involved is too large to be dealt with in a general article of this kind, but a new approach to the subject and many interesting technical developments are to be found. Cf. Royston (1967), Curtis (1968) for recent developments in brewing. In the case of effluent disposal the Annual Reports of the Water Pollution Research Laboratory, Stevenage, Herts. (H.M. Stationery Office, London) summarize recent publications which often contain much of interest on techniques and instrumentation. The Proceedings of the International Congresses on Water Pollution are also of interest.

5. *Surface cultures*

Surface culture has been widely used for growing micro-organisms. It is only necessary to provide an incubator or incubation room and suitable culture bottles. The type of incubation room and other facilities required for shaken cultures is generally suitable for surface culture work. Shelving should be open and the layout should allow good circulation of air among the culture flasks so as to avoid local heating. On the whole, less heat is given off by surface cultures, but in a confined space in a heated building some cooling may be necessary. It is assumed that only a few hundred bottles are likely to be used at once. With very large numbers, special ventilation is required. An absence of vibration is necessary and the flasks should be arranged so that they are not disturbed during incubation. It is important that incubation rooms should be kept clean and tidy and preferably shelving should be of metal, so as to reduce the risk of contamination and prevent attack by mites. Most types of micro-organisms can be grown in surface culture.

Many types of flasks and bottles can be used for surface culture. A number of special bottles have been devised. Some of these are included in the following list which includes some other possible fermentation flasks—

TABLE VII

Flasks and bottles for surface culture

Culture bottle	Approx. price (£)	Volume of medium (litres)
1-litre conical flask	0·35	0·35
Roux bottle (offset neck)	0·64	0·4
"Glaxo" culture bottle	0·7	0·5
Thompson bottle ("double Roux")	0·9	1·0
Fernbach flask (2·8 litres)	1·3	0.8
Quart bottle (round, narrow neck)	0·05	0·35

The volume of medium quoted gives a depth of about 2 cm, this being suitable for most purposes. The total output of product usually depends on the volume of medium up to a depth of 2–3 cm, after which it declines.

The choice of bottle or flask depends on circumstances. With a given shelf area the size of the flask does not make a great deal of difference to the actual area of medium on the shelf, but bottles that can be stacked allow several layers of medium to be used and are thus more economical in space. Larger bottles such as Thompson bottles mean less bottles to inoculate and handle, but the individual bottles are heavier to wash-up and transport. It may also be difficult to get good cover on larger areas of medium. When only a small quantity of metabolism solution is needed, the type of bottle is less important, but if large quantities are needed the cost of special bottles is considerable and this capital is permanently tied up. Conical flasks or ordinary bottles of suitable shape, such as the quart bottle mentioned (used on its side) may be used for other purposes or even discarded. The handling and storage of special bottles can itself be quite a problem. All these points should be borne in mind in the selection of fermentation bottles. Flasks and bottles are conveniently plugged with cotton wool or special plugs of aerated polystyrene (Scientific Furnishings Ltd., Poynton, Cheshire, England) may be used. For some purposes trays, fitted with suitable lids, are useful. Ordinary enamel-ware may be used. Such fermenters are certain to become infected, but if for instance, very acid conditions are being used and the fermentation lasts only 2–3 days, satisfactory results may be obtained. Fermentation bottles are sterilized in the autoclave in the usual way, inoculated and placed or stacked on the shelves of the incubator. Growth usually requires 1–3 days for bacteria and yeasts and 6–10 days with mycelial cultures.

Inoculation is carried out with a suspension of cells or spores. While it is not usually difficult to get satisfactory results some attention may have to be given to inoculation in order to obtain good surface cover. With inadequate inoculation cover is only partial or may be semi-submerged. Mould spores may sink or float depending on their surface properties. If they sink, as many as 99·9% may be lost, while others stick to the sides. Often moulds are grown on bran, the particles float long enough to allow a surface cover to form. The bran culture should not be allowed to be too wet when growth is complete. When only a few bottles are in use, inoculation is not a problem, but with a hundred or so, especially with large bottles, it may be difficult to obtain even, healthy growth. Growth tends to be outward from the walls and is rapid for 3–5 cm and then slows off. With large flasks this can leave a large uncovered area in the centre of the medium. If the bottles are subject to vibration or are disturbed, many spores sink or the early growth may sink. For this reason surface-culture fermenters must be kept as still as possible.

6. *Comparison of solid and liquid medium cultures*

Culture methods using solid media may be divided into three main categories. The first and in some ways the most important uses media solidified with agar, a method of the widest application in microbiological work. The second uses bran and other similar materials for producing spore inocula for laboratory and other fermentations. Lastly there is the use of various solid medium processes for preparing foods, especially in the Far East, and for cheese manufacture; bran cultures were also used at one time for producing enzymes.

Growth on solid agar medium is analogous to that on stationary liquid media, though with a greater degree of restriction of growth. As a rule a layer of medium only a few millimeters thick is used, so that colonies are prevented from becoming too large. Characteristic features of the method are the local exhaustion of the medium around the colonies and interference between colonies due to the production of toxic substances. When an agar plate is spread with identical spores a uniform population of colonies usually arises consisting of colonies all about the same size. If, however, the growth of any of the colonies is delayed, for instance because of uneven germination, the slower growing colonies are at a disadvantage and are smaller and often of a slightly different appearance. This effect is particularly marked with actinomycetes, when the colonies may appear most varied. Careful consideration should be given to these factors when plating new cultures of micro-organisms.

Colonies of all types of organisms, on agar, show characteristic shapes. Bacterial colonies may be rough or smooth, raised, rounded or flat. Mould colonies are characteristically folded or wrinkled. These effects are due to differential growth between different parts of the colonies, as well as to drying out of parts of the agar which changes the shape of the bed on which the colonies lie.

These types of behaviour occur on agar in much the same way as on the surface of liquid medium. Another similarity was noted by Frank *et al.* (1951), who observed that the number of spores produced by *Penicillium notatum* depended not on the surface area but on the volume of agar medium. Differences from liquid medium arise when molten medium is inoculated and allowed to set in deep layers. Growth may occur near the top (aerobic), in a layer some way below the surface (microaerophilic) or in the lower parts (anaerobic) or in some cases bacteria may grow throughout the layer showing their adaptability to all conditions.

Diffusion effects involved in the use of agar media have already been mentioned. These have been put to use in many different ways, such as in assaying antibiotics and in isolating deficient strains and so on.

Using bran for producing mould spores is a well established practice. It is only necessary to moisten the bran and sterilize it in test-tubes or flasks. The best degree of moisture should be found by trial, it is usually 15–25%. The bran is inoculated, shaken to mix and incubated. Usually a good layer of spores forms over the bran flakes. The spores may be washed off the bran and used as a suspension, or portions of bran may be scooped out aseptically into surface-culture flasks, when the floating bran particles give a good start to the formation of a felt. The best alternative is pearl barley which is a soft grain giving good yields of spores, but care must be taken in sterilizing it as it may soften and form into masses. Other grains are sometimes referred to. Russian workers apparently use millet, but the writer found that millet has too hard a cortex and spore formation was not good. Bran is not really very convenient for large scale use as it involves a good deal of labour in preparing culture-bottles, and the bran suspensions obtained are not easy to handle. It is however an easily applied stock method for making inocula, and its use can save a lot of experimentation. Rarely the culture liquifies the bran, forming a soggy mass which will, however, dry out if given long enough. In the Far East a number of foods and sauces are made by inoculating piles of rice, soya beans or other materials with suitable fungi. These processes, and other methods of treating foods, have been reviewed by Hesseltine and Wang (1967).

F. The conduct of fermentation experiments

Many types of experiment are involved in fermentation work and it is not possible to describe them in detail. The present section is only intended to gather together a number of points which experience suggests are important in the conduct of experiments, and it is hoped that they may be of use to anyone starting work in the field. Fermentation work is not especially difficult but requires more planning and is more laborious than some other fields of study. The various topics are dealt with in a general way, in each case the details must be arranged to suit the particular culture being studied.

Full details of working methods are given in the cited papers by Chain *et al.* (1954), McCann *et al.* (1961), Elsworth and Stockwell (1968) and Sargeant (1968b). The paper by Hosler and Johnson (1953) and its references also give many interesting details of fermentation work.

1. *The culture*

The organism will probably be received as a slope or as a lyophilized preparation. It is important to make sure that it is carefully handled and stored otherwise it may change and cause a variety of troubles. It is desirable to set up a master culture as soon as possible and to organize a system of sub-culturing so as to ensure a regular supply of sub-cultures for experimental work. It is highly undesirable to rely on slopes that are held in the laboratory or refrigerator for indefinite periods—it is surprising how weeks turn into months and the cultures become more and more decrepit and unreliable. The master culture should preferably be lyophilized in serum or stored in soil or as an oiled culture. Methods of culture preservation have been described by the author (Calam, 1964) and by LaPage *et al.*, this Series, Vol. 3A.

On receiving the culture, slopes should be set up. One of these should be covered with sterile medicinal paraffin and held in the refrigerator, or transplanted into a stab or meat culture with bacteria, to serve as a temporary master culture. This may suffice for a short programme. The culture should be streaked or plated and examined for variability. In many cases these tests will be satisfactory and it is possible to make a permanent master culture immediately. If variation is observed or suspected, colonies should be picked off on to slopes for examination and classification. Often variation on streak plates is more apparent than real. If several variants are found they should be re-plated and the most stable forms selected, the variants should also be tested for productivity. In this way a satisfactory culture can very quickly be chosen for experimental work. It is still necessary to look out for mites which can cause trouble if they are accidentally introduced.

Once the master culture has been established, a regular system for producing working slopes can be set up. Slopes will probably keep for one to three months in the refrigerator and should be regularly replaced by fresh sub-cultures from the master. Slope to slope transfers over several generations usually lead to trouble with variability. At all times slopes should be carefully examined, and any signs of variation noted.

2. *Inoculum cultures*

Production cultures are usually inoculated with a specially grown inoculum culture. It is quite possible to start directly from a slope culture, but this may be unreliable, leading to slow initial growth and poor reproducibility, due to too small a quantity of organisms being present to give a good start. It is usually advisable to add 1–5% of well grown inoculum to the production fermenter and in many cases 10% or more is used. Each culture is however a law unto itself.

Inoculum cultures are often grown in shaken flasks. When fermenters are being used and 0·5–1 litre of inoculum is required, it is best prepared in another fermenter. Fermenters are also convenient for inoculum production as often several fermenters can be inoculated from a single batch of culture, thus reducing variation. The seeding of inoculum cultures can itself require care, and rather than use slopes it may be better to use bran cultures or large spore preparations grown on agar flats in bottles, or to grow small shaken cultures in flasks. These and other details are a matter of experience. Usually a well grown slope will inoculate 1–3 litres of medium.

The criterion used for judging inocula is the results they give in the fermenter. In practice inocula are usually satisfactory if they give rapidly increasing, even growth in 1–2 days with moulds and actinomycetes, or ½–1 day with yeasts or bacteria. Growth should be free from lumps or large pellets which are caused by inadequate seeding. Microscopically the inoculum culture should be full of healthy living cells. The decision as to whether the inoculum is ready may be based on appearance, quantity of growth on standing, changes in pH or exhaustion of sugar: for detecting the latter "Clinitest" tablets (Ames Coy. Slough, Bucks.) and the same company's "Clinistix" are invaluable. Inoculum should not be allowed to grow too long as it may lose its vigour.

With care there should be no risk of infection at the inoculum stage. A test for infection may be made during the growth of the inoculum, using a rapid bacterial growth test which will give a result before the inoculum matures. Unfortunately light infections with bacteria are not usually visible under the microscope so that a microscopical examination is not much help. The main defence against infection is good technique.

For growth of inocula rich media are usually used, such as one containing glucose and corn-steep liquor. The medium used for the production stage is often suitable.

3. *The main growth or production culture*

The operation of the main fermentation stage usually involves setting up the flasks or fermenters in a routine manner. The only problem is to make sure that all the mechanical devices are operated correctly. The following notes draw attention to points which should be kept in mind.

The apparatus should be assembled and checked well in advance. With new apparatus a trial run of a day or two is advisable so that all the automatic controls may be adjusted and any missing parts or defects made good. This should include a thorough test for freedom from infection. All the ancillary equipment such as air-filters, feed-pumps and sampling units should be carefully checked. When the fermentation is about to start time should be allowed to make sure that all is in order and that all the solutions, pipettes and so on are sterilized and ready. The volume of medium should be carefully checked, if necessary by weighing the fermenter, as losses sometimes occur during sterilization due to boil-outs. The air-inlet system should be checked by momentarily closing the inlet cock and observing that the air-flow ceases; if there is a leak in the filter or inlet-pipe the lack of air may pass unnoticed. It is convenient to have a large pressure cooker handy in which small objects can be sterilized at a moment's notice.

Especially in a laboratory where several groups of workers are using similar media it is best to have loose-leaf reference books containing all the recipes, which should be individually numbered. This is to prevent confusion as to what recipes are being used. Thus a medium can easily be modified by one of the groups, so that a name like "Czapek" can cover a multitude of variations. It is better that each variant should have a number of its own. If A.R. chemicals are being used there is little trouble, but where works materials are used care should be taken over specifications and it is advisable to retain reference samples. Thus names like "starch" or "chalk" can cover a range of substances which can be changed without the fact being noticed. Care should be taken with some media when recipes are first used. Thus media containing starch can become very thick and make-up involves knowing the correct way to deal with this. In some cases it is better to measure an ingredient separately (by calibrated scoop) into each flask rather than to try to pour out separate lots of a thick suspension from a make-up vessel. The meaning of terms should be clear, for instance whether "5% glucose" means 5% before or after inoculation and whether anhydrous, mono-hydric or crude glucose is meant. Details of this kind can cause confusion especially when analytical studies are involved.

Fermentation experiments should be planned well in advance. This seems obvious enough, but as a rule there are more ideas than fermenters, and unless the object is clearly defined and the methods clarified the best results may not be obtained. Planning should include the programme of additions and sampling and any other variations that are to be made. As fermenters are operated continuously several workers may be involved, and it is important that all should be clear as to what should be done.

It is important to use a reliable system of recording. This should give all details of culture, medium and fermentation conditions as well as any additions, and the final volume of culture. The system should be simple, often standard forms can be used. For records during the fermentation, a foolscap note-book with the pages ruled into four columns is convenient. In these can be written the age of the culture, description at the time, any changes made or additions and the "status" of the culture. In the latter column can be written in a word or two whether the culture was thought to be satisfactory or not. It is as well to make a full record of all aspects of the culture, including density of growth and colour. If possible, objective standards (such as colour patches) should be used, as people vary in their opinions about these matters. The records should include data from chart and instrument readings. In addition to these records it is as well to have a day-to-the-page diary so that informal notes can be freely jotted down at any time. Temporary notes of pH readings, mycelial weight measurements and the like should be written in handy notebooks rather than on bits of paper, so that they will not be lost. Dates should also be noted. This emphasis on recording arises from the fact that fermentation work is relatively slow and it is easy for information to be lost. If this occurs, experiments may have to be repeated with considerable loss of effort and time.

Especially in planning shaken flask work, including large numbers of flasks, it is well to keep in mind the possibility of using statistically designed experiments such as fractional factorial designs (Davies, 1954) to obtain maximum use of labour and equipment.

Finally, in planning fermentation work it is essential to keep in mind the time factor, especially when the results will be needed. Sometimes unlimited time is available, but often time is short and the programme must be limited so that useful (if incomplete) results will be obtained at a time when they can be applied to some other part of the work of the laboratory.

G. Examples of the development of fermentation processes

In the present section accounts are given of the development of some well-known fermentation processes. The object is neither to give a historical description nor to discuss theoretical aspects of fermentation. It is rather to give an impression of the lines which have been followed in development work and the way in which microbiologists have approached these problems; it is also hoped to indicate some of the newer ideas which are now developing. This Chapter deals with theoretical and practical aspects of fermentation, and it is hoped that this final section may indicate the way in which these two sides of the problem can be drawn together and reduced to practice.

Each fermentation is different, yet all have many features in common. Describing the development and optimization of the process is more difficult. Some workers proceed scientifically, some intuitively, the best use a combination of both processes.

The fermentation processes chosen for review are all well known. The fact that they relate to manufacturing methods is not in itself of importance, since the principles of process improvement may be applied on any scale. Continuous fermentation is not discussed.

1. *Citric acid*

The modern commercial process arose from the work of Currie (1917) who obtained high yields of citric acid with *Aspergillus niger* when using a medium with a very low pH (2·0). It is necessary to keep sugar always slightly in excess so as to avoid decomposition of the acid by reversed metabolic reactions. Basically the process consists of culturing *A. niger*, a fast growing profusely sporing organism, on a molasses medium (10–15 % sugar) at an initial pH of 2·0 to 4·0, with a low concentration of nitrogen. A felt of the organism rapidly forms on the medium, which is held in large trays. A yield of about 70% of the sugar is obtained in 5–10 days. A leaflet issued by the Selby firm of J. Sturge and Co. shows a view of a citric acid plant with storage tanks for molasses, low brick buildings with ventilation cowls where the culture trays are incubated, together with other buildings for isolating and purifying the product.

Foster (1949) refers to the possibility that a shift would soon be made to a submerged culture process. Martin (1963), writing fourteen years later indicates that this had now started and mentions the use of 30,000-gallon stainless steel tanks for this purpose. It is clear that in spite of intense research in many laboratories and numerous claims of success, a reliable submerged culture process is difficult to achieve, depending as it does on the careful adjustment of the content of trace elements in the medium. Miall (1966) summarizes some recent developments. These include an increase in yield following a change from molasses to glucose, and the addition of phenols to check the growth of *A. niger* and enable less pure starting materials to be used, an effect already observed with some of the lower alcohols. The morphology of the organism has been improved by adding surface active agents to the medium to encourage the formation of mycelial growth rather than pellets. For an account of current techniques, see Miles Laboratories (1968).

Thus the development of the citric acid process continues, still clearly proving difficult to standardize and evoking every kind of skill to bring success. In the meantime, the surface culture process is still holding its own on the manufacturing scale. Foster's comment is interesting. He points out that as citric acid was the first fungal product to be made commercially it had a tremendous effect on the outlook of microbiologists as it brought out so many points of importance, such as (a) strain specificity and the need for careful strain selection, (b) physiological degeneration, (c) control of the process to enhance the production of a particular product, (d) the remarkable sensitivity of the fungus to small changes (trace elements); additionally it aroused great interest in the metabolic pathways involved in product formation. At the same time it created the feeling that fungi were adapted for surface growth. There is no doubt, wrote Foster, that the history of mould metabolism would have been very different if the citric acid process had been started in submerged culture. This instance exemplifies how stubborn and complex a fermentation project can be, and also indicates what can happen when a particular technique becomes established that wider knowledge and experience would suggest as being ultimately inferior.

2. *The griseofulvin fermentation*

Griseofulvin, a valuable antifungal antibiotic, was discovered by Oxford *et al.* (1938). The development of the commercial process has been described by Rhodes (1963). Griseofulvin was first obtained by surface culture but yields were low. Extensive research by Glaxo Ltd. led to the discovery of a mould (*Penicillium patulum*, C.M.I. No. 39,809) which would produce griseofulvin under submerged conditions, and the development of media for shaken flask fermentations. The medium eventually chosen had the recipe, corn-steep liquor to give 0·15% N, lactose, 7%; KCl, 0·1%; KH_2PO_4, 0·4% and limestone, 0·8%. It was inoculated with a well-grown culture of the mould and incubated on the shaker for 10 to 12 days at 25°C. Purely synthetic media were unsatisfactory since they gave uncontrollable fluctuations of pH (Rhodes *et al.* 1955).

Mutation and selection resulted in the isolation of a strain of *P. patulum* S 152, which gave twice as much griseofulvin as the original culture. Later Hockenhull (1959) developed the fermentation process for use on the large scale. This involved the inoculation of a corn-steep plus chalk medium and starting a feed of glucose which is carefully varied to allow the pH to rise to 6·8–7·2 by 20 hours and to remain at that level for the rest of the fermentation. Considerable skill and experience were necessary to do this as the requirement for sugar depended on various factors such as type of medium and fermenter, rate of aeration and age of culture. An excess of carbohydrate rapidly lowered the pH. In this way griseofulvin production was increased to about 7 g/litre in 12 days.

An interesting feature of the griseofulvin fermentation is the great difficulty of working it successfully in stirred laboratory fermenters. Possibly because of the thick, viscous type of culture produced, turbulence is reduced, or perhaps for some other reason, the process only works well in shaken flasks or on the large scale. Fortunately it seems possible to carry forward shaken flask results to the full scale, but experience with the griseofulvin fermentation casts doubt on a good deal of what is written about the practice of scaling up. It is a good thing that those engaged in developing the process did not allow themselves to be put off by an apparently insoluble problem.

3. *Glutamic acid and lysine*

Glutamic acid is an important amino-acid which finds wide application in food preparation for improving flavour. Formerly produced mainly from gluten, it is now being made by bacterial fermentation. Lysine is required as a fodder supplement, as many grasses used for feeding animals are deficient in lysine and have therefore a very poor food value. Lysine is also made by bacterial fermentation, using biochemically deficient mutants of the glutamic acid producing organisms. Developments in this field have been led by Japanese workers, and sharply emphasize the contribution being made by Japan in microbiological research and manufacture. A valuable review of the methods of producing glutamic acid and lysine has been given by Kinoshita (1963) in whose laboratory most of the initial work was done. Kinoshita describes the bacteria used and fermentation conditions (see also Kyowa, 1957 and Huang, 1964).

A typical glutamic acid fermentation is described by Kinoshita (*loc. cit.*). The medium contained glucose (10%), urea (0·5%), trace elements and small quantities of corn-steep liquor and casein hydrolysate. During the first 48 hours cell growth was rapid, reaching 8 g/litre, and frequent additions of urea were made to control pH at 7·5–8·0. Production of glutamic acid was maximal between 12 and 48 hours. Commercially, cheap sources of sugar are used (molasses or hydrolysed starch), yields are in the range 30–50 g/litre. Recently glutamic acid has been obtained from fermentations based on hydrocarbons (Kyowa, 1968). An account of a search for suitable organisms has been given by Chen *et al.* (1959) who screened 2522 strains from 458 soil samples. Of these 94 gave 10% of glutamic acid from glucose and 5 gave 20%.

A direct fermentation for the production of lysine was described by Kyowa in 1960, using auxotrophic mutants of *Micrococcus glutamicus* requiring homo-serine. Other auxotrophs producing lysine required a combination of threonine and methionine, a combination of threonine and cystathione or of threonine homo-cystine. It is considered that accumulation of lysine occurs, instead of glutamic acid, because the mutations block the pathway to methionine, thus directing biosynthesis into the desired direction, but it is not clear why this should also stop formation of glutamic acid. The media used for lysine production are generally similar to those used for glutamic acid, and adjustments in the growth factors are important. Yields of about 25 g/litre can be obtained; higher values than this having been reported in some cases. According to Brecka *et al.* (1966) lysine-producing bacteria tend to lose their auxotrophy, with a corresponding reduction in lysine yield. This can be corrected by adding erythromycin to the medium which selectively destroys the prototrophic revertants. Legchilina and Shishkina (1965) have achieved the same result by introducing a second deficiency, thus increasing the stability of the organisms and reducing production of unwanted by-products.

In the previously mentioned article, Kinoshita gives an interesting discussion of the biochemistry of the glutamic acid and lysine fermentations. The reaction network involved in glutamic acid formation uses both the glycolysis of sugar via triose-phosphate, and the pentose shunt, the proportion being determined by cultural conditions. Formation of glutamate is increased by the presence of ammonia which couples the isocitrate and L-glutamate dehydrogenases so as to trap the oxidation product in the form of L-glutamate. A balance between aerobic and anaerobic conditions is also needed, a specific oxygen diffusion-rate in the actively growing culture of $3–5 \times 10^{-6}$ moles O_2/atm/min./ml being optimal.

Biotin has been shown to have a critical effect on the level of glutamic acid production by *Micrococcus glutamicus*. The best glutamate-producing organisms require biotin for growth in any case, its absence leading to distortions of the morphology of the cells. Biochemically, biotin is involved in the NADP-regenerating system which plays an important part in glutamate biosynthesis and in the carbon dioxide fixation step which is also involved. Biotin also affects oxidation of organic acids. With too little biotin the products of glucose metabolism would flow too freely into the T.C.A.-cycle, while with too much there would be excessive formation of lactate from pyruvate, with diminished production of glutamic acid.

As mentioned above, lysine is produced by auxotrophic mutants of organisms which form glutamic acid. In this case high concentrations of biotin are desirable together with an optimal balance between the required amino-acids. Kinoshita (*loc. cit.*) stresses the fundamental importance of the glutamic acid and lysine processes, because of their introduction of auxotrophic organisms into commercial fermentations, and the use of a knowledge of the mechanisms involved for the optimization of yields. This is an example of the type of thinking about cell-reactions which Japanese workers and others are beginning to develop, and which are coming to play a fundamental role in the design of fermenters.

REFERENCES

Aiba, S., Humphrey, A. E., and Millis, N. F. (1965). "Biochemical Engineering." Tokyo University Press.

Ajinomoto (1961). Br. Pat. 876,943.

Ajinomoto (1968). Br. Pat. 1,104,355.

Bent, K. J. (1964). *Biochem. J.*, **92**, 280–289.

Blakeborough, N., Shepherd, P. G., and Nimmons, I. (1967). *Biotechnol. Bioengng*, **9**, 77–89.

Borrow, A., Jeffreys, E. G., Kessel, R. H. J., Lloyd, E. C., Lloyd, P., and Nixon, I. S. (1961). *Can. J. Microbiol.*, **7**, 227–276.

Brecka, A., Plachy, J., and Kalina, V. (1966). *Process Biochem.*, 359–365.

Brookes, R., and Hedén, C.-G. (1967). *Appl. Microbiol.*, **15**, 219–223.

Bu'Lock, M. J. (1965). "The Biosynthesis of Natural Products." McGraw-Hill, London.

Calam, C. T. (1964). "Progress in Industrial Microbiology", Vol. 5. Heywood, London.

Calam, C. T. (1965). *J. gen. Microbiol.*, **41**, i.

Calam, C. T. (1967). *Process Biochem.*, **2** (No. 6), 19–000.

Calam, C. T., and Davies, A. (1961). *Abstr 5th Int. Cong. Biochem.*, Moscow, 282.

Calam, C. T., Driver, N., and Bowers, R. H. (1951). *J. appl. Chem.*, **1**, 209–216.

Calam, C. T., Oxford, A. E., and Raistrick, H. (1939). *Biochem. J.*, **33**, 1488–1495.

Chain, E. B., Paladino, S., Ugolini, F., Callow, D. S., and Sluis, J. Van der, (1954). *Rc. Ist. sup. Sanità*, **17**, 61–86.

Chatigny, M. A. (1961). *Adv. appl. Microbiol.*, **3**, 131–192.

Chen, C. T., Tu, T. M., and Chen, L. T. (1959). *J. Ferment. Technol. Osaka*, **37**, 29.

Currie, J. N. (1917). *J. biol. chem.*, **31**, 15–37.

Curtis, N. (1968). *Process Biochem.*, **3** (No. 4), 17–19.

Davies, O. L. (1954). "The Design and Analysis of Industrial Experiments." Oliver and Boyde, Edinburgh.

Demain, A. L. (1966). *Adv. appl. Microbiol.*, **8**, 1–27.

Dicks, J. W., and Tempest, D. W. (1966). *J. gen. Microbiol.*, **45**, 547–557.

Dworkin, M., and Foster, J. W. (1956). *J. Bact.*, **72**, 646–659.

Elsworth, R., and Stockwell, F. T. E. (1968). *Process Biochem.*, **3** (No. 3), 15–18.

Fiechter, A. (1965). *Biotechnol. Bioengng*, **7**, 101–128.

Filosa, J. A. (1963). U.K. Pat. 914,568.

Filosa, J. A. (1967). U.K. Pat. 1,089,887.

Foster, J. W. (1949). "Chemical Activities of the Fungi." Academic Press, New York.

Frank, M. C., Calam, C. T., and Gregory, P. H. (1948). *J. gen. Microbiol.*, **2**, 70–79.

Gaden, E. L., Jr. (1962). *Biotechnol. Bioengng*, **4**, 99–103.

Hamer, G., Hedén, C.-G., and Carenberg, C.-O. (1967). *Biotechnol. Bioengng*, **9**, 499–000.

Harrison, D. E. F., and Pirt, S. J. (1967). *J. gen. Microbiol.*, **46**, 193–211.

Hepple, J. R. (1965). *J. appl. Bact.*, **28**, 52–55.

Hepple, J. R. (1968). *Chemy Ind.*, 670–674.

Hedén, C.-G. (1958). *Nord. Med.*, **60** (No. 31), 1090–1098.

Hesseltine, C. W., and Wang, H. L. (1967). *Biotechnol. Bioengng*, **9**, 275–288.

Hickey, R. J. (1964). "Progress in Industrial Microbiology", Vol. 5, Heywood, London.

Hockenhull, D. J. D. (1959). Br. Pat. 868,958.
Hockenhull, D. J. D., and Faulds, W. F. (1955). *Chemy Ind.*, 1390.
Hockenhull, D. J. D., and MacKenzie, R. M. (1968). *Chemy Ind.*, 607–610.
Hosler, P., and Johnson, M. J. (1953). *Ind. Engng Chem.*, **45**, 871–874.
Huang, H. T. (1964). "Progress in Industrial Microbiology", Vol. 5. pp. 55–92. Heywood, London.
Humphrey, A. E. (1967). *Biotechnol. Bioengng*, **9**, 3–24.
Imeida, Y., Takahashi, J., Yamada, K., Ucheda, K., and Aida, K. (1967). *Biotechnol. Bioengng*, **9**, 45–54.
Jensen, A. L., and Schultz, J. S. (1966). *Biotechnol. Bioengng*, **8**, 539–548.
Kluyver, A. J., and Perquin, C. H. C. (1933). *Biochem. Z.*, **266**, 68–81.
Koga, S., Burg, C. R., and Humphrey, A. E. (1967). *Appl. Microbiol.*, **15**, 683–689.
Kyowa Hakko Kogyo Kabushi Kaisha (1960). Br. Pat. 851,396.
Kyowa Hakko Kogyo Co. (1967). Br. Pat. 1,095,724.
Laine, B. M., Vernet, C., Champagnat, A., and Vinh, K. C. (1967). Br. Pat. 1,059,888.
Legchilina, S. P., and Shishkina, T. A. (1965). *Genetica, Moscow*, (No. 3), 182–189.
Leudeking, R. (1967). *In* "Biochemical and Biological Engineering Science", (Ed. N. Blakeborough.) Academic Press, London.
McCann, E. P., Parker, A., Pickles, D., and Wright, D. G. (1961). *Chem Engr., Lond.*, **157**, A61–A71.
Martin, S. W. (1963). "Biochemistry of Industrial Microorganisms." Academic Press, London.
Martin, S. W. (1956). Can. Pat. 526,987.
Matilova, V., and Brecka, A. (1967). *Appl. Microbiol.*, **15**, 1079–1082.
Maxon, W. D., and Chen, J. W. (1966). *J. Ferment. Technol. Osaka*, **44**, 255–263.
Miall, L. M. (1966). *Rep. Prog. appl. Chem.*, **51**, 452–459.
Miles Laboratories (1961). U.S. Pat. 2,970,084.
Miles Laboratories (1968). Neth. Pat. 67,08573; U.S. Pat. 2,494667 (1950).
Monod, J. (1942). "Recherches sur la Croissance des Cultures Bacterienne", Thesis of 1942. Hermann, Paris, 2nd ed. 1958.
Moraine, R. A., and Rogrovin, P. (1966). *Biotechnol Bioengng*, **8**, 511–524.
Moyer, A. J., and Coghill, R. D. (1946). *J. Bact.*, **51**, 57–78.
Nicholson, D. E. (1967). "Metabolic Pathways" (chart). Koch-Light Laboratories, Colnbrooke, Bucks.
Nixon, I. S., and Calam, C. T. (1968). *Chemy Ind.*, 604–606.
Oldshue, J. Y. (1966). *Biotechnol. Bioengng*, **8**, 3–24.
Oxford, A. E., Raistrick, H., and Simonart, P. (1938). *Biochem. J.*, **33**, 240–248.
Paladino, S., Ugolini, F., and Chain, E. B. (1954). *Rc. Ist. sup. Sanità*, **17**, 87–144.
Paladino, S. (1954). *Rc. Ist. sup. Sanità*, **17**, 145–148.
Parker, R. B. (1966). *Biotechnol. Bioengng*, **8**, 473–488.
Pirt, S. J. (1965). *Proc. R. Soc. B.*, **163**, 224–231.
Pollock, M. R. (1950). *Br. J. exp. Path.*, **31**, 739–753.
Puziss, M., and Hedén, C.-G. (1965). *Biotechnol. Bioengng*, **7**, 355–366.
Pyke, M. (1958). *In* "Chemistry and Biology of Yeasts", (Ed. J. H. Cooke). Academic Press, London.
Ramkrishna, D., Fredrikson, A. G., and Tsuchiya, H. M. (1967). *Biotechnol. Bioengng*, **9**, 129–170.
Rhodes, A., Cross, R., Ferguson, T. P., and Fletcher, D. L. (1955). Br. Pat. 784,618.

Righelato, R. C. (1967). Ph.D. Thesis, University of London.
Righelato, R. C., Trinci, A. P., Pirt, S. J., and Peat, A. (1968). *J. gen. Microbiol.*, **50**, 399–412.
Royston, M. G. (1966). *Process Biochem.*, **1** (No. 4), 215–221.
Sargeant, K. (1968a). *Chemy Ind.*, 85–88.
Sargeant, K. (1968b). *Process Biochem.*, **3** (No. 4), 51–53.
Schultz, J. S. (1964). *Appl. Microbiol*, **12**, 305–310.
Smith, C. G., and Johnson, M. J. (1954). *J. Bact.*, **68**, 346–350.
Solomons, G. L. (1967). *Process Biochem.*, **2**, (No. 3), 7–12.
Spencer, J. F. T., Tulloch, A. P., and Gorin, P. A. J. (1962). *Biotechnol. Bioengng*, **4**, 271–279.
Vernet, C., and Filosa, J. A. (1967). Br. Pat. 1059,885.
Walsby, A. E. (1967). *Biotechnol. Bioengng*, **9**, 443–447.
Wolnak, B., Andreen, B. H., Chisholm, J. A., and Saddeh, M. (1967). *Biotechnol. Bioengng*, **9**, 57–76.
Yamashita, S., and Hoshi, H. (1969). Reports of 4th International Fermentation Symposium (Academic Press.) In press.

CHAPTER IX

Methods for Studying the Microbial Division Cycle

CHARLES E. HELMSTETTER

Radiation Physics Section, Roswell Park Memorial Institute, Buffalo, New York, U.S.A.

and

Department of Biophysics, State University of New York, Buffalo, New York, U.S.A.

I. INTRODUCTION

Studies on the course of biochemical events during the microbial division cycle generally require techniques for identifying or isolating large numbers of cells of the same age. During the past 15 years, two fundamentally different experimental approaches have been developed to meet this requirement. In one approach (synchronization techniques), a culture is treated to yield a synchronously dividing population either by withdrawing a group of cells of the same age, or by inducing phased growth in the entire culture through a programme of environmental changes. Repetitive analysis of a synchronously dividing culture can provide considerable information concerning the sequence of biochemical events during the division cycle. However, treatments that yield synchronous growth may also disturb the course of these events, and this consideration led to the development of the second approach. In methods of the second type (classification techniques), the cells in the culture are sorted out according to age (or size) so that biochemical analyses can be performed directly on cells in various interdivisional stages in the original culture.

The technique an investigator might choose for a particular experiment would depend upon the organism and the problem to be investigated. Many cell cycle studies are concerned with what Maaløe (1962) has described as the "normal" division cycle. A culture of "normal" cells can be considered to be a culture undergoing balanced growth, i.e., a culture in which every extensive property increases by the same factor in the same time interval (Campbell, 1957). For studies of this kind, techniques that do not introduce transient distortions in biochemical processes must be employed (Abbo and Pardee, 1960; Maaløe, 1962). However, objections to the concept of a "normal" division cycle have been voiced, in regard to its definition and on whether normality is a necessary requirement for investigating the process of division. It has been suggested that balanced growth is not a requirement for providing an insight into the essential activities for division, since cells continue to divide in populations that have been forced to divide synchronously, and an understanding of the distortions that have occurred can increase our knowledge of this process (Padilla and James, 1964). Clearly, both considerations are valid and their relevance to the problem depends entirely on the intent of the particular investigation.

The ideal method for studying the cell cycle would be one that allowed the investigator a completely free choice of the growth conditions and physiological state of the cells. We wish to study the course a cell travels from its last division, or between divisions, in various states, such as in exponentially growing cultures, early stationary-phase cultures, or during a shift in growth conditions. Cells undergoing balanced growth are in only one physiological state of interest. With these considerations in mind, any method that introduces a change in the physiological state chosen for investigation is undesirable. As an example, if a property of cells as a function of age is to be examined in a population that is entering the stationary phase of growth, any method that introduces an alteration of this state, or permits the cells to leave this state before the analysis has been completed, would be unacceptable.

This report is divided into two main Sections, corresponding to the two experimental approaches described above. In each case, those methods that most closely approach the ideal are described in detail and those that probably cause alterations in physiological state are referred to briefly.

II. CULTURING AND ASSAY PROCEDURES

A. Media

Culture media of special or unusual constitution are not required with most methods, but for binding–elution techniques (Sections III C and IV B) and probably for filtration techniques (Section III A), a very rich medium, such as nutrient broth, is unsuitable, since it interferes with the response of the cells in these procedures. If an enriched medium is needed, it can be prepared by adding supplements, such as amino-acids, nucleosides and vitamins, to a minimal salts medium.

In some methods, cells are collected and resuspended in new medium, and this can be either fresh medium identical to that in which the cells were grown, or medium that had previously supported growth to the same concentration of cells (conditioned medium). To reduce the possibility of disturbing the cells by changing the medium, it is preferable to use conditioned medium. Conditioned medium can be prepared by clarifying a portion of the experimental culture (or an equivalent culture) by filtration. When synchronized cells are resuspended in conditioned medium rather than fresh medium, the degree of synchrony is usually somewhat better and the doubling time is shorter.

B. Growth conditions

Most cell-cycle studies are performed on exponentially growing cultures. These are usually obtained by inoculating fresh medium with cells from a culture that has grown to completion in medium containing a limiting quantity of carbon source(s), or from an exponentially growing culture. The time between inoculation and exponential growth depends on the inoculum, organism and growth conditions. For bacteria, a minimum of 2 h is required, and it is often convenient to incubate overnight (about 15 h) to yield an exponentially growing culture containing 10^8 bacteria/ml. From a physiological standpoint, a lower concentration of cells would be preferable (ca. 10^6 bacteria/ml) in order to minimize interactions between cells and changes in environmental conditions during incubation. However, a fairly high concentration of cells (10^8/ml in the case of bacteria) is needed for selection–synchronization procedures, since only a small fraction of the cells is selected from the culture. As a general rule, the concentration of cells in the starting culture should be as low as is experimentally feasible.

C. Counting

Since cell-cycle studies involve relatively small changes in the number of cells (a factor of 2 or less) during a fraction of a generation time, cell numbers should be measured with the most accurate and rapid means available. Electric sensing zone instruments have proved most satisfactory (Harvey, 1968). The use of the Coulter Electronic Particle Counter, the first and most commonly used instrument of this type, has been described by Kubitschek (This Volume, p. 593). Because of its relative accuracy, speed and capacity for comparative estimation of cell sizes, the Coulter Counter is a very useful instrument in experiments involving synchronous growth or selection of cells by size.

D. Evaluation of synchrony

Experiments on the cell cycle should always be carried through at least two, and preferably three division cycles. In this way it can be determined if a pattern observed in the first cycle repeats in the second and third cycles or if it is transient, thus suggesting a distortion. The existence of repetitive patterns does not rule out the possibility of a distortion, but a rapidly changing pattern usually indicates an effect caused by the treatment. Also, in evaluating a new technique it is instructive to determine if biochemical properties that change both continuously and discontinuously during the division cycle can be found. If in a given experiment all parameters studied are either continuous or periodic, there is a possibility that, in the first case, the cells were not actually synchronized, and in the second case, the synchronization treatment induced the periodicities. Observations of both types, in conjunction with an acceptable synchronous growth curve measured with an electric sensing zone instrument, provide strong support for a good experimental technique.

III. SYNCHRONIZATION TECHNIQUES

A. Size selection by filtration

1. *Maruyama–Yanagita technique*

Maruyama and Yanagita (1956) were first to describe a successful method for obtaining synchronously dividing populations by selecting cells of similar sizes from an exponentially growing culture. A culture of *Escherichia coli* B was flushed through a stack of filter papers, and since small cells presumably travelled through the pile more easily than large cells, an abundance of large cells remained near the top of the pile and small cells were found in the filtrate. The cells in the filtrate, or those obtained by eluting selected papers, grew synchronously during subsequent incubation. The degree of synchrony achieved with this method depends on the extent of size separation and the effect of the filtration process on cellular metabolism. To improve the former, minimize the latter and adapt the method for use with other organisms, numerous modifications of the original filtration procedure have been reported. The original technique will be presented first and this is followed by a description of the important modifications.

Figure 1 shows a sketch of the paper pile filtration apparatus used by Maruyama and Yanagita for fractionating cells by size (Maruyama, 1964).

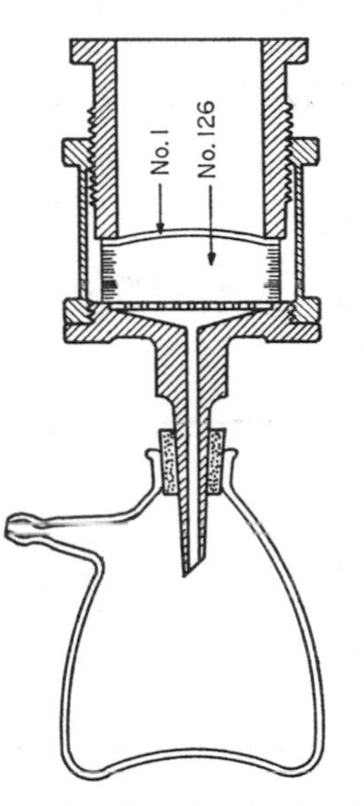

FIG. 1. Paper pile apparatus for fractional filtration of cells. Inside diameter, 55 mm. (From Maruyama, 1964).

The paper pile consists of two sheets of Toyo No. 1 filter paper (a fine paper equivalent to Whatman No. 1) on top of 18 sheets of Toyo No. 126 paper (a coarse paper equivalent to Eaton–Dikeman No. 623 or 624). The filter papers are cut into 55-mm diameter circles and screwed tightly into the filtration apparatus. A Büchner-type funnel with a coarse fritted-glass disc can also be used by clamping the paper pile in place with a ring or by taking care that cells do not seep down the sides of the pile. Piles consisting of 15-cm diameter papers have been used, and larger sizes might be used with larger cultures. The pile is wetted by passing 1 litre of water and then 50 ml of elution medium through it by suction. The elution medium used by Maruyama and Yanagita consisted of fresh minimal salts medium lacking glucose to prevent growth during filtration. Since filtration can be performed rapidly, conditioned medium might be used to avoid a period of starvation.

A 1-litre culture of *E. coli* B containing about 10^8 cells/ml growing exponentially in minimal medium is centrifuged as rapidly as possible and resuspended in 2 ml of medium to yield a suspension containing approximately 10^{11} cells. This thick suspension is placed uniformly on the paper pile and drawn into the upper sheets by suction. Filtration is then performed by pouring 50 ml of elution medium on the pile and filtering by suction at a flow rate of about 2·5 ml/sec. As emphasized by Abbo and Pardee (1960), the entire procedure should be performed at constant temperature (usually 37°C). The increase in number of cells in the filtrate during incubation is shown in Fig. 2. Although the small cell fraction is

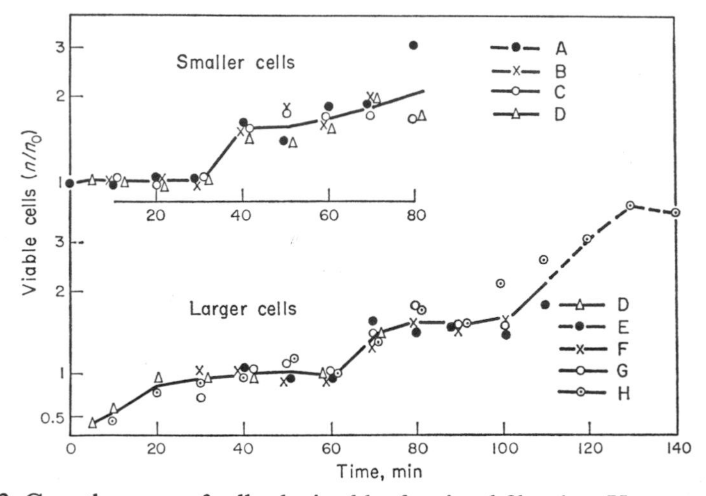

FIG. 2. Growth curves of cells obtained by fractional filtration. Upper and lower curves represent growth curves obtained after inoculations of smaller cells and larger cells, respectively. All experiments from A to H were carried out under the same experimental conditions using different batches of cells. Average numbers (n_0) of cells during the first flat phase in each experiment were as follows: A, 0·65; B, 0·57; C, 0·72; D, 2·33 (larger cells) and 0·83 (smaller cells); E, 1·28; F, 1·00; G, 3·02; and H, 4·57 cells/ml ($\times 10^8$). In each experiment the ratio of the number (n) of cells to the average number, n/n_0, was plotted logarithmically against time. Curves were drawn through the average points at successive time intervals. (From Maruyama, 1964).

usually used in experiments on the cell cycle, large cells can be collected by removing the third, fourth and fifth sheets of filter paper from the top of the pile and eluting these with minimal medium. Growth of the large-cell fraction is also shown in Fig. 2. The yield of both small and large cells is about 5% of the original culture.

2. *Modifications*

(a) *Escherichia coli*. Abbo and Pardee (1960) reported a modification of the filtration technique which they considered less likely to disturb cellular metabolism than the original technique. In their modification, a culture is filtered through a paper pile without concentrating the cells in advance. The Abbo–Pardee paper pile consists of, from top to bottom, two No. 1, one No. 42, twenty No. 1, and one No. 42 Whatman filter papers. A 200-ml culture of *E. coli* B growing exponentially in minimal medium at a concentration of 1 to 2×10^8 cells/ml is poured directly on top of the paper pile and filtered under vacuum. Culture medium is poured on top of the pile to elute the bacteria and filtration is stopped when the filtrate becomes very slightly turbid. Abbo and Pardee reported a relatively high degree of synchronous division for 4 or 5 generations, but the intervals between the increases in cell number varied over a range of approximately 40–65 min.

Nagata (1963) has developed a modification of the Maruyama–Yanagita technique that includes the step in which concentrated cells are placed on top of the pile, plus an additional step in the pre-filtration culturing procedure which was designed to ensure a relatively homogeneous and healthy population of cells. A culture of *E. coli* is grown overnight with aeration at 37°C and then washed twice and re-inoculated into fresh medium. The culture is incubated at 37°C for more than 10 h until it reaches stationary phase, after which incubation is continued for an additional 120 min. This growth into early stationary phase was considered necessary for successful synchronization. The culture is then treated essentially as in the Maruyama–Yanagita technique, except that filtration is performed in two or three successive elutions with 10 ml of culture medium over a 1-min period.

(b) *Alcaligenes faecalis*. The filtration technique has been modified for use with *A. faecalis* (Lark, 1958; Lark and Lark, 1960). Seven-litre cultures containing 2×10^8 cells/ml growing exponentially are sedimented as rapidly as possible (4 min/litre in a continuous flow centrifuge in the case of Lark and Lark, 1960), resuspended in 15 ml of nitrogen-free medium at 37°C and homogenized for 45 sec in a blender. This suspension is absorbed into the top layer of a filter paper pile consisting of 9 sheets of 15-cm diameter Eaton-Dikeman No. 624 paper clamped in a funnel. The pile is then flushed with 300 ml of nitrogen-free medium under vacuum. Synchronously dividing cells are obtained by eluting the middle three papers within 1 min into 300–400 ml of culture medium. The amount of flushing should be determined empirically, and the authors suggest that a dye be used to trace the path of the organisms through the pile to ensure that they filter through to the bottom layers and to determine the area of the paper in the direct flow of the wash fluid.

(c) *Bacillus megaterium*. Imanaka *et al.* (1967) have reported a modification of the filtration procedure for use with *B. megaterium*. Nine sheets of Whatman No. 40 paper are placed in a 9·0-cm diameter Büchner funnel, moistened with distilled water and packed down on the fritted-glass disc with a blunt glass rod. Before autoclaving, 500 ml of distilled water are passed through slowly under suction. Just before an experiment, 50 ml of sterile medium is passed through and discarded. A 100-ml culture in the late log phase of growth at 30°C is filtered by suction in about 1 min and the cells in the filtrate are used for synchronous growth studies.

3. *Applications*

To date, more biochemical studies of the bacterial division cycle have been performed on synchronous cells obtained with various modifications of the filtration procedure than with any other technique. For example, it has been used to study nucleic acid synthesis (Maruyama, 1956; Maruyama and Lark, 1959, 1961, 1962; Lark, 1960, 1961; Abbo and Pardee, 1960; Nagata, 1963; Masters *et al.*, 1964; Rudner *et al.*, 1964; Rudner *et al.*, 1965), protein synthesis (Maruyama, 1956; Abbo and Pardee, 1960; Masters *et al.*, 1964; Kuempel *et al.*, 1965; Nishi and Kogoma, 1965), the sequence of genome replication in *E. coli* (Nagata, 1963) and the effects of various deleterious agents such as radiation and chemical mutagens (Yanagita *et al.*, 1958; Rudner, 1960; Anderson, 1961; Helmstetter and Uretz, 1963; Ryan and Cetrulo, 1963).

Although the filtration procedure has been widely used, it has two disadvantages. First, it has proved difficult to reproduce in different laboratories, and a great deal of trial and error is usually necessary to achieve even a small degree of synchronous growth. Second, it appears that this seemingly gentle technique can cause significant distortions in the sequence of biochemical events during the division cycle. That is, different investigators using the filtration technique to measure the same biochemical events in the same organism have obtained entirely different results. For instance, Maruyama (1956) found a step-wise increase in DNA content during the division cycle of glucose-grown *E. coli* synchronized by filtration, whereas Abbo and Pardee (1960) and Nagata (1963) found a continuous increase. The extent to which any filtration technique might affect cellular metabolism is unknown, but cells which have been concentrated by centrifugation or grown into the stationary phase are probably not typical of those in exponentially growing cultures. When stationary-phase cultures are used in the filtration procedure, the achievement of synchronous growth may be due partly to the resuspension of these cells in fresh medium, and the experimental results obtained would reflect the transition from stationary phase to exponential phase. Considering the difficulty in reproducibility and the extra treatments required to obtain good results, some of the newer techniques to be described in later Sections of this Chapter appear to be superior to the filtration procedures.

B. Size selection by centrifugation

1. *Mitchison–Vincent technique*

Methods for separating cells by size from exponentially growing cultures by centrifugation have been reported by Maruyama and Yanagita (1956), Corbett (1964) and Mitchison and Vincent (1965, 1966). The Mitchison–Vincent technique has proved most successful and has been used to obtain synchronously dividing cultures of *Schizosaccharomyces pombe*, *Saccharomyces cerevisiae* and *E. coli*. The method consists of layering a concentrated sample from an exponentially growing culture on top of a linear sucrose gradient, and centrifuging until the cell layer has moved two-thirds of the way down the tube. Small cells sediment slower than large cells, and they can be separated from the top of the gradient to obtain a synchronous culture.

Linear sucrose gradients are prepared in 16 × 150 mm (15 ml) or 25 × 170 mm (80 ml) centrifuge tubes. The gradient should be 10–40% sucrose (w/v) from top to bottom in the large tubes and 2–12% in the small tubes. The function of the gradient is to stabilize the liquid in the tube and facilitate layering the cells on top of the liquid.

An exponentially growing culture is concentrated rapidly by continuous-flow centrifugation or by collecting the cells on a membrane filter. This suspension, 0·5–1 ml for the small tubes and 2–5 ml for the large tubes, is then layered on top of the gradient. The number of cells that can be loaded depends on their size, and examples for yeast and bacteria are given in the legends to Figs 4 and 6, respectively. The suspension is stirred into the sucrose with a thin rod until the layer of cells is about twice its original thickness. The tubes are centrifuged in a swinging-bucket rotor at a speed and for a duration that are functions of the type of cells and the temperature (see Figs. 4 and 6). As centrifugation progresses, the layer of cells broadens and moves down the tube. Sedimentation is stopped when the layer has moved about one-half to two-thirds of the length of the tube. The top portion of the cell layer is removed with a syringe and re-suspended in culture medium. A syringe with a long needle should be used, and the tip of the needle is placed at the bottom of the portion to be removed, since in the gradient the liquid that is drawn into the syringe comes from above the point of the needle. If 5–20% of the length of the layer is removed this would correspond to 1–5% of the total number of cells in the tube.

Figure 3 shows the frequency distribution of the lengths of the cells from the top of the layer compared to cells in an exponentially growing population. In each case the cells from the top of the gradient are distributed at the lower end of the size distribution of the exponential phase culture. This degree of size separation is sufficient to produce the synchronous

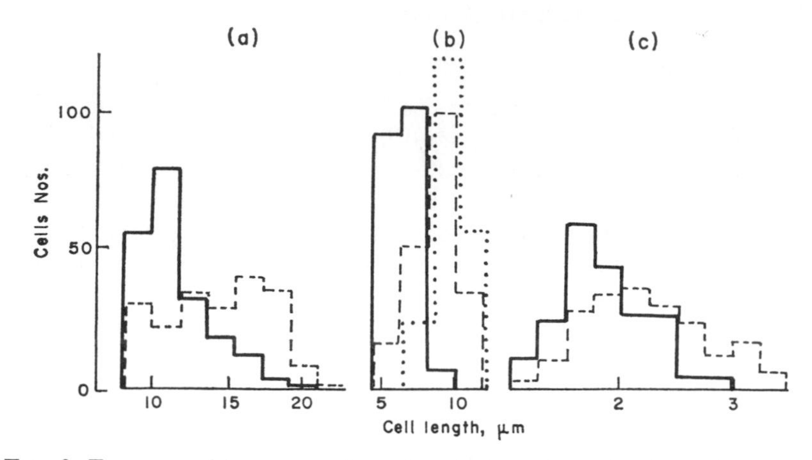

FIG. 3. Frequency histograms of cell length from 200 cells. (a). *Schizo. pombe*, 10–40% gradient in 80-ml tube loaded with 2×10^{10} cells from exponential culture ($8 \cdot 6 \times 10^6$ cells/ml). Centrifuged at 500 *g* for 10 min at 20°C: ———, top sample (0·5% of cells loaded); – – –, control. (b). *S. cerevisiae*, 10–40% gradient in 15 ml tube loaded with 6×10^8 cells from stationary culture (11×10^6 cells/ml). Centrifuged at 200 *g* for 6 min at 20°C: ———, top sample (top 10% of cell layer); . . . ., bottom sample (bottom 20% of cell layer); – – –, control. (c). *E. coli*, 10–40% gradient in 15-ml tube loaded with 2×10^{10} cells from exponential culture (6×10^8 cells/ml). Centrifuged at 2500 *g* for 20 min at 12°C: ———, top sample (2% of cells loaded); – – –, control. (From Mitchison and Vincent, 1965).

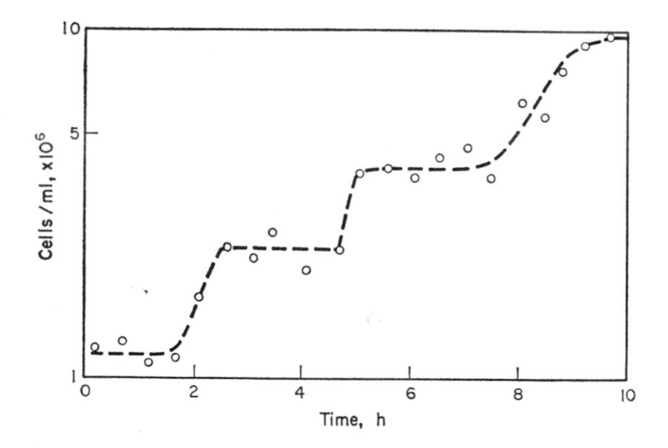

FIG. 4. Cell numbers in synchronous culture of *Schizo. pombe*. 10–40% gradient in 80-ml tube loaded with 3×10^9 cells from exponential culture ($3 \cdot 2 \times 10^6$ cells/ml). Centrifuged at 500 *g* for 10 min at 25°C: top sample suspended in fresh medium. (From Mitchison and Vincent, 1965).

growth shown in Figs. 4, 5 and 6. Figure 4 shows the increase in cell number of *Schizo. pombe* taken from the top of a gradient and cultured in fresh medium. Experimental details are given in the legend to the Figure. Figure 5

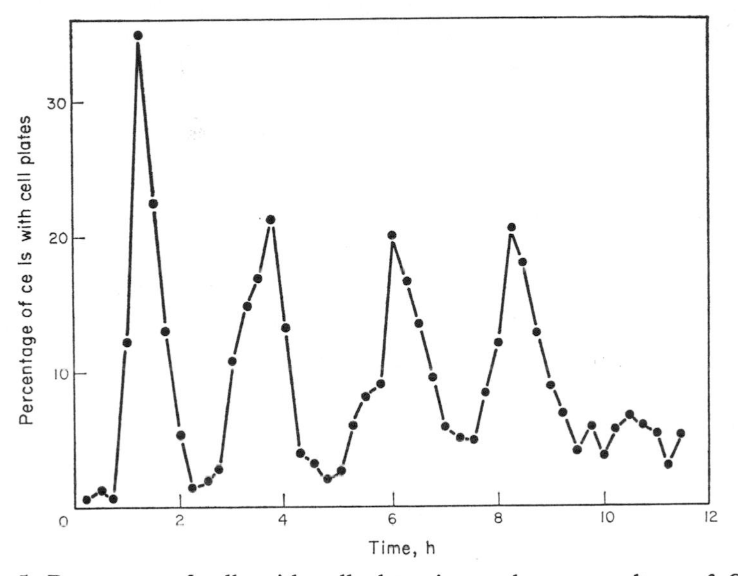

FIG. 5. Percentage of cells with cell plates in synchronous culture of *Schizo. pombe*, 10–40% gradient in 80-ml tube loaded with 3×10^9 cells from exponential culture (2×10^6 cells/ml). Centrifuged at 500 *g* for 9 min at 25°C: top sample (4% of cells loaded) suspended in fresh medium. (From Mitchison and Vincent, 1965).

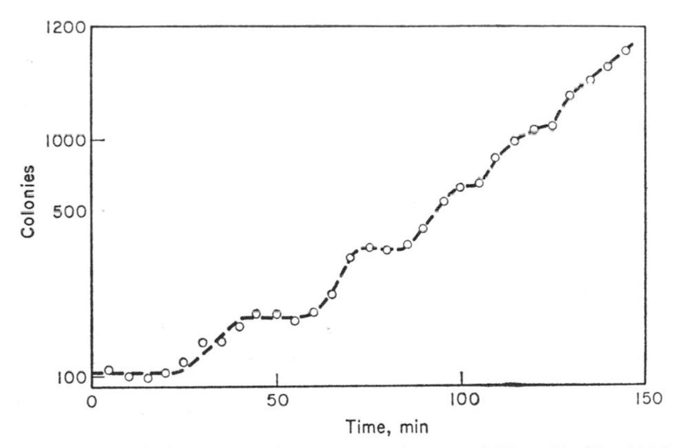

FIG. 6. Colony counts from synchronous culture of *E. coli*, 10–40% gradient in 15-ml tube loaded with 2×10^{10} cells from exponential culture (6×10^8 cells/ml). Centrifuged at 2500 *g* for 20 min at 12°C: top sample (2% of cells loaded) suspended in fresh medium. (From Mitchison and Vincent, 1965).

shows quite dramatically the degree of synchronous division obtained with *Schizo. pombe* in terms of the percentage of cells with cell plates as a function of incubation time. In recent experiments, with better selection of the top of the layer, the first peak of cells with cell plates was 60% and the second 30% (J. M. Mitchison, personal communication). Figure 6 shows the result of a similar experiment with *E. coli*. Again, synchronous growth is quite evident, and it is not unlike that shown in Fig. 9 which was obtained with the membrane elution technique (Section III C).

2. *Applications*

The gradient–centrifugation technique has been used to demonstrate both periodic and continuous enzyme synthesis and step-wise DNA synthesis during the division cycle of yeast (Bostock *et al.*, 1966; Tauro and Halvorson, 1966; Halvorson *et al.*, 1966), and the polarity of chromosome replication in F^- mating types of *E. coli* (Donachie and Masters, 1966). The advantages of this method are that it can be used with a wide variety of organisms and it gives a reasonably large yield of cells. Also, in common with other selection techniques, it is less likely to cause a pronounced disturbance in cellular metabolism than the induction–synchronization techniques. However, the requirement for concentrating the cells before layering on the gradient could result in a distortion. To minimize any disturbances, the procedure should be performed rapidly with as little change in temperature as possible, particularly with bacterial cultures.

If the cells are osmotically damaged by suspension in the sucrose solution, an effective gradient could be made with protein or dextran (Mitchison and Vincent, 1966). Also, Bostock *et al.* (1966) recently used glucose gradients instead of sucrose gradients with *Schizo. pombe* in order to avoid causing any complications by changing the carbon source. The extent to which the experimental manipulations influence biochemical processes can be estimated by using the remainder of the gradient for a control, asynchronous culture. Bostock *et al.* (1966) found periodic enzyme synthesis during incubation of the small cell fraction of *Schizo. pombe* and no steps in the control, and they concluded that the periodic enzyme synthesis was not a consequence of the treatment.

Corbett (1964) has used a sedimentation technique to obtain synchronous populations of the protozoan *Tetrahymena pyriformis*. His technique involves centrifuging exponentially growing cultures at 550 *g* for 6 min. It was found that the cells remaining in the supernatant after centrifugation grew synchronously. For studies on cells in known physiological states, this technique may prove to be a promising partner to the more popular temperature-induction techniques for *Tetrahymena*.

C. Age selection

1. *Experimental procedure*

In a third type of selection technique, synchronously dividing populations of *E. coli* B/r can be obtained by selecting cells of uniform age from a growing population (Helmstetter and Cummings, 1963, 1964). In this technique, newly formed cells are eluted from a culture that is bound to the surface of a membrane filter, and the cells in a sample of the effluent collected during a small fraction of a generation time grow synchronously. Presumably, each bacterium that is eluted from the membrane is that sister of a new sister pair that does not retain the parental attachment to the membrane. Figure 7 shows a schematic illustration of this principle. If the

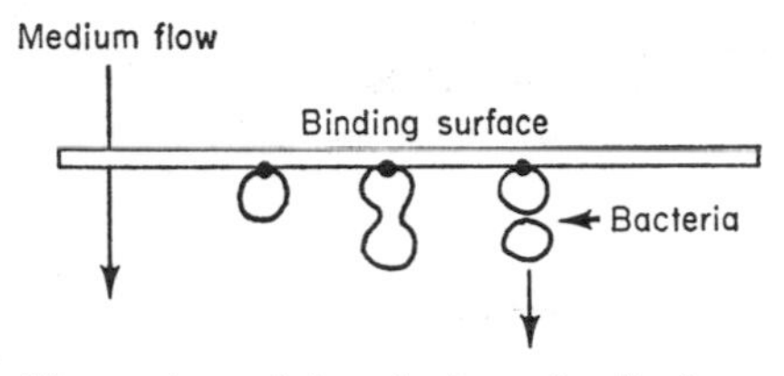

FIG. 7. Schematic illustration of the elution of cells from a culture bound to a membrane. Three successive stages in the division cycle of an individual cell are indicated. If the attachment to the membrane remains fixed, then the daughter cell indicated by the arrow is the only cell that would be eluted from the surface as culture medium is passed through.

attachment were irreversible, only new-born cells could elute from the surface, and if occasionally both daughters remain attached to the surface this would only effect the concentration of new-born cells in the effluent.

Most of the filter holders that have been used in this technique have been complex and expensive (Helmstetter and Cummings, 1963, 1964; Cummings, 1965; Clark and Maaløe, 1967). However, recently a simplified and inexpensive arrangement has been developed (Helmstetter, 1967). Figure 8(a) shows a sketch of the apparatus. A 15-cm diameter porcelain filter holder (Schleicher & Schuell Co., Keene, New Hampshire, type PA 15) is used as the holder for the membrane. The support plate supplied with the filter holder is replaced with a stainless-steel screen (Millipore Corp., Bedford, Mass., Cat. No. YY–2214264) and the porcelain upper section is replaced with seven rubber gaskets supplied with the apparatus. A 15-cm diameter, type GS Millipore filter is clamped in the holder before the experiment begins and it remains in place throughout the entire experiment. A type GS membrane is used because its pores are small enough so that the bacteria remain on the surface of the membrane, yet large enough so that medium flows easily through the membrane during elution. The stem of the bottom section is cut off level with the lower surface of the stopper for ease of handling. The entire experiment is performed in a 37°C incubator. A Full View incubator (Precision Scientific Co., Chicago, Illinois) is convenient for this purpose, since it is the right size and shape to contain the apparatus, and it has an air-curtain double door so that manipulations can be performed within the incubator with the lower door open. After the cells are bound to the membrane, the apparatus is inverted and elution is performed by connecting the top of the filter holder to a reservoir of conditioned medium

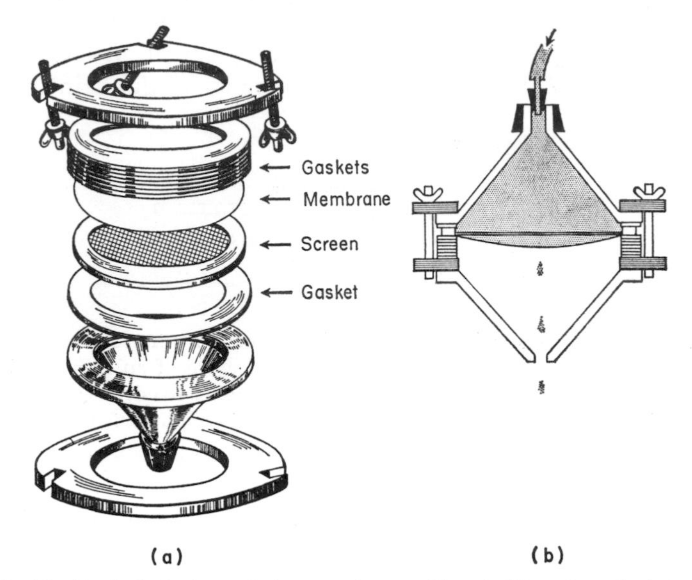

FIG. 8. (a). Exploded view of the membrane filter holder described in the text. (b). Configuration of the holder during elution.

through a Model AL–4–E Sigmamotor pump (Sigmamotor, Inc., Middleport, New York).

A 100-ml culture of *E. coli* B/r is incubated at 37°C with shaking until a concentration of 1×10^8 cells/ml in the exponential phase of growth is reached. The entire culture is then poured on top of the membrane, through which the medium is drawn by suction. In early experiments, filtration was performed by pressure, but identical results are obtained with vacuum filtration and it is considerably more convenient. Filtration is stopped when a small amount of the culture remains above the membrane, in order to avoid drying the cells. The remaining fluid is poured off, the entire filter holder is inverted, the top is filled with medium (approximately 300 ml) and it is connected to the pump. In order to reduce evaporation and a consequent reduction in temperature on the membrane, the inverted filter holder is placed on top of a second porcelain bottom section from a type PA 15 filter holder. The stem of this bottom section is removed completely so that the effluent can drop freely through the hole. Figure 8(b) shows a sketch of the apparatus as it appears during the elution procedure. The pump is operated at a rate of 15 ml/min for about 2 min to hasten wash-out of unbound or weakly bound cells, and then the rate is reduced to the value desired for the particular experiment.

After the first 20 min of elution, most of the cells in the effluent are new-born. Figure 9 shows the synchronous growth of a sample collected between 38 and 40 min of elution from a population growing in glucose minimal medium.

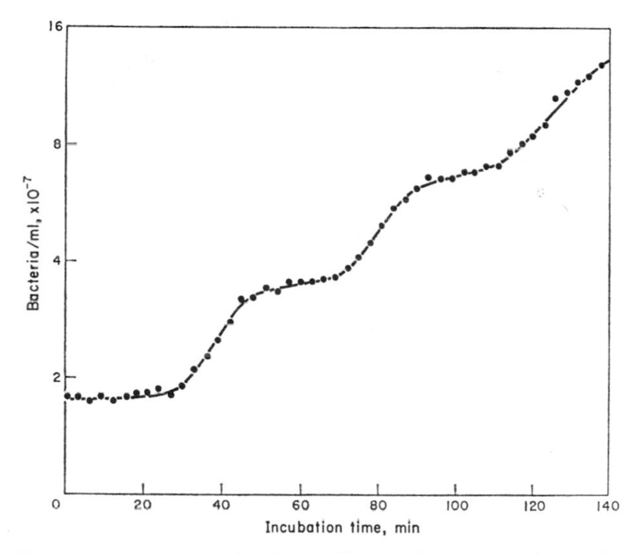

FIG. 9. Cell numbers in a sample of the effluent from a membrane-bound population of *E. coli* B/r growing in glucose minimal medium. The sample was collected between 38 and 40 min of elution and incubated at 37°C with shaking. Cell numbers were determined with a model B Coulter Counter.

The number of cells in the effluent varies with elution time and reaches a maximum of from 5×10^7 to 1×10^8 cells/min. The concentration can be varied by adjusting the flow rate, since the degree of selection of new-born cells is not critically dependent upon the elution rate over a wide range. Essentially the same results have been obtained with flow rates between 1 and 100 ml/min. Lower rates can be used over short periods of time to yield more concentrated samples. However, a single sample containing a total of about 10^8 cells is frequently not sufficient for biochemical analysis. If more cells are needed, replicate samples can be collected and incubated for different intervals to obtain cells of different ages for analysis.

Cummings (1965) and Clark and Maaløe (1967) increased the concentration of cells in the effluent by increasing the surface area. Cummings constructed a stainless-steel filter holder that supported a 293-mm diameter membrane filter, and Clark and Maaløe used an apparatus that supported five 142-mm diameter membrane filters. In both cases the yield of new-born cells in the effluent was about $4–5 \times 10^8$ cells/min. However, the results indicate that the quality of selection of newly formed cells was inferior to that achieved with a single 142-mm diameter membrane, probably owing to less efficient flow of medium across the membrane. Of the two methods for obtaining large numbers of cells, i.e., collecting multiple samples from a small membrane or a single sample from a large membrane, the former appears to be superior, since the quality of selection is better and concentrated samples can be collected by decreasing the flow rate.

2. *Applications*

Using this technique, Cummings (1965) demonstrated that DNA and RNA are synthesized continuously during the division cycle of glucose-grown *E. coli* B/r, and that β-galactosidase can be induced throughout the division cycle. Clark and Maaløe (1967) measured the rate of DNA synthesis and the extent of DNA synthesis in the presence of chloramphenicol during the division cycle of *E. coli* B/r. They concluded that rounds of DNA replication are initiated during the latter half of the division cycle in glucose-, glycerol- and succinate-grown cells. Harvey *et al.* (1967) used the technique to measure the size distribution of newly formed cells, and Schleif (1967) measured the kinetics of ribosomal protein synthesis as a function of cell age.

The major disadvantage of this procedure is that it does not work satisfactorily with all micro-organisms or even with all strains of *E. coli*. The best results are obtained with *E. coli* B/r, and the quality of selection of new-born cells from membrane-bound populations of other strains tested was variable, but always poorer than with strain B/r. For instance, most of the cells in the effluent from a membrane-bound population of strain B/r are new born, but only about one-half of the cells from populations of strain K12 are new born.

A second criticism of this technique is that, as with the size-selection techniques, cellular metabolism may be disturbed somewhat by the procedure. Although the initial trauma of filtering the cells on the membrane can be minimized by delaying collection of new-born cells until the cells have adapted to growing on the membrane, the trauma of transferring the new-born cells from membrane culture to batch culture is more difficult to avoid. It may be possible to avoid this problem by collecting and growing the synchronous cells on individual membrane filters, but this would be a complex and delicate procedure.

With larger cells (e.g., protozoa) synchronous populations are easily obtained by picking dividing cells from a growing culture. Synchronous populations have been obtained in this way by Prescott (1955, 1960, 1966), and a thorough description of the methods for collecting protozoans with a braking pipette has been reported by Stone and Cameron (1964).

D. Induction techniques

1. *Temperature*

The goal in the development of new procedures for studying the division cycle has been to devise methods that involve a minimum of disturbances to the cells. In this Chapter, emphasis is placed on these newer techniques. The induction–synchronization techniques have either been described in detail elsewhere or they have been supplanted by the newer methods. For these reasons they will be reviewed briefly and reference will be made to more complete descriptions.

One of the earliest means used to synchronize growth of micro-organisms was a single shift in temperature. Hotchkiss (1954) found that a single low-temperature treatment could synchronize growth in a culture of pneumococci. In his experiments the temperature of an exponentially growing culture was reduced from 37° to 25 °C for 15 min and then returned to 37°C. Similar temperature shocks have been shown to synchronize growth in various micro-organisms (Lark and Maaløe, 1954; Maaløe and Lark, 1954; Falcone and Szybalski, 1956; Hunter-Szybalska *et al.*, 1956; Prescott, 1957; Roslansky *et al.*, 1958; Scott and Chu, 1958; Stárka and Koza, 1959; Doudney, 1960; Zeuthen, 1964). However, the most successful methods for synchronizing bacterial growth with temperature shifts have involved repeated changes in temperature at intervals of a generation time. Lark and Maaløe (1954) described a method for synchronizing *Salmonella typhimurium* by alternating 28-min periods at 25°C with 8-min periods at 37°C. In these experiments, essentially all of the cells divided during the brief period at 37°C. Similar periodic cold- or heat-shock treatments have been used to synchronize *T. pyriformis* (Zeuthen and Scherbaum, 1954; Zeuthen, 1958; Padilla and Cameron, 1964). This organism has also been synchronized by exposure to repeated temperature shocks applied at shorter intervals (Scherbaum and Zeuthen, 1954; Zeuthen and Scherbaum, 1954). For instance, if cells are subjected to alternating exposures to 34°C and 28°C, with the period at 28°C being sufficiently short so that the cells do not divide, synchronous divisions ensue when the cells are finally placed at 28°C and allowed to grow unrestrictedly. At this point, they are 2–4 times their normal size, and the next two or three divisions occur more rapidly than the ordinary maximal growth rate. Techniques for achieving synchrony in this manner have been described by Plesner *et al.* (1964), Scherbaum and Jahn (1964), Moner (1965) and Moner and Berger (1967), and reviewed by Scherbaum (1960) and Zeuthen (1964). Temperature-induced synchronization of the alga *Astasia longa* has also been reported (James, 1959, 1964; Padilla and James, 1960), and a detailed description of a technique for continuous maintenance of synchronized growth of *Ast. longa* by periodic temperature fluctuations once per generation has been reported by Padilla and James (1964). Finally, Neff and Neff (1964) have reviewed the techniques for inducing synchronous growth in *Amoeba*.

2. *Illumination*

Photosynthetic cell types can be induced to divide synchronously by means of a repetitive light–dark cycle. Various techniques have been developed for synchronizing *Chlorella* in this way (Tamiya *et al.*, 1953; Lorenzen, 1957; Sorokin, 1957; Schmidt and King, 1961), and these have been described in detail by Kuhl and Lorenzen (1964), Lorenzen (1964), Tamiya (1964), Tamiya and Morimura (1964), and Schmidt (1966). Repetitive light–dark cycles have also been used to synchronize *Euglena* (Cook and James, 1960; Edmunds, 1964; Padilla and James, 1964). Howell *et al.* (1967) have reported a method for the continuous synchronous culture of photosynthetic micro-organisms, and Lafeber and Steenbergen (1967) and Kates and Jones (1967) have described recent developments in the synchronous culture of photosynthetic algae.

3. *Starvation*

Starvation treatments of various forms have been used extensively for synchronizing growth in cultures of micro-organisms. Barner and Cohen (1956) observed a phasing of cell division in cultures of *E. coli* $15T^-$ following withdrawal and re-addition of thymine to the cultures. Starvation of *E. coli* for glucose, either alone or in conjunction with a cold shock, has also been reported to phase cell division (Campbell, 1956; Scott *et al.*, 1956; Scott and Chu, 1958; Perry, 1959). Starvation of cells for required deoxyriboside also introduces a degree of synchronous growth (Burns, 1959, 1961, 1964).

Along the same line, some of the earliest attempts to synchronize micro-organisms involved the dilution of stationary phase cultures into fresh medium (Browning *et al.*, 1952; Houtermans, 1953). Yoshikawa and Sueoka (1963a) have suggested that stationary phase cells of *Bacillus subtilis* contain complete, non-replicating chromosomes and they found (Yoshikawa and Sueoka, 1963b) that stationary-phase cultures showed some synchrony of chromosome replication after dilution. Masters *et al.* (1964) also found that re-suspension of stationary-phase cells of *B. subtilis* in fresh medium resulted in synchronous division and stepwise enzyme synthesis. The most successful procedure of this type was reported by Cutler and Evans (1966, 1967a, b). They found excellent synchronization of *Proteus vulgaris* and *E. coli* by dilution of stationary-phase cultures with fresh medium. Cells were grown in 125 ml of minimal medium to early stationary phase, harvested by centrifuging at 3000 **g** for 9 min at 37°C, and inoculated into 350 ml of fresh medium. After two generations of synchronous growth (growth into early stationary phase), the cells were harvested again and inoculated into 5500 ml of medium at 37°C. This treatment yielded very dramatic synchrony of cell division for several generations. The optimal harvest time for a particular organism must be determined empirically, but in their experiments it was found to be about one-half generation after the cultures entered stationary phase, as determined by the optical density.

The explanation for the achievement of synchronous growth by dilution of stationary phase cells is as yet unclear. It is not surprising that stationary-phase cells might contain complete, non-replicating chromosomes, since it has been suggested by Maaløe and Hanawalt (1961) that protein synthesis is required to initiate chromosome replication, but not to maintain it. If growth into stationary phase is similar to starvation, chromosome replication would be expected to continue to completion, but new rounds of replication would not be initiated. However, synchrony of cell division need not be a direct consequence of terminalization of the chromosomes, since re-initiation of chromosome replication may not occur in synchrony.

There have been two recent reports of bacterial synchronization by starvation for required amino-acids. Stonehill and Hutchison (1966) reported the synchronization of a methionine and threonine auxotroph of *Streptococcus faecalis* by withdrawal and re-supplementation of the required amino-acids. Matney and Suit (1966) described a method for synchronizing amino-acid-requiring mutants of *E. coli* by depletion and re-supplementation of required amino-acids. Cells were grown in 1 μg/ml of the amino-acids until this was depleted as determined turbidimetrically; incubation was then continued for 3 additional hours, after which time the required amino-acids were added at normal supplement concentrations. They suggested that the synchronization of cell division, which was not observed following abrupt withdrawal of required amino-acids (Maaløe and Hanawalt, 1961), was due to the gradual depletion of a small amount of the amino-acids. Based on a similar idea, Altenbern (1966) reported the synchronization of *Staphylococcus aureus* by treatment with phenethyl alcohol, which was assumed to result in terminalization of the chromosomes in the culture, similar to amino-acid starvation (Treick and Konetska, 1964). The usefulness of these starvation procedures will have to await further investigations.

Until recently, the most popular methods for synchronizing yeast involved starvation techniques (Sylvén *et al.*, 1959; Williamson and Scopes, 1960, 1962; Sando, 1963; Williamson, 1964). In the Williamson–Scopes technique for *S. cerevisiae*, large cells are collected by sedimentation and subjected to a regime of alternate feeding and starving. The starvation methods have been used extensively by Halvorson and co-workers to investigate the regulation of enzyme synthesis during the division cycle of yeast (Halvorson *et al.*, 1964, 1966; Gorman *et al.*, 1964; Tauro and Halvorson, 1966).

In summary, various methods are available for inducing synchronized growth, and the degree of synchrony is usually as good or better than that achieved with selection techniques. However, there can be little doubt that induction techniques disturb some metabolic processes, and this is a problem with which the investigator must contend in every experiment.

IV. CLASSIFICATION TECHNIQUES

To avoid the possibility of introducing distortions in the physiological state of cells during studies on their division cycle, it would be preferable to do the experiments on cultures which have not been treated to obtain synchronous growth. There are two ways in which an investigation of the division cycle of cells in untreated cultures might be accomplished. First, the cells could be analysed as a function of their interdivisional stage by sorting them out according to their age. Once they were separated into age classes, standard biochemical procedures could be applied. Second, the cell cycle could be studied by continuously examining cells of a single, known age in a growing culture. In this type of experiment, the response of cells of different ages to an analytical treatment could be determined by measuring the effect of the treatment on the cells, or their progeny, at a later time in their growth cycle. The usefulness of these approaches depends on the development of methods, in the first case, for sorting out cells as a function of their age, and in the second case, for identifying and examining cells of a specific age.

Techniques such as these have been used for many years to study the division cycle of higher organisms. For the most part, they have been used to investigate the period of DNA synthesis during the mitotic cycle (see reviews by Taylor, 1960, and Sisken, 1964). In an experiment of the first type, a population is pulse labelled with radioactive thymidine and examined by autoradiography to determine the extent of incorporation as a function of cell size or nuclear volume (Woodward *et al.*, 1961). The period of DNA synthesis (S period; Howard and Pelc, 1953) can be determined directly as a function of cell age if time-lapse motion pictures are taken of the cells before the labelling. The age of the cells at the time of pulse labelling is found by measuring the time on the film since their last mitosis. The length of the S period can also be determined approximately by measuring the fraction of cells that incorporate label during a brief exposure, if the intermitotic time and the position of the S period in the mitotic cycle are known.

Most experiments on the mitotic cycle have involved techniques of the second type, i.e., continuous observation of one age (Painter and Drew, 1959; Stanners and Till, 1960). The stage in the mitotic cycle that is usually observed continuously is metaphase, since it is easily recognizable and brief. In this case, the S period and the gaps in DNA synthesis before and after it (G_1 and G_2) are measured by pulse labelling a population with radioactive thymidine and measuring the number of labelled metaphase figures as a function of time subsequent to the labelling period (Painter and Drew, 1959). The time required for labelled metaphase chromosomes to appear is a measure of the length of the G_2 period; labelled metaphases continue to appear for a time equal to the S period, and finally, the number of labelled metaphases again decreases corresponding to the appearance in mitosis of cells that were in the G_1 period at the time of pulse labelling. By measuring the number of grains per cell in the autoradiograms, the relative amount of radioactive thymidine incorporated during the pulse can be determined, and this gives an estimate of the rate of DNA synthesis during the S period. This experiment can also be done by exposing the cells continuously to radioactive thymidine. The time required for the appearance of labelled metaphase chromosomes again equals the G_2 period. The length of the S period is determined by the time for the grain count over metaphases to increase from zero to a maximum value, and the intermediate grain counts give the manner in which DNA synthesis proceeds with time over the S period (Stanners and Till, 1960). As a final example of techniques of the second type, methods have been developed for life-cycle analysis based on the pattern of accumulation of cells at a particular point in the cycle (Evans *et al.*, 1957; Neary *et al.*, 1958; Puck and Steffen, 1963; Puck *et al.*, 1964; Tobey *et al.*, 1966). These methods consist of inhibiting cells at a specific point in the division cycle and measuring the relative number of cells accumulating at that point as a function of time. The pattern in which cells accumulate can be used to determine the effects of various treatments on cell growth as a function of age.

Although these techniques have been used extensively with various plant and animal cells, they had not been applied to micro-organisms until very recently. This delay was probably due to the absence of an easily recognizable stage in the division cycle of micro-organisms similar to metaphase in higher organisms. During the past few years, several techniques for studying the division cycle of micro-organisms in asynchronous cultures have been described. Some of these are simply alternative applications of techniques that have been used previously to select synchronous populations, but the experimental details are slightly different and they are significantly different conceptually. The ideal method would be one that permitted cells to be sorted out directly as a function of age. Such a method has yet to be developed, but methods that permit fractionating cells as a function of size and continuous withdrawal of cells of a particular age are now in use.

A. Classification by size

1. *Experimental procedures*

In Section III B the gradient centrifugation technique of Mitchison and Vincent (1965) for obtaining synchronously dividing populations by separating small cells from a culture was described. Manor and Haselkorn (1967) and Kubitschek *et al.* (1967) have used similar gradient-centrifugation procedures for fractionating cells by size. Large cells sediment more rapidly than small cells, so that after a period of centrifugation a gradient of cells is formed. By collecting consecutive fractions from the centrifuge tube, measurements of interdivisional events can be performed as a function of size.

These procedures have wide application in studies of the biochemistry of the cell cycle. For measurement of the rate of synthesis of a macromolecule, a culture is pulse labelled with a radioactive precursor of the macromolecule, growth is stopped, and the cells are fractionated for measurement of the amount of label incorporated in relation to size. The quantity of the macromolecule in cells of different sizes can be found either by long-term labelling followed by fractionation, or by fractionating the cells followed by chemical analysis of each size class. The rate of induced enzyme synthesis can be found by briefly exposing a culture to the inducer, fractionating the cells and measuring the total enzyme per cell in each size class. This is the general way most experiments would be performed, and only the fractionation procedures themselves will be described here. Since the techniques are very new and somewhat different, both will be presented.

(a) *Manor–Haselkorn technique*. In this technique, 20-ml cultures of *E. coli* (58–161) are grown into the exponential phase and growth is terminated by pouring the suspension onto a half volume of ice-cold culture medium. The culture is centrifuged at 0°C and re-suspended in 1 ml of ice-cold medium in preparation for layering on the gradient. All subsequent operations are performed at 0°–4°C. Manor and Haselkorn layered the re-suspended bacteria on top of a 63-ml linear caesium chloride gradient (0·3–0·5 mole/litre) in a 6-in long centrifuge tube and centrifuged at about

1000 **g** for 25 min at 4°C in a swinging-bucket rotor. The details of the centrifugation procedure depend upon the organism and a certain amount of trial and error would no doubt be necessary.

Figure 10 shows the percentage of cells in various fractions taken from the bottom of the tube. The size-distribution of cells in an exponentially growing population and in different fractions from the gradient are shown

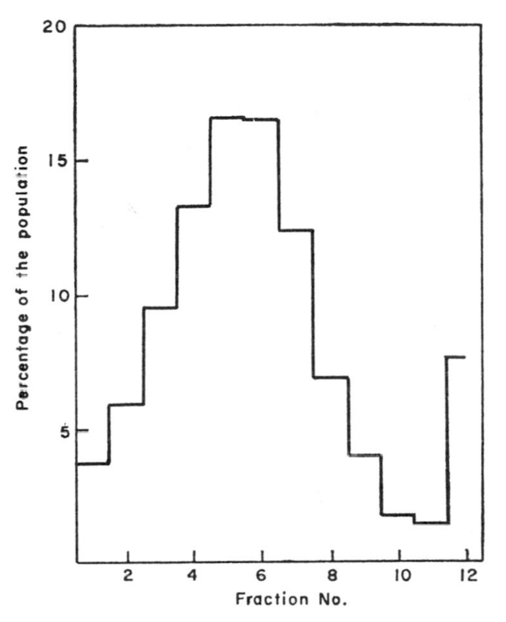

FIG. 10. Zone centrifugation of an exponentially grown population of *E. coli* 58–161 $RC^{relaxed}$. Centrifugation was at about 1000 *g* and 4°C for 25 min and twelve fractions were collected. The total number of bacteria was determined in each fraction with a Coulter Electronic Counter. The sedimentation profile is normalized (total analysed population, 100%). (From Manor and Haselkorn, 1967).

in Figs. 11 and 12, respectively. The selection of small cells in fraction 10 appears quite good, whereas the spread in size in the fraction near the bottom of the tube, which should contain large cells, is fairly wide.

(b) *Kubitschek–Bendigkeit–Loken technique.* The method of Kubitschek *et al.* (1967) was used with *E. coli* $15T^-$ growing slowly in a minimal medium (1–12 h per doubling) in batch cultures or in a chemostat. To obtain generation times between 80 min and 12 h, cells were grown in chemostat cultures in which growth was limited with 100 μg/ml of glucose and the growth rate was controlled by the rate of addition of the nutrient. Chemostat cultures contained $1–2 \times 10^8$ cells/ml. For generation times less than 100 min, cells were grown in batch cultures in which the glucose concentration was

1 mg/ml. These cultures were used when a concentration of 1×10^8 cells/ml in exponential growth was reached. Growth was stopped by the addition of formaldehyde to a final concentration of 0·4% and about 25 ml of the suspension was centrifuged and resuspended in 0·1 ml of a 2·5% sucrose solution.

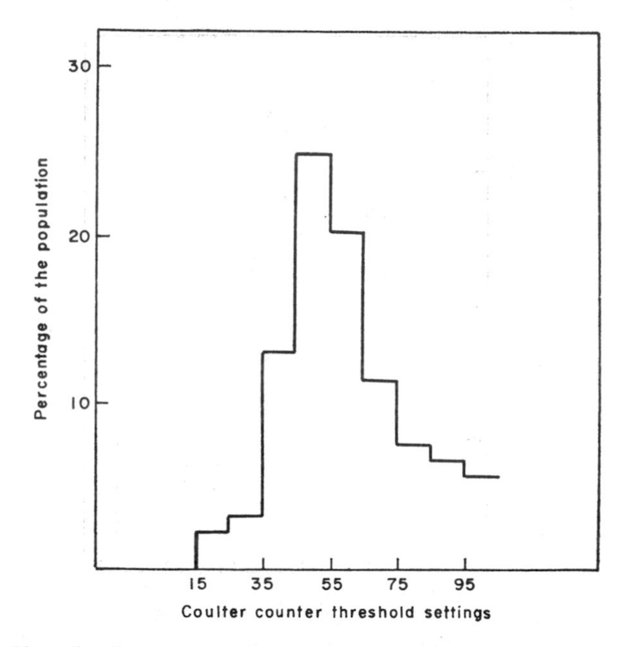

FIG. 11. Size distribution of exponentially grown population of *E. coli* 58–161 $RC^{relaxed}$. A portion was withdrawn from a logarithmic phase culture of bacteria, and the size distribution of the cells was measured with a Coulter Electronic Counter; the distributions were analysed differentially, that is the number of cells was measured between pairs of upper and lower electronic thresholds, as recorded. The sizes of the electronic signals are roughly proportional to the volumes of the cells. The distribution is normalized (total population, 100%). 5·6% of the cells were larger than the uppermost threshold specified and are not included in the figure. (From Manor and Haselkorn, 1967).

Sucrose gradients (5–15%) are prepared in 13-ml volumes in 12×100 mm cellulose nitrate centrifuge tubes. In early experiments, approximately 0·05 ml of the concentrated cell suspension was layered on top of the gradient, but concentration and sample size are not very critical, and in recent experiments 0·2–0·3 ml samples have been used (H. E. Kubitschek, personal communication). The sample is placed within a smaller cylinder of polycarbonate (12 mm i.d.) that extends about 2 mm into the gradient and reduces impingement of cells upon the walls of the tube during centrifugation. The cell suspension is placed on the gradient with a hypodermic needle that is bent to provide an inverted point. Centrifugation is at 3000 r.p.m. (approximately 2500 *g*) in a swinging-bucket rotor until the band of cells extends from 2 to 4 cm below the meniscus. This takes 6 min for

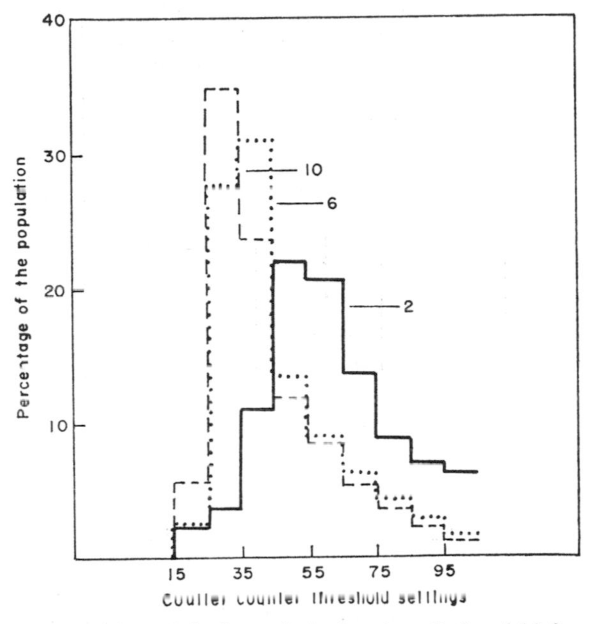

FIG. 12. Size distribution of the bacteria in fractions 2, 6 and 10 from the gradient shown in Fig. 10. The size distributions were analysed as described in the legend to Fig. 11. The distributions are normalized (total population, 100%). 5·3, 0·7 and 3·4% of the cells in fractions 2, 6 and 10, respectively, were larger than the uppermost threshold specified above, and are not included in the figure. (From Manor and Haselkorn, 1967).

freely growing cells and up to 11 min for chemostat cells. After centrifugation, 0·1-ml samples are removed from the gradient with a bent hypodermic needle at intervals of approximately 2·5 mm along the tube.

Figure 13 shows the distribution of cell volumes in various fractions from the gradient, compared to the size distribution in the initial population. As with the previous technique, the best separation is obtained with the small-cell class.

2. *Applications*

Both techniques are recent developments, and therefore, they have not been used extensively for studies on the microbial division cycle. Manor and Haselkorn (1967) used their technique to measure the rate of RNA synthesis as a function of size of *E. coli*, and Kubitschek, *et al.* (1967) measured the period of DNA synthesis during the division cycle. The latter authors found that DNA synthesis begins late in the division cycle

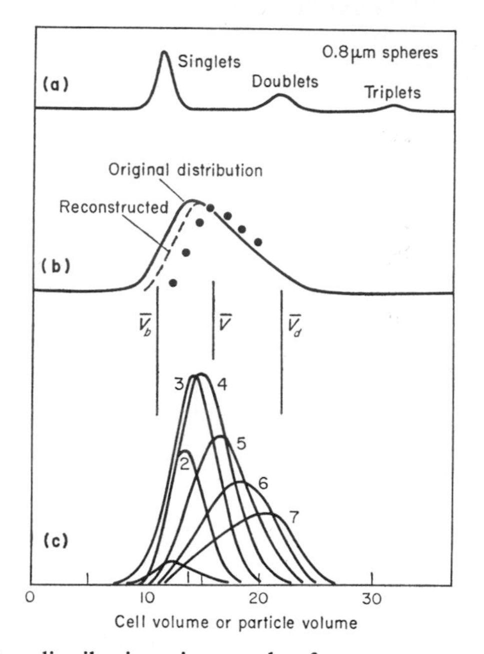

FIG. 13. Cell-size distributions in samples from a sucrose gradient. Particle concentrations are shown as a function of volume. (a). Calibration with 0·8 μm polystyrene latex spheres. Doublets and triplets refer to the volumes measured when two or three particles pass through the counting aperture simultaneously. (b). Size distributions for bacteria grown in a chemostat culture at a generation time of about 8 h. Measured distribution, ———. Distributions reconstructed from samples shown below in (c): ●, number of cells versus mean cell volume for each sample; – – –, sum of distributions of individual samples. (c). Cell-size distributions for seven successive samples taken from a sucrose gradient. (From Kubitschek *et al.*, 1967).

of slow-growing cells, i.e., in chemostat cultures with generation times longer than 2 h DNA synthesis does not begin until the cells have passed through two-thirds of the division cycle.

The centrifugation techniques are close to the ideal method for investigating the microbial division cycle, since the physiological state of the cells under examination can be precisely and reproducibly defined. A relatively minor disadvantage of the techniques is that measurements are made as a function of size rather than age, and a relationship between size and age must be determined or assumed. This is not a serious disadvantage, since a method that accurately determines a property as a function of size is far superior to methods (i.e., synchronization techniques) that determine a property as a function of age plus an unknown distortion.

As shown in Figs. 12 and 13, the size distributions in each fraction are broad and overlap considerably, especially for rapidly growing, unlimited cultures. It is to be hoped that the resolution of size classes will be improved, so that the finer features of the division cycle will become amenable to study.

Finally, the caesium chloride technique is not suitable for studies requiring viable cells after fractionation, since it disrupts ribosomes inside the cells and is probably lethal (H. Manor, personal communication).

B. Classification by age

1. *Experimental procedure*

The membrane-elution procedure for obtaining synchronously dividing populations of *E. coli* B/r (described in Section III C) can also be used for studying the division cycle of cells in asynchronous cultures. The principle behind this application of the technique corresponds to the second type of classification method, since the division cycle is studied by continuously observing a single age class in the population. The advantage of this method for observing a single age class is that the cells are *withdrawn* from the culture, and therefore, can be analysed with standard biochemical techniques (Helmstetter, 1967). A schematic outline of the procedure is shown in Fig. 14. The rate of synthesis of a macromolecule is determined by pulse labelling a culture with a radioactive precursor of the macromolecule (Step I), binding the labelled cells to a membrane filter (Step II), and measuring the radioactivity in the new-born cells, which are eluted continuously from the membrane filter (Step III). The new daughter cells, which elute at four distinct times during each generation of elution, are shown, and the amount of label in each of these cells reflects the amount of label incorporated into their ancestors during the pulse labelling period. For instance, the cells that elute from the membrane during the first generation are daughters of the oldest through to the youngest cells initially attached, and each of these contains half of the label in its parent. Thus, the rate of incorporation of a labelled compound into cells of different ages in a population is determined by pulse labelling the cells and measuring the amount of label in their progeny. In a similar way, the initial rate of induced enzyme synthesis or mutation induction, the quantity of a macromolecule, or the effect of inhibitors is found by pulse labelling the culture with an inducer or mutagen, by long term incorporation of a radioactive precursor, or by exposing the culture to an inhibitor, respectively, and measuring the effect in the new-born cells.

The organisms, culturing conditions, apparatus, and binding-elution procedure are identical to those described in Section III C. Again, the cells are bound to a grade GS, 152-mm diameter Millipore filter, but in this case the cells are washed after binding to remove compounds added to the culture before filtration. A 100-ml culture at a concentration of 10^8

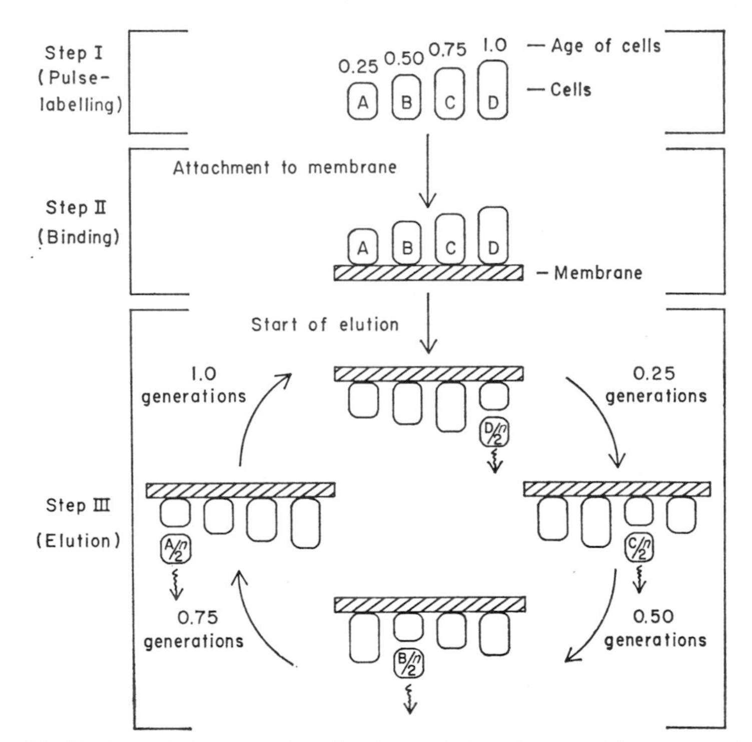

FIG. 14. Outline of the procedure for determining the rate of incorporation of a labelled molecule into cells of different ages in an exponentially growing culture. The ages of the cells at the time of pulse labelling are shown at the top of the figure, and the letter in each cell indicates the amount of label assumed to be incorporated during the exposure. (From Helmstetter, 1967).

cells/ml is filtered by suction, filtration is stopped when almost all of the fluid has passed through the filter, and the remaining portion is poured off. The cells are washed by passing 100 ml of conditioned medium through the filter in the same manner. The filter apparatus is then inverted and elution is begun by pouring conditioned medium into the top and connecting it to the pump. The total time required for these operations is about 1 min.

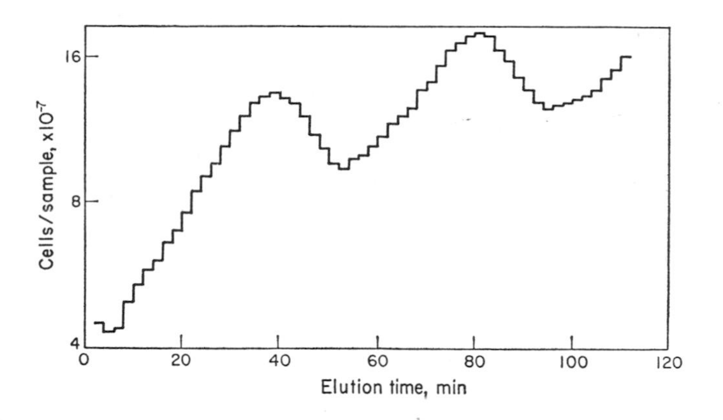

FIG. 15. Number of cells in consecutive samples of the effluent from a membrane-bound culture of *E. coli* B/r growing in glucose minimal medium. A 100 ml culture containing 10^8 cells/ml was filtered onto the membrane by suction. Elution was with conditioned medium at a rate of 5 ml/min. Samples were collected over 2 min intervals and the cell number was measured with a Coulter Counter.

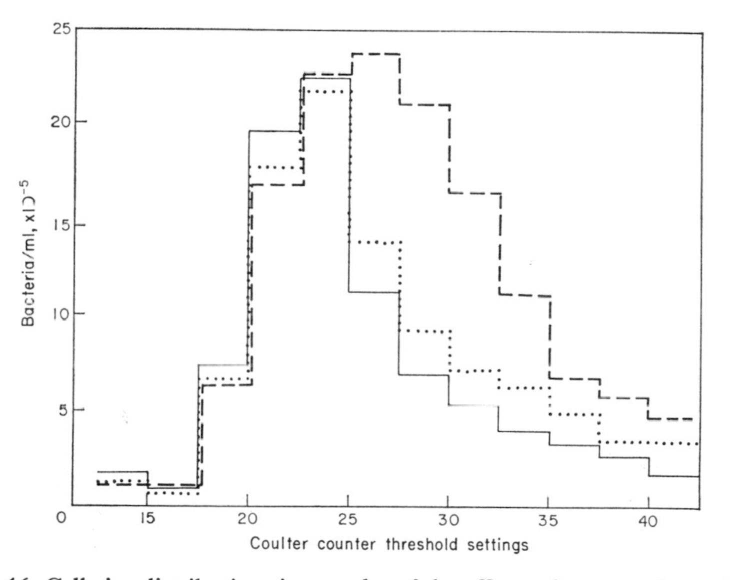

FIG. 16. Cell-size distributions in samples of the effluent from membrane-bound *E. coli* B/r, and in an exponentially growing culture. The distributions are shown as the number of cells between pairs of threshold settings in a model B Coulter Counter: ———, size distribution in a sample of the effluent collected between 38 and 40 min of elution; . . . , size distribution in a sample of the effluent collected between 118 and 120 min of elution; – – –, size distribution in a sample from an exponentially growing culture. Cells were grown in glucose minimal medium at 37°C.

The flow rate is adjusted to the desired value (usually 1–15 ml/min) by setting the pumping rate and samples of the effluent are collected consecutively. A portion of each sample is removed for measurement of the cell number, and the remainder is used for the experimental measurement.

Figure 15 shows the concentration of cells in the effluent from a membrane-bound population of *E. coli* B/r growing in minimal medium containing glucose as the carbon source. After the first 20 min of elution, most of the cells in the effluent are new-born, and Fig. 16 shows the size distribution in samples of the effluent compared to the distribution in an exponentially growing culture. The shape of the elution curve is a consequence of the exponential age distribution of the bound population (Powell, 1956) which contains twice as many newly formed cells as cells about to divide.

2. *Applications*

This technique can be used to study various problems, as long as these problems can be investigated by introducing an environmental change, e.g., radioactive precursors, inducers, radiation, chemical mutagens, antibiotics and nutritional shifts. It has been used to determine the relationship between chromosome replication and the division cycle (Helmstetter, 1967; Helmstetter and Cooper, 1968; Cooper and Helmstetter, 1968).

The major advantage of this technique, like those described in the previous Section, is that it allows analysis of untreated cultures in known physiological states. A second advantage is that the resolution is quite high and it is limited only by the dispersion in generation times of the cells on the membrane.

The major disadvantage of this technique is that at present its usefulness is limited to strain B of *E. coli*. In addition, there is an uncertainty of about 0·1 generations in the ages of the ancestors of the eluted cells. It must also be assumed that the cells divide on the membrane in the same relative progression as in batch cultures, but there is no evidence to suggest otherwise. These questions have been discussed in detail previously (Helmstetter, 1967; Helmstetter and Cooper, 1968).

C. Specialized techniques

A few classification techniques have been developed expressly for studying particular aspects of the division cycle of cells in asynchronous cultures. Although the methods are only applicable to certain problems, these problems can be studied extensively by experimenting on cells in various physiological states.

1. *Autoradiography*

Williamson (1965) has used autoradiography to measure the period of DNA synthesis during the division cycle of *S. cerevisiae* in exponentially growing cultures. A culture is briefly exposed to radioactive adenine, the cells are treated to remove virtually all of the radioactive constituents other than DNA, and autoradiographs are prepared. In the autoradiographs, only those cells which were synthesizing DNA during the period of exposure to the precursor are labelled, and their ages can be estimated by measuring their sizes. Cell age can be determined by measuring the ratio of the bud length to the length of the parental part of the cell, and converting this ratio to an estimate of age by means of a calibration curve. The calibration curve is obtained by following individual cells photographically. Relative bud length was considered a valid criterion of age, since the bud increases in size during growth whereas the parental part of the cell does not. This procedure for estimating age does not account for the possibility of a time interval between the separation of parent and daughter cells and the appearance of new buds. However, Williamson suggested that the average length of this lag period was less than 6% of the generation time, and therefore, little error is involved in neglecting this period.

The results indicate that DNA is synthesized during the first quarter of the cell cycle in exponentially growing cultures. This result agrees with that observed in cells growing in artificially synchronized cultures (Williamson and Scopes, 1962).

2. *Marker-frequency analysis*

Yoshikawa and Sueoka (1963a) and Yoshikawa *et al.* (1964) have investigated the mode of replication of the *B. subtilis* chromosome by measuring the relative frequency of genetic markers in exponentially growing cultures.

The theory of marker-frequency analysis in asynchronous populations has been described in detail by Sueoka and Yoshikawa (1965). The total length of the chromosome is considered to be one unit, so that the origin is at position 0 and the terminus at position 1. In an exponentially growing population, the chromosomes are in various stages of replication, and the frequency distribution of the position of the replication points can be given by the cellular age distribution function derived by Powell (1956). Under this condition, and assuming that replication proceeds at a constant rate per replication point, the frequency $g(x)$ of a genetic marker located at x on the chromosome in an exponentially growing population can be given by:

$$g(x) = 2^{n(1-x)}$$

where n represents the average number of replication positions per chromosome. When n is less than 1 there is a resting period between rounds of chromosome replication, and when it is greater than 1 there are multiple replication points per chromosome. When n equals 1, the average number of copies per chromosome for a genetic marker located at the origin, $g(0)$, is 2, and the average number for a marker located at the terminus, $g(1)$, is 1 with intermediate values between. When n equals 2, i.e., when the chromosomes contain 3 replication points, the ratio of the frequency of a marker located at the origin to one located at the terminus would be 4. This analysis assumes that the replication origin and direction of replication are fixed in the population.

This theoretical analysis of marker frequency has formed the basis of the experiments of Yoshikawa, Sueoka and co-workers. They analysed the marker frequency in cultures of *B. subtilis* using genetic transformation. Since the efficiency of transformation varies for different markers, the actual frequency of different markers in a population cannot be determined by direct comparison of the numbers of transformants for the markers. Rather, the marker frequencies in an exponentially growing culture are determined by comparing the frequency of transformants in the population with that in a standard culture in which all markers are known to be equally frequent. Stationary-phase cells and spores (Sueoka and Yoshikawa, 1963; Yoshikawa *et al.*, 1964) have been used as standards. A genetic map of *B. subtilis* has been constructed based on the relative frequency of transformants for various markers from samples of DNA from exponential- and stationary-phase cultures. It was as a result of these experiments that the concept of multiple-fork replication was developed based on the observation of marker-frequency ratios greater than 2.

Berg and Caro (1967) used the principle of marker-frequency analysis to investigate the mode of chromosome replication in *E. coli* K12. Instead of transformation, they used generalized transduction with the bacteriophage P1. Exponentially growing cultures were infected with P1 and the progeny phage were used to transduce markers into a multiply-marked recipient. The fractional number of transductions for each marker relative to the total number of transductants was determined, and these values were compared for different strains. They used this technique to determine the dependence of the replication origin on the position of the incorporated F factor in Hfr mating types. Nagata (1963) proposed that in Hfr strains chromosome replication starts at the F factor and proceeds opposite to the direction of transfer during conjugation. If this were the case, different marker frequencies would be expected in strains containing the incorporated F factor at different loci on the chromosome. Berg and Caro found that the relative marker frequencies were the same in different strains, and they concluded that the position and orientation of the F factor does not determine the origin and direction of replication.

3. *The kinetics of cell growth*

In a population of growing micro-organisms there is a distribution of cell sizes, and the shape of the distribution depends, in part, on the kinetics of growth of the individual cells. Collins and Richmond (1962) and Koch and Schaechter (1962) have derived theoretical expressions for the distribution of cell sizes, and Harvey *et al.* (1967) have used the equation of Collins and Richmond, which relates the kinetics of growth and the size distribution, to study the growth of individual cells. The general conclusion from all of these studies is that the specific growth rate of individual cells of *B. cereus*, *E. coli* and *Azotobacter agilis* increases between divisions.

Studies of this type on the kinetics of growth require very accurate measurements of size distributions (Koch, 1966). With the development of electric sensing zone instruments, measurements of cell size have become fairly routine procedures. However, there are problems associated with obtaining accurate volume distributions with these instruments, and some of these problems, along with possible solutions, have been discussed recently by Harvey (1968).

V. CONCLUSIONS

The choice of technique for an investigation on the cell cycle is dictated by the organism and type of experiment. In general, however, classification techniques are preferable to synchronization techniques. The fundamental difference between these two approaches to cell-cycle studies is that with synchronization techniques the age of the cells to be analysed is known, but their physiological state is somewhat uncertain, whereas with classification techniques the age is somewhat uncertain, but the physiological state of the cells is predetermined. In the latter case the precise age of the cells to be analysed may be uncertain, but it can usually be determined within certain experimental limits. On the other hand, the extent to which a synchronization treatment causes a change in state is difficult to estimate without reference to an untreated control.

In either case, the centrifugation and membrane-elution techniques are recommended for most experiments, since they combine minimum distortion with reasonably good resolution. Unlike some of the earlier methods, both of these procedures are reliable and easily performed.

REFERENCES

Abbo, F. E., and Pardee, A. B. (1960). *Biochim. biophys. Acta*, **39**, 478–485.

Altenbern, R. A. (1966). *Biochem. biophys. Res. Commun.*, **25**, 346–353.

Anderson, P. A. (1961). *Biochim. biophys. Acta*, **49**, 231–232.

Barner, H. D., and Cohen, S. S. (1956). *J. Bact.*, **72**, 115–123.

Berg, C. M., and Caro, L. G. (1967). *J. molec. Biol.*, **29**, 419–431.

Bostock, C. J., Donachie, W. D., Masters, M., and Mitchison, J. M. (1966). *Nature, Lond.*, **210**, 808–810.

Browning, I., Brittain, M. S., and Bergendahl, J. C. (1952). *Tex. Rept. Biol. Med.*, **10**, 794–802.

Burns, V. W. (1959). *Science, N.Y.*, **129**, 566–567.
Burns, V. W. (1961). *Expl Cell Res.*, **23**, 582–594.
Burns, V. W. (1964). *In* "Synchrony in Cell Division and Growth" (Ed. E. Zeuthen), pp. 433–439. Wiley, New York.
Campbell, A. (1956). *Bact. Proc.* 40–41.
Campbell, A. (1957). *Bact. Rev.*, **21**, 263–272.
Clark, D. J., and Maaløe, O. (1967). *J. molec. Biol.*, **23**, 99–112.
Collins, J. F., and Richmond, M. H. (1962). *J. gen. Microbiol.*, **28**, 15–33.
Cook, J. R., and James, T. W. (1960). *Expl Cell Res.*, **21**, 583–589.
Cooper, S., and Helmstetter, C. E. (1968). *J. molec. Biol.*, **31**, 519–540.
Corbett, J. J. (1964). *Expl Cell Res.*, **33**, 155–160.
Cummings, D. J. (1965). *Biochim. biophys. Acta*, **95**, 341–350.
Cutler, R. G., and Evans, J. E. (1966). *J. Bact.*, **91**, 469–476.
Cutler, R. G., and Evans, J. E. (1967a). *J. molec. Biol.*, **26**, 81–90.
Cutler, R. G., and Evans, J. E. (1967b). *J. molec. Biol.*, **26**, 91–105.
Donachie, W. D., and Masters, M. (1966). *Genet. Res.*, **8**, 119–124.
Doudney, C. O. (1960). *J. Bact.* **79**, 122–124.
Edmunds, L. N. Jr. (1964). *Science, N.Y.*, **145**, 266–268.
Evans, H. J., Neary, G. J., and Tonkinson, S. M. (1957). *J. Genet.*, **55**, 487–502.
Falcone, G., and Szybalski, W. (1956). *Expl Cell Res.*, **11**, 486–489.
Gorman, J., Tauro, P., LaBerge, M., and Halvorson, H. O. (1964). *Biochem. biophys. Res. Commun.* **15**, 43–49.
Halvorson, H. O., Gorman, J., Tauro, P., Epstein, R., and LaBerge, M. (1964). *Fedn Proc. Fedn Am. Socs exp. Biol.*, **23**, 1002–1008.
Halvorson, H. O., Bock, R. M., Tauro, P., Epstein, R., and LaBerge, M. (1966). *In* "Cell Synchrony-Studies in Biosynthetic Regulation" (Ed. I. L. Cameron and G. M. Padilla), pp. 102–116. Academic Press, New York.
Harvey, R. J. (1968). *In* "Methods in Cell Physiology" (Ed. D. M. Prescott), Vol. 3, pp. 1–23. Academic Press, New York.
Harvey, R. J., Marr, A. G., and Painter, P. R. (1967). *J. Bact.*, **93**, 605–617.
Helmstetter, C. E. (1967). *J. molec. Biol.*, **24**, 417–427.
Helmstetter, C. E., and Cooper, S. (1968). *J. molec. Biol.*, **31**, 507–518.
Helmstetter, C. E., and Cummings, D. J. (1963). *Proc. natn. Acad. Sci. U.S.A.*, **50**, 767–774.
Helmstetter, C. E., and Cummings, D. J. (1964). *Biochim. biophys. Acta*, **82**, 608–610.
Helmstetter, C. E., and Uretz, R. B. (1963). *Biophys. J.*, **3**, 35–47.
Hotchkiss, R. D. (1954). *Proc. natn. Acad. Sci. U.S.A.*, **40**, 49–55.
Houtermans, T. (1953). *Z. Naturf.*, **8B**, 767–771.
Howard, A., and Pelc, S. R. (1953). *Heredity, Suppl.* 6, 261–273.
Howell, J. A., Tsuchiya, H. M., and Fredrickson, A. G. (1967). *Nature, Lond.*, **214**, 582–584.
Hunter-Szybalska, M. E., Szybalski, W., and DeLamater, E. D. (1956). *J. Bact.*, **71**, 17–24.
Imanaka, H., Gillis, J. R., and Slepecky, R. A. (1967). *J. Bact.*, **93**, 1624–1630.
James, T. W. (1959). *Ann. N.Y. Acad. Sci.*, **78**, 501–514.
James, T. W. (1964). *In* "Synchrony in Cell Division and Growth" (Ed. E. Zeuthen),. pp. 323–349. Wiley, Interscience, New York.
Kates, J. R., and Jones, R. F. (1967). *Biochim. biophys. Acta*, **145**, 153–158.
Koch, A. L. (1966). *J. gen. Microbiol.*, **45**, 409–417.
Koch, A. L., and Schaechter, M. (1962). *J. gen. Microbiol.*, **29**, 435–454.

Kubitschek, H. E., Bendigkeit, H. E., and Loken, M. R. (1967). *Proc. natn. Acad. Sci. U.S.A.*, **57**, 1611–1617.
Kuempel, P. L., Masters, M., and Pardee, A. B. (1965). *Biochem. biophys. Res. Commun.*, **18**, 858–867.
Kuhl, A., and Lorenzen, H. (1964). *In* "Methods in Cell Physiology" (Ed. D. M. Prescott), Vol. 1, pp. 159–187. Academic Press, New York.
Lafeber, A., and Steenbergen, C. L. M. (1967). *Nature, Lond.*, **213**, 527–528.
Lark, K. G. (1958). *Can. J. Microbiol.*, **4**, 179–189.
Lark, K. G. (1960). *Biochim. biophys. Acta*, **45**, 121–132.
Lark, K. G. (1961). *Biochim. biophys. Acta*, **51**, 107–116.
Lark, K. G., and Lark, C. (1960). *Biochim. biophys. Acta*, **43**, 520–530.
Lark, K. G., and Maaløe, O. (1954). *Biochim. biophys. Acta*, **15**, 345–356.
Lorenzen, H. (1957). *Flora (Jena)*, **144**, 473–496.
Lorenzen, H. (1964). *In* "Synchrony in Cell Division and Growth" (Ed. E. Zeuthen), pp. 571–578. Wiley, New York.
Maaløe, O. (1962). *In* "The Bacteria" (Ed. I. C. Gunsalus and R. Y. Stanier), Vol. IV, pp. 1–32. Academic Press, New York.
Maaløe, O., and Hanawalt, P. C. (1961). *J. molec. Biol.*, **3**, 144–155.
Maaløe, O., and Lark, K. G. (1954). *In* "Recent Developments in Cell Physiology" (Ed. J. A. Kitching), pp. 159–169. Butterworths, London.
Manor, H., and Haselkorn, R. (1967). *Nature, Lond.*, **214**, 983–986.
Maruyama, Y. (1956). *J. Bact.*, **72**, 821–826.
Maruyama, Y. (1964). *In* "Synchrony in Cell Division and Growth" (Ed. E. Zeuthen) pp. 593–598. Wiley, New York.
Maruyama, Y., and Lark, K. G. (1959). *Expl Cell Res.*, **18**, 389–391.
Maruyama, Y., and Lark, K. G. (1961). *Expl Cell Res.*, **25**, 161–169.
Maruyama, Y., and Lark, K. G. (1962). *Expl Cell Res.*, **26**, 382–394.
Maruyama, Y., and Yanagita, T. (1956). *J. Bact.*, **71**, 542–546.
Masters, M., Keumpel, P. L., and Pardee, A. B. (1964). *Biochem. biophys. Res. Commun.*, **15**, 38–42.
Matney, T. S., and Suit, J. C. (1966). *J. Bact.*, **92**, 960–966.
Mitchison, J. M., and Vincent, W. S. (1965). *Nature, Lond.*, **205**, 987–989.
Mitchison, J. M., and Vincent, W. S. (1966). *In* "Cell Synchrony—Studies in Biosynthetic Regulation" (Ed. I. L. Cameron and G. M. Padilla), pp. 328–331. Academic Press, New York.
Moner, J. G. (1965). *J. Protozool.*, **12**, 505–509.
Moner, J. G., and Berger, R. O. (1967). *J. Cell Physiol.*, **67**, 217–223.
Nagata, T. (1963). *Proc. natn. Acad. Sci. U.S.A.*, **49**, 551–559.
Neary, G. J., Evans, H. J., and Tonkinson, S. M. (1959). *J. Genet.*, **56**, 363–394.
Neff, R. J., and Neff, R. H. (1964). *In* "Synchrony in Cell Division and Growth" (Ed. E. Zeuthen), pp. 213–246. Wiley, New York.
Nishi, A., and Kogoma, T. (1965). *J. Bact.*, **90**, 884–890.
Padilla, G. M., and Cameron, I. L. (1964). *J. Cell Physiol.*, **64**, 303–307.
Padilla, G. M., and James, T. W. (1960). *Expl Cell Res.*, **20**, 401–415.
Padilla, G. M., and James, T. W. (1964). *In* "Methods in Cell Physiology" (Ed. D. M. Prescott), Vol. 1, pp. 141–157. Academic Press, New York.
Painter, R. B., and Drew, R. M. (1959). *Lab. Invest.*, **8**, 278–285.
Perry, R. P. (1959). *Expl Cell Res.*, **17**, 414–419.
Plesner, P., Rasmussen, L., and Zeuthen, E. (1964). *In* "Synchrony in Cell Division and Growth" (Ed. E. Zeuthen), pp. 543–563. Wiley, New York.

Powell, E. O. (1956). *J. gen. Microbiol.*, **15**, 492–511.
Prescott, D. M. (1955). *Expl Cell Res.*, **9**, 328–337.
Prescott, D. M. (1957). *J. Protozool.*, **4**, 252–256.
Prescott, D. M. (1960). *Expl Cell Res.*, **19**, 228–238.
Prescott, D. M. (1966). *J. Cell Biol.*, **31**, 1–9.
Puck, T. T., and Steffen, J. (1963). *Biophys. J.*, **3**, 379–397.
Puck, T. T., Sanders, P., and Peterson, D. (1964). *Biophys. J.*, **4**, 441–450.
Roslansky, J. D., Branchflower, N. H., and Huckman, M. S. (1958). *Anat. Rec.*, **132**, 498.
Rudner, R. (1960). *Biochem. biophys. Res. Commun.*, **3**, 275–280.
Rudner, R., Prokop-Schneider, B., and Chargaff, E. (1964). *Nature, Lond.*, **203**, 479–483.
Rudner, R., Rejman, E., and Chargaff, E. (1965). *Proc. natn. Acad. Sci. U.S.A.*, **54**, 904–911.
Ryan, F. J., and Cetrulo, S. D. (1963). *Biochem. biophys. Res. Commun.*, **12**, 445–447.
Sando, N. (1963). *J. gen. appl. Microbiol.*, **9**, 233–241.
Scherbaum, O. (1960). *Ann. Rev. Microbiol.*, **14**, 283–310.
Scherbaum, O., and Jahn, T. L. (1964). *Expl Cell Res.*, **33**, 99–104.
Scherbaum, O., and Zeuthen, E. (1954). *Expl Cell Res.*, **6**, 221–227.
Schmidt, R. R. (1966). *In* "Cell Synchrony—Studies in Biosynthetic Regulation" (Ed. I. L. Cameron and G. M. Padilla), pp. 189–235. Academic Press, New York.
Schmidt, R. R., and King, K. W. (1961). *Biochim. biophys. Acta*, **47**, 391–392.
Schleif, R. (1967). *J. molec. Biol.*, **27**, 41–55.
Scott, D. B. M., and Chu, E. C. (1958). *Expl Cell Res.*, **14**, 166–174.
Scott, D. B. M., DeLamater, E. D., Minsavage, E. J., and Chu, E. C. (1956). *Science, N.Y.*, **123**, 1036–1037.
Sisken, J. E. (1964). *In* "Methods in Cell Physiology" (Ed. D. M. Prescott), Vol. I, pp. 387–401. Academic Press, New York.
Sorokin, C. (1957). *Physioligia Pl.*, **10**, 659–666.
Stanners, C. P., and Till, J. E. (1960). *Biochim. biophys. Acta*, **37**, 406–419.
Stárka, J., and Koza, J. (1959). *Biochim. biophys. Acta*, **32**, 261–262.
Stone, G. E., and Cameron, I. L. (1964). *In* "Methods in Cell Physiology" (Ed. D. M. Prescott), Vol. I, pp. 127–140. Academic Press, New York.
Stonehill, E. H., and Hutchison, D. J. (1966). *J. Bact.*, **92**, 136–143.
Sueoka, N., and Yoshikawa, H. (1963). *Cold Spring Harb. Symp. quant. Biol.*, **28**, 47–54.
Sueoka, N., and Yoshikawa, H. (1965). *Genetics*, **52**, 747–757.
Sylvén, B., Tobias, C. A., Malmgren, H., Ottoson, R., and Thorell, B. (1959). *Expl Cell Res.*, **16**, 75–87.
Tamiya, H. (1964). *In* "Synchrony in Cell Division and Growth" (Ed. E. Zeuthen), pp. 247–305. Wiley, New York.
Tamiya, H., and Morimura, Y. (1964). *In* "Synchrony in Cell Division and Growth" (Ed. E. Zeuthen), pp. 565–569. Wiley, New York.
Tamiya, H., Iwamura, T., Shibata, K., Hase, E., and Nihei, T. (1953). *Biochim. biophys. Acta*, **12**, 23–40.
Tauro, P., and Halvorson, H. O., (1966). *J. Bact.*, **92**, 652–661.
Taylor, J. H. (1960). *Adv. biol. med. Phys.*, **7**, 107–130.
Tobey, R. A., Petersen, D. F., Anderson, E. C., and Puck, T. T. (1966). *Biophys. J.*, **6**, 567–581.
Treich, R. W., and Konetzka, W. A. (1964). *J. Bact.*, **88**, 1580–1584.

Williamson, D. H. (1964). *In* "Synchrony in Cell Division and Growth" (Ed. E. Zeuthen), pp. 589–591. Wiley, New York.
Williamson, D. H. (1965). *J. Cell Biol.*, **25**, 517–528.
Williamson, D. H., and Scopes, A. W. (1960). *Expl Cell Res.*, **20**, 338–349.
Williamson, D. H., and Scopes, A. W. (1962). *Nature, Lond.*, **193**, 256–257.
Woodard, J., Rasch, E., and Swift, H. (1961). *J. biophys. biochem. Cytol.*, **9**, 445–462.
Yanagita, T., Maruyama, Y., and Takebe, I. (1958). *J. Bact.*, **75**, 523–529.
Yoshikawa, H., and Sueoka, N. (1963a). *Proc. natn. Acad. Sci. U.S.A.*, **49**, 559–566.
Yoshikawa, H., and Sueoka, N. (1963b). *Proc. natn. Acad. Sci. U.S.A.*, **49**, 806–813.
Yoshikawa, H., O'Sullivan, A., and Sueoka, N. (1964). *Proc. natn. Acad. Sci. U.S.A.*, **52**, 973–980.
Zeuthen, E. (1958). *Adv. biol. med. Phys.*, **6**, 37–73.
Zeuthen, E. (1964). *In* "Synchrony in Cell Division and Growth" (Ed. E. Zeuthen), pp. 99–158. Wiley, New York.
Zeuthen, E., and Scherbaum, O. (1954). *In* "Recent Developments in Cell Physiology" (Ed. J. A. Kitching), pp. 141–156. Butterworths, London.

CHAPTER X

Methods of Microculture

LOUIS B. QUESNEL

Department of Bacteriology and Virology, University of Manchester, Manchester, England

I. INTRODUCTION

All the methods of culture considered in this Chapter are aimed at the study of single cells and their reactions to one another and to their environment. To record their behaviour microscopic examination is necessary and, in the absence of biochemical techniques for investigating living single cells, comparatively few properties can be measured. Ideally a microcultural technique should be capable of maintaining the cells under the required growth conditions while allowing for continued or repeated observation over long periods of time. There are however applications of microcultural techniques where it is not essential that the cells be alive at the time of observation. In the latter cases no exceptional forms of instrumentation are usually required, but some types of microcultural studies present formidable although generally unappreciated technical difficulties.

II. THE PROPERTIES OF CELLS AMENABLE TO MEASUREMENT

A. Physical dimensions

If the individual cells can be viewed optically with sufficient definition, it is possible to make "direct" measurements of cell lengths and widths by the use of some form of micrometer. Alternatively by taking photographs lengths and widths as well as shape can be permanently recorded. Assuming the three-dimensional geometry of the cells to be cylinders, spheres, ellipsoids, helices or modifications of these forms it is possible to calculate volumes.

B. Cell structures and inclusions

Any structure that can be made visible or photographable can be recorded. There are not many inclusions that can be resolved in living bacteria, but it is possible under suitable conditions (which are described later) to record such structures as nuclei, spores, end-bodies, fat globules, cross walls, flagella.

The use of fluorescence techniques that allow bacteria to be "stained" without inhibition of division has led to interesting discoveries on the growth and division of cell walls and the disposition of newly formed wall material (Chung *et al.*, 1964a, b, 1965); the primulin-fluorescence technique has led to an extensive study of the budding of yeasts, their scars and ageing (see review by Beran, 1968).

C. Growth parameters

Either directly or by the use of time-lapse cinematography it is possible to record individual generation-times and from these to construct genealogies and to assess rates of reproduction and clonal development. Definitions of cell viability are precarious and demand a specific and restricted statement for any particular experimental situation, but with this realization viabilities can be obtained and the antimicrobial effects of drugs can be usefully estimated even e.g., on intracellular bacteria (Showacre *et al.*, 1961).

D. Behavioural reactions

A varied assortment of other phenomena can also be recorded, such as motility (Weibull, 1960) and other movements such as palisading and snapping (Hoffman and Frank, 1965), agglutination, aggregation and colonial structure and even the formation of eco-systems (Ware and Loveless, 1959). Phototaxis, aerotaxis and chemotaxis have also been investigated (Engelmann, 1894; Sherris *et al.*, 1957; Clayton, 1958; Baracchini and Sherris, 1959).

III. METHODS OF MICROCULTURE

In a previous discussion of microcultural techniques applicable to the study of tubercle bacilli, Pryce (1941) divided the available methods into two main classes: "open" methods, in which the medium is exposed to air; and "closed" methods, in which the medium is sealed with little or no air. Hoffman (1964) has modified this by calling "open" methods "all those which interpose an air break in the optical path through the microculture preparation", and "those which have no air breaking the optical path" are "closed". While considerations such as whether or not an air space is interposed in the optical path through the microculture chamber may have technical and observational consequences of interest to the experimenter, the "air break" is not necessarily of great consequence to the bacteria and a more fundamental division of microcultural types should probably be made between those in which environmental conditions are continually varying and those in which the environmental factors are maintained constant, or very nearly so. The need to consider this basic problem has become increasingly evident with the many recent reports of the vast changes that occur in microbial composition or behaviour consequent upon quite minor (to us) changes in the environment.

For the purpose of this article, all the methods that have been used without particular regard to the control of the micro-environment will be described in sections based on the nature of the preparation, while those methods that are aimed at maintaining environmental control will be described under the general heading "perfusion chambers".

A. Wet film

Undoubtedly the simplest form of microcultural chamber is the wet film, prepared by enclosing the organisms in a layer of liquid between slide and coverslip. In such a liquid layer there is ample room for small mobile organisms to move about while being observed microscopically, and their movements can be observed and recorded by film. Such preparations very rapidly dry out unless the edges of the coverslip are sealed by means of

Vaseline or silicone grease to prevent evaporation. They have several advantages: they are simple, they are thin and they are "optically clean". Their thinness permits microscopy at its very highest resolving power, while deformation of the light path is minimal, as there need be only one layer of differing refractive index if oil immersion objective and condenser are used, *viz*, the suspending fluid. Even this latter can be made to have a refractive index similar to that of glass by the addition of suitable compounds, such as glycerine or gelatin, or bovine plasma albumin, but they obviously then form part of the micro-environment and may influence the cells in undesired ways.

The limitations of the simple wet film used as a microcultural chamber are equally obvious. Since the suspending menstruum is of a fixed small amount, conditions of limiting nutrient may soon arise. For the same reason rapid changes of environmental chemistry will take place as ingredients of the environment are utilized, adsorbed or absorbed, and as products of metabolism begin to accumulate, resulting in changes of pH, toxic effects, etc. Aeration is minimal, and oxygen starvation is a fairly immediate effect upon obligate aerobes. Also, the mobility of the cells in free suspension, due either to Brownian motion or their own motility, makes accurate records of cell parameters extremely difficult.

B. Hanging drop

The simple hanging drop preparation is probably best assembled by applying the "drop" to a coverslip already having four small smears of Vaseline, one at each corner. The hollow-ground slide is then pressed on to the coverslip so that the Vaseline makes contact with the flat of the slide while the dome of the drop is accommodated by the "hollow" of the slide (see Fig. 1a). The preparation can then be easily inverted and more Vaseline

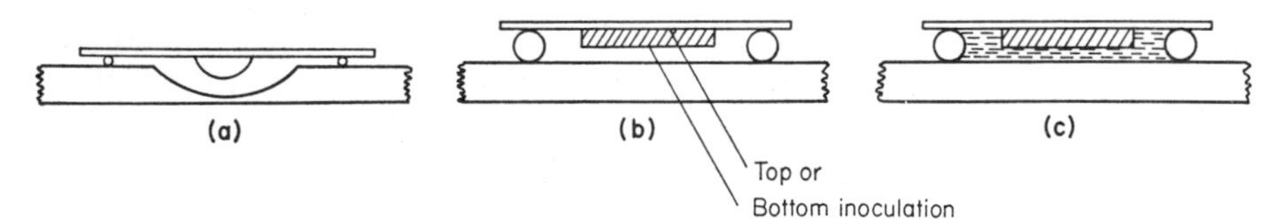

FIG. 1. Diagramatic representations of simple microcultural preparations: (a), hanging drop (Section III B); (b) hanging block (Sections III D and E); (c), hanging block with liquid well (Section III D).

added to the contact edge of the coverslip to make a wholly enclosed chamber if required.

The advantage of the hanging drop over the wet film is the increased capacity for aeration of the preparation as the film is surrounded by a comparatively large air space enabling gaseous exchange to a limited extent; also, individual microdrops can be used to isolate selected organisms. All the other disadvantages mentioned in the previous Section apply here, with the added interference with the optical path, which results from the inclusion of an air layer in the sandwich, so that there are at least three different refractive indices involved. The situation is made worse by the curvature of the "hollow" of the ground slide and the curvature of the droplet. Really good microscopy is not possible with hanging-drop preparations utilizing a hollowed slide. It is better to replace the hollow-ground slide by a standard flat slide and mount the coverslip on a platform of grease or Plasticine. Alternatively, a hollowed slide with flat floor is recommended.

In spite of the unsuitability of hanging-drop preparations for most purposes other than the observation of motility, an early and ingenious application of this microcultural method to the study of the relationship between cell generation times and temperature was made by Barber in 1908, and his results are still much quoted in modern textbooks. Later Allen (1923) used the same technique to estimate the generation times of bacteria growing in milk, and Kahn (1929) used it to study mycobacterial life cycles. Barber's method was as follows. By means of a specially designed micropipette and micromanipulator used in conjunction with a microscope, he transferred single cells from an inoculum hanging drop to other coverslips to form single-cell hanging drop cultures, which were then maintained at different temperatures for given periods of time. The cells were then killed and lightly stained by the injection of a microdrop of methylene blue and KOH. The organisms produced in a given time were counted and from the number and size of the cells the generation times at different temperatures were calculated.

Barber's work is worth special attention as it is one of the earliest attempts at a quantitative assessment of the behaviour of individual bacteria in microculture. His experimental approach could well be emulated today. (An ingenious device for isolating single bacteria, which utilized a cannibalized microscope as a micromanipulator for use with a second microscope, was described by Malone (1918), inspired by the paper in which "Barber . . . described a machine devised for picking out single organisms" (see Johnstone, this Volume, p. 455).

C. Hanging drop with oil bath

In 1949, de Fonbrune (see Johnstone, this Volume, p. 455) published a sophisticated version of Barber's original method in which high-quality micromanipulators were used in conjunction with an oil chamber for isolating micro-droplets on separate parts of a coverslip. His method has been used and slightly modified by various workers, e.g., Lederberg (1956) and

Nossal (1958b), for various studies on the behaviour of single cells in antigen-antibody reactions involving bacteria and for the study of the kinetics of the release of poliomyelitis virus from single cells by Lwoff *et al.* (1955).

The adaptation used by Nossal (1958a) is described by him as follows: "A new 2 × 1 in. glass microscope coverslip was washed in distilled water and dried with a Kleenex tissue. The coverslip was divided into 9 rectangles by indian-ink lines. It was then placed on an oil chamber (Fig. 2). This

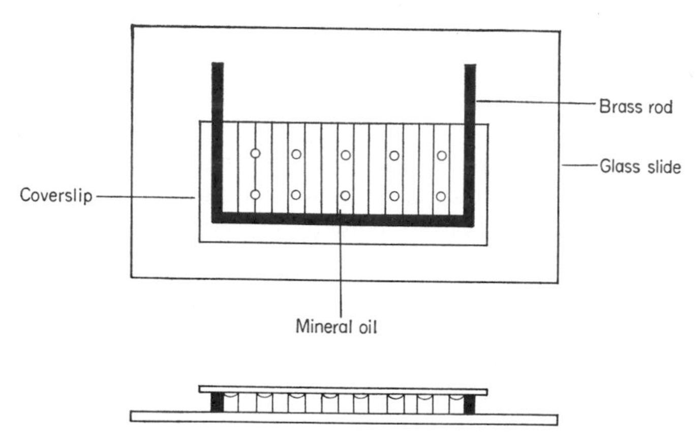

FIG. 2. de Fonbrune chamber method used by Nossal. The upper surface of the coverslip is ruled with indian-ink lines. The droplets of cell suspension are deposited on the unruled surface under a layer of oil and then inverted over the oil chamber. (From Nossal, 1958a.)

consisted of three brass rods glued on to a glass slide in the form of a rectangle with one side missing. A thin layer of the mineral oil Flozene (H. C. Sleigh & Co.) was spread over the unmarked surface of the coverslip. A glass Pasteur pipette was drawn out to a diameter of about 5 μm in a small flame and attached to one end of a rubber tube, the other end of which was held in the operator's mouth. Droplets of the suspension were then deposited on the surface of the coverslip, under the layer of oil which minimized evaporation. If the volume of the droplets was kept relatively constant in the vicinity of 10^{-7} ml, they contained from 0 to 6 cells (of lymph node tissue prepared from immunized rats as described in Nossal's paper).† Larger droplets containing up to 100 or 1000 cells could also be dispensed. In all experiments, many droplets containing no cells were deposited for control purposes. The coverslip was then inverted and the space beneath it filled with mineral oil.

"The chamber was placed on a microscope and the droplets surveyed at

† Parentheses mine.

100-fold magnification using dark-ground illumination. It was found useful before incubation of the microdroplets to add a small constant quantity of suspending medium to all microdroplets using the micropipette and manipulator. Within the first 5 min after deposition of the original droplets, there was a tendency for them to spread somewhat and flatten out into discs convenient for microscopic observation. If a little further fluid was added at this stage, little or no further spreading occurred, even after some hours of incubation. The addition of a little fluid increased the ratio of volume to surface area and thereby reduced the total loss of water through the paraffin during the hours of incubation".

The chamber with its inoculated lymph node cells was then incubated at 37°C for 4 h, after which it was returned to the microscope stage and about ten motile bacteria were added by micropipette to each droplet. The microdroplets were then surveyed at appropriate intervals, and the cell content as well as the motility of the bacteria was recorded. Total loss of motility of all the bacteria was recorded as "inhibition". If even one of the inoculated bacteria remained motile, this was recorded as "no inhibition".

A series of publications by Nossal (1958a, b, 1959, 1960), by Nossal and Makela (1962a, b) and by Makela and Nossal (1961a, b, 1962) illustrate the variety of immunological phenomena that can be studied by this technique. Their results up to 1962 are reviewed by Nossal and Makela (1962b), who point out that "just as immobilization is a slightly more sensitive method for H-antibody assay than is H-agglutination, so chaining appears to be slightly more sensitive than O-agglutination in assaying anti-O activity of the droplets. Again, no more than a semiquantitative assessment of the antibody content of positive droplets is possible".

Attardi *et al.* (1959) used a similar microdrop method for measuring the production of antibody by single cells, isolated from rabbits hyperimmunized with bacteriophage, using a virus neutralization technique. In order to transfer individual microdrops from the oil chamber to recovery media, they removed the coverslip containing the microdrops to the stage of a stereomicroscope and then absorbed the selected droplet on the tip of a fine pointed strip of sterile filter paper (Fig. 3). The filter paper was then dropped into broth, the virus extracted and the broth plated with bacteria suitable for the detection of the relevant phages. Marked decrease in the expected plaque count was taken as evidence for the formation of antiphage antibody in the relevant microdrop. This phage inactivation technique has the advantage that a better expression of the quantitative relationships of the antigen–antibody response is possible, but it still does not allow the determination of antibody yield per cell in absolute terms.

Attardi and co-workers reported an alternative micropipette technique in the same paper. In this method individual lymph node cells from a

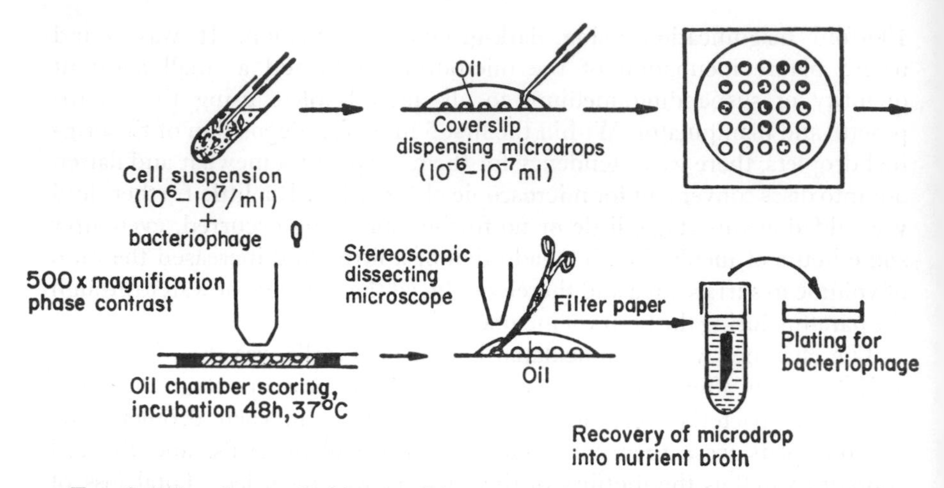

FIG. 3. Microdrop technique used by Attardi *et al.* (1959) to measure antibody production by single cells against bacteriophage. The microdrops are removed by absorbing on to filter paper from beneath the oil layer. (From Attardi *et al.*, 1959).

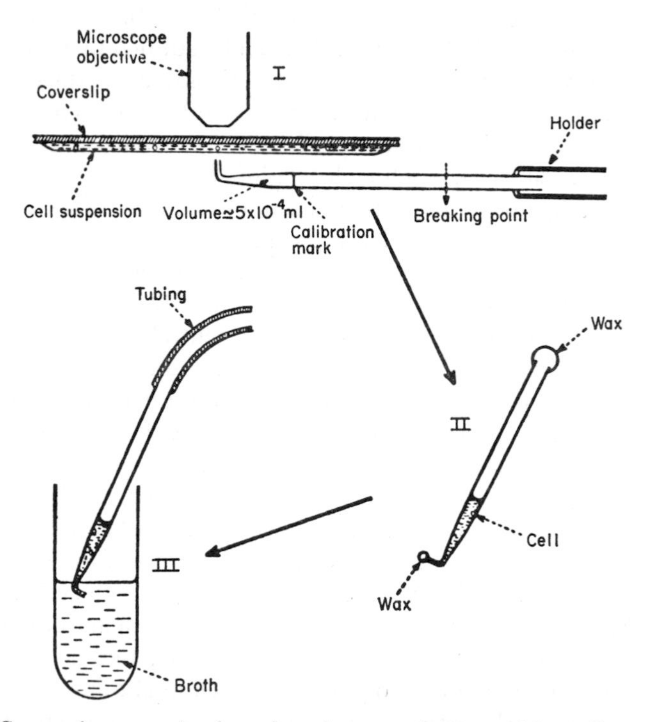

FIG. 4. Successive steps in the micropipette technique of Attardi *et al,* (1959); for details see text. (From Attardi *et al.*, 1959)

dilute suspension containing both lymphoid cells and the phages T_2 and T_5 at known concentrations, were drawn up into carefully calibrated micro-pipettes (Fig. 4). The micropipettes were broken at the point indicated in the Figure and the ends sealed with wax. After incubation for the required time, the contents of each were blown out into broth and the phage assay performed. Attardi *et al.* used phase microscopy for their observations, a method that has several advantages over dark-field illumination as used by Nossal.

The de Fonbrune method has also been adopted for studying the kinetics of the release of poliomyelitis virus from single cells by Lwoff *et al.* (1955). The oil-bath technique has been found suitable for animal-cell culture, since paraffin oil, as well as being biologically "neutral," is permeable to gases and allows the exchange of O_2 and CO_2 between the suspending nutrient droplet and the atmosphere. Lwoff *et al.* have tested several paraffins and recommend Bayol F (Penola Oil Co.) and also the silicone derivative silicone dimethylsiloxane (Dow Corning Fluid 200) as suitable. Their paper records many useful details of technique, and should be consulted; only a summary of the method will be given here.

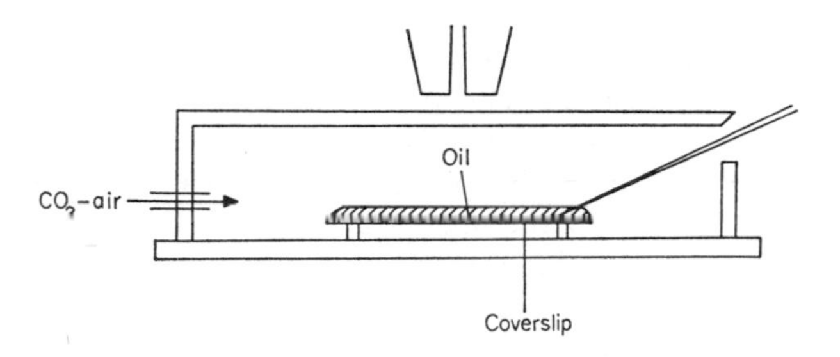

FIG. 5. Use of Lucite box containing coverslip with microdrops submersed in paraffin oil; method of Lwoff *et al.* (1959).

Monolayer cultures of monkey kidney cells were prepared and suspensions of infected cells were made. A suitably cleaned coverslip placed on the base of a 6 cm Petri dish was layered with oil to a depth of 3–4 mm and the dish placed in a Lucite box, $3\frac{1}{2} \times 4 \times 1$ in. having an inlet on one side for the CO_2–air mixture and a horizontal slit, 30×5 mm, on the other to allow the manipulation of a fine pipette (Fig. 5). The Lucite box was placed on the stage of a dissecting microscope which was kept in a thermostatically regulated box (with hand ports) maintained at 37°C. By means of a pipette with attached rubber tubing held in the mouth of the operator, a drop of nutrient medium and a separate drop of cell suspension containing from 200–400 cells, were deposited on the coverslip. Fresh medium was then sucked into a pipette and the tip placed near a cell in the inoculum droplet so that it was aspirated by capillarity along with a small volume of medium. The cell plus medium, to a drop volume of approximately 4×10^{-6} ml, was deposited on another part of the coverslip and rows of such droplets were deposited by the same process.

To estimate the virus liberated by particular cells during a given time period, the fluid of the droplet was removed insofar as possible, without damaging the cells, the original droplet replenished with fresh medium and the droplet again extracted and pooled with the first extract. About 95–99% of the original fluid was thus withdrawn and this was mixed with 0·2 ml of Earle's saline containing 20% chick embryo extract and frozen. These samples were later assayed for plaque-forming particles by "plating" on monolayer cultures of Rhesus monkey kidney cells.

The above free-hand manipulation under the dissecting microscope does not permit the observation of the cells under high magnification, but this can be carried out by using the de Fonbrune chamber. In this case the coverslip was inoculated with several droplets containing single cells, each of these droplets being surrounded by a further 10 droplets of fresh medium alone. After 1 h at 37°C in the humidified CO_2–air atmosphere, the coverslip was inverted and placed over the de Fonbrune chamber and the chamber space filled with oil. The chamber was then placed on the stage of a phase–contrast microscope that was enclosed in a Lucite box with controlled temperature and atmosphere as before. The preparation could then be examined and photographed by phase contrast. After appropriate times samples were removed to the nutrient droplets by means of micro-manipulators, their new positions marked on a chart, the coverslips removed and the samples withdrawn by pipette (as before) to chick embryo extract and subsequently assayed.

D. Hanging block

The hanging-block technique appears to have been originated by Hill (1902), and a clearer description than his own could hardly be given. "Pour melted agar into a Petri dish to the depth of about $\frac{1}{8}$–$\frac{1}{4}$in. Cool this agar and cut from it a block about $\frac{1}{4}$–$\frac{1}{3}$ in. square and of the thickness of the agar layer in the dish. This block has a smooth upper and under-surface. Place it, under-surface down, on a slide and protect it from dust. Prepare an emulsion in sterile water of the organism to be examined if it has been grown on a solid medium or use a broth culture; spread the emulsion or broth upon the upper surface of the block as if making an ordinary coverslip preparation. Place the slide and block in a 37°C incubator for 5 or 10 min to dry slightly. Then lay a clean sterile coverslip on the inoculated surface of the block in close contact with it, usually avoiding air bubbles. Remove the slide from the lower surface of the block and invert the coverslip so that the agar block is uppermost. With a platinum loop run a drop or two of melted agar along each side of the agar block, to fill the angles between the sides of the block and the coverslip. This seal hardens at once, preventing slipping of the block. Place the preparation in the incubator again for 5 or 10 min to dry the agar seal. Invert this preparation over a moist chamber and seal the coverslip in place with white wax oı paraffin. Vaseline softens too readily at 37°C, allowing shifting of the coverslip. The preparation may then be examined at leisure."

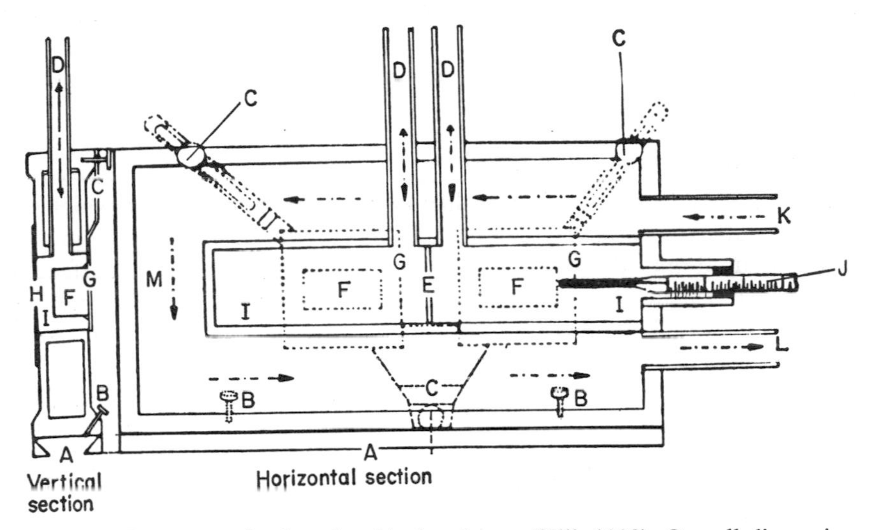

FIG. 6. Warm stage for hanging block cultures (Hill, 1902). Overall dimensions $3\frac{1}{2}\times 2$ in., made of brass. A, dovetail for attaching warm stage to mechanical stage; B, fixing screws; C, stage clamps for holding coverslips; D, pipes connecting with moist chamber; E, partition between chambers; F, hanging agar block; G, coverslip; H, slide cemented to base of chamber; I, moist chambers; J, thermometer; K, L, inlet and outlet tubes for warm water; M, water space. Note: the "vertical section" is a composite of several section planes. (From Hill, 1902).

The moist chambers used by Hill were an integral part of two cleverly designed heated stages that are described and illustrated in the original paper. However, because of the thickness of the preparation, he appears not to have been able to use a condenser during microscopic examination. The more sophisticated chamber allowed for the circulation of water, air, etc, around the suspended hanging block (see Fig. 6). "If anaerobic conditions are desired, a small cup containing pyrogallol is placed in the moist chamber and the moist chamber is made air-tight." He mentions the use of "very thin films of agar", but only so as to enable bacteria to "be stained under the observer's eye". He describes post-fissional movements and palisading in "*B. diphtheriae*" and "*B. typhosus*."

The hanging-block technique used in the manner described by Hill has the serious drawback that the optical conditions are very poor unless modifications are made. The first obvious one is to reduce greatly the depth of the agar block to little more than 1 mm and at the same time to reduce the depth of the air space beneath it. Eliminating the air space by filling it with water or nutrient medium (colourless if possible) to form a "hanging block with liquid in-fill," will also improve optical conditions (Fig. 1c). In these chambers there will also be the problems of constancy of environment mentioned in Sections IIIA and IIIB, but these can be greatly improved if there is a facility for the replacement of liquid in the chamber, for which Hill's heated stage provides.

E. Hewlett's coverslip method

Hewlett's method (1918) is in reality a form of hanging block in which the block is reduced to a thin layer of agar. In this method a thin layer of molten agar is poured onto the surface of a coverslip. When gelled the surface of the agar is inoculated by spreading over it a small quantity of diluted culture. The coverslip is then inverted over a moist chamber and examined with a dry lens. The net result is a thin hanging block with the lower or "free" surface of the agar inoculated. The air space can be filled with liquid, but bacteria may be washed off as a result. The great drawback of this method is the inability, usually, to view the cells under the 100× oil-immersion objective as the cell layer is usually beyond the working focal distance of the lens (but see next paragraph). The method has recently been modified by Hoffman and Frank (1961) for studies on cellular aggregations. The technique is as follows. A No. 1 coverslip 22×30 mm is immersed in 95% (v/v) ethanol in water and sterilized by flame after draining off excess on to blotting paper. The coverslip is then inclined at approximately 45°C and a drop of hot molten 2% water agar allowed to run down the surface, leaving a thin streak of agar in the centre. After drying for 1 min, a drop of inoculum is allowed to run down the streak and excess is drained off by standing on edge in a Columbia jar (A. H. Thomas Co., Philadelphia, U.S.A.) for 2–5 min. It is then removed, and both ends of the streak are excised by means of a sharp blade to leave a block about 3 mm square in the centre. The agar block thus has two "square" edges and two "free-flow" edges (Fig. 7a).

Inoculated streaks like these were used by Hoffman and Frank in one of two ways. In the first (Fig. 1b) the coverslip was inverted over a chamber formed by placing two coverslips 1 cm apart on a glass slide. The chamber was then sealed with paraffin to prevent drying. In the preparation the agar layer was thin enough not to make contact with the upper surface of the slide and to enable the cells to be focused upon by a 100×, N.A.† 1·30 Zeiss oil-immersion objective. See Section III F.2 for the second method of use.

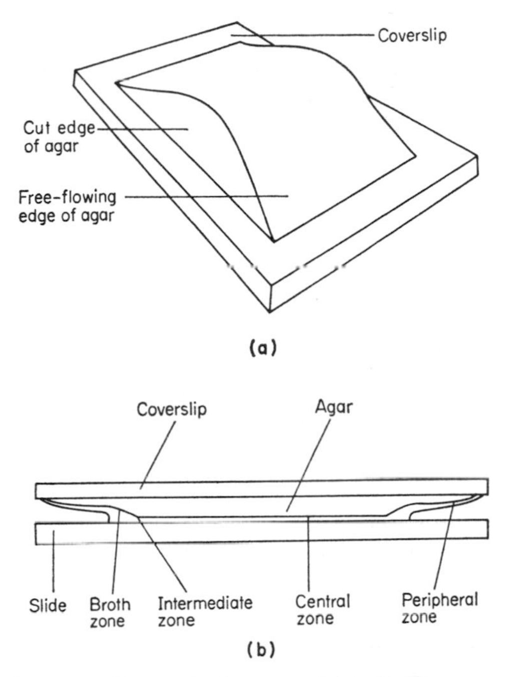

FIG. 7. Hewlett's coverslip method as used by Hoffman and Frank (1961): (a) coverslip with agar "wedge"; (b) mounted with agar in contact with slide; (From Hoffman and Frank, 1961). (Reproduced with permission from *J. Gen. Microbiol.*)

F. Agar block microcultures

These preparations could be more exactly described as agar sandwich cultures. If we regard the agar as the slice of meat and the inoculum as the butter, the differences arise mainly in the buttering. Methods 1, 2, 3 and 4 below can be described as buttered-beef methods, whereas 5 and 6 are buttered-bread methods.

1. *Ørskov's (1922) method*

Molten agar is poured into a sterile Petri dish to a depth of 2 or 3 mm. A drop of the inoculum is placed on the centre of the plate and spread outwards by means of a bent glass rod so as to give a graded inoculation. Small

† N.A. = numerical aperture = $n \sin\alpha$, where n is the refractive index of the immersion medium and α is half the angular aperture of the objective lens.

blocks of agar are excised by sterile scalpel, or other suitable instrument, and placed on sterilized slides. A coverslip is then applied to the upper, inoculated surface of the agar.

2. *Hoffman and Frank's (1961) method*

This is the same as the Hoffman and Frank preparation described in Section IIIE above, except that in this case the inoculated preparation is inverted directly on to a slide so that the agar surface is in contact with the slide. The preparation is sealed as before and is similarly thin (Fig. 7b).

3. *Knaysi's (1940) method*

This is the reverse of the method described by Hoffman and Frank, in that the agar is applied to the slide between two coverslips to form a small mound. The top of the mound is inoculated and the covering slip is applied to make contact with this. The chamber is then sealed by a mixture of paraffin and Vaseline. This type of preparation has been successfully used for dark-ground photomicrographic studies of growing bacteria and yeasts.

4. *Koopman's (1960) method*

Slides with sealed-on glass rims are used. The rimmed off space is filled with a clear solid nutrient medium (1·5–2 ml) to form a thin layer. The film is inoculated with a horsehair (to avoid damage to the surface) and the sites of inoculation are marked on the slides with a glass cutter. The slides are stored in well sealed flat boxes for incubation. This technique was used by Koopman to study the effect of incorporated antibiotics on colony formation by various bacteria. It was suggested as an alternative to more conventional routine antibiotic inhibition tests.

5. *Hort's (1920) method*

Here the layer of agar is formed by pouring molten agar directly on to the slide and allowing it to set; excess can be trimmed off. A small droplet of inoculum is then deposited on the coverslip (within the confines of an etched circle) and the coverslip is inverted and pressed on to the surface of the agar before the inoculum has dried. Sealing can be performed as before.

6. *Method of Fleming* et al. *(1950)*

Fleming *et al.* give the following description of the method. "The culture is spread on a sterile coverslip and allowed to dry. Then penicillin agar (1–1·5% agar) at a temperature under 50°C is dropped on the coverslip and as soon as it has set the coverslip is inverted on a slide. When the coverslip is picked up with a pair of forceps the whole of the agar adheres to it and there is almost no disturbance of the culture. The coverslip with the agar can be fixed in formalin and when fixation is complete the agar can be gently removed leaving on the coverslip an almost undisturbed pattern of the bacilli in the culture which can be stained to give a beautiful permanent record of the culture."

Pulvertaft (1952) used this technique with great effect in several studies by phase-contrast cinemicrography to record the reactions of various species of bacteria to penicillin, streptomycin, chloramphenicol, aureomycin and terramycin. Pulvertaft and Haynes (1951) also used it to study the germination of spores that had been formed in the microculture itself. Slide cultures of 24 h growths of *Bacillus cereus* and *Bacillus subtilis* were prepared and sealed with wax. "Within a week in most cases very large numbers of spores had formed . . ." The coverslip was then removed with a scalpel bringing the adhering agar with it. One half of the agar was then cut away and the coverslip replaced on a clean slide and sealed on three sides. "Hartley's broth was run in on the fourth side, where the agar had been removed, and this side was also then sealed. Germination was observed under a $\frac{1}{12}$ in. objective by phase contrast, the microscope being mounted in an incubator at 37°C."

Of the methods described above, the one most injurious to the bacteria is Fleming's, as this involves two sources of serious shock. Firstly, the simple process of air drying is now well recognized as a source of cell damage leading to death (Hewitt, 1951; Webb, 1959; Webb and Malina, 1967), and secondly, the drying film is then subjected to the further hazard of having hot agar at 48°–50°C poured upon it. Such temperatures are beyond the maximum for growth of many species and can kill a proportion of the cells.

The method may be improved by applying the agar before drying of the culture droplet, and the preparation should be inverted on to a slide, before the agar has set. In this way better contact is made with the slide and the heat of the agar can be disseminated more quickly. Hewitt (1951) recommends, where possible, the use of 30% gelatin in place of agar because of its lower melting point and high refractive index, but this implies a lower experimental temperature. Hoffman and Frank (1963) have found that the incorporation of 6% inositol provides a high degree of protection to *Escherichia coli* in drying suspensions, and this protection has been investigated in some detail by Webb (1967).

G. Postgate slide chamber

This culture chamber is in effect a miniaturization of the common Petri plate and was first described by Postgate *et al.* (1961). Its preparation and use are described in detail by Postgate (this Volume, p. 611) and need not be repeated here.

It is often necessary to filter hot agar medium through a membrane filter to free it from debris and dead bacteria which abound in some samples of dried media. This is especially necessary where one is interested in counting single organisms as in the estimation of viabilities. Some experiments may involve the use of considerable numbers of slide cultures and where this is the case or where there is only one experimenter examination and counting of the bacteria may not be possible immediately after the growth period required by the experiment. In the author's experience one of the great advantages of this method is the ease with which the results can be put into cold storage for assessment when time permits. My own procedure is to place the slide cultures into Petri dishes with a piece of filter paper, remove the coverslips, apply 3 or 4 drops of chloroform to the filter paper and close the Petri dishes. The cells are instantly killed and the slide cultures can then be kept in the refrigerator to await examination. If precautions are taken to prevent the agar layer from drying out, e.g., by replacing the coverslips and resealing, the bacteria can still be counted under phase-contrast illumination after several days of storage at 4°–10°C.

H. Jebb and Tomlinson's wire-loop method

A method that enables the assessment of various media as well as the assessment of growth rate in developing mycobacterial colonies was devised by Jebb and Tomlinson (1960). For this method wire loops are prepared by twisting together 3 strands of stainless-steel wire, 40 s.w.g., and forming loops about $\frac{3}{8}$ in. in diameter which are then "flattened" to ellipses. The free ends of each loop are cut off fairly short and fused into the narrowed end of a piece of glass rod. The rod is pushed through a hole punched in the cap of a 1 oz "Universal" (wide-necked vials; J.669, United glass Bottles Ltd., St. Helens, England) and the assembly autoclaved. Purified watery agar, melted and cooled to 46°C, is mixed with the cell suspension (in the case of mycobacteria considerable processing is necessary in order to obtain a suspension of isolated cells) and the prepared loops are dipped into this, removed and gently rocked to obtain an even thickness of film, then returned to their bottles and allowed to set. The charged loops are transferred to Universals containing the selected media and the cultures incubated as required. After various periods of time, loops are withdrawn and the films of agar transferred to slides, fixed and stained. Fixation was performed by Jebb and Tomlinson using abolute alcohol to dry out the films. (For some types of experiment it may be possible to examine the agar films directly by phase-contrast microscopy).

I. Petri plate methods

The advent of plastic Petri plates with translucent, flat, thin walls has greatly facilitated the microscopic observation of micro-organisms growing in or on solid media without the need for preparation of slide chambers or the techniques described under Section IIIF. However, long before their introduction, the ordinary glass Petri plate had been used for microcultural studies. Basically, the Petri plate culture is simply a larger type of agar block culture allowing quick preparation and a large area so that many differently pre-treated specimens can be allowed to grow in the same environment.

Microcolonies developing on the surface of an agar plate can be viewed directly by phase-contrast microscopy with dry high power lenses or under oil immersion by first covering the growth with a coverslip. If required, small blocks can be cut from the agar and impression preparations made on coverslips, fixed and stained to reveal structures, e.g., nuclei (Stempen, 1950; Hillier *et al.*, 1949), colonial morphology, etc. (see Bissett, 1950).

Mahoney and Chadwick (1965) have described a rapid method for measuring antibiotic sensitivity based upon the inhibition of microcolony formation on the surface of agar plates containing antibiotic. They have found the correlation with the results of conventional disc tests to be good. They investigated the use of the method for direct testing of "raw" pathological specimens (Chadwick and Mahoney, 1966), and by combining this technique with specific fluorescent antibody staining were able to determine the antibiotic sensitivity of enteropathogenic *E. coli* in artificially mixed cultures within 5 h (Mahoney and Chadwick, 1966).

A method that enables developing microcolonies to be studied by electron microcscopy following a period of observation by light microscopy has been described by Hillier *et al.* (1948) (Fig. 8). The surface of a thinly poured agar plate is (a) flooded with sterile distilled water from a free-flowing pipette and immediately afterwards (b) a drop of collodion (0·5–1% in amyl acetate) is applied and allowed to spread. Speed is essential as material dissolved from the agar layer may prevent spreading of the collodion and lead to the formation of a thick film. The solvent is allowed to evaporate completely (it is growth inhibitory), the plate is tilted and the water layer removed completely (c) to leave the collodion film lying on the surface of the agar. A suitably thin membrane is, at this stage, invisible except for a thickening at its edges. A small droplet of suspension in water or saline of the test organism is applied (d) and the preparation incubated as required (e). When the growth has reached the desired stage, as determined by light microscopy (f), a selected area of the agar and membrane in excised (g) and the slab of agar then slid under the surface of clean distilled water (h), held in another dish, in such a way that the membrane gently floats off on to the surface, bearing the organisms on its upper side (see Fig. 8). The floating membrane is then recovered on a mesh screen suitable for electron microscopy (i), and any excess water removed from beneath by absorption with filter paper. It can now again be examined under the light microscope (j) before treatment for observation in the electron microscope.

Pearce and Powell (1951) have investigated the use of oblique and vertical illumination, in the manner used by metallurgists, for studying micro-colonial development. They described a means of making a Petri dish moist

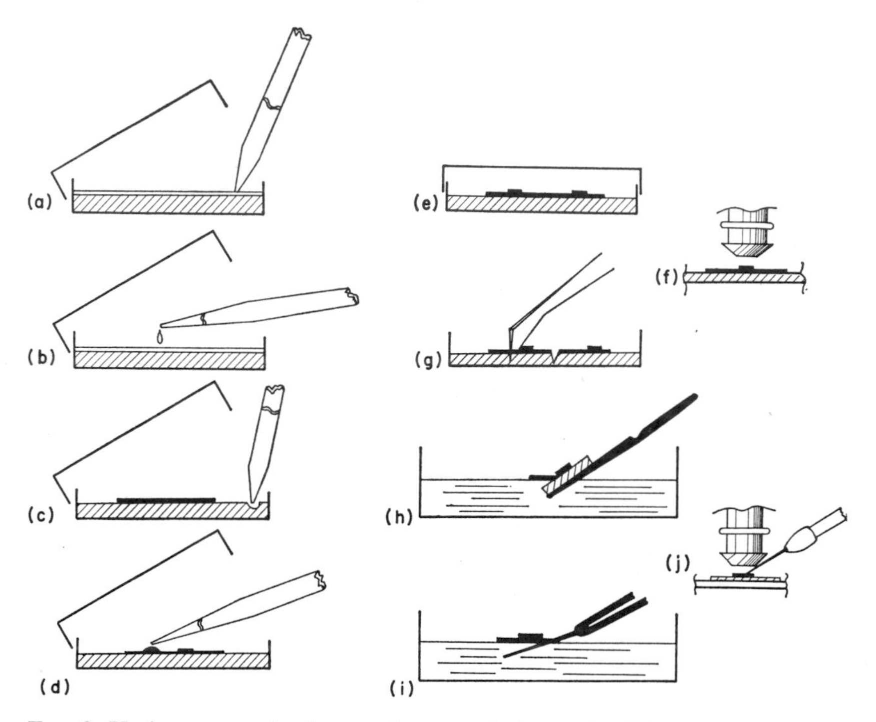

FIG. 8. Various stages in the membrane technique of Hillier *et al.* (1948): (a), flooding the agar plate; (b), applying drop of collodion; (c), withdrawal of water layer; (d), application of test organisms; (e), growth period followed by (f) microscopic examination; (g), excision of sample; (h), flotation of membrane with sample; (i), recovery on electron microscope grid; (j), microscopic examination.

chamber so as to avoid "hardening of the surface layers through concentration of the agar, and a consequent progressive change in the appearance of bacterial growths". The growths were observed at magnifications of 400–600. Also described is a metal chamber that encloses a Petri plate, designed so that the atmospheric environment of the culture can be controlled, e.g., by flushing with O_2-free N_2 which enabled them to follow the growth of anaerobes such as *Clostridium welchii* on a nigrosin tryptic meat agar plate.

"Oblique incident illumination is the same as ordinary dark ground, except that the hollow cone of light is applied from above instead of below." The arrangement of the microscopic elements is complicated, and there seems little to be gained and much to be lost in the use of the "over-head" illumination system. "In vertical illumination, light from an illuminated iris, brought nearly parallel by means of a lens, is directed towards a thin glass plate lying on the axis of the microscope immediately behind the objective. The plate is tilted so as to reflect part of the light directly downwards through the objective, which then forms an image of the iris on the surface of the specimen. If the surface reflects specularly, only a brightly illuminated circle is seen—irregularities of contours appear dark wherever the surface departs sufficiently from normality to the axis, or scatters some of the incident light." Photographs taken of bacteria illuminated in this fashion show dark "walls" lightening towards the centre of the cell (i.e., the part of wall normal to the axis) in a 3-D type effect, against a light background. Such techniques would now appear to be of little value since the advent of machines such as the scanning electron microscope.

It is also possible to grow bacteria on Cellophane films laid on the surface of an agar plate. However, it is necessary if the growth of the colony is to be more or less "normal," that the relative humidity of the micro-environment should be kept high, otherwise many species tend to form a "rough" -type, "tough" colony. The Cellophane-disc technique enables one to transfer the organisms from one type of medium to another with comparatively little disruption of the cells. "Grids" may be scored on the Cellophane by means of the razor-blade stack device of Powell (1956c).

J. Flat capillary method

Sherris *et al.* (1957) have described a method for following the tactic responses of motile bacteria enclosed in flat capillaries, and used it to study the effects of oxygen and arginine. The capillaries are made in the way that Pasteur pipettes are drawn from glass tubing, except that, after heating, the free ends of the glass tube are turned inward through 90° and then drawn apart to give finally a ┌┐-shaped tube; the flat capillary forms the "cross-bar" of the ┌┐. With practice one can control the thickness of the wall and the width of the "flat" capillary. Pieces of capillary at least 50 mm long were used and "gases other than air were introduced by the following method. One end of the capillary tube was attached by rubber tubing to a cylinder of the appropriate gas and the other end held beneath the surface of the bacterial suspension. After allowing sufficient time for the air in the tube to be completely replaced by the gas, the cylinder was closed and the rubber tubing squeezed and then released to withdraw a column of suspension into the capillary tube. The tube was then sealed at both ends leaving the bacterial suspension in contact with the gas at the end which had been attached to the cylinder."

As used by Sherris *et al.* the capillaries were mounted on slides and fixed with plasticine for examination by a 4 mm objective. Improved optical conditions can be obtained, however, by mounting the capillary in a layer of immersion oil between slide and coverslip and viewing by, for example, the Wild 50 × oil-immersion phase objective.

K. Perfusion chambers

These can be defined as microcultural methods in which nutrient is provided by a solution that can be changed, either intermittently or continuously, by a flow system. There are several reasons why perfusion chambers are more useful than other forms of microculture and the considerable attention that they now receive has stemmed from the growing awareness of the necessity of defining the micro-environment as precisely as possible if meaningful statements are to be made concerning the morphology and physiology of microbial cells. It is also a fundamental consideration in studying the effects of chemical compounds on living cells that the most information is usually obtained if all features of the experimental situation are retained constant while the factor whose effect is to be studied is varied in a known way, and various methods designed to achieve this have been introduced in recent years. Operationally two broad categories of chamber can be recognized; (1) those in which intermittent changes only are easily realized; (2) chambers designed to allow liquid to flow continuously through the chamber. From the design point of view there is the difficult problem of a support for the cells (see Section IVA). In the case of tissue cells that adhere to a glass surface the problems are not nearly so great as they are with chambers that must be designed to retain cells that do not easily adhere to surfaces.

1. *Chambers permitting intermittent change of medium*

(a) *Mackaness chamber.* Mackaness (1952) designed a chamber that permitted the observation of a population of macrophages; the medium could be changed without loss of cells and permanent-stained preparations could be easily made. The chamber is made from a piece of Perspex 2 in. × $1\frac{1}{4}$ in. × 2·2 mm thick with a central hole 12 mm in dia. The surface on each side of the aperture is countersunk to a depth of 0·35 mm to accommodate standard $\frac{7}{8}$ in. coverslips, leaving a space between coverslips of 1·5 mm. (Fig. 9a.) A small perspex ring (1 mm section, 6 mm i.d.) is fixed to the base coverslip by paraffin wax (m.p. 56°C). Two holes are drilled in the side to communicate with the chamber, and these are plugged by two stainless-steel pins fixed into a wedge of Perspex that abuts on the "slide" when the pegs are pushed home (Fig. 9b, 9c). For some applications the central ring would be unnecessary.

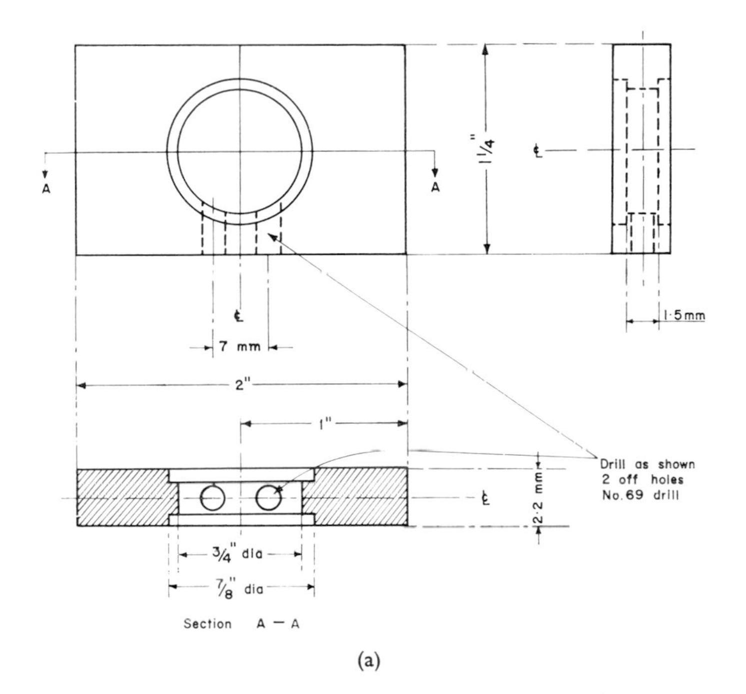

(a)

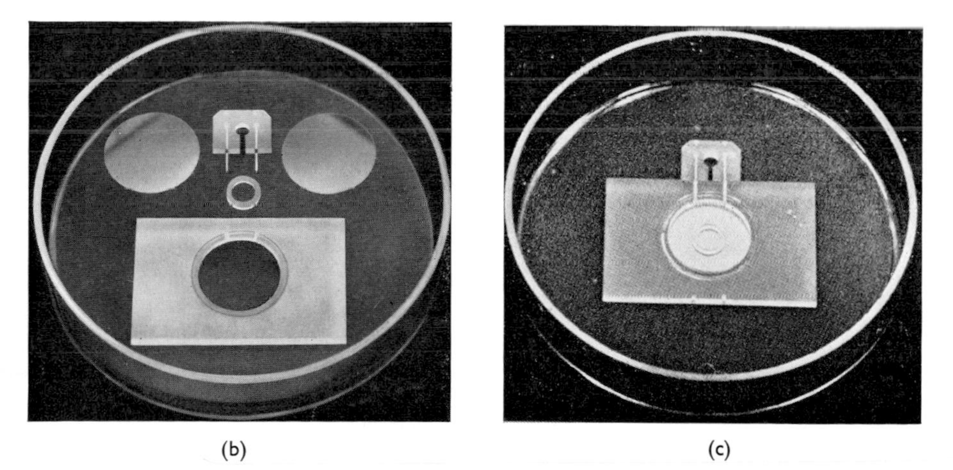

(b) (c)

FIG. 9. (a), Details of construction of Mackaness "perspex" slide chamber; (b), view of the components; (c), assembled chamber. (From Mackaness, 1952.)

(b) *Harris modification.* Harris (1955) modified the Mackaness chamber to "permit adequate oxygenation of the cells over several days". In effect the preparation is a double layered chamber involving three coverslips; the measurements are clearly given in Fig. 10. The cell suspension is introduced into the lower compartment, formed by A and B, so that the cells settle out and become adherent to A. The medium is then changed, coverslip B removed and coverslip C sealed. More medium is introduced through the drill holes until the cavity of the chamber is filled except for a large air bubble, which is of such a size that it does not encroach upon coverslip A.

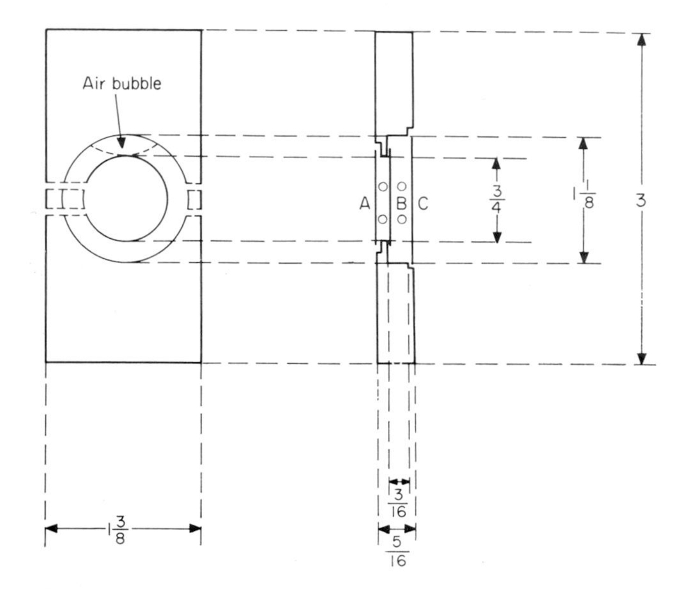

FIG. 10. Details of construction of Harris chamber; dimensions in inches. (From Harris, 1955.)

The air bubble is gassed with 95% O_2/5% CO_2 and the drill holes sealed. The chamber is attached to a vertical turntable (set at 1 rev/min) so that the gas bubble circles round the peripheral shelf as the turntable revolves.

(c) *Perspex slide for roller culture.* Pulvertaft *et al.* (1956) constructed a simple chamber from a Perspex slide by recessing a moat 3 mm deep in one surface, leaving a central pillar 6 mm in dia. Two holes bored through the edges connected with the moat (Fig. 11). "Cells for culture may be mounted in culture fluid directly on to the pillar; a coverslip is applied and sealed on with paraffin wax. One of the orifices is then plugged with a wooden spiggot impregnated with wax, and the moat is half filled with culture medium. The other hole is then plugged. It is important to wait until the wax is firmly set before running in fluid; otherwise an emulsion of wax is formed. For the same reason silicone seals are useless. Cultures are fed at any required interval after removing both spiggots and withdrawing the old medium.

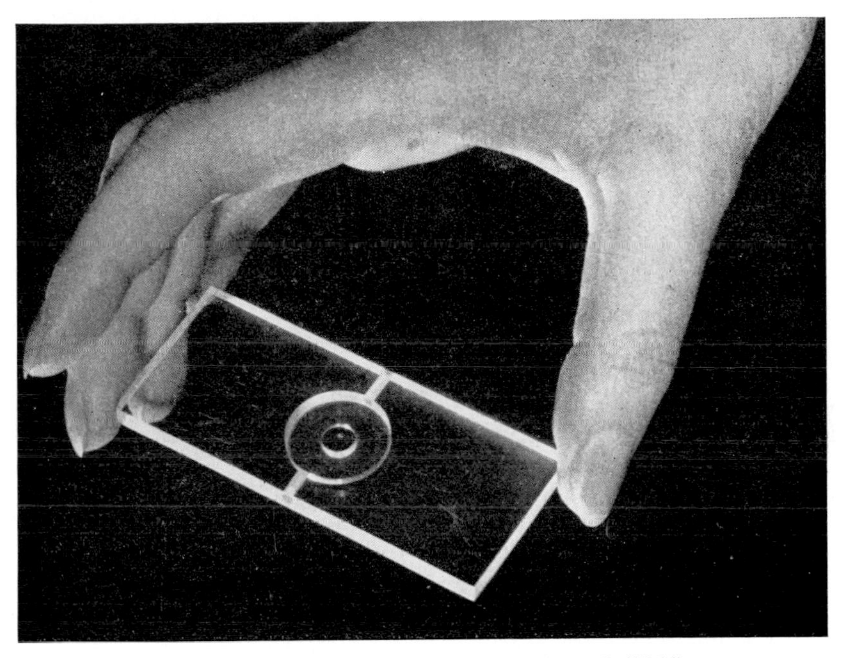

FIG. 11. Perspex chamber of Pulvertaft *et al.* (1956).

"We find it preferable in many cases to have an agar base. For this purpose the pillar is recessed 1 mm below the slide level, and molten serum agar is dropped on to it. A coverslip is immediately applied and the agar sets with a plane surface. With care and practice the coverslip can be slid off; the standard technique is then followed." The slides can also be made thinner than the one shown here. A very similar chamber is described by Cruickshank *et al.* (1959), who recommend beeswax as a sealing agent (see p. 425).

2. *Chambers designed for continuous perfusion*

(a) *Pomerat chamber.* A simple but rather fragile chamber has been described by Pomerat (1951) consisting of two coverslips sealed together by wax with a thin space between them. Perfusion is made possible by sealing two fine glass tubes, serving as entry and exit ducts into the wax on opposite sides of the chamber. The bottom coverslip is considerably larger than the top slip, and the overlap is cemented to the base of a suitably shaped block of aluminium or stainless steel to give the assembly greater rigidity and to facilitate handling (Fig. 12). This chamber has been used e.g., by Hu *et al.* (1951) in a study of the effect of ^{32}P on living adult human epidermal cells.

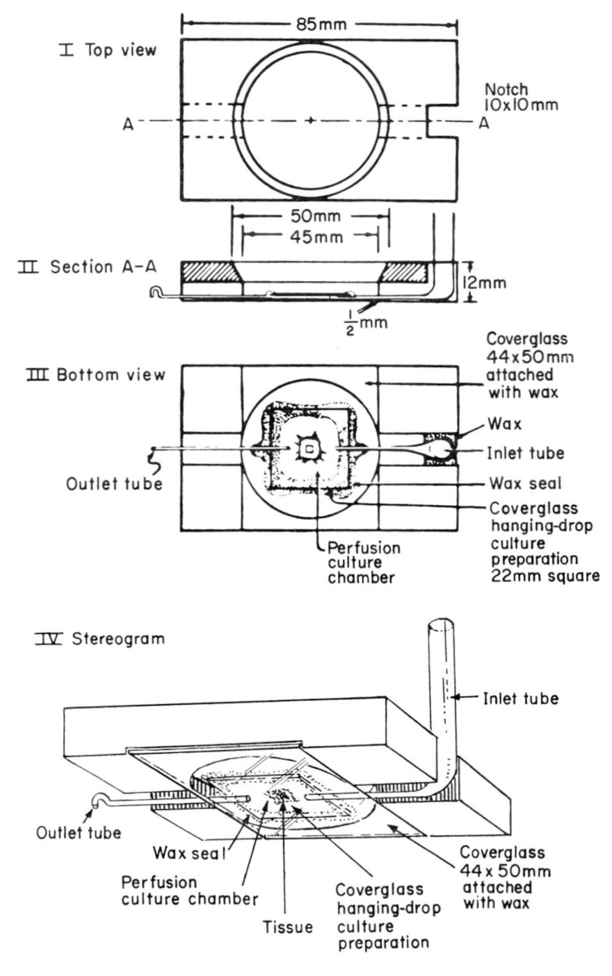

FIG. 12. Pomerat perfusion chamber: I, II, details of metal supporting frame; III, bottom view of assembled chamber; IV, perspective view of assembly. (From Pomerat, 1951)

(b) *Gustafson and Kinnander's "microaquaria" (1956)*. This preparation for restraining the movements of sea-urchin larvae provides an interesting modification of the wet-film technique. The device consists of a strip of nylon net inserted between the coverglass and the slide. The mesh size used should be chosen with regard to the size of the object to be "trapped". In Gustafson and Kinnander's work they chose a mesh opening of 130μm which "was wide enough to permit the slight expansion of the blastula during gastrulation." In order to prevent swimming movements of the larvae, crystals of $CaCO_3$ were deposited on the mesh by boiling the fabric in, e.g., "a solution of pure $CaCl_2$ and $NaHCO_3$." The crystals caused slight local depressions in the blastula without seriously disturbing its volumetric expansion, and kept the larvae in more or less fixed position for time-lapse cinemicrography during several hours. A continuous supply of O_2 and drainage of waste products were achieved by gently flowing seawater between the slide and coverglass using filter paper strips as siphoning devices and a Mariotte flask as a continuous feeding reservoir.

(c) *Rose chamber* (*1954*). This chamber is made from two stainless-steel plates $2 \times 3 \times \frac{1}{8}$ in., each with a centre hole $1\frac{1}{16}$ in. dia. bevelled at a 45° angle. The hole in the bottom plate has a countersink $\frac{1}{16}$ in. deep extending for $\frac{1}{8}$ in. from the perimeter of the hole. Each plate has four holes, the top plates recessed to fit the heads of flat topped Allen bolts that screw into the threaded bottom plate holes.

The centre of the chamber is made from gum latex $2 \times 1\frac{3}{4} \times \frac{1}{8}$ in., having a central hole $\frac{13}{32}$ in. in dia. When assembled, the steel plates press two coverslips (50×43 mm) on to the top and bottom surfaces of the latex centre piece. The final tightening of the Allen screws must be done after insertion of a No. 25 $\frac{1}{2}$ in. needle through the edge of the gasket to allow equilibration of pressure as the air within the chamber is compressed. The inoculation of the already assembled chamber can be achieved in the same way by injection through the gasket wall. Alternatively the cell explant may be placed in the partly assembled chamber before the top coverslip is positioned and the chamber assembly then completed.

"The fact that a perfusion system can be plugged into this chamber by mere insertion of a few needles that are attached to tubes and supply bottles distinguishes this chamber as something different in the tissue-culture field."

Essentially the same chamber has been described by Richter and Woodward (1955) and a refinement using round coverslips and gasket (of silicone rubber) has been suggested by Sharp (1959).

(d) *Sykes and Moore chamber*. Sykes and Moore (1959, 1960) have published descriptions of two simple chambers that permit medium to be changed. Since the principles underlying both are the same, only the later one will be described. The chamber consists basically of two metal rings that mate by a screw thread. When screwed together they compress a silicone-rubber ring between two coverslips in the manner of the Rose chamber. The thicker metal (base) ring has four diametrically opposed holes drilled in the circumference so that the rubber gasket can be pierced. The outside diameter of the rubber ring and of the coverslips is 25 mm (Fig. 13). The methods of use are similar to those described for the Rose chamber.

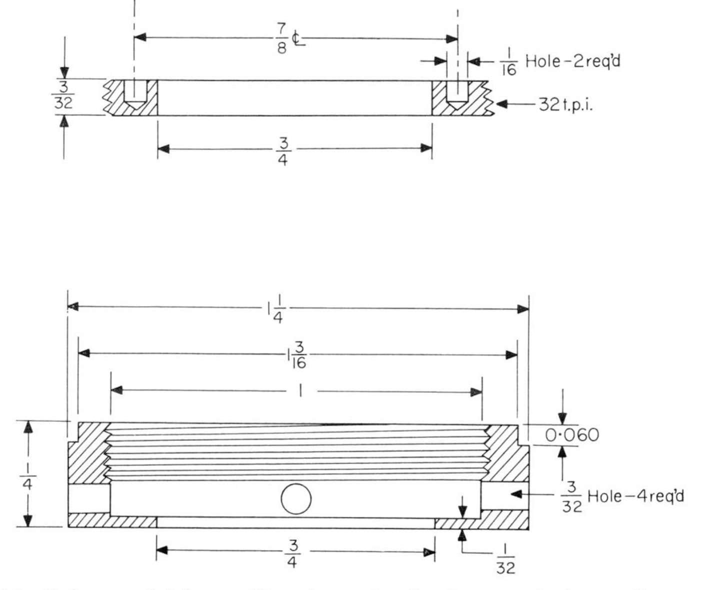

Fig. 13. Sykes and Moore Chamber: details for metal rings: dimensions in inches. (From Sykes and Moore, 1960.)

(e) *Barski and Robineaux's chamber* (*1956*). This chamber is very similar to that of Rose, but instead of having a rubber core, there is an Araldite core. The inlet and outlet tubes are stainless-steel fine-bore tubing embedded in the Araldite. The seal between Araldite core and coverslips is effected by means of Vaseline.

(f) *Toy and Bardawil chamber* (*1958*). Toy and Bardawil make the following six minimal requirements for a suitable perfusion chamber—

1. Simple design and inexpensive fabrication, approaching the ideal of a single-shot, disposable unit.
2. Rapid loading by minimally skilled personnel.
3. Convenient exchange or perfusion of medium.
4. Freedom from toxicity or contamination.
5. Simultaneous microscopic observation and photography, with a short optical path for phase microscopy.
6. Easy demounting, cleaning and sterilization.

Their chamber consists of three pieces (Fig. 14): (a) a roof plate with openings for the upper window and the two duct tubes through which medium is perfused; (b) a core plate serving as a support for the coverslips; (c) a base plate with an opening for the lower window. The ducts are cemented

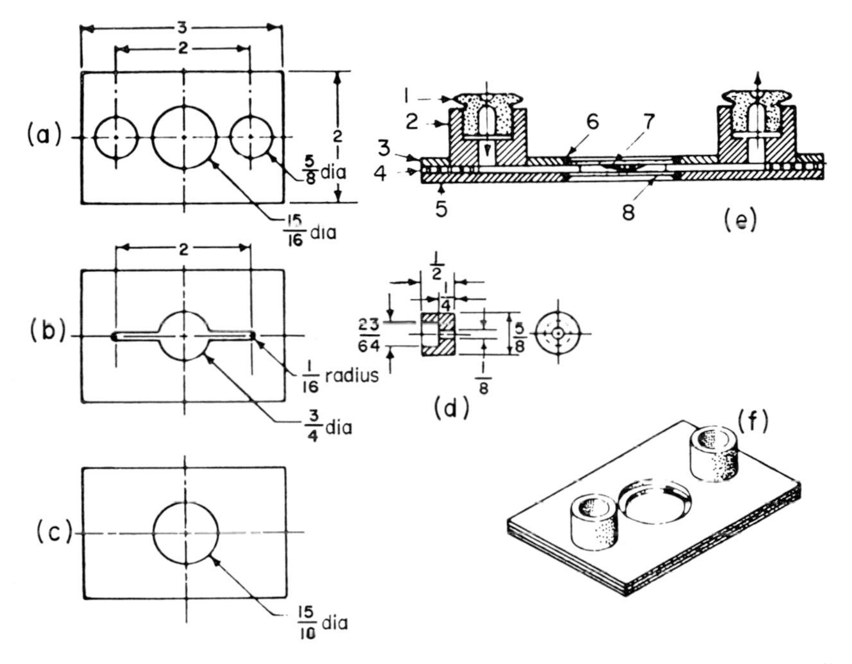

Fig. 14. Toy and Bardawil chamber. Basic chamber design for $\frac{7}{8}$ in. dia. coverslips showing assembly details (dimensions in inches): (a), roof plate; (b), core plate; (c), base plate; (d) duct (2 needed); (c), assembled chamber, section [1, rubber plug: 2, duct; 3, as (a); 4, as (b); 5, as (c); 6, wax seal; 7, tissue explant; 8, coverslip] (f), perspective view of assembly. (From Toy and Bardawil, 1958.)

in place and the coverslip windows are fixed in place by a mixture of paraffin and beeswax. The inoculum is placed on either window of the chamber as required for microscopy. Two variations of the basic design are also given; two of the three are shown in Fig. 15. The various layers of the chamber are made from Plexiglas. The surfaces are cemented together by first applying ethylene dichloride to form a superficial film of partially dissolved Plexiglas, then parts (a), (b) and (c) are clamped together under pressure until the primary set has taken place. After a few minutes the two duct tubes may be cemented to the frame. The superfluous solvent is volatized off in an incubator. Sterilization may be carried out by X-rays or ultraviolet rays.

(g) *Chamber of Christiansen* et al. (*1953*). This chamber designed for use with a stage incubator, may be made either of plastic or glass. The plastic chamber (Fig. 16) is milled from $\frac{1}{4}$ in. acrylic plastic sheet (Plexiglas) to the dimensions shown in the Figure. (This material must be water cooled during working to prevent heat distortion.) The dimensions given are critical if it is to be used with the microscopic electrical incubator for oil-immersion objectives for which it was designed (C.S. and E., A.S. Aloe Co.,

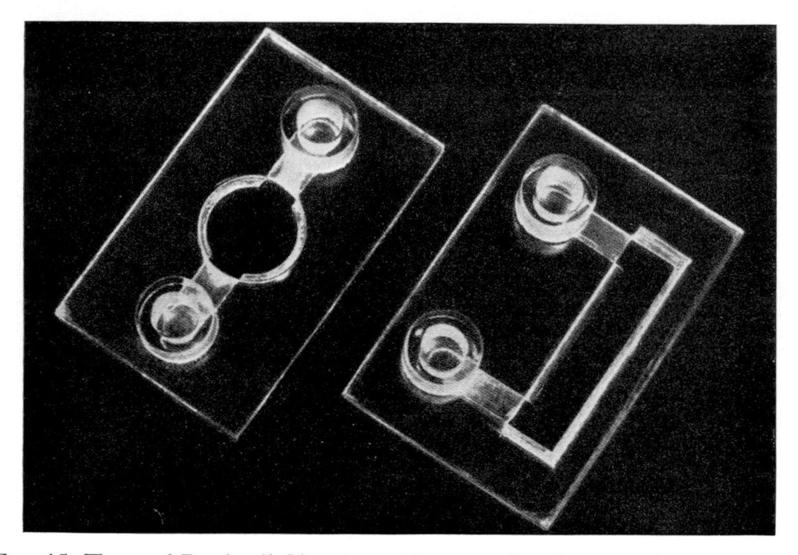

FIG. 15. Toy and Bardawil Chambers. Photographs of frames designed for $\frac{7}{8}$ in. coverslips, left, and for 11×50 mm coverslips, right. (From Toy and Bardawil, 1958.)

St. Louis, U.S.A.). Two grooves are engraved on the underside of the block extending from the circular opening to within 20 mm of each end of the block. At these points each groove joins a $\frac{1}{16}$ in. hole drilled upwards and outwards through the block. Two flow channels are cemented on the upper surface. The bottom of the chamber is formed by attaching a thin coverglass (0·24 mm $\times$ 50 mm long) to the base so as to enclose the grooves and holes cut in the plastic block. The culture adheres to the top coverslip which covers the entire circular opening.

A glass chamber to the same design can be made by grinding the holes and grooves in a 1 mm thick slide. "This slide was then ground to a thickness of 0·8 mm. The inlet and outlet tubes were set in thicker glass blocks which were then attached to the slide by sodium silicate cement." The glass chamber may be sterilized by autoclaving; the plastic one by ultraviolet irradiation.

(h) *Schwöbel chamber.* Schwöbel (1954) described a perfusion chamber (made from stainless steel) and medium reservoir system that he used for studying tissue cells. In an improved model of the same basic design

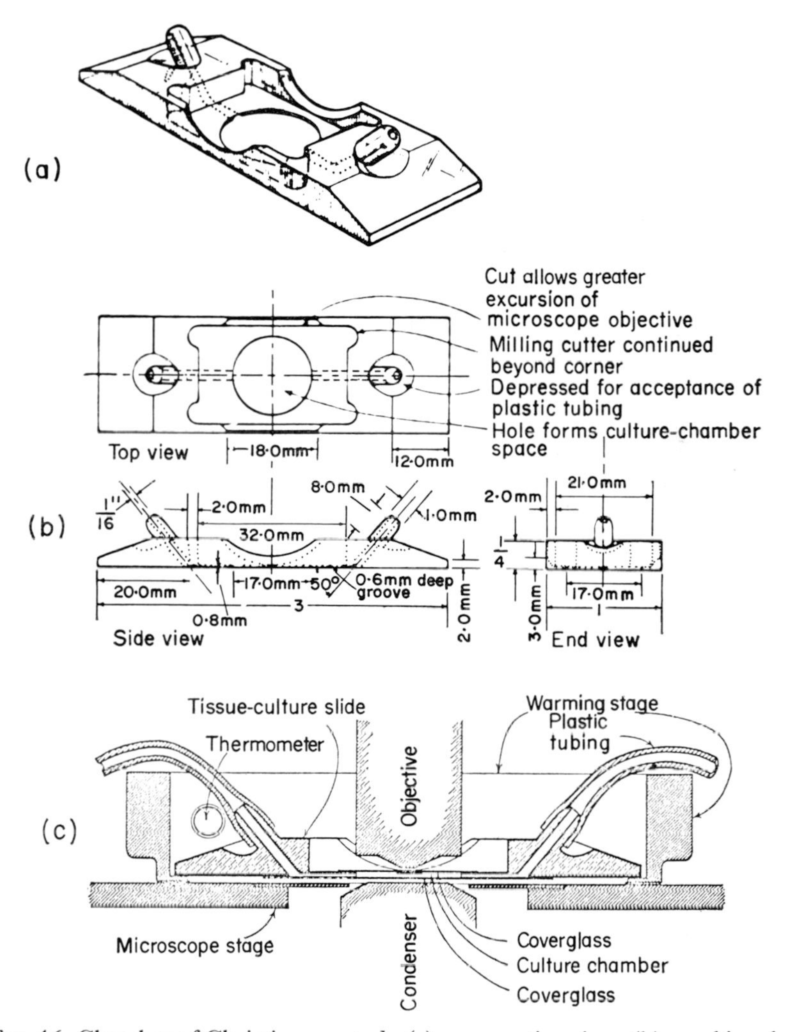

FIG. 16. Chamber of Christiansen *et al.*: (a), perspective view; (b), working drawings; (c), cross-section of assembled chamber in place on microscope stage. (From Christiansen *et al.*, 1953.)

(Schwöbel, 1955) the entire apparatus was made from glass. The glass version (Fig. 17) consists essentially of a culture chamber (U) formed from two coverglasses sealed to the top and bottom of a 6 mm high section of a glass tube of 30 mm dia. (The space between the two coverglasses is packed with synthetic fibre). Fused into the wall of the chamber on one side is a tube of 4 mm i.d. (R) by which the specimen can be introduced, and opposite are two smaller tubes (L) of 1·5 mm i.d. that provide for influx and efflux of medium from a reservoir (V). The rate of flow of the medium is determined

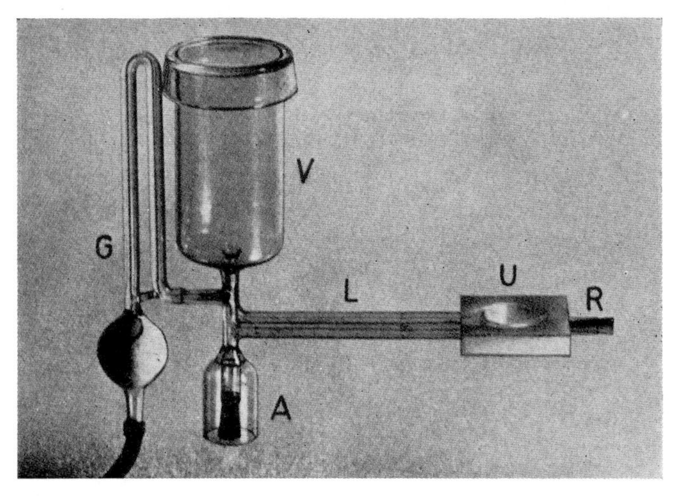

FIG. 17. Schwöbel chamber and medium supply system: U, culture chamber; R, tube for introduction of explant; L, influent and effluent tubes; V, medium reservoir, G, gas inlet tube with filter; A glass bell protecting end of medium effluent tube. (From Schwöbel, 1955.)

by the rate at which gas is allowed to collect in the reservoir, and this is controlled by a simple pressure clip in the gas line (G). Effluent medium can be collected aseptically under the glass bell (A).

(i) *Ware and Loveless chamber*. For an investigation of the construction of biological film in a sewage filter Ware and Loveless (1959) designed a perfusion chamber in which the organisms were allowed to flow through a shallow channel formed by slide and coverslip. This "microfilter" was constructed by cementing two thin strips of glass to a microscope slide with a $2 \times \frac{7}{8}$ in. No. 0 coverglass as the roof of the chamber. Glass rings were cemented over each open end of the channel to form reservoirs for adding and removing the perfusion liquid (Fig. 18). The rate of flow through the chamber was determined by the head of fluid in the reservoir, which was itself controlled by an electromagnetic valve operated through a proximity switch with electrodes in the reservoir. Liquid was removed from the effluent reservoir by suction. A record was made by means of a 35 mm time-lapse camera and Leitz microscope set up for phase contrast.

(j) *Thomas and Cramer chamber* (*1966a*). The construction of this chamber is illustrated in Fig. 19. Two aluminium plates ($\frac{1}{8}$ in. thick, 57×57 mm) screwed together press on silicone-rubber gaskets that themselves press on

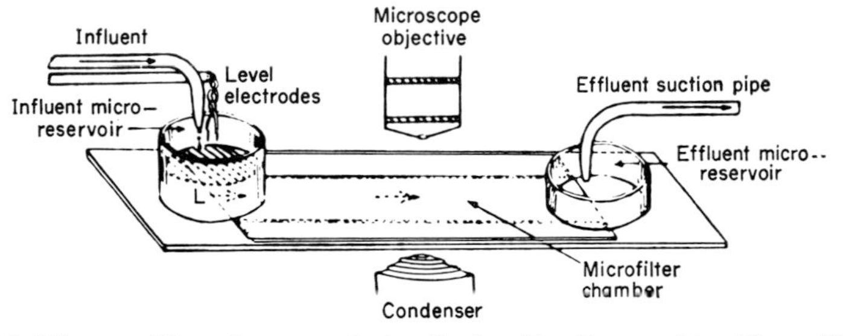

FIG. 18. Ware and Loveless percolating "microfilter" assembly. (From Ware and Loveless, 1959.)

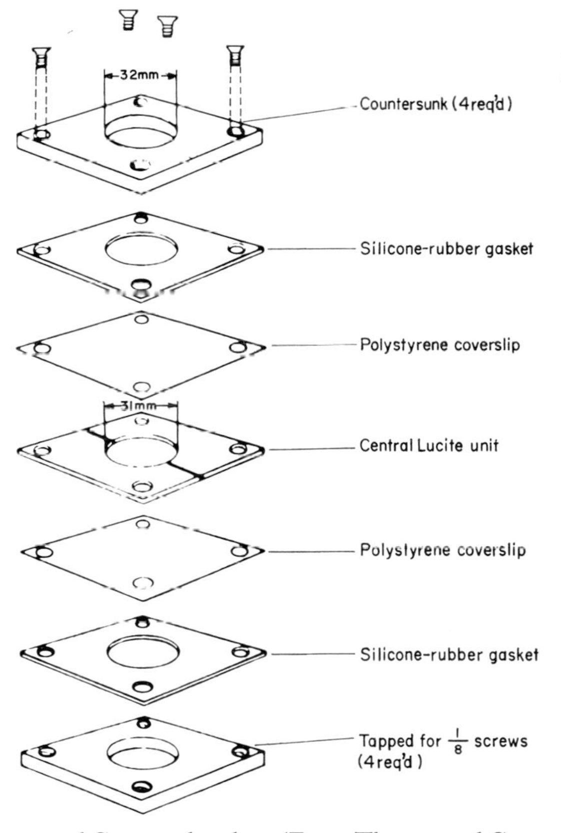

FIG. 19. Thomas and Cramer chamber. (From Thomas and Cramer, 1966a.)

polystryene coverslips which enclose a central Lucite unit $\frac{1}{16}$ in. thick bored to take 22 s.w.g. hypodermic needles as inlet and outlet ports to the central chamber. The chamber was used in conjunction with an air-displacement system for maintaining a regulated flow of nutrient. The system consisted of a pump that forced air or CO_2 through a Swinney filter with Millipore membrane to feed sterile gas to an enclosed medium reservoir thus displacing medium, which flowed to the perfusion chamber. The rate of gas flow, and hence of medium flow, was controllable. The chamber effluent was collected in sterile vented tubes. The authors give examples of easily compiled multiple chambers, e.g., by introducing a second Lucite unit into the "pile" separated from the first by a dialysis membrane enabling "interchamber membrane dialysis and diffusion studies and control of pH with CO_2."

(k) *The Carter Chamber.* A beautifully simple and efficient chamber has been designed by Carter (1966) which he describes as follows. "The chamber consists of a double ring of stainless steel, permanently bonded to a glass base plate with epoxy resin (Araldite AT. 1, Ciba Ltd.). The two concentric rings are separated by an annular channel. Before the cover glass is positioned, a heavy silicone oil (viscosity 10,000 cs) is applied by syringe to the outer ring. The coverglass is square so that its overlapping corners make it easy to place in position and remove as required. The spring clips are made of plated beryllium copper. Each clip is fitted by depressing the central tongue and locating it in a recess in the side of the chamber. The two arms of the clip then press the cover glass firmly against the inner ring. A short length of stainless steel tube gives access to the chamber". Also described in the paper is a means of agitating the contents electromagnetically, and a gravity-feed liquid circulation system.

3. *Chambers with membrane cell support*

As previously stated, the continuous observation of microbial cells is almost impossible unless some form of support for the cells can be provided. While it is possible for cells to remain comparatively immobile lying on the base of a culture chamber, observation under oil immersion (usually a pre-requisite where bacterial cells are concerned) may not be possible unless an "inverted" microscope is used. The design of perfusion chambers for studying growing bacteria has as a result centred round the problem of a suitable support for the cells. Comparatively few chambers that successfully overcome this difficulty, have been described.

(a) *The Hartman chamber.* Hartman and Hartman (1962) have described a way of converting an electron-microscope specimen grid into multiple microchambers using Formvar (polyvinyl formal dissolved in ethylene dichloride or dioxan) as restraining membranes. The specimen grid is dipped into a solution of 0·2% Formvar in ethylene dichloride, touched briefly to filter paper and placed on a coverslip. A drop of the inoculum is placed on the grid and excess withdrawn so that the level of fluid is just below the surface of the grid. If the grid surface is wet the second film will not adhere properly. "The floor of the chamber is then formed by dipping the coverslip through a film that has been previously cast on glass and stripped off on to a water surface; casting, stripping and dipping of Formvar films were described in detail by Oster and Pollister ('Physical Techniques in Biological Research,' Academic Press Inc., New York, 1956, Vol. 3, p. 181)". The coverslip is then inverted over the well of a special slide and the edges sealed with wax except at the side channels. The slide used is the Fisher–Littman wellslide, the bottom of the well being optically flat. Input and output channels are ground into the surface to communicate with the well. The means of connecting these channels to input and outflow tubes is not described, but could probably be achieved in the manner of Pomerat (1951). Or the grid assembly could easily be applied to one or other of the flow chambers already described.

There are several problems associated with this technique. Firstly, one must ensure that no ethylene dichloride remains in the Formvar films before contact with the inoculum; secondly, it must be shown that the solutes to be perfused are capable of passing through the plastic film so that they can come into contact with the cells, and thirdly there are optical deficiencies that result from the curvature of the Formvar films. The curvature of the films makes them act as lenses and they refract light unless the surrounding medium is of the same refractive index as the film. Diffraction of light also occurs where light impinges on the walls of each microchamber. The effect of these two features on definition and on field brightness is clearly shown in the published photographs.

(b) *The Vischer chamber*. Vischer (1956) has described a perfusable chamber made from Plexiglas that contains an inner ring, upon which the coverglass carrying the inoculum is placed, and this communicates with an outer reservoir chamber which surrounds it (Fig. 20). The cells are retained on the underside of the coverglass by means of a Formvar membrane. As described by Vischer the cell inoculum is placed on the coverslip and incubated for 1 h to allow the cells to adhere to the glass surface. The coverglass with cells is then dipped under a Formvar film floating on the surface of a saline solution and drawn upwards through the film so that the glass becomes enfolded by the Formvar membrane. Membrane which adheres to the obverse side is removed mechanically, leaving the cells trapped between coverglass and membrane. The preparation is then seated on the inner ring of the chamber, membrane side down.

No doubt this method is very useful for tissue cells. There is the possibility, however, that if bacteria are used the entire inoculum may be washed off in the process of applying the membrane. The earlier remarks made in

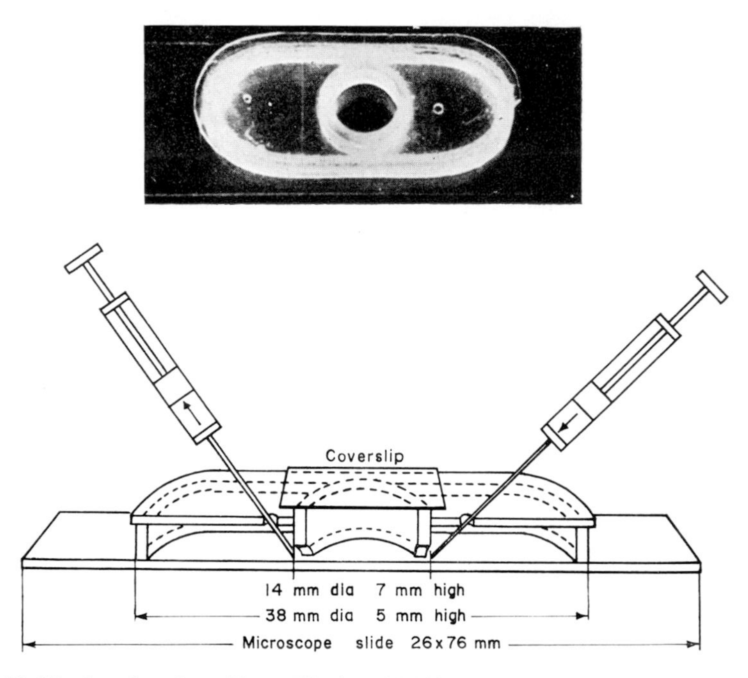

FIG. 20. Vischer chamber. (From Vischer, 1956.)

regard to the use of Formvar apply here also, and, of course, these processess should be carried out in a temperature-stabilized environment to eliminate heat shock effects.

(c) *Powell chamber*. In 1951 Harris and Powell published the design for a chamber which utilized a cellulose film (Cellophane) as the semipermeable membrane support for bacteria in a continuous-flow system. The organisms resting on the upper surface of the membrane were fed from below and vertical illumination was used for observation. There were serious limitations to this chamber pointed out by Powell (1956a) such that only dry objectives could be used, vertical illumination was inadequate, and some evaporation occurred at the Cellophane surface giving rise to local changes in the concentrations of dissolved solutes.

Later, however, this model was superseded by an elegant but sophisticated design that met all these objections and opened up a whole new field of microbial investigation, to date only little ploughed. This chamber is described by Powell (1956a) in the following paragraphs.

"The somewhat complicated mechanical design of the chamber is dictated by two conditions. First, it must be thin enough to work between a condenser and objective of standard pattern and permit an adequate

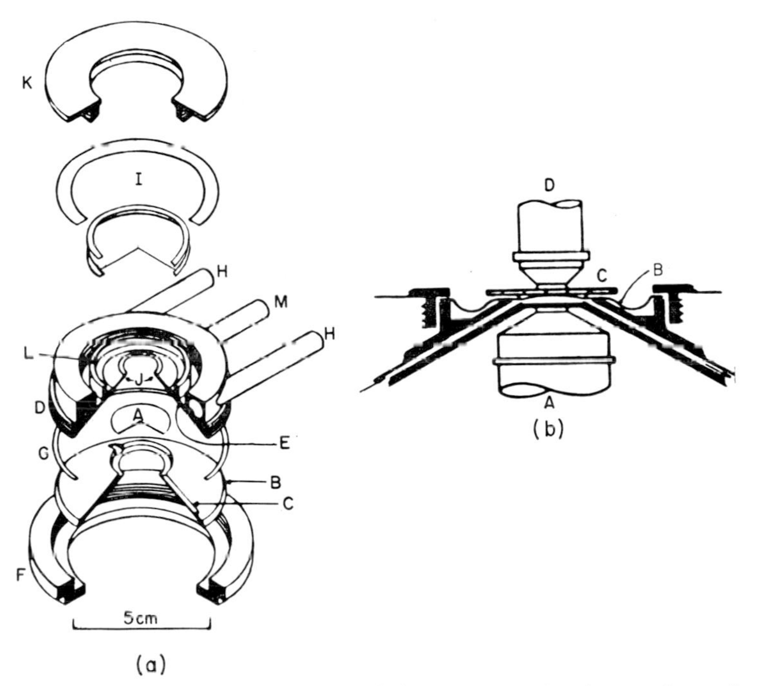

FIG. 21. Powell chamber: (a), exploded quarter-section isometric projection; lettering explained in the text. (b), schematic section of central portion in use with oil-immersion condenser (A) and objective (D). The Cellophane membrane (B) is surmounted by the coverslip (C). Arrows indicate the medium flow path. (From Powell, 1956a.)

illuminating aperture (e.g., 0·9). Second, the Cellophane membrane must be taut and plane, to prevent distortion of the illuminating beam; this is especially important for phase-contrast illumination.

"The component parts of the chamber are shown in 'exploded' isometric projection quarter section [in the Figure reproduced here as Fig. 21]. The floor of the chamber is formed by a coverglass (A) cemented with shellac or Bakelite† into a conical stainless-steel shell (B). On opposite sides of the

† Araldite is preferable to both.

outer surface of the shell are channels (C) along which medium flows into and out of the chamber. This shell is inserted from below into the main body of the chamber (D), also of stainless steel, so that the channels (C) are in register with corresponding channels (E) in the undersurface of the body. (It is not necessary for the two conical surfaces to form a liquid-tight joint.) The shell is held in position by a locking ring (F). At its lower rim the outer surface is relieved to a depth of 0·005 in., and a narrow annular washer of polyvinyl chloride (G), 0·007 in. thick, is empressed in this gap when the locking ring is tightened. At their lower ends, the channels (E) communicate with tubes (H) for connection to the medium circulating pump. The sides and floor of the chamber, with passages for ingress and egress of liquids, are thus determined.

"The chamber is closed by a Cellophane membrane (I) stretched over the cone (J); the working volume is then about 3 mm deep and 15 mm dia . . . The membrane is clamped between polyvinyl chloride washers at its edge by the ring (K) bearing on the surface (L). This surface is on the same plane as the upper surface of the cone (J). The annular gap between (J) and (L) is connected to a water vaccum pump via the tube (M). The Cellophane is drawn down into the gap to an extent depending on its Young's modulus in any direction; the radial stress is nearly uniform and the surface of the membrane within (J) is taut and very nearly plane."

The rather complex arrangement for stretching the membrane is necessary because Cellophane is elastically highly anisotropic and corrugates instantly upon wetting. Once it has been satisfactorily stretched it will remain so almost indefinitely so long as it is not allowed to dry out again. The stainless-steel tubes (H) and (M) are soft-soldered into the body; silver-soldered joints corrode very rapidly in contact with bacteriological media.

The author's experience with this chamber since 1957 has confirmed Powell's caution that it requires skill and care in use; (a modified design has been described; Quesnel, 1961). Powell describes an air-driven pump constructed from glass as a means of supplying well aerated media. With the advent of numerous peristaltic "low-flow" pumps on the commercial market, the air driven pump can be dispensed with for the growth, certainly at any rate, of non-fastidious aerobes. There are many "tricks-of-the-trade" to be learnt from experience in the setting up of the chamber, and since details of this process have not before been published this would seem to be an appropriate Volume in which to do so.

SETTING UP THE POWELL CHAMBER

The description given here relates to the use of a peristaltic pump of variable speed and the letters quoted refer to the items of apparatus indicated in Fig. 22. All pieces of the apparatus that come into contact with the nutrient solution are sterilized in advance and maintained so at the temperature to which the test organisms will be subjected during the experiment. Firstly, the glass influent tube (A) is inserted in the neck of the reservoir bottle (B) and the bung (C) fitted in place. The silicone-rubber section (D) of the influent tube is then seated in the appropriate position under the pressure rollers (E) of the pump (P).

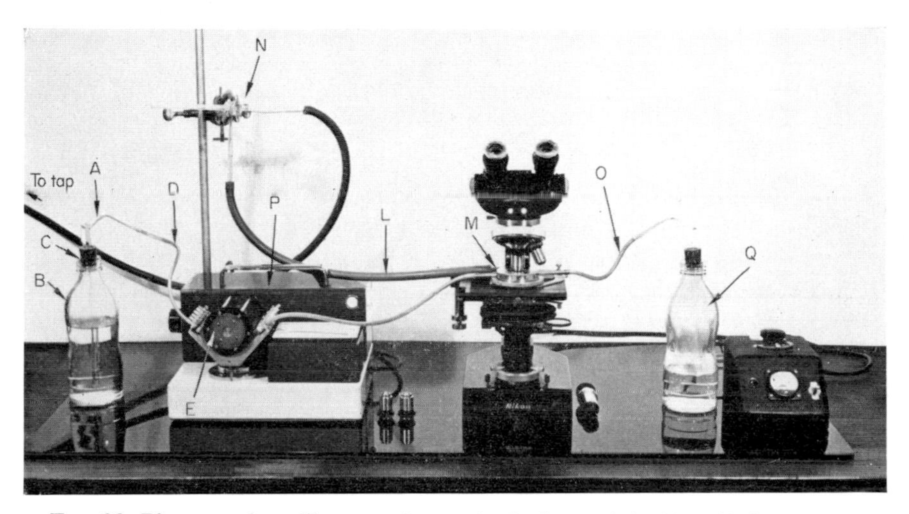

FIG. 22. Photograph to illustrate the method of use of the Powell chamber. The letters refer to the explanation in the text.

The culture chamber is removed from the container in which it has been kept sterile after autoclaving, and the rubber vacuum tubing (L) is connected to the chamber vacuum pipe (M). The vacuum line (L) is connected via a one-way valve to a Geissler pump (water tap). Inserted in the vacuum line is a three-way glass stop-cock (N) lightly greased and capable of effectively regulating the pressure level in the line (L). The third part of the three-way cock serves as a "bleed" when suction is not being applied to the chamber; the water runs continuously to avoid "kick-back" which occurs when operating the water tap.

MOUNTING THE MEMBRANE

The vacuum line (L) is blocked by turning the cock to the "bleed" position, and the chamber supported on the work bench in a suitable position. A single Cellophane disc is carefully placed by forceps, centrally on the top of the chamber. The PVC washer is immediately placed upon it (before it curls) and the top locking ring of the chamber screwed tight. At this stage the Cellophane disc should be unwrinkled, as wrinkling usually indicates faulty seating of the membrane. If seated correctly, the stop-cock is opened fractionally until the membrane is stretched taut but only just so.

ESTABLISHMENT OF CONTINUOUS FLOW

The pump motor is now switched on and medium pumped along the influent tube until it reaches about 1 in. from the distal end. It is then switched off and the tube (translucent) is carefully examined for bubbles, which may completely ruin an experiment if subsequently carried along and into the chamber. They can be eliminated by holding the tube taut and at an angle of about 60° to the bench and then plucking the stretched tube like a harp-string. The vibrations cause the bubbles to rise quickly to the surface at the open end of the tube. The open end of the influent tube is now connected to the chamber at its inlet metal tube and the effluent tube (O) connected to the outlet metal tube of the chamber while the glass exit tube is placed in the effluent reservoir bottle (Q). The flow path is now complete. The next operation requires dexterity for success.

STRETCHING THE MEMBRANE

The pump is re-started at a flow rate of about 100 ml/h and the glass stop-cock held in readiness. As soon as the inflowing medium makes contact with the Cellophane, the stop-cock is opened gradually during the course of about 2 sec. If this operation is correctly performed the "slack" due to the stretching of the membrane that occurs as it is wetted is taken up smoothly and evenly under the increasing force of the vacuum in the annular trough resulting in a smooth, taut membrane. If the stop-cock is opened too slowly, corrugation of the membrane takes place, leading to leakage of the medium, or, alternatively to a membrane that cannot be stretched flat even after maximum vacuum is applied. If the stop-cock is opened too suddenly the sudden drop in pressure in the trough often tears the membrane or wrenches it from its seating.

In the event of an unsuccessful attempt to stretch the membrane correctly the pump must be stopped, the chamber dried out and the operations repeated with a fresh membrane. However, once the membrane is stretched correctly it will remain taut unless dry-out occurs or the vacuum line is broken. The chamber must now be carefully inspected for trapped air bubbles and every step taken to ensure that they are cleared from the chamber before the membrane is inoculated.

It is a useful precaution to allow medium to flow slowly under the membrane for a half-hour or so to leach out plasticizers that are introduced in the final stages of the manufacture of Cellophane (usually urea and glycerol). Inoculation is then performed by allowing a very small droplet on the end of a pipette to touch the membrane. The pipette itself must not be allowed to touch the membrane as it is easily scratched and scratches greatly impair the phase performance of the preparation. A coverslip of appropriate size is applied by forceps and the chamber can now be mounted on the microscope stage.

IV. SPECIAL PROBLEMS ASSOCIATED WITH MICROCULTURAL METHODS

A. Retention of cells

Obviously if cells are to be observed for extended periods of time their location must be comparatively constant so that there is no problem in identifying specific cells at any particular time. This restraint would not apply, however, where continuous microscopic examination was not a necessary feature of the particular experiment but was required, say, only to determine the end point of a particular reaction, or final numbers of cells after a particular period of time. Such would be the case for example in the determination of generation times by the hanging-drop method of Barber (1908), or in the work of Nossal and Makela (1962a, b) in the study of various immunological phenomena, such as the immobilization of motile bacteria by lymph node cells, or chaining before O-agglutination.

Continuous records of individual motile cells are possible by means of cinematographic technique, (which may, however, obscure phenomena discernible by direct observation; Quesnel 1966). In practice good quality photographs of motile cells require short-time exposures not normally possible at high magnification because of problems of inadequate illumination, but it is possible to achieve this by the use of synchronized high-intensity flash. Alternatively one may escape the complexity of electronics by using other high-intensity sources of light, such as the sun, so effectively utilized by Pijper (1949; Pijper and Abraham, 1954) in numerous studies on the motility of bacteria by dark-field illumination. Despite 20 years of experiments, Pijper (1951) found it "impossible for me to regard flagella as motor organs." The problems that arise here stem from our lack of detailed knowledge of the effect of light on bacteria and on other cells and it has been shown, for example, that visible wavelengths can induce mutations in bacterial cells (Webb and Malina, 1967). It is clear, also, that it would be only by luck that any particular motile cell remained in the field of observation for an extended period of time, so that hanging-drop or wet preparations in which motile cells are freely suspended usually preclude the possibility of longterm records of such individual cells.

It is, however, possible to "trap" motile organisms in certain cases, as has been neatly achieved by Gustafson and Kinnander (1956) in micro-aquaria formed from nylon mesh of a size capable of retaining individual highly motile sea-urchin larvae, so that a photographic record could be made of embryological development.

More usually the solution to the problem of unwanted mobility in small cells that suffer the bombardments of molecular activity in Brownian motion is to provide some sort of substratum to which the cells can adhere.

The situation for the tissue-culturist is comparatively simple. Metazoan cells are by microscopic standards large, heavy objects falling easily under gravity. Moreover, they "stick" to glass and form, as is well known, highly adhesive monolayers. Bacteria, by comparison, seem to bounce off glass surfaces, probably because their surface-tension forces are high as a result of their wall structure, but no doubt there may be some encapsulated forms with a slime layer that "sticks" to glass. In this connection it is interesting to note that bacteria stripped of their walls form protoplasts that appear to adhere to glass, as evidenced by the comparative absence of Brownian motion once they have been allowed to "settle."

Claims have been made by Barski (1950) that glass is not the best surface upon which to grow tissue cells and he describes methods for using PVC films to produce enhanced cultural growths; and Thomas and Cramer (1966b) strongly recommend polystyrene.

For microbial cells, the simplest means of cell retention have been the use of solidifying or gelling agents, such as agar, gelatine or silicate. No doubt some of the new thixotropic agents may one day be found suitable for the same purpose. Agar or the equivalent can be used as the support in all the techniques described in Sections III D, E, F, G, H and I. When a perfusion chamber is employed the problem is more serious, as the irrigation of the chamber by liquid will cause displacement unless the cells are firmly adherent to the substratum. Here again tissue cell layers can be perfused with comparative ease, a fact that explains the rapid proliferation of perfusion chambers for the microscopic study of such cells. Alas, the evolution of perfusion chambers for studying individual bacteria has been sluggish by comparison. Very few practicable chambers have so far been devised, as few suitable retaining membranes have been discovered. The problems here are twofold: the membrane must be permeable to the nutrient molecules required by the cells and to the products of metabolism, and its tensile strength must be sufficiently great for it to be "stretched" flat by some mechanical means. Hartman and Hartman (1962) used Formvar (PVC) membranes in a perfusion chamber for bacteria, but this technique has deficiencies that were pointed out earlier (Section III J.3). Vischer (1956) also used Formvar.

The search for a suitable membrane support led Harris and Powell (1951) to the use of Cellophane. Their chamber required the use of vertical illumination, the bacteria were subjected to drying effects and only dry objectives could be employed. These problems were solved in a later more complex chamber (Powell, 1956a) which utilized a Cellophane membrane that was stretched and made slightly convex to press the bacteria lightly against the coverslip. The retention of bacteria under pressure in this way was found to be unnecessary, at least in the case of *E. coli* cells that adhered sufficiently well to Cellophane to resist dislocation by the perfusing liquid, and the Powell chamber could be used with a concave membrane and no overt constraint upon the cells. (Quesnel, 1960, 1962, 1963).

B. Nutritional problems

All the microcultural methods already described, with the exception of the perfusion chambers, have inherent problems of environmental inconstancy. Unless there are the means for adding fresh nutrient and removing metabolic products, as is achieved in perfusion chambers, the system is provided with a limited source of nutrient and cells will sooner rather than later suffer the effects of pH or other changes due to metabolic activity. Clearly, the smaller the volume of suspending menstruum the more rapidly will these effects be observed, so that good optical conditions (thinness of preparation, etc.) limit good nutritional conditions in the simple forms of microcultural preparation.

Although bacteria are minute compared to the agar film upon which they grow, e.g., in Ørskov's method, the limitations upon growth are clearly demonstrated in the work of several authors. Bayne-Jones and Adolph (1932b) investigating the generation times of *E. coli* cells grown on agar block cultures state: "no reproduction occurred in the first hour of culture—this lag period being occupied by growth in size and not in numbers. The maximal rate of reproduction occurred at 2 h, and after 3½ h reproduction was not observed to occur." In other words, development of the micro-colonies was inhibited beyond the fourth generation. On the other hand, Hoffman and Frank (1963) were able to photograph the development of *E. coli* cells for up to nine generations in a "Fleming closed chamber" preparation. Here it is evident that by the eighth generation there is serious disruption of the division synchrony evident in the earlier generations and lengthening of generation times, almost certainly the result of environmental deficiency. In their earlier published genealogy (Hoffman and Frank, 1961) these phenomena are well marked even at the fourth generation.

There is now a deepening realization of the very fundamental changes that can be induced in bacteria as a result of comparatively minor changes in the environment, so that it is extremely important to know what the environmental changes are or alternatively to aim at keeping the environment constant. Thus we find that cells grown at different growth rates have completely different macromolecular compositions, and that deficiencies in certain substrates can lead to the extremely rapid accumulation of storage products or decrease of extracellular secretions; see for example, the reviews by Kjeldgaard (1967), Herbert (1961) and Wilkinson (1959). There are also unexpected phenomena, such as substrate-accelerated death and various other distortions of physiology resulting from minimal stresses upon bacteria, recently reviewed by Postgate (1967).

C. Optical problems

Since all the methods involve microscopic observation, then the best optical conditions must be sought. Basically, where very small objects like bacteria are involved, one aims for maximum resolution and maximum magnification without loss of image definition. Many of the cultural methods described militate against really critical microscopy. Usually one must be prepared to compromise since many of the features that enhance the microscopic possibilities, restrict the growth properties or increase the technical problems of chamber handling and construction to too high a degree.

1. *Choice of microscope*

The microscope of choice will have high-quality optics and an efficient light source. For the best quality this usually means short focal length objective and condenser systems of high N.A. This may well mean that chambers of certain thicknesses, even as little as 3 mm may be too thick to work satisfactorily with a particular lens system. As a general rule it is better to work at lower magnification but maintain critical definition, rather than to use oil-immersion optics in a system where it is impossible to obtain the correct relative focussing distances of condenser and objective. It is a simple matter to produce highly magnified images from good negatives, it is quite impossible to obtain good results, enlarged or otherwise, from poor negatives. A similar caution applies in the choice of eyepiece lenses for a particular optical system. No increase in resolution can be obtained by eyepieces, and in general definition falls off as the magnifying power of the eyepiece is increased. Increased enlargement does not increase definition, so that in practice there is nothing gained in the use of eyepieces of power above 10× and often much lost. (See Quesnel this Series, Vol. 5.)

The advent of long working distance objectives and condensers has greatly alleviated this difficulty. Obviously, long working distance lenses cannot be used under homogeneous immersion, so that a lower effective N.A. must be tolerated. Alternatively if the chamber is too thick it is possible to modify the optics and yet obtain excellent results. Powell (1956a) had to make modifications to the condenser system (Watson Universal No. 1 condenser in conjunction with Watson phase-contrast system of that date) in order to obtain the 3 mm working distance required for the establishment of the phase optical paths. A larger illuminating annulus was used, brought into focus by an auxilliary lens—an objective of 75 mm focal length—mounted in place of the condenser iris and carefully centred. This modification enabled the use of double immersion and the image quality remained extremely good in spite of the lack of theoretical perfection. Quesnel (1961) using a Watson Bactil Binocular microscope with the same phase optics used an 8 cm focal length wide-diameter (5 cm) bi-convex lens cemented to the condenser housing below the iris diaphragm, while the condenser itself was raised by the insertion of a tubular sleeve 19 mm in height to enable the top element to be racked through the stage. Good contrast was obtained over the whole field of view.

2. *The optical path through the chamber*

Apart from being thin, the chamber used should be selected with a view to the simplicity of the optical path through the chamber. Ideally the space between the condenser and objective faces should, with the exception of the specimen, be optically homogeneous, that is, there should be no changes in refractive index†. In practice this may be impossible, but, at least, the number of changes in refractive index, and thus light-path deviations, should be kept to the minimum. For example, a simple wet film between coverslip and slide used with dry optics involves media of three different refractive indices; air between preparation and objective and preparation and condenser; glass, top and bottom layers of the preparation; and (say) water, the middle layer. This means refraction of the light rays at four interfaces excluding the lens faces themselves. By oil-immersing the condenser and objective only two refracting interfaces are left, i.e., from glass to water and water to glass. If a hanging agar block with a liquid well is used then there are two additional interfaces between media of different refractive index, and so on. Also, the greater the difference in refractive index between contiguous layers, the greater will be the angle of refraction, which may lead to either a greatly diminished effective N.A. and to a loss of light intensity owing to the refraction of light beyond the angular aperture of the objective or to total internal reflection at one or other interface within the sandwich, or both.

In addition to the distorting effect of refractive-index changes there are likely to be numerous diffractions that scatter light and limit good definition. These are caused as a rule by scratches on any of the glass surfaces in the light path, the presence of dust particles, microscopic crystals, various inclusions, such as dead bacteria usually found in agar layers, etc. If any layer in the sandwich is coloured, then absorption of certain wavelengths will occur.

3. *Image contrast*

In bright-field microscopy, objects become visible only if they can alter the path of light impinging upon them. If light is totally absorbed then no image is formed at that point in the field and it is therefore represented as a black area. On the other hand, with most material suitable for microscopic examination some light is transmitted through the specimen, and often nearly all, so that contrast is poor. To obtain good contrast images of living cells the most useful instrument is the phase-contrast microscope, which is described in detail by Quesnel, this Series, Vol. 5. Suffice it to say here that it is possible to select the degree of contrast required in two ways; light passing

† But see 3 below.

through the phase-displacing ring can be displaced by selected fractions of a wavelength or alternatively the phase displacement produced at the specimen can be altered by altering the refractive index of the suspending fluid.

The different effects that may be obtained are shown by comparing, for example, the photographs of Knöll and Zapf (1951) with those of Mason and Powelson (1956) and Adler and Hardigree (1965). Knöll and Zapf studied the behaviour of nuclei during growth of microcolonies on agar-block microcultures using bright-phase contrast. In their photographs the nuclei appear as bright areas in a darker background; the phase-bright areas were shown to stain Feulgen positive. They claim to have discerned in *E. coli* a division rhythm for the nuclei in that post-fissional cells contain two nuclei while pre-fissional cells contain four. In one clone, a 4–8 rhythm was established. By changing the refractive-index properties of the supporting medium, Mason and Powelson (1956) were able to increase and reverse the contrast of the image. In their system, the best results were obtained with a gelatin–brain–heart infusion broth containing from 20–27% gelatin for *E. coli* and from 28–35% gelatin for *B. subtilis*. In their preparations, the nuclei appeared dark against a lighter background of cytoplasm. Their results indicate a uni-nucleate/bi-nucleate rhythm for *E. coli* and a 2–4 rhythm for *B. subtilis*. They agreed in general with Lark and Maaløe (1954) that the separation of the daughter nuclei follows rather than precedes wall fission.

The excellent contrast shown in the photographs of Adler and Hardigree (1965) (reproduced here as Fig. 23) was obtained by mounting the cells in a layer containing 20% gelatin and 1% agar in nutrient broth. The nuclear bodies appear as bright areas against a very dark cytoplasm under the phase-contrast microscope. The principles underlying such refractrometric tricks have been described by Barer *et al.* (1953) and Hancox and Kruszynski (1956).

4. *Intensity of illumination*

Because of the great difficulty of obtaining good contrast in living specimens by simple bright-field illumination the vast majroity of such observations are now performed by phase contrast. The intensity of illumination required for phase-contrast observations is very considerably greater than for ordinary microscopy, since only a fraction of light emitted by the source can be used in image formation. This is so because the substage annulus immediately eliminates most of the light and a further proportion is absorbed by the phase plate in order to "balance" the proportions of transmitted and diffracted radiation. In addition it is common practice to use a filter, usually green, which helps to eliminate chromatic aberration and improves image definition and is especially useful in photography on panchromatic

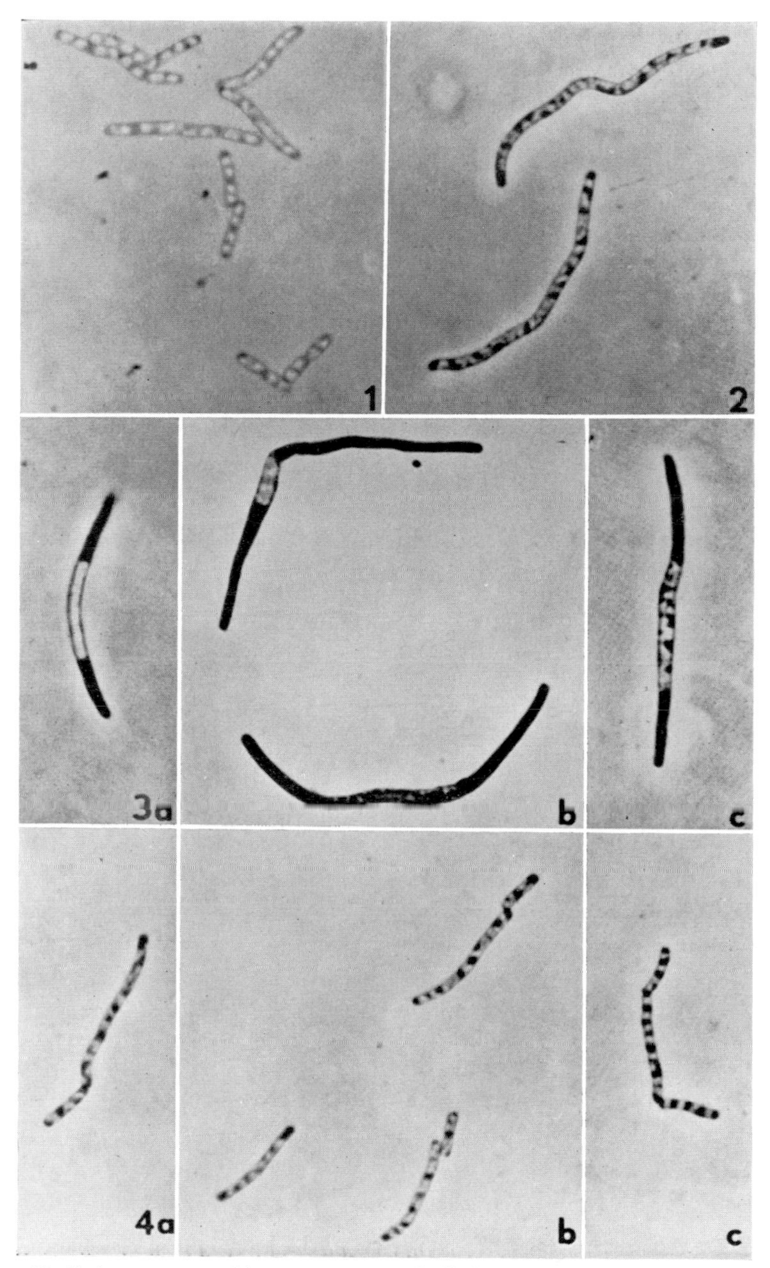

FIG. 23. Enhancement of image contrast. Cells "suspended" in 20% gelatin, 1% agar in nutrient broth, photographed by phase illumination. Nuclear material bright: 1 unirradiated cells of *E. coli* K-12 AB 1899; 2, filaments formed by growth of irradiated cells on nutrient agar; 3, examples of filaments formed by growth of irradiated cells on nutrient agar containing 10μg of mitomycin C per ml for 3 h; 4, divisions induced by exposure to pantoyl lactone. Magnification, 1650X. (From Adler and Hardigree, 1965.)

emulsions. This will further reduce the light intensity by 50% or more. Indeed, high-intensity sources are necessary for good phase contrast if photographs are to be made easily. High-intensity illumination generates considerable quantities of heat and the specimen must be protected by heat-absorbing filters introduced into the light path. (The heat generated by the lamp may be enough to change significantly the temperature of the room.) Wherever light is used its temperature effects must be borne in mind.

In addition to heating effects, there is a growing literature on the effects of visible wavelengths on microbial cells. As early as 1907 Thiele and Wolf pointed out that visible light kills *E. coli* and other species at 30°–40°C but not at 14°–20°C in the presence of oxygen. More recently Mackie and van den Lingren (1924) have claimed that *E. coli* at 37°C was definitely inhibited in the red region of the visible spectrum; *Staphylococcus aureus*, *Staphylococcus albus* and *Pseudomonas pyocyanea* were inhibited by red, yellow and blue wavelengths. Pneumococci and *Streptococcus pyogenes* showed "complete inhibition in all colours even with inoculation from dense emulsions". *Staph. albus*, *Vibrio cholerae*, *Neisseria meningitidis*, *Neisseria gonorrhoeae* and *Corynebacterium diphtheriae* were markedly inhibited by (artificial) white light. In addition it has now been shown that visible light is capable of significantly increasing the mutation rate of resistance to T_5 bacteriophage in *E. coli*, the mutation rates being proportional to irradiance (Webb and Malina, 1967).

Under oil-immersion optics, the light available for photography of the specimen is extremely limited and beyond the limit of sensitivity of most conventional light meters. There is equipment available that can amplify the output from the photocell and allow accurate measurements to be made. These are dealt with in a later Chapter on photomicrography (Quesnel, this Series, Vol. 5). Even so one finds that the exposure required may run to several seconds, and it is often extremely difficult to maintain precise focus during such extended exposures unless special precautions are taken (see below).

D. Temperature control

"Very few investigations concerned with microcultural problems have been sufficiently aware of the importance of holding temperature variations down to a few hundredths or thousandths of a degree" (Hoffman, 1964). This may be an overstatement of the case but emphasizes the caution required if observed variations in behaviour are to be attributed to causes other than temperature variations. Certainly one should aim at a restriction of the order of $\pm 0{\cdot}1$°C.

Few precise studies have been made of the responses in bacterial composition and growth rate to temperature. At higher temperatures at or just beyond the temperature for maximum division rate, a difference of 2° or 3°C can be the difference between maximum multiplication and death (Ingraham, 1958). Temperature shifts below the maximum have been for long used as a means of synchronizing cell divisions, which of necessity implies that the temperature shift affects cells of different individual ages in quite different ways. Each biological process is characterized by a definite temperature characteristic, and a change of temperature may mean a change in a controlling "master reaction." This may lead to a rate difference in an overall process such as growth; it may also result in a difference of metabolic pathway and a changed end product. The need for exogenous sources of previously non-essential nutrilites by heat-injured cells, whether the injury is from too high or too low a growth temperature (Campbell and Williams, 1953), and the disruption of cytokinesis without restraint on total growth rate at elevated temperatures (Hoffman and Frank, 1963) are examples of such shifts in reaction control.

1. *The heated stage or stage incubator.*

Heated stages are made by several manufacturers. They have the advantage of being simple to use and do not limit the operations of the experimenter. They may be heated by circulating hot liquid, or electrically, depending on design. Among the earliest described are two due to Hill (1902). The more complex of these chambers (Fig. 6) was made from brass and was designed to accommodate hanging-block preparations. A large coverglass formed the bottom of the chamber, and the coverglass from which the hanging block was suspended formed the roof. Warm water was circulated through an outer jacket and liquid or gas could be passed through the inner chamber in direct contact with the suspended block. The depth of the device was too great to permit high-power microscopy. The simpler chamber was made from glass slides and glass tubing and provided a moist chamber for a hanging-block preparation. Heating was by circulation of warm water. Both devices incorporated thermometers.

For some purposes a simple water heated stage such as that shown in Fig. 24 may be found adequate (Hambleton, 1964). The inoculum on a small plug of agar is fixed between two coverslips sealed with wax to the top and bottom surfaces of the base plate so as to enclose the central aperture. This stage may be conveniently made of brass, but must be nickel plated for use, since brass may prevent growth or inhibit the germination of spores. The germination rates of spores of *Bacillus megaterium* have been studied with the aid of this device (Hambleton and Rigby, 1966).

A very simple electrically heated stage is made by C. Reichert, Optische Werke AG, Vienna, in which a simple slide platform receives heat by conduction from an electric heating element housed in a metal block on one side of the platform and good temperature control at the site of growth is not possible. The same manufacturers supply a more complex instrument, called the Kofler Micro Cold-Stage, with a temperature range from $-50°C$ to $+80°C$ (as distinct from a similar hot stage with a range from 20°C to 350°C). The Micro Cold-Stage consists of a metal base heated by an electric element while the top of the stage forms a heating chamber in which the sample is placed on a half slide and covered with a glass lid. The half slide is

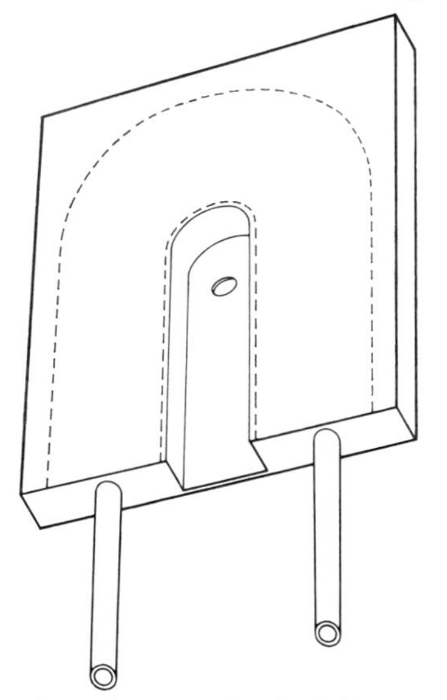

FIG. 24. Simple water-heated stage. Dashed lines indicate internal channel. The specimen is mounted between coverslips fixed by wax to either side of the base plate, to cover the central aperture.

held in a sample-shifter operated by a screw which enables it to be positioned. A central aperture in the base permits illumination from below to be focused on the specimen. Heating is controlled by a regulating transformer and temperature is measured by a thermometer with precision-ground bulb built into the stage. There is in addition a CO_2 inlet (for cooling), regulating valve and expansion chamber in the body of the stage. While this stage is designed for the accurate determination of melting points and severely limits the use of high-power optics, it could probably be modified for microcultural use. Messrs Wild, Heerbrugg of Switzerland manufacture a hot-and-cold stage, in which the heating is performed electrically and cooling is achieved by circulating cold water through a system of tubes built into the body of the chamber. The chamber itself sits high on the microscope stage and must be used in conjunction with a long working distance condenser. There is no thermostat. "It heats or cools regularly and is not subject to periodical fluctuations in temperature." Temperature selection is by means of a transformer rheostat with continuous setting on a scale 1–10 which covers a range from ambient to ca. 50°C. A warming-up time of at least 1 h is recommended to ensure constant temperature. The performance of this instrument is not known to the author, but there is the interesting possibility that the water system might be used in conjunction with the electric system for heating purposes. The Wild stage incorporates a thermometer calibrated in $\frac{1}{2}$° steps.

Schmitt and Schmitt (1931) have described a method by which very precise temperature control of liquids may be achieved. In this design a toluol–mercury regulator is inserted in the grid circuit of a Thyratron regulator tube (General Electric Co. of America, Syracuse, type FG–27) which is a mercury-vapour tube capable of controlling a peak current of 5 A by means of a grid in which a current of less than 0·1 mA may flow. The designers claim to have obtained temperature control to within 0·001°C for an insulated water bath.

Zeiss (Oberkochen) supply two heating stages, one for temperatures from 35°–43°C and the other covering the range 30°–60°C. Both devices are basically a metal plate to which a current is supplied and which is heated as a result of the inherent resistance. The specimen slide sits on the plate and is warmed by conduction and the whole is enclosed by a plastic cover with an aperture through which the objective works. Temperature regulation on the 35°–43°C model is by means of a simple manually variable resistance and the stage temperature is read by an attached thermometer. The 30°–60°C model incorporates a contact thermometer, on which the desired temperature can be set with a magnet, and this regulates the heating current via a relay. The makers claim a temperature constancy of ±0·2°C at the specimen after a warm-up period of 15 min. Up to 40°C the setting accuracy is ±0·5°C. An earlier version of the Zeiss heating stage was tested critically by Engel and Zerbst (1960) who found that even with well made instruments such as this, there arise a variety of heat flow gradients as a result of uneven heating, which leads to unstable temperatures at the locus of growth. The author has not tested the newer instruments and is unable to comment on the precision of their control.

A "heating and cooling microscope stage 80 with automatic thermoregulation" is made by Ernst Leitz (Wetzlar, Germany) which enables observations to be made at constant or varying temperatures between −20° and +80°C. Heating is achieved by an electric heating element built into the stage plate linked with an automatic thermoregulator and thermometer for setting and controlling the stage temperature. This thermometer is curved and mounted in a groove at the outer edge so that there is a flush finish to the top of the stage. A round metal plate positioned over the slide helps to stabilize the temperature at the growth site. The stage is operated in conjunction with a variable transformer. There are built-in channels for water circulation (for cooling and heating) and freezing of the specimen *in situ* (by CO_2) is also possible. Condensation deposits on the glass surfaces at low temperatures can be avoided by the use of an accessory N_2 compartment that fits snugly around the objective and over the slide. The precision of temperature control is not given, and the author has not tested this stage.

Whatever the merits of any particular heated stage, there are a number of problems associated with their use. In the first place a heated stage does not heat the microscope to the working temperature and the material of the microscope is therefore subject to the ambient fluctuations in temperature which cause loss of focus. Depth of focus at high magnifications is so minute (less than 0·25 μm above N.A. 1·0) that the slightest variations in temperature necessitate re-adjustment of the focus. On the other hand Wyckoff and Lagsdin (1933) found that precise control of the temperature of the instrument enabled precise focus to be maintained without attention over long periods; Ware and Loveless (1959) found that these variations could be compensated for by the insertion of a small piece of $\frac{3}{4}$ in. dia Perspex tube between a steel bracket attached to the body tube and the limb of the microscope. The correct length of Perspex tube for use with the 40× objective used, was found by trial and error.

In addition to focussing problems, which are of great importance if a good photographic record is required, there is need to consider the serious temperature shock to the organisms when they are first placed on the warm stage. Preparations made up at room temperature and subsequently transferred to a stage at 37°C (say) are undergoing the equivalent of a temperature shift synchronization procedure (Maaløe, 1962) with effects that may last 2 or 3 h. If one is to be stringent in the use of heated stages, a considerable re-adjustment time must be allowed before the culture can be considered free from the effects of the temperature shift.

It is also clear from the work of Engel and Zerbst (1960) that with heated stages there are numerous heat-flow gradients which make it unlikely that there is a stable, uniform and definable temperature at the actual locus of growth. The steepness of such heat gradients would be somewhat buffered where flowing perfusion chambers were used, so long as great care was taken to equate the temperature of the feed liquid to that of the stage.

In general terms, heated stages have the advantage of being easily available, comparatively cheap, and usually do not hamper the activities of the experimenter. Their disadvantages are the poor temperature control at the growth site, and the fact that stabilization of the temperature of the microscope is not achieved. In addition, some prevent the use of high resolution optics.

2. *The incubator box*

To stabilize the temperature of the whole culture and instrument system, an incubator box can be used. Such boxes usually take the form of rigid plastic structures that envelope the immediate environment of the microscope leaving only the light source and the oculars protruding from the chamber so formed. Here the intention is to heat the air inside the box rather than the apparatus directly, and the need for a long heat-up time to bring all the experimental components to the temperature of the heated air before the start of the experiment is obvious. The air itself may be heated in different ways.

Wyckoff and Lagsdin (1933) used an incubator box which had (incredibly), "a brass frame, fibre sides, a sliding aluminium top and a glass front," in which the air was heated by means of low-wattage blackened bulbs. The temperature was controlled by a glass–mercury regulator operated through the thyratron relay circuit of Schmitt and Schmitt (1931). Electrical resistances controlled by a thermoregulator have been used as the heat source by Deschiens *et al.* (1945) for an incubator box constructed of wood and glass. The box was so designed that focussing and stage movements could be manipulated from outside the enclosure of the box. By this means they were able to maintain temperatures of 37° ± 0·25°C for extended experimental periods.

Despite the stability of temperature claimed by some workers, significant temperature gradients will be established in still air. If the sensing probe of the control circuit is attached to the culture chamber or stage, by the time the rise in temperature is registered and the heating elements inactivated, the lag may result in considerable subsequent overshoot of the temperature as the over-heated air around the source spreads through the box. This problem has been realized and overcome in several incubator boxes of more recent design. The solution in general terms is to blow heated air at a carefully controlled temperature through the chamber continuously, so that an even temperature is maintained throughout the enclosed space. Sage Instruments Inc. (White Plains, New York) make a wide range of custom-built incubator boxes employing this principle. The Plexiglas incubator has a thermostatically controlled heat source and circulating filtered air system, enabling the selection of any temperature from ambient to 40°C. The thermostat, heat source, blower and filters are all outside the box and air is delivered through a duct. A precision thermometer, indicating temperature at the microscope stage, is incorporated. The temperature variation is claimed to be less than 1°C.

Hoffman and Frank (1963) used a similar system with better temperature control to limit variation to less than 0·1°C. In this design the heating element with blower is monitored by a thermistor control, and temperature overshoot is minimized by a rheostat that maintains the heating element "at a temperature that will give heated air at a level just slightly above that desired for the experiment."

Nippon Kogaku (Tokyo) supply an incubator chamber for use with their inverted research microscope. With the exception of the fine-focus control, access to the movements must be made via curtain ports in the front of the chamber. Heating is performed by "monitored" warmed air fed into the chamber from an external heater–blower unit linked to a sensor placed within the chamber. A similar system manufactured by the Union Optical Co. Ltd (Tokyo) for use with their microscopes has the advantage that all the microscope movements can be made from outside the incubator box by means of extension control levers, which are supplied.

An incubator bath designed to maintain a constant temperature at the site of the specimen has been described by Gittens and James (1960). They have controlled the temperature of the circulated water within $\pm 0{\cdot}02^\circ C$ by means of a Shandon Circotherm II Constant Temperature Unit (Shandon Scientific Co. Ltd, London). Not least among the drawbacks of the design, however, is the need to waterproof the optical system as the lenses are used submerged in water.

Various other workers have designed incubator boxes, e.g., Lwoff *et al.* (1955); the Pulvertaft incubator modified by Ross (1967), which used heating coils; Pederson *et al.* (1933), who demonstrated the benefit of forced circulation of the heated air.

The use of an incubator box of sufficient size will solve the problem of culture preparation at the experimental temperature. Unless, as mentioned before, the culture apparatus and inoculum and all the preparatory manipulations are subject to the experimental temperature, the cells will suffer a heat shock on being introduced into the incubator box. These preparations must therefore be made inside the incubator and all the materials must be placed in it long in advance of the start of the experiment. Where a continuous flow microcultural method is being used, involving reservoir bottles, pump, etc., working conditions inside the box may become very difficult.

In general, incubator boxes have the advantage that a high degree of temperature control is possible with the right sort of heating system, *viz*, a thermistor-regulated blown hot-air supply. They also provide for stable microscope temperatures, thus minimizing the loss-of-focus problem; and enable the cultures to be warm while the experimenter is cool. The box itself can be used as the camera support as described by Ross (1967), so that camera vibrations are not transmitted to the microscope. The limitations to the use of the incubator box are the restraints upon manipulations around the microscope, and of the microscope itself, and the need to prepare for the experiment "within the box".

3. *The incubator room*

While tissue culturists may be prepared to tolerate temperature fluctuations of the order of $\pm$ 1·0°C, there is a growing fastidiousness among students of bacterial growth that demands a much closer control of temperature at the growth site. For them the final solution appears to be the incubator room or HADES (heated and darkened experimental situation). For some microcultural techniques, a closely controlled incubator box may be adequate, but in continuous-flow technique demanding medium reservoirs, pumps, etc., the numerous manipulations involved dictate a free access to the apparatus only possible in an incubator room.

The advantages of working in HADES are several. Primarily, it provides for the proper temperature stabilization of all the instruments and implements, cultures, etc., required in the experiment, and because of the larger volume of air involved a higher degree of stability at the growth site is possible. The obvious disadvantage is the fact that the experimenter has to operate at the uncomfortable optimum growth temperatures of the more popular microbes. However, since temperature control can be of a high order it is possible to set up a time-lapse photomicrographic system to record the experiment in the absence of the operator and this is a considerable advantage in experiments of extended duration.

HADES should have well insulated walls and an air-lock double door entrance so that large volumes of hot air are not lost suddenly. The hot air should enter at low level and be vigorously circulated by a fan so that there is thorough mixing and a minimum of temperature gradients in the room. Even so, the temperature-sensing device should be as near as possible to the growth site and the thermo-regulating system should be designed to give a minimum of overshoot. In a continuous-flow system where the cells are bathed in liquid, because of the heat capacity of the liquid its temperature fluctuations will be appreciably less than the fluctuations in the temperature of the circulated hot air.

Temperature sensing can be achieved by several devices such as thermocouples, resistance thermometers and thermistors. Thermocouples are comparatevely slow to react to temperature change and comparatively inaccurate as temperature measurers. Resistance thermometers utilize the change in electrical resistance of a conductor to determine the temperature. The most sensitive are platinum, and changes of temperature of less than 0·05°C can readily be detected. By means of a suitable bridge circuit and strip-chart recorder, a detailed record of temperature variation can be obtained.

Temperature control is probably best achieved by means of a thermistor. Thermistors are sintered mixtures of metallic oxides that have a high negative temperature coefficient of resistance, the specific resistance falling a hundredfold or more for a temperature change from $-50°$ to $+50°C$. With the appropriate resistor in series with the thermistor and applied voltage, an appreciable change in current with temperature can be obtained in milliseconds. By incorporating the thermistor in a bridge circuit with a relay in which the variable resistor determines the temperature at which the relay will operate, a temperature control circuit can be made; or better, a continuously compensating heat-control system may be made by linking the bridge into a regulatory circuit via a transformer as shown in Fig. 25.

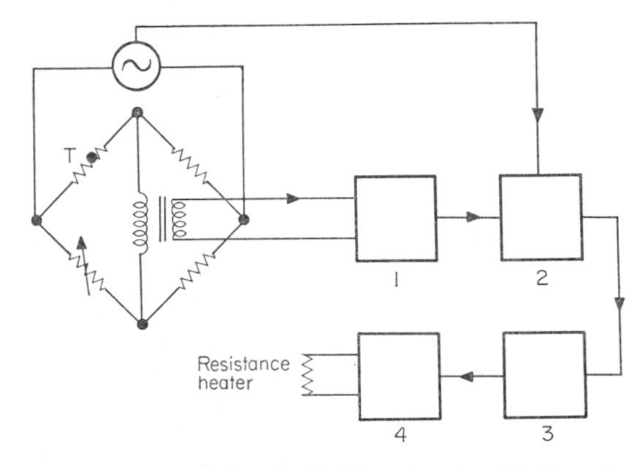

FIG. 25. Temperature control circuit. Bridge circuit with thermistor, T, feeding: 1, amplifier; 2, phase-sensitive rectifier; 3, stabilizing network; 4, power amplifier.

In the circuit outlined in Fig. 25 minute changes in temperature at the thermistor (T) would lead to a potential difference across the arms of the bridge and a signal would be put out to the amplifier (1). The amplified signal from this and a signal from the a.c. source are then fed to a phase-sensitive rectifier (2) in order to distinguish whether correction of temperature upwards or downwards is required. The signal from this stage is then passed to a stabilizing network (3), but, provided that the thermal time constant of the controlled area is sufficiently long, this stage could probably be omitted and the stage 2 output fed directly to a power amplifier (4). The power amplifier controls the resistance heater (situated in the duct through which air is blown into the incubator room) continually and smoothly, activating it to a greater or less degree by changing the power output to the resistance in response to the rectified signal from 3. Since the heater is continually activated, this system requires that external air at a lower temperature be fed continually to the air circulating through the room, thus supplying fresh air to the operator.

Such a circuit could only be used successfully for the control of temperature within a narrow range by means of a low-power heater unit, to complement the action of a less finely controlled main heat source which would cease to operate once the room was brought from ambient to within perhaps 1°–2° of the required experimental temperature.

The high sensitivity of the thermistor controlled circuit enables it to be applied in unconventional ways such as the thermistor-controlled air-curtain incubator available from Sage Instruments, Inc. (White Plains, New York). The makers claim that this instrument totally eliminates the need for enclosures of any kind. In operation a stream of air at a specified temperature, closely controlled by the thermistor, is blown over the specimen continually, while the temperature at the stage is monitored by a suitable probe which feeds back information to the controller. It is claimed to be able to control the temperature of a stage specimen to an accuracy of ±0·2°C, within the range 30°–40°C. This instrument provides a highly sensitive and convenient form of local heating very suitable for enclosed stage specimens, but not as dependable where continuous flow techniques involving pumps and delivery tubing are used.

E. Recording methods

While it is possible to make direct measurements of cell dimensions by micrometer, the "filar" micrometer being particularly versatile and accurate (Quesnel, this Series, Vol. 5), even with this instrument each individual measurement requires a comparatively long period of time, and this rules it out as a means of following growth in cell size for more than a few cells in any experiment. For a complete record of the changes in cell dimensions and cell inclusions during an extended experiment a permanent time-lapse photographic record is required.

In the pioneering experiments of Adolph and Bayne-Jones (1932) and Bayne-Jones and Adolph (1932a, b), measurements were made by projecting images of the cells on to a screen to a magnification of 30,000 × and measurements were made of the images. Knaysi (1940) used a similar method to study the rate of increase in substance during growth of *Schizosaccharomyces pombe* and *B. cereus*, but instead of using bright-field microscopy he used dark-field illumination. Nowadays, one would use phase-contrast microscopy (Hoffman and Frank, 1963, 1961; Durham *et al.*, 1967; Collins and Richmond, 1962), while Harvey and Marr (1966) measured the dimensions of *E. coli* from electron micrographs printed to a final magnification of 10,000.

While photographs provide the best method of permanent recording, it is still necessary to have a code system for identifying the individual cells in any particular clone if growth parameters, such as generation times and genealogies, are to be easily reported. Several different codes have been used. Kelly and Rahn (1932) used a simple system in which cells were identified by a Roman numeral denoting the generation of offspring to which they belonged and the time in minutes of the "birth." Bayne-Jones and Adolph (1932a, b) utilized a form of binary coding in which each clone was given a code capital letter and each cell of the clone was identified by a permutation of code letters a and b in a binary system.

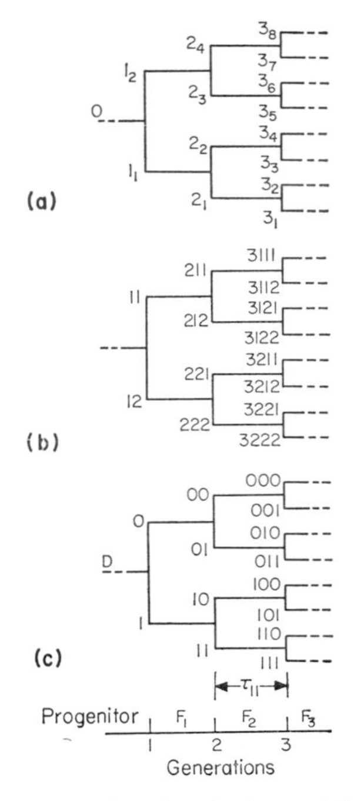

FIG. 26. Comparison of systems of coding for bacterial cells in genealogical relationships. For explanations, see text.

Hoffman and Frank (1961) identified cells in their genealogical charts by designating the original organisms as 0 (0th generation) and organisms in each subsequent generation bore the number 1, 2, 3, etc., to denote the generation to which they belonged. The bacteria within a generation were given numerical subscripts from 1 to n, where n was the maximum possible number of viable organisms that could be formed for any particular generation ($n = 2g$, where g is the generation). The subscripts were given in order of the chronological times of division, the first to divide receiving the lowest subscript (Fig. 26a). It should be noted that the first divider in chronological time does not necessarily result from the shortest mother generation time. The coding system used by Powell (1958) is slightly different (Fig. 26b). Here generation times of individuals (and so, the individuals themselves) were coded by an initial number that denoted the generation to which the cell belonged, as in the previous system, but the individuals of any particular generation were identified by the addition to this of a binary code using the digits 1 and 2 as the code symbols. Thus the code for a third-generation cell always starts with 3, and the remainder of the code number is derived from the full permutation of 1 and 2 in groups of three. On Powell's system the generation time of cell 211 would be denoted by τ_{211} and on termination it gives rise to cells 3111 and 3112. The system of binary coding used by Quesnel (1963) is illustrated in Fig. 26(c), about which the following points should be noted. The progenitor D has an unknown generation time, but a first division time d_D representing the time from inoculation to first fission can be allocated. The code numbers of members of different generations have different numbers of digits, and the number of digits corresponds to the cardinal number of the generation completed by their inception. Thus, the F_1 offspring formed at first division have single digit code names; F_2 offspring have two-digit code names, etc. Since bacterial multiplication is a binary process, the code system is a complete permutation in each generation so that no surplus or "stray" code numbers are possible. The code numbers for sisters are always identical except in the last digit; those for first cousins differ only in the last two digits, etc. The notation is easily applicable even to large numbers of generations. The generation time of any particular cell is denoted by prefixing the symbol τ to the code number of the cell; thus, τ_{11} is the time interval between the inception and the termination of cell 11, the termination resulting in the sister pair, 111, 110.

The survival and viability of micro-organisms under conditions of minimal stress have recently been reviewed by Postgate (1967). He defines viability as the ratio of the number of viable units in a microbial population to the total number of microbes present. A viable microbe is one "capable of dividing to form one or more live daughter cells when provided with a favourable environment." Statements about viability should be accompanied by a specification of the environment and also the time allowed for division before the assessment was performed. The method of determining viabilities by the slide chamber technique has been described in detail and evaluated by Postgate *et al.* (1961) and by Postgate in this Volume, p. 611.

Where growth and development of colonies is being studied continuously, it is not possible to assess viabilities unless genealogical identification of the cells is possible (see above), since non-viable cells formed within the clones are not distinguishable as non-viable, as for example, single-cell sites representing non-viable cells and microcolonies representing viable transplant cells are distinguishable in the Postgate method. Powell (1958) and Quesnel (1963) used similar criteria for assessing the viability of cells within growing clones, in the log phase in the former case, and in the lag phase in the latter study. These criteria were that under the conditions of culture a cell was considered non-viable if it became dimly visible under phase-contrast and then showed no sign of growth during about 1 h, or, failing this if it showed no sign of growth during about 3 h (Powell, 1958).

ACKNOWLEDGMENT

I would like to thank Prof. John R. Postgate for reading the manuscript and making a number of useful suggestions.

REFERENCES

Adler, H. I., and Hardigree, A. A. (1965). *J. Bact.*, **90**, 223–226.
Adolph, E. F., and Bayne-Jones, S. (1932). *J. cell. comp. Physiol.*, **1**, 409–427.
Allen, P. W. (1923). *J. Bact.* **8**, 555–566.
Attardi, G., Cohn, M., Horibata, K., and Lennox, E. S. (1959). *Bact. Rev.*, **23**, 213–223.
Baracchini, O., and Sherris, J. C. (1959). *J. path. Bact.*, **77**, 565–574.
Barber, M. A. (1908). *J. infect. Dis.*, **5**, 379–400.
Barer, R., Ross, R. F. A., and Tkaczyk, S. (1953). *Nature, Lond.*, **171**, 720–724.
Barski, G. (1950). *Annls Inst. Pasteur, Paris*, **78**, 666–670.
Barski, G., and Robineaux, R. (1956). *Annls Inst. Pasteur, Paris*, **90**, 514–517.
Bayne-Jones, S., and Adolph, E. F. (1932a). *J. cell. comp. Physiol.*, **1**, 387–407.
Bayne-Jones, S., and Adolph, E. F. (1932b). *J. cell. comp. Physiol.*, **2**, 329–348.
Beran, K. (1968). *Adv. Microbiol. Physiol.*, **2**, 143–171.
Bissett, K. A. (1950). "The Cytology and Life History of Bacteria". Livingstone, Edinburgh.
Campbell, L. L., and Williams, O. B. (1953). *J. Bact.*, **65**, 141–145.
Carter, S. B. (1966). *Expl Cell Res.*, **42**, 395–398.
Chadwick, P. and Mahoney, D. H. (1966). *Can. J. Microbiol.*, **12**, 683–690.
Christiansen, G. S., Daines, B., Allen, L., and Leinfelder, P. (1953). *Expl Cell Res.*, **5**, 10–15.
Chung, K. C., Hawirko, R. Z., and Isaac, P. K. (1964a). *Can. J. Microbiol.*, **10**, 43–48.
Chung, K. C., Hawirko, R. Z., and Isaac, P. K. (1964b). *Can. J. Microbiol.*, **10**, 473–482.
Chung, K. C., Hawirko, R. Z., and Isaac, P. K. (1965). *Can. J. Microbiol.*, **11**, 953–957.
Clayton, R. K. (1958). *Arch. Mikrobiol.*, **29**, 189–212.
Collins, J. F., and Richmond, M. H. (1962). *J. gen. Microbiol.*, **28**, 15–33.
Cruickshank, C. N. D., Cooper, J. R., and Conran, M. B. (1959). *Expl Cell Res.*, **16**, 695–698.
Deschiens, R., Jouan, C., and Lamy, L. (1945). *Annls. Inst. Pasteur, Paris*, **71**, 220–222.
Durham, N. N., Noller, E. C., Burger, M. W., and Best, G. K. (1967). *Can. J. Microbiol.*, **13**, 417–421.
Engel, H. J., and Zerbst, E. (1960). *Z. wiss. Mikrosk.*, **64**, 384–394.
Engelmann, T. W. (1894). *Pflügers Arch. ges. Physiol.*, **57**, 375–386.
Fleming, A., Voureka, A., Kramer, I. R. H., and Hughes, W. A. (1950). *J. gen. Microbiol.*, **4**, 257–269.
de Fonbrune, P. (1949). "Technique de Micro-manipulation". Masson, Paris.

Gittens, G. J., and James, A. M. (1960). *Analyt. Biochem.*, **1**, 478–485.
Gustafson, T., and Kinnander, H. (1956). *Expl. Cell Res.*, **11**, 36–51.
Hambleton, R. (1964). M.Sc. Thesis, University of Manchester.
Hambleton, R., and Rigby, G. (1966). *J. Pharm. Pharmac.*, **18** (suppl.), 30S–32S.
Hancox, N. M., and Kruszynski, J. (1956). *Expl. Cell Res.*, **11**, 327–339.
Harris, H. (1955). *Br. J. Exp. Path.*, **36**, 115–127.
Harris, N. K., and Powell, E. O. (1951). *Jl. R. microsc. Soc.*, **71**, 407–420.
Hartman, R. S., and Hartman, R. E. (1962). *J. Bact.*, **84**, 595–596.
Harvey, R. J., and Marr, A. G. (1966). *J. Bact.*, **92**, 805–811.
Herbert, D. (1961). *Symp. Soc. gen. Microbiol.*, **11**, 391–416.
Hewitt, L. F. (1951). *J. gen. Microbiol.*, **5**, 287–292.
Hewlett, R. T. (1918). "Manual of Bacteriology", 6th Edn, Churchill, London. quoted by Hort (1920).
Hill, H. W. (1902). *J. med. Res.*, **7**, 202–212.
Hillier, J., Knaysi, G., and Baker, R. F. (1948). *J. Bact.*, **56**, 569–576.
Hillier, J., Mudd, S., and Smith, A. G. (1949). *J. Bact.*, **57**, 319–338.
Hoffman, H. (1964). *A. Rev. Microbiol.*, **18**, 111–130.
Hoffman, H., and Frank, M. E. (1961). *J. gen. Microbiol.*, **25**, 353–363.
Hoffman, H., and Frank, M. E. (1963). *J. Bact.*, **85**, 1221–1234.
Hoffman, H., and Frank, M. E. (1965). *J. Bact.*, **90**, 789–795.
Hort, E. C. (1920). *J. Hyg. Camb.*, **18**, 361–368.
Hu, F. N., Holmes, S. G., Pomerat, C. M., Livingwood, C. S., and McConnell, K. P. (1951). *Tex. Reps. Biol. Med.*, **9**, 739–748.
Ingraham, J. L. (1958). *J. Bact.*, **76**, 75–80.
Jebb, W. H. H., and Tomlinson, A. H. (1960). *J. gen. Microbiol.*, **22**, 93–101.
Kahn, M. C. (1929). *Am. Rev. Tuberc.*, **20**, 150–200.
Kelly, C. D., and Rahn, O. (1932). *J. Bact.*, **23**, 147–153.
Kjeldgaard, N. O. (1967). *Adv. microbiol. Physiol*, **1**, 39–95.
Knaysi, G. (1940). *J. Bact.*, **40**, 247–253.
Knöll, H., and Zapf, K. (1951). *Zentbl. Bakt. Parasit Kde, Abt I. Originale*, **157**, 389–406.
Koopmans, R. K. (1960). *Antibiotics Chemother.*, **10**, 612–622.
Lark, K. G., and Maaløe, O. (1954). *Biochem. biophys. Acta*, **15**, 345–356.
Lederberg, J. (1956). *Genetics*, 41, 845–871.
Lwoff, A., Dulbecco, R., Vogt, M., and Lwoff, M. (1955). *Virology.*, **1**, 128–139.
Maaløe, O. (1962). *In* "The Bacteria" (Ed. I. C. Gunsalus and R. T. Stanier), Vol. 4, pp. 1–32. Academic Press, New York.
Mackaness, G. B. (1952). *J. Path. Bact.*, **64**, 429–446.
Mackie, T. J., and van der Lingen, S. (1924). *J. Path. Bact.*, **27**, 134–136.
Mahoney, D. E., and Chadwick, P. (1965). *Can. J. Microbiol.*, **11**, 829–836.
Mahoney, D. E., and Chadwick, P. (1966). *Can. J. Microbiol.*, **12**, 699–702.
Makela, O., and Nossal, G. J. V. (1961a). *J. Immun.*, **87**, 447–456.
Makela, O., and Nossal, G. J. V. (1961b). *J. Immun.*, **87**, 457–463.
Makela, O., and Nossal, G. J. V. (1962). *J. exp. Med.*, **115**, 231–243.
Malone, R. H. (1918). *J. Path. Bact.*, **22**, 222–223.
Mason, D. J., and Powelson, D. M. (1956). *J. Bact.*, **71**, 474–479.
Nossal, G. J. V. (1958a). *Br. J. exp. Path.*, **39**, 544–551.
Nossal, G. J. V. (1958b). *Aust. J. exp. Biol. med. Sci.*, **36**, 235–244.
Nossal, G. J. V. (1959). *Br. J. exp. Path.*, **40**, 301–311.
Nossal, G. J. V. (1960). *Br. J. exp. Path.*, **41**, 89–96.

Nossal, G. J. V., and Makela, O. (1962a). *J. Immun.*, **88**, 604–612.
Nossal, G. J. V., and Makela, O. (1962b). *A. Rev. Microbiol.*, **16**, 53–74.
Ørskov, J. (1922). *J. Bact.*, **7**, 537–549.
Pearce, T. W., and Powell, E. O. (1951). *J. gen. Microbiol.*, **5**, 91–103.
Pederson, C. S., Yale, M. W., and Eglinton, R. (1933). *N. Y. St. agric. exp. Stn. tech. Bull.*, No. 213.
Pijper, A. (1949). *Symp. Soc. gen. Microbiol.*, **1**, 144–179.
Pijper, A. (1951). *Nature, Lond.*, **168**, 748–749.
Pijper, A., and Abraham, G. (1954). *J. gen. Microbiol.*, **10**, 452–456.
Pomerat, C. M. (1951). *In* "Methods in Medical Research" (Ed. Visscher, M. B.), Vol. 4, pp. 275–277, Year Book Publishers, Chicago.
Postgate, J. R. (1967). *Adv. microbial Physiol.*, **1**, 1–23.
Postgate, J. R., Crumpton, J. E., and Hunter, J. R. (1961). *J. gen. Microbiol.*, **24**, 15–24.
Powell, E. O. (1956a) *Jl. R. microsc. Soc.*, **75**, 235—243.
Powell, E. O. (1956b). *J. gen. Microbiol.*, **15**, 492–511.
Powell, E. O. (1956c). *J. gen. Microbiol.*, **14**, 153–159.
Powell, E. O. (1958). *J. gen. Microbiol.*, **18**, 382–417.
Pryce, D. M. (1941). *J. Path. Bact.*, **53**, 327–334.
Pulvertaft, R. J. V. (1952). *J. Path. Bact.*, **64**, 75–89.
Pulvertaft, R. J. V., and Haynes, J. A. (1951). *J. gen. Microbiol.*, **5**, 657–663.
Pulvertaft, R. J. V., Haynes, J. A., and Groves, J. T. (1956). *Expl Cell Res.*, **11**, 99–104.
Quesnel, L. B. (1960), *J. appl. Bact.*, **23**, 99–105.
Quesnel, L. B. (1961). Ph.D. Thesis, University of Bristol.
Quesnel, L. B. (1962). *J appl Bact.*, **25**, iii.
Quesnel, L. B. (1963). *J. appl. Bact.*, **26**, 127–151.
Quesnel, L. B. (1966). *Nature, Lond.*, **211**, 659–660.
Richter, K. M., and Woodward, N. W. (1955). *Expl. Cell Res.*, **9**, 585–587.
Rose, G. (1954). *Tex. Rep. Biol. Med.*, **12**, 1074–1083.
Ross, K. F. A. (1967). "Phase Contrast and Interference Microscopy for Cell Biologists". Edward Arnold, London.
Schmitt, F. O., and Schmitt, O. H. A. (1931). *Science, N.Y.*, **73**, 289–290.
Schwöbel, W. (1954). *Expl. Cell Res.*, **6**, 79–81.
Schwöbel, W. (1955). *Expl. Cell Res.*, **9**, 583–585.
Sharp, J. A. (1959). *Expl. Cell. Res.*, **17**, 519–521.
Sherris, J. C., Preston, N. W., and Shoesmith, J. G. (1957). *J. gen. Microbiol.*, **16**, 86–96.
Showacre, J. L. Hopps, E. H., Du Buy, H. C., and Smadel, J. E. (1961). *J. Immun.*, **87**, 153–161.
Stempen, H. (1950). *J. Bact.*, **60**, 81–87.
Sykes, J. A., and Moore, E. B. (1959). *Proc. Soc. exp. Biol. Med.*, **100**, 125–127.
Sykes, J. A., and Moore, E. B. (1960). *Tex. Rep. Biol. Med.*, **18**, 288–297.
Thiele, H., and Wolf, K. (1907). *Arch. Hyg. Bakt.*, **60**, 29–39.
Thomas, F. L. X., and Cramer, L. M. (1966a). *Tex. Rep. Biol. Med.*, **24**, 700–711.
Thomas, F. L. X., and Cramer, L. M. (1966b). *Tex. Rep. Biol. Med.*, **24**, 712–719.
Toy, B. L., and Bardawil, W. A. (1958). *Expl. Cell. Res.*, **14**, 97–103.
Vischer, W. A. (1956). *Schweiz. Z. Path. Bakt.*, **19**, 560–566.
Ware, G. C. and Loveless, J. R. (1959). *J. appl. Bact.*, **21**, 309–312.
Webb, R. B. and Malina, M. M. (1967). *Science., N.Y.*, **156**, 1104–1105.

Webb, S. J. (1959). *Can. J. Microbiol.*, **5**, 649–669.
Webb, S. J. (1967). *Can. J. Microbiol.*, **13**, 733–742.
Weibull, C. (1960). *In* "The Bacteria" (Ed. I. C. Gunsalus and R. Y. Stanier), Vol. 1, pp. 153–205. Academic Press, New York.
Wilkinson, J. F. (1959). *Expl Cell Res.*, **7** (suppl.), 111–130.
Wyckoff, R. W. G., and Lagsdin, J. B. (1933). *Rev. scient. instrum.*, **4**, 337–339.

NOTE ADDED IN PROOF: An improved design of the chamber of Cruickshank *et al.* (1959) (see this Volume, p. 387), moulded from "Nunclon" plastic, is readily available from Stenlin Ltd, Richmond, Surrey, England, at very low cost.

CHAPTER XI

Solid and Solidified Growth Media in Microbiology

R. C. CODNER

School of Biological Sciences, Bath University of Technology, Bath, England

In a liquid environment, micro-organisms may either grow as single cells and small aggregates of cells dispersed in the liquid, or as films and colonies on the interfaces bounding the liquid, including the solid-liquid interfaces and particularly the gas-liquid interface. If solid materials capable of supplying the nutrient requirements of micro-organisms contain more than a critical proportion of water they become subject to microbial attack and may support the growth of micro-organisms on surfaces exposed to the surrounding gas phase. Solid substrates may also support the growth of anaerobic organisms mixed with or occluded by the solid. It is not usually considered that micro-organisms multiply in a purely gaseous environment, though the existence of an air-inhabiting "plankton", as opposed to a fortuitously airborne microbial flora, cannot be entirely ruled out (Gregory 1961). It is largely as a result of the manipulation of these environments for the growth of micro-organisms that the science of microbiology has advanced.

One of the early major steps in this manipulation was the introduction of methods for the controlled growth of micro-organisms on or in solid or gelled substrates. This enabled progeny from the divison of a single micro-cell to develop in spatial isolation from the progeny of other cells in the environment. By such spatial separation on air/gel interfaces, Robert Koch was able to obtain pure cultures with facility, first using cut surfaces of boiled potato and later (1881) by using flat glass plates onto which thin layers of meat bouillon containing dissolved gelatin were poured and allowed to set. Serum coagulated by heat was introduced by Koch in the following years for cultivating the tubercle bacillus while a little later, at the suggestion of Frau Hesse, agar-agar was adopted as a gelling agent by Koch. In this early work the layers of gels on flat glass sheets were covered by bell jars to exclude airborne dust particules but in 1887 at the suggestion of R. J. Petri the flat glass sheets were replaced by the familiar shallow covered glass dishes which today bear his name. The use of agar gels in Petri dishes is the basis of a high proportion of microbiological methods in use today. Accounts of this early work may be found in Bullock (1938) and Brock (1961), the modern preparations of agar and gelatin and their use is described by Bridson and Brecker (this Series, Vol. 3A). The formation of colonies from individual micro-organisms present in liquids or gases enables these viable micro-organisms to be enumerated without resort to tedious and often difficult microscopic observation. The examination of the colonial characters and morphology on solid surfaces provides useful features for use in the identification and classification of micro-organisms, while solid media may be designed which throw into prominence colony characteristics of certain organisms or which selectively permit colony formation by certain groups of organism. The reaction of the metabolic products of one micro-organism on another may have either stimulatory or antagonistic effects which can be directly observed and measured on solid media.

The growth rate of micro-organisms on solidfied substrates may be compared by direct measurement of colony dimensions or, with fungi, the extension of individual hyphae growing on a plane surface may be measured microscopically. In the early stages of colony formation by bacteria and yeasts, growth rate may be followed by estimating numbers of cells present.

The use of cultures on solidified media for inoculum production has found favour particularly with fungi and actinomycetes in which sporulation is stimulated by the emergence of the hyphae from the submerged to the aerial condition. Some microbial products have been prepared by using solid particulate substances.

Some of these aspects of growth on solid substrates are discussed below while other areas such as the development of selective media are dealt with elsewhere in this Series (Veldkamp, Vol. 3A).

I. INTERFACIAL ENVIRONMENTS

A. Effects of interfaces on growth

Aside from the purely technical aspects of using solid or solidified substrates, the interfaces between liquid and gas or liquid and solid or the lines at which all three phases meet represent very special microbial environments. The effect of solid surfaces on bacterial activity has been discussed by ZoBell (1943) who pointed out that the rate of bacterial multiplication and the rate of oxygen uptake was higher in samples of sea water in contact with a greater surface area of glass than in samples in contact with smaller areas of glass. He concluded, however, that these beneficial effects of solid surfaces are usually only evident in environments low in nutrient substrates (less than 10 mg/litre). One explanation put forward by ZoBell for these observations was the absorption and concentration of limited nutrients on the solid surfaces where they become more readily available to the micro-organisms. He was able to demonstrate that from 2% to 27% of the oxidizable organic matter in sea water could be absorbed onto glass surfaces varying from 2 to 200 cm^2/ml of sea water. It was also clear from this work that more micro-organisms can occur on the glass surfaces of containers than occur free in the water occupying them. The distribution of the exposed surface in the dilute medium affected the efficiency of the growth promotion; thus glass tubes placed in water in such a way that no part of the water was more than 3 mm from a glass surface were more effective in promoting bacterial growth in sea water than a layer of glass balls 2 mm in diameter, having an equivalent surface area to the tubes, lying on the bottom of the container. Plastics were reported to vary in the rate at which bacteria became attached to them in sea water and to vary in the accumulation of organic matter on their surfaces. Other reports quoted by ZoBell indicate that glazed and unglazed porcelain, asbestos and emery grit promote bacterial activity in dilute nutrient solutions, while other reports, reviewed by ZoBell, indicate stimulation of bacterial metabolic activity with various clays.

The beneficial effects of solid surfaces in bacterial cultures, it is suggested, arise not only from the absorption of substrates but also from the retention of exoenzymes and hydrolysis products produced by the cultures at a position close to the bacterial cells.

In soil, the presence of strongly absorbing surfaces on which microbial activity takes place is likely to be of even greater significance than in the situations discussed by ZoBell. Enzymes, bacteria and enzyme or bacterial substrates become absorbed to soil colloids and soil surfaces. This type of situation has been examined by Eastermann and McLaren (1959) who point out that added absorbents may have a variable effect on the metabolic activities of bacteria. Clay minerals may absorb proteins in their lattice layers thus preventing proteolysis by enzymes or bacterial cells. Similarly, phosphatases may be inhibited by the presence of clay minerals in the system. However, in other reports quoted and in the experimental work described by Eastermann and McLaren the rate of proteolysis is shown to be enhanced in a system where protein is adsorbed on Kaolinite. This effect occurs both with organisms which adsorb strongly to the Kaolinite (such as *Flavobacterium*) or which are not adsorbed. (*Pseudomonas* sp.) The enhancing effect does not occur with protein substrates modified so that they are not adsorbed to the clay. With dilute inocula of *Flavobacterium* the rate of proteolysis was enhanced when the substrate was adsorbed on Kaolinite but with concentrated inocula the enhancement due to adsorption disappeared.

Rem and Stotzky (1966) confirmed the stimulation of respiration of bacteria by Montmorillonite. They suggest that while the mechanisms involved are complex, one factor involved is the maintainence of pH at levels adequate for sustained growth by ion exchange mechanisms. The effect of Montmorillonite does not appear to be due to any extent to the supply of inorganic ions since Kaolinite is effective for this purpose and the related Montmorillonite shows a stimulation greater than does Kaolinite. Montmorillonite shows its effect on a wide range of bacterial species and at various stages in the growth cycle but the effect is particularly marked in the shortening of the lag phase. Stotzsky (1966a) showed that the maintenance of pH by clay particles is dependent both on the initial pH of the system and the buffering activity of the clay particles. Certain other particulate materials possessing ion exchange properties such as polystyrene latex, glass beads and ion exchange resins did not affect respiration of *Agrobacterium radiobacter* as did clays of the Montmorillonite type.

Stotzsky (1966b) showed that stimulation of the respiration of *A. radiobacter* by clays may be related to the cation exchange capacity of the clay. The particle size of the clay did not appear to be a critical factor. Correlation between surface area of the clay and bacterial stimulation was not established unequivocally.

The use of Zeolites to promote an increase in the attack of micro-organisms on hydrocarbon substrates is described in a U.S. Patent (1965). The hydrocarbon is finely dispersed in the aqueous medium to which zeolite with a pore size of 3–10 Å is added.

The author (unpublished observation) has shown that the sporulation of *Streptomyces* cultures on complex agar media is stimulated by the inclusion of magnesium or calcium trisilicates.

B. Observation of growth and aggregation at interfaces

Hoffman (1964) discusses the morphology of bacterial aggregates at interfacial environments with particular reference to dental plaque, the thin film of zoogleal material on the surface of teeth. This review points out that the study of such microbial aggregates presents problems not only in the methods of observation but also in examination of biological factors concerned in the environment and the characters of the micro-organisms, involved in formation of the aggretates.

Methods of observation of microbial growth at interfaces show an increasing degree of complexity from the hanging agar block method of Hill (1902) who used a small cube of agar in place of a hanging drop suspended from a cover slip in a moist chamber. The inoculum was placed on the surface of the agar block which adheres to the cover slip.

Fleming, *et al.* (1950) allowed a drop of molten agar to set on a previously inoculated sterile cover slip. The agar covered surface of the cover slip is then sealed onto a slide for observation. Pearce and Powell (1951) used a metallurgical microscope providing both oblique and vertical illumination for observation of microbial growth on agar surfaces either in Petri dishes or in Conway diffusion units in which growth on agar in the centre well is kept humid by water placed in the surrounding annulus. This paper also discusses the growth of micro-organisms on a stretched Cellophane membrane which, during growth of the culture but not during observation is in contact with nutrient solution. Developing further the application of Cellophane in this work, Powell (1956) described a micro-culture chamber consisting of a stretched Cellophane film on the upper surface of which cells forming the inoculum are placed and beneath which nutrient solution is circulated by a micro-pump. The Cellophane presses the cells against a cover glass thus keeping the cells co-planar. While this restriction is advantageous for observation of the developing cells, it may impose restraint on the growth and post-fissional movement of the cells and the developing micro-colony (see Quesnel, this Volume p. 365).

Campbell (1968) gives details of an apparatus for the observation of fungal development based on the device of Powell (1956). In this modification an interfacial system consisting of an Oxoid MF 50 cellulose acetate membrane filter is perfused on the lower side with a flow of liquid medium while the fungus growth on the upper surface of the membrane has a suitable gas stream continuously drawn past it at a measured rate, thus controlling both the liquid and gaseous environments.

The importance of emergence from the liquid phase into the gas phase as the most powerful stimulus to sporulation in the strains of *Penicillium* and *Aspergillus* studied so far has been emphasised by Morton (1961). It is stated that, "whereas submerged sporulation occurs, if at all, only in response to narrowly defined conditions of the composition of the medium aerial sporulation takes place readily and abundantly irrespective of the nature and composition of the medium in which the fungus has been growing". The stimulus to sporulation provided by breaking through the air–liquid interface is reversed if the mycelium is transferred back from aerial to submerged conditions. The emergence of the mycelium into the air phase causes physiological changes of which one manifestation is the ability of the aerial mycelium to assimilate ammonia in the absence of an external carbohydrate source. Even in aerial conditions the mycelium is surrounded by a water film but it is suggested that this air–water interface may be sufficiently close to the cell wall to allow the unfolding of globular protein molecules of the cell wall in the air–water interface. Morton further tentatively postulated that an initial stimulus might be transmitted from unfolding molecules near the interface through molecules in the cell wall to initiate metabolic changes in the cell itself.

C. Adhesion and attachment of micro-organisms to interfaces

Interfacial phenomena governing the adhesion of the unicellular alga *Chlorella* to glass have been examined by Nordin *et al.* (1967). The adhesion of this alga to glass does not occur under most conditions as both alga and glass have negative zeta potentials, but if ferric chloride was added to the algal suspension in 0·05 M sodium chloride, the alga and the glass acquire different zeta potentials and adhesion increases as the difference in zeta potentials increases. Adhesion is stated to be strongest at between 2 and 3×10^{-5} M ferric chloride.

It is also suggested that other "local conditions" on the glass surface are involved in adhesion and that the cell surface may not have uniform properties of adhesion since algal cells were occasionally observed to oscillate about a single point of attachment. This phenomenon of the adhesion of algal cells is of practical importance in water treatment for domestic water supplies and in photosynthetic gas exchangers.

The holdfast structure of the caulobacters represents a peculiar adaptation for attachment of these organisms at solid–liquid interfaces where nutrient concentration is increased by adsorption. It is also suggested that the holdfast adhesion is the enabling mechanism for the ectocommensal existence lead by these organisms in the natural environment (Poindexter, 1964).

The attachment of caulobacters to solid surfaces has been exploited for their demonstration in the floated cover slip technique of Board (1967). Poindexter (1964) pointed out that the holdfast structure of the stalked bacteria is also involved in the mutual adhesion of cells of these organisms to form pellicles at the air–liquid interface where cells congregate as a result of aerotactic movement during the motile phase of growth. Bacterial fimbriae, of widespread occurrence in the Enterobacteriaceae, are similarly involved in the adhesion of these organisms to many substrates. In suspensions of red blood cells most fimbriate strains cause haemagglutination. Strains of *Klebsiella* (*Aerobacter*) *aerogenes* which have thick fimbriae with an adhesive property which is subject to inhibition by D-mannose, will also adhere to cells of *Candida albicans*, to mycelium of fungi including *Aspergillus niger*, to plant root hairs and to glass and cellulose fibres (Duguid, 1959). The association of the presence of fimbriae with the ability to form pellicles at the air liquid interface in cultures of *Shigella flexneri* has been discussed by Duguid and Wilkinson (1961). The advantage of this property to shigellas, which are strict parasites and which do not normally encounter the stagnant poorly aerated aqueous environments occupied by the saprophytic enterobacteria is not clear.

Haemagglutinating fimbriate salmonellas, in which adhesion is inhibited by mannose have been shown by Old *et al.* (1968) to produce surface pellicles in static liquid culture after the cessation of the log phase and to give rise by multiplication of organisms in the pellicle to a prolonged phase of renewed growth. Non-fimbriate forms and forms showing type 2 non-haemagglutinating fimbriae did not produce pellicles nor did they show a marked post-logarithmic growth phase. Cultures of *Salmonella typhimurium* bearing mannose-sensitive fibriae show a 24 h delay in the formation of a pellicle and the onset of the secondary growth phase in the presence of 0·2% mannose and methyl mannoside which is not metabolized. A much shorter delay in pellicle formation of 2–3 h was shown in a mannose utilizing strain in the presence of mannose than in non-mannose utilizing strains. In anaerobic culture or in shaken aerobic cultures where pellicle formation was suppressed there was no secondary phase of growth and it was concluded that in pellicle forming fimbriate salmonellas the active secondary phase of growth was due to the increased availability of atmospheric oxygen to organisms supported in the pellicle.

II. COLONY CHARACTERISTICS

The characteristic features of colonies play a considerable part in the identification of bacteria and moulds. The description of bacterial colonies may be standardized by using the terms recommended by the Society of American Bacteriologists (1957) or by Wilson and Miles (1962). Standardized descriptive charts for fungus colonies have been set out for aspergilli (Raper and Fennell, 1965) and for penicillia (Raper and Thom, 1949). For yeasts many workers give descriptions of colony forms either from streaks and giant colonies on malt agar or malt gelatin, but Lodder and Van Rij (1952) state, "In contrast to many earlier investigators we attach little value to this character for differentiating purposes".

The hyphal structures and modes of branching of the hyphae which contribute to the colonial form of filamentous fungi have been discussed by Butler (1966). Many of the factors contributing to the morphology of microbial colonies are little understood. Even where colonies are grown on a homogeneous substrate such as an agar gel the situation during growth is made complex since the early stages of the growth of the colony modify the surrounding medium over which later growth will take place. This occurs both by depletion of nutrients by diffusion towards the centre of the colony and by diffusion outwards from the colony of products of metabolism. Such considerations may give rise to the concentric zonation often seen in mould colonies even where diurnal fluctuations of light and temperature are as far as possible eliminated. This formation of concentric zones in colonies may be analagous to the physico-chemical phenomenon of the formation of banded precipitates in agar, gelatin or silicate gels known as Liesegang rings (see Glasstone, 1962, for a discussion on this phenomenon).

Attempts have been made by examining bacterial colonies using the culture techniques outlined above (IB) to correlate the mode of cell division, the post fissional movements of the cells and the early stages of microcolony formation so adequately described by Graham-Smith (1910), with the ultimate macroscopic appearance of the colonies. The medusa head colonies of *Bacillus anthracis* and the rhizoidal colonies typical of *Bacillus mycoides* depend on the cells remaining in chains and showing simulataneous growth in the entire chain. This growth produces looped chains which influence the ultimate surface configuration of the macro colony. (Bissett, 1963). Similar chainlike growth of the cells may give rise to rough colonial forms in streptococci though it must be borne in mind that the presence or absence of capsular substances is a factor of fundamental importance in determining the appearance of colonies of streptococci and pneumococci. Hoffman and Frank (1961) discuss the formation of microcolonies in relation to the effect of the palisading movement of one cell of a daughter pair alongside another. They conclude that for *Escherichia coli* in both rough and smooth phases the orderly arrangement of the cells in the microcolonies is established primarily by palisading due to the sliding of one cell over another in smooth colonies while in rough forms palisading due to buckling of cell chains is occasionally found. When growth of *E. coli* takes place at 44°C palisading may be inhibited with a resulting decrease in the degree of linear arrangement of cells within the colony and the formation of an irregular network of loosely packed cells.

A further contribution to the colonial character of micro-organisms is sometimes made by the formation of sectors and papillae. It was pointed out by Dean and Hinselwood (1957) that in the light of a study of factors influencing papilla formation this phenomenon could not be explained by mutation changes. The formation of sectors however may be satisfactorily accounted for in terms of genetic mutation.

The smooth and rough colonies encountered in yeasts appear, at cellular level, to be associated in some instances with the degree to which the cells remain attached together after the bud is fully formed, and also with the extent to which the yeast forms pseudomycelial cells.

The situation may be further complicated by differences in behaviour between haploid and diploid colonies of a single yeast species. This and other aspects of colonial morphology in yeasts have been discussed by Morris (1966) and by Ingram (1955).

The arrangement of cells in microcolonies has been discussed in relation to a possible rapid diagnostic method for *B. anthracis* by Chadwick (1963) who describes the characteristic microcolonial forms of the related *Bacillus cereus*, *Bacillus subtilis*, *Bacillus subtilis var. mycoides* and *Bacillus megaterium* and speculates on the causes of the differences in microcolonial structure in these organisms.

Finkelstein and Punyashthiti (1967) discuss colonial forms of enteric organisms on McConkey's agar as an aid to diagnosis using oblique illumination for examination under a low power stereoscopic microscope. These workers suggest that colonial morphology can be used as an adjunct to conventional methods of diagnosis and as a quick guide to the application of appropriate definitive diagnostic tests.

Fowell (1965) applied a similar approach to the identification of "wild yeasts contaminating baker's yeast using lysine agar. Fowell however, did not resort to oblique illumination though Kulka *et al.* (1951) had earlier recommended the use of oblique illumination in the photography of giant yeast colonies.

The general contours of colonies and the degree to which the centre of a colony may be raised above the surface of the solid medium on which it is growing depends to a large extent for aerobic organisms both on the diffusion of oxygen into the colony and the transport of nutrient into the colony from the substrate in relation to the requirements of the organisms. In this connection Pirt (1967) discusses a mathematical model, which accounts for the limited thickness to which bacterial and yeast colonies grow on solidified media. Predictions for the thickness attained may be made either from the oxygen requirement of the organism or from the diffusion of a limiting nutrient. Work of Lindegren and Hamilton (1944), suggests that in yeasts the translocation of nutrients into the colonies may be complicated by the presence of pseudomycelial threads which extend from the colony into the medium below. They also indicate that the surface layers of old colonies of *Saccharomyces cerevisiae* is composed of autolysed cells which surround a core of actively growing cells.

Morris and Hough (1956) discuss the structure of the yeast colony in relation to the type of budding shown by the cells in the colony. They point out that cells in colonies on malt gelatin show a budding form which corresponds with the type of budding found in liquid malt medium. Penetration of the agar by pseudomycelial strands was again reported, confirming the earlier work of Lindegren and Hamilton (1944). All these factors must bear on the ultimate morphology of the colony and would complicate any mathematical approach to the formation of giant colonies in yeasts.

In *Neurospora crassa*, mutants known as "clock" mutants have been described by Sussman *et al.* (1964) which show bands of regular periodicity, when grown on agar gel surfaces, formed by changes in the branching pattern of the mycelium from sparse monopodial branching of the wild type to alternating bands of dense cymose growth. Another similar rhythmic phenomenon in *Neurospora* is the "patch" mutant which gives rise to the formation of bands of sporulating growth as the colony extends. Berliner and Neurath (1965) suggest that the rhythm in *Neurospora* mutants has unusual non-circadian characteristics. The period of the rhythm, depending on the strain and environment, may vary from 18–110 h. The phase of the rhythm is not related to time of day and is not phased by light and dark cycles. Both the period of the rhythm and the growth rate of the organism are influenced by temperature but are different functions of temperature. The phase of the rhythm is set by transfer to fresh medium and in a later paper Berliner and Neurath (1966) report on the control of the periodic rhythm by the nature of the carbohydrate source; thus a "clock" mutant grown on 1% sucrose showed no zonation though this same strain showed well defined bands on dextrose and fructose and maltose.

Another morphological mutant of *N. crassa* demonstrable on a solid substrate is the "colonial" form which gives dense restricted colonies on the surface of agar while the normal "wild type" gives sparse loosely spreading mycelium on solid medium. It is reported that these "colonial" mutants, which are similar to the wild type when the latter is grown on sorbose-containing medium, show differences from the wild type in the composition of their cell walls. de Terra and Tatum (1963) showed significantly less glucose and more glucosamine per unit weight of cell wall in "colonial" than in "wild type" mycelium. This work was extended by Mahadevan and Tatum (1965) to an examination of the structural polymers of the cell walls of "wild type", "colonial" and sorbose induced colonial growth of *Neurospora*. It was concluded that changes in the proportion in the cell wall of a peptide–polysaccharide complex and a β-1, 3 glucan were involved in the modification of the overall morphology of the colonies.

Other facets of microbial growth which only become apparent when micro-organisms are grown on solid surfaces, are the formation of migrating and rotating colonies associated with some members of the genera *Proteus*, *Clostridium*, and *Bacillus*. These colonial movements are more obvious in colonies growing on agar media which have been incompletely dried. Murray and Elder (1949) point out that the colonies do not move at random, but show a predominantly counter-clockwise rotation. The terminal filaments of colonies of *Bacillus mycoides* may indicate either predominantly clockwise or predominantly counter-clockwise colony rotation. Predominantly clockwise strains outnumber counter-clockwise strains by 3 to 1. Yuill and Yuill (1954) gave an account of a similar phenomenon which occurs rarely in strains of *Aspergillus niger*. Among spiral colonies in this organism counter-clockwise left-handed forms are the more usual, though occasional clockwise strains were encountered. One such clockwise strain showed counter-clockwise growth on Cellophane laid on the surface of the agar and became clockwise as the hyphae grew off the edge of the Cellophane onto the exposed agar surface. Similar spiralling of the mycelial strands in *Cephalosporium* strains has been observed fairly frequently by the author (unpublished).

III. RELATION OF GROWTH RATE AND COLONY SIZE

The extent to which the measurement of growth by determining linear increase in colony dimensions of micro-organisms on solid media is justifiable, has been examined by a number of workers. This approach has been applied more widely to rapidly growing fungi than to bacteria, where the rate of colony extension is usually more limited and consequently less readily measured. With fungi it is also possible to obtain information on growth rate by direct measurement of the extension of individual hyphae under the microscope. Both these methods have the advantage of being simple and non-destructive. It is, however, well-known that the radial extension of fungal colonies on media in which one or more ingredients are deficient may be as rapid as on complete medium, but that the growth is more profuse, and branching is more extensive, on complete medium. This lack of agreement between colony extension and total growth is shown clearly in the work of Brancato and Golding (1953) in which colony diameter and colony dry weight were determined on agar of different depths with the results shown in Table I, from which it is clear that the use of colony diameter does not give a true indication of the relative growth when certain factors are varied.

However from other experiments with penicillia and aspergilli on the effect of temperature, pH, relative humidity and the concentration of sucrose, sodium chloride and ethylene gycol in the medium on colony extension, Brancato and Golding concluded that the increase in the diameter of the colony gives a satisfactory measurement of growth rate and pointed out that soon after germination a characteristic radial growth rate is established at optimum temperature.

Grogan and Purdy (1954) performed experiments on the nutrition of *Sclerotinia sclerotiorum* in which growth on deficient liquid media measured by fungal dry weight was compared with growth on deficient media solidified with agar measured by colony diameter. It was shown that while radial growth of colonies on agar deficient in K, Mg, P, and Fe was as rapid as on complete medium, the mycelium on the deficient media was sparse and thin. This situation was truly indicated by the lower dry weights of mycelium harvested from the deficient media without agar. No assessment was made in this work of the possible contribution to the nutrient status of the medium by the agar used.

TABLE I

Relation of thickness of agar to the diameter and dry weight of colonies on malt agar.

°C	Organism	Depth of Medium			
		4 mm		2 mm	
		Av. dia. mm	Av. dry wt. mg	Av. dia. mm	Av. dry wt. mg
35	*Aspergillus niger*	45·9	19·7	46·3	11·0
32·2	*Aspergillus flavus*	52·4	33·7	52·7	3·6
26·7	*Penicillium notatum*	27·8	14·3	28·5	12·4

From Brancato and Golding (1964).

Plomley (1959) discussed the growth of the fungus *Chaetomium globosum* in terms of hyphal growth, increase in colony size and change of density of the hyphae. Hyphal extension was measured using photographs taken at various time intervals. Plomley in agreement with the work of Smith (1923) found that the growth of hyphae takes place only at the tip ahead of the first cross wall, and that the apical segment is longer than segments behind it.

Hyphal growth rate remains constant for an initial period followed by a phase in which linear hyphal growth rate and colony extension become exponential. The mass of mycelium also increases exponentially as shown by an increase in hyphal density. In the third phase of colony development marginal growth occurs in such a way that colony extension rate remains constant, due to branching, while total hyphal extension is still exponential.

The phase of constant rate of extension of fungal colonies was also observed by Fawcett (1925) and by Beadle *et al.* (1943) who both measured colony extension in straight parallel sided tubes. It was again suggested that while total hyphal extension is intrinsically exponential, colony extension remains linear by virtue of branching. In circular colonies branches spread to fill all available agar surface behind the leading hyphae, but in parallel sided tubes it is suggested (Beadle *et al.*, 1943) that after attainment of a certain hyphal density at the advancing colony margin, a substance, inhibitory to branching, accumulates and suppresses branching thus causing the fungus to advance at constant hyphal density and at constant rate.

Bertrand *et al.* (1968) examined the problem of indefinite uninterrupted growth of *N. crassa* in continuous growth tubes to determine whether uninterrupted hyphal extension was possible and whether spontaneous mutants with supressive phenotypes could be expressed. It was found that two cultures from a common origin both showed stop-start growth behaviour. In one culture there were frequent stops of long duration (average 16 days) while in the second culture there were infrequent stops of shorter duration (average 2 days). The rate of growth between stops was also found to be extremely variable. Many of the significant increases and decreases in growth rates noted in these prolonged experiments lasting over a four year period, were not reproduceable in sub-cultures. However, certain of the growth rate changes were reproduceable and had a genetic basis with a suppressive phenotype. In the long term vegetative growth of *Neurospora* the experiments indicate that phenotypic changes are due to genetic changes at both cytoplasmic and nuclear levels.

Pirt (1967) proposed a mathematical model to account for the constant rate of increase in colony radius on solid media and measured the effect of several environmental factors on colony radial growth rate for various bacteria. Accurate measurement of the bacterial colonies was carried out using an enlarger and projection screen (Shadowmaster Buck and Hickman Ltd. Otters Pool Way, Watford, Herts; a device for measuring small engineering components). The colonies magnified to ten times were measured on the screen with a transparent ruler. It was shown that reducing the agar concentration from 1% to 0·6% did not affect growth rate but that increasing the depth of agar from 1 mm to 3·44 mm caused an increase in initial radial growth rate of *E. coli* but that further increase in depth was without effect. It was demonstrated that the initial radial growth shows a linear relationship with the square root of the initial glucose concentration over a considerable range which differed markedly for the three organisms tested from 0–1·25 g/litre for *Streptococcus faecalis* and 0–2·5 g/litre for *E. coli* to 0–5 g/litre for *K. aerogenes*. For *E. coli* the initial radial growth rate of colonies in air at 1 atm pressure bears a linear relationship to the square root of the glucose concentration over the range of glucose concentration up to 2·5 g/litre. When the same organism is grown in 100% oxygen at 1 atm pressure the initial radial growth rate is directly proportional to the square root of the glucose concentration over the range 1·7–10 g/litre suggesting that at higher glucose concentrations the availability of oxygen is the growth limiting factor. In this experiment oxygen at 1 atm. inhibited colonies at low glucose concentrations. It is also of interest that initial radial growth of colonies reaches zero at a low but finite glucose concentration termed the lag concentration which varies between 0·090 g/litre for *E. coli* and 0·005 g/litre for *St. faecalis* growing in air. At 10 g/litre glucose both *E. coli* and *St. faecalis* showed strong inhibition of colony growth due presumably to the formation of toxic products and it is suggested that the level of glucose in routinely used laboratory media may be limiting the growth of colonies of these organisms. It is also pointed out that the very minute colonies formed by lactic acid bacteria are usually attributed to their intrinsically lower growth rate but when compared at an initial glucose concentration of 1 g/litre the initial radial growth rate of *St. faecalis* is 23 μm/h whereas that of *E. coli* is 20 μm/h. This is taken as indication that the lower colony growth rate of *St. faecalis* is due to inhibition by the high glucose concentrations used.

Using *E. coli*, the effect of sulphanilamide and sulphanilamide plus para-aminobenzoic acid on the specific growth rate measured in liquid culture and the initial colony growth rate were determined and it was established that, within the limits of experimental error, the ratio of the initial colony growth rate to the square root of the maximum specific growth rate was a constant. Changes in maximum specific growth rate (α_m) brought about by change in temperature of incubation of glucose limited cultures were not, for *E. coli* and *K. aerogenes*, always reflected by changes in colony radial growth rate K_r though for *St. faecalis*, the ratio $K_r : \sqrt{\alpha_m}$ did not vary by more than 5% from the mean.

IV. MATERIALS OTHER THAN AGAR AND GELATIN FOR SUPPORT OF INTERFACIAL GROWTH

Among the inert gelling and support materials used for the growth of micro-organisms, to which the usual gelling agents agar and gelatin are inimical, the most widely used is silica gel. This material was used with success by Winogradsky (1891) in the study of the nitrifying bacteria. The method of preparing silica gels used by Winogradsky and other workers involved the addition of hydrochloric acid to sodium or potassium silicate solution followed by the removal of the salt formed by dialysis until chloride was no longer detectable in the dialysate. The silica sol which was stable to sterilization was then gelled by the addition of a small amount of concentrated sodium chloride solution and at the same time the other medium ingredients were added at suitable concentration.

Subsequently a method of forming the gel by adding acid to sodium or potassium silicate to give a pH close to 7 was used by several workers. The gel was formed in a Petri dish and freed of salt by diffusion into water, followed by infiltration with nutrient solution by diffusion. The nutrient gel could be sterilized by autoclaving. The use of *ortho*-silicic acid tetraethyl ether as a starting material for the preparation of a silica sol by hydrolysis has been suggested by Ingleman and Laurell (1947).

More recent techniques for the preparation of silica gels have used either Ludox (E. I. Dupont de Nemours and Co. Inc. Wilmington, Delaware), a colloidal solution containing some 30% of hydrated silica with a small amount of alkali for stabilization, or sodium silicate as starting materials from which the purified silica sol is obtained by ion exchange processes. (Taylor, 1950; Kingsbury and Baghoorn, 1954; Pramer, 1957).

Kingsbury and Baghoorn (1954), starting with Ludox as silica source, used a mixed bed resin system, composed of the strong base exchange resin. Amberlite IR–120 (Rohm and Haas Co. Philadelphia, Pennsylvania) and the weak anion exchange resin Amberlite IR–45, in the preparation of the silica sol. The column was regenerated to get the resins in the appropriate form before application of the Ludox. An excess of the anion exchange resin was used so that the cation exchange resin become saturated first whereupon the pH of the effluent showed a sharp change from pH 2–3 to above pH 7 as the sodium ions broke through. This change was readily detected by checking the eluate with Congo Red. Pramer (1957), starting with reagent grade *meta*-sodium silicate, produced the silica sol by using a cation exchange column of Amberlite IR–120. Stable silica sols were prepared from the sodium silicate solution containing 1·5% SiO_2 by collecting fractions with a pH of 3·4 or less or with silicate solutions containing 3% SiO_2 by collecting eluate fractions with a pH or 3·0 or less.

Pramer examined the influence of a number of factors on the sol–gel transformation in relation to the time required for gelation and teh extent of syneresis during subsequent incubation. Factors investigated included the concentration of silica, pH, temperature, varying sodium chloride concentration and the effect of a number of single salts at a single concentration (M/30). Silica sol containing 1·5% SiO_2 with M/30 sodium chloride gelled most rapidly at pH 6–7. Gelation was retarded at pH values outside this optimum range, the retarding effect being greater at high than at low pH. The time required for gel formation varied inversely with temperature and directly with sodium chloride content. Syneresis, which is minimal at 1% SiO_2 increases with increasing silica concentration up to 3%. Syneresis also increases slightly with increasing pH between 4 and 7. The ability of sols to withstand autoclaving without gelling or becoming opalescent during treatment is related to the pH of the silica sol. Sols containing 3% SiO_2 at pH 2·0 remain clear and do not gel too rapidly for subsequent manipulation when autoclaved for 15 min. at 15 lb/sq. in. After autoclaving, the pH of the sol is adjusted with sterile caustic soda and plates are prepared from equal volumes of silica sol (3% SiO_2) and a separately sterilized solution containing appropriate medium ingredients. The strength or hardness of the gel may be increased by using increased sodium chloride concentrations, by using the lowest acceptable pH in the final gel and by using high silica content sols (Kingsburg and Baghoorn 1954). With stronger gels the problem of cracking of the gel increases.

In addition to studies on nitrifying organisms, silica gel has been used *inter alia* in the cultivation of halophils (Moore, 1940) in the demonstration of cellulose decomposers (Waksman and Carey, 1926) and in a viable count technique for *Mycobacterium tuberculosis* (Mitchison and Selkon, 1957).

Other materials used to provide physical support for the growth of micro-organisms include amorphous diatomaceous silica (Meyer, 1966) used for the growth of fungi by placing 10 Hyflo Supercel (Johns Manville Ltd) in a 10 cm plate together with 20 ml aqueous medium. Excess water is removed under a heat lamp until the surface of the slurry assumes a dull matt appearance. The use of vermiculite as a support material for the growth of *Aspergillus oryzae* in the production of amylase has been described by Meyrath (1965) while Bindal and Sreenivasaya (1945) tested asbestos powder as a support medium in the production of diastase.

In earlier methods for the induction of sporulation in yeasts a variety of solid support substrates were used including blocks of gypsum, vegetable wedges, filter paper and blocks of wood and fire-brick. These methods have been reviewed by Mrak and Phaff (1949).

Polypropylene moulding granules that float on aqueous media and that withstand autoclaving at 15 lbs/sq. in. have been used for supporting surface pellicles of fungi on liquid media. The use of polyacrylamide gels to support the formation of microbial colonies has been shown to be feasible (J. R. Norris, personal communication). These gels withstand autoclaving and may be freed of soluble impurities by diffusion into water prior to soaking in the required nutrient solution.

V. PRODUCTION OF METABOLITES ON MOISTENED NATURAL SUBSTRATES

While the present trend in fermentation technology is towards the production of metabolic products in sumberged aerated culture it has for certain products been found convenient to use particulate natural materials as the growth medium. The production of fungal amylases is one important example of the use of a solid particulate substrate, moistened bran, for the growth of selected strains of *Asp. oryzae* strains. Processes have been devised for the production of the mould bran on laboratory, pilot plant and full plant scale. Small quantities of mould bran may be prepared for laboratory use with full aseptic control in plugged conical flasks. Wheat bran, moistened with approximately its own weight of water, is placed in layers about 1 cm deep in the flasks which are sterilized by autoclaving. The sterilized bran is inoculated either with a spore suspension or by loop, and incubated. This method has been scaled up using covered aluminium trays in a process described by Roberts *et al.* (1944) who used layers of moist bran up to 1 in. deep, for *Asp. oryzae*. On a larger scale, where gas exchange problems involve the use of forced aeration and where strict aseptic precautions are more difficult, the problem of contamination arises. Growth of contaminants may be suppressed to a large extent by acidifying the bran before sterilization to pH 3·5–4 with 0·1 N to 0·3 N hydrochloric acid. This acidification technique described by Underkofler *et al.* (1939) was also covered by patents by Underkofler (1942) and by Christensen (1944). Using acidified sterile bran, attempts were made to scale up the production of mould bran by using rotating drums through which humidified air was passed thus supplying oxygen to the mass of tumbled bran. While this method was satisfactory at the 5 gallon laboratory drum scale, the process proved unsatisfactory on any larger scale due to mechanical damage to the young mycelium. (Underkofler *et al.*, 1947).

A method of growing the mould on sterile acidified bran in modified aluminium saucepans with the base perforated with a number of $\frac{1}{8}$ in. dia holes and with an air inlet in the lid, proved satisfactory (Hao *et al.*, 1943) provided that the temperature rise could be controlled by adjusting the air flow rate, and kept below 45°C. Aeration was continued throughout the growth period which varied from 12–24 h. After removal from the pan, the mould bran was air dried at room temperature.

The perforated saucepan method was modified for pilot plant and plant scale operation, by using incubation cells described by Underkofler *et al.* (1947) in which sterile acidified bran, inoculated with the fungus, was confined in vertical or inclined boxes, two sides of which were made from woven wire mesh (8 mesh/in) through which air could be forced in either direction. The direction of flow of the air through the bran was controlled by the manipulation of sliding "dampers" in the appropriate air inlets and outlets. For maximum enzyme production using incubation cells of this type temperatures were held between 32° and 38°C, and temperature and moisture content of the bran were interdependent variables. For the above temperature range optimum moisture content for enzyme production was obtained using 8 parts by weight 0·1 N hydrochloric acid and 10 parts by weight of bran. This gave a final moisture content of 51% taking account of the 12% moisture content of the bran used. Incubation cells of varying sizes were tested and it was considered impracticable to use cells larger than 6 × 6 × 1 ft thick. For a 1 ft thick bran layer the air volume required reached a peak of 52 cu. ft/h/cu. ft. of bran at about 15 h with an average of 32 cu. ft/h/cu. ft. of bran. 1 cubic ft of bran represents about 12 lbs of dry bran processed. The inlet air was 85–90% saturated with moisture at a temperature of 30°–32°C. Higher amylase activity could be produced in the bran by secondary incubation after removal of the mould bran from the incubation cells. The material was broken into pieces not more than 1 in. in diameter and placed in containers with perforated bottoms for secondary incubation for a further period of 18–20 h in a current of air in layers not more than 2–3 ft deep. For this secondary incubation stage humidified air at 26°–30°C was used.

Using wire mesh sided incubation cells on the large scale for primary incubation was not entirely satisfactory, owing to the difficulties encountered in temperature control and the need for very critical adjustment of the air flow which was often made difficult by shrinking of the bran. With improved design of handling equipment and methods Underkofler *et al.* (1947) reported the successful operation of a large scale tray unit with trays 5 × 14 ft hinged along one edge for ease of discharge, with bran layers up to 2 in. deep. The trays were set up in tiled ventilated incubator rooms. One such plant at Eagle Grove, Iowa had an output of 10 tons per day. Jefferys (1948) also described a mechanized tray production unit in which growth of the mould took place in an aeration tunnel housing some 40 trucks holding 5000 lbs of substrate.

One problem common to solid substrate large scale mould bran production units is the preparation of sufficient inoculum and the development of suitable methods of applying the inoculum to the large quantities of moist sterile bran. (Underkofler *et al.*, 1947).

Inoculum for amylase production by *Asp. oryzae* on bran was prepared by growing spore cultures on a mixture of 10 g ground maize and 100 g wheat bran moistened by 60 ml of 0·2 N hydrochloric acid, containing 0·62 ppm $ZnSO_4.7H_2O$; 0·63 ppm $FeSO_4.7H_2O$ and 0·08 ppm $CuSO_4.5H_2O$. In laboratory scale studies this material was distributed in 10 g quantities in 250 ml conical flaskes and sterilized by autoclaving. After inoculation, the flasks were incubated on their sides to expose a large surface of the bran to the air. The sporing culture dries out during incubation and for laboratory enzyme production cultures, the sporing culture was used at a level of 1% for inoculating the moist sterile bran. On the pilot plant and plant scale inoculum was produced using the above substrate in covered trays with a central air distributor in the lid, and air outlets at each corner of the lid. On the larger scale of working it was possible to reduce the inoculum level to 0·10% or even to 0·04% by applying the dry spore inoculum to the bran used in the enzyme production stage with an insecticide duster while the moist bran was agitated in a mechanical mixer.

Bindal and Sreenivasaya (1945) reported on the use of malt residues as a substrate for the production of diastase. After extraction of the malt, the grain residues were broken up by grinding and used as a substrate for *Asp. oryzae*. Of a number of malted cereal residues tested, maize gave the highest yield of enzyme. The yield could be improved by the addition of groundnut (*Arachis hypogea*) cake.

Fungal proteases have been produced by similar techniques to those used for the production of amylases, the mould strains being selected strains of *Aspergillus flavus*, *Asp. oryzae* and *Aspergillus wentii*. Wheat bran, soya bean cake, alfalfa meal, broken grain, middlings and brewers grain have been reviewed as possible substrates by Hoogerheide (1954).

The pectic enzymes represent another example of a group of enzyme preparations derived from the growth on solid substrates of selected mould strains While the processes for the production of pectic enzymes are less well documented than those for fungal amylase production, it is clear that most commercial preparations are derived either from *Penicillium* strains or from the lighter coloured members of the *Asp. niger* group grown on bran, on bran admixed with fruit pomace, the pectin rich press cake after the juice has been expressed, or on sugar beet residues. The enzymes present in the pectolytic complex include pectin methylesterase which removes methoxyl groups from the polygalacturonic acid chains prior to random or terminal cleavage of the β-1,4 glycosidic linkages between the galacturonic acid units.

It is likely that the use of solid substrates for enzyme production will gradually be replaced by submerged culture processes as the submerged culture techniques are improved. Recovery of the enzymes from submerged cultures, in convenient form for storage, is rather more difficult than in the case of bran culture techniques where the mould bran may be conveniently stored in the air dry state.

It is not intended here to discuss food products prepared by microbial action on solid natural substrates which are so commonly used in the Far East. This subject has been most adequately examined by Hesseltine (1965), however, arising out of the use of Tempeh-Bongkrek a fermented food prepared domestically in Java from coconut press cake by the growth of a *Rhizopus* probably *Rhizopus oligosporus*, a number of cases of food poisoning have occurred which were traced to the presence of toxins produced by a pseudomonad in imperfectly prepared Tempeh. This organism, subsequently known as *Pseudomonas cocovenenans*, was cultured on defatted shredded copra for the experimental production of the toxin Bongkrekic acid. The method described by Nugteren and Berends (1957) involved the use of 2·5 kg of commercial shredded copra. The oil, amounting to some 65% of the dry matter, was partially removed by soaking the material for several hours in water at 60°C. The shredded copra was then squeezed out in small batches in cloth and the soaking and pressing process repeated with fresh water. The defatted damp copra now showing a lumpy porous consistency was placed in shallow layers 1–1·5 cm deep in a glass box and inoculated with a broth culture of the pseudomonad.

The copra was agitated daily during incubation and at the end of the incubation period, some 4 days, the covers of the boxes were removed and the material was allowed to air dry before extraction of the Bongkrekic acid. The Bongkrek toxins have recently been discussed by van Veen (1966). Certain food-borne mycotoxins have been produced initially, for convenience and simplicity, on solid natural substrates while submerged culture production methods were being evolved. Thus the aflatoxins have been produced on sterile moistened groundnuts by Sergeant *et al.* (1961), De Iongh *et al.* (1962) and Codner *et al.* (1963). The last group pointed out that more uniform yields are obtained if the flasks containing the nuts are shaken briefly each day, to prevent the nuts becoming caked together by the growing mycelium. Other workers including Asao, T. *et al.* (1963) and Asao, Y. *et al.* (1966) used wheat for the production of aflatoxin. The wheat was crushed and moistened with an equal weight of water before autoclaving. Purchase and Nel (1966) initially prepared ochratoxin from *Aspergillus ochraceus* on sterilized corn meal though this mycotoxin was also later prepared in larger quantities in submerged culture.

These few examples serve to show that it is often convenient to obtain microbial products from simple solid substrate systems while more sophisticated submerged culture techniques are being developed.

VI. TRICKLING FILTER TECHNIQUES

In considering the growth of micro-organisms at interfaces, it is possible to trace among the techniques used in aerobic microbiological processes a trend towards the most highly aerated and agitated submerged cultures and consequently a trend away from the use of unsupported surface pellicles of microbial growth at the air-liquid interface. In the manufacture of vinegar the Orleans process using a surface pellicle of *Acetobacter* was replaced by processes in which the medium is trickled over a microbial film supported on inert materials presenting a very large surface. Coiled beechwood shavings have traditionally been used for this purpose but many other support materials including wood wool, grape stems, corn cobs, a basket work made of rattan, coke, wood charcoal and ceramic materials have also been used (Vaughn, 1954). In the vinegar industry there is an increasing trend towards the development of continuous aerated cultures using turbine type mixing devices (Ebner *et al.*, 1967) or cylcone type fermenters and further development of support materials for *Acetobacter* films seems unlikely.

However, in the field of effluent disposal where an analogous biological oxidation takes place, either by the use of a zoogloeal film supported on inert material or by a zoogloeal floc dispersed throughout the aerated liquid as in the activated sludge process, the present trend does not appear to favour the latter to the exclusion of the use of biological filtration beds. Consequently much work in this country and elsewhere has been directed to the development of suitable media for the support of the zoogloeal film in biological filters. The traditional beds packed with an aggregate of rock, slag or clinker are now being challenged by the use of towers filled with specially designed plastic support media. A number of such support media are now available and references to them can be found in Noble (1966).

Some of the advantages claimed for one such medium, the I.C.I. product "Flocor", which may be taken as an example, include low weight, a surface area comparable with that obtainable with rock aggregates, a large void volume giving adequate ventilation without forced draught. The high void volume also provides adequate accommodation for film growth and permits high hydraulic loading without ponding. The material is so designed that the applied liquid cannot fall freely through vertical channels and is distributed in such a way as to cause the liquid to flow in a thin film over the corrugated surfaces without channelling and without accumulating in drops in the angles of the corrugations.

The investigation of the activities of microbial films in percolating filters packed with aggregate or other randomly shaped material is unsatisfactory and simplified model systems for this purpose based on the development of a film on the inner surface of a rotating inclined cylinder have been devised by Comstock *et al.* (1952). A film flow reactor has been described by Maier (1968) using a flat inclined bed on which zoogloeal film is supported and enclosed by a layer of fibre-glass.

Metabolic activities of micro-organisms in soil have been studied by the soil perfusion technique in which certain elements of the microflora are caused to show increased activity and perhaps to become dominant by modifying the environment of the soil by use of a suitable perfusion solution. This technique was originally used by Lees and Quastel (1946) in the study of the biochemistry of nitrification in soil. Their apparatus consisted of two parts. The reciprocator (Fig. 1a) consisting of a 300 ml Buchner flask fitted with two tubes as shown into which a slow flow of water, controlled by a capillary tube, enters via the side arm. The flask alternately fills up and empties via the siphon tube to waste. At each cycle air is slowly forced out of the reciprocator to the perfusion unit and then sucked in from it. In the perfusion unit proper (Fig. 1b), the sieved soil (20–50 g) is held in a glass tube (25 × 2·5 cm) on a glass wool plug with a second glass wool plug

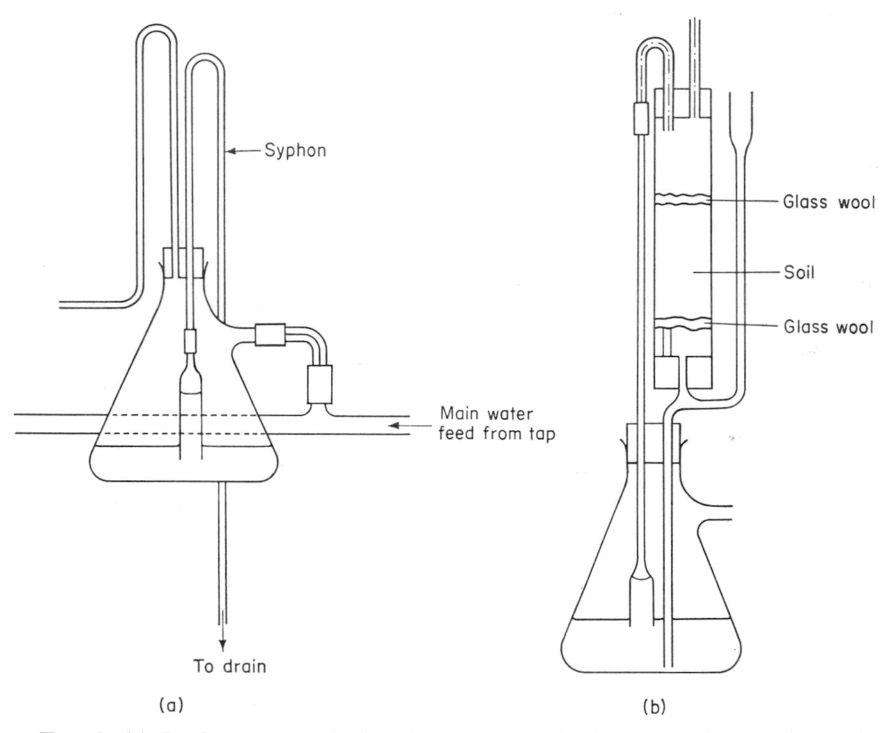

FIG. 1. (a) Reciprocator unit for circulation of air through soil perfusion unit (b). (From Lees and Quastel, 1946).

on top to avoid puddling. The bung below the soil column connects to the side arm of a T-piece the main arms of which run from a wide mouth above the level of the soil through a bung in the mouth of the 300 ml Buchner flask and almost to the bottom of this flask which acts as a reservoir. A lift tube passes through the second hole in the reservoir bung. The wide lower end of the lift tube is adjusted by means of the rubber connection so that it just touches the surface of the liquid in the reservoir when the soil is saturated with the solution. The upper end of the lift tube leads through a capillary tube in the bung at the top of the soil tube into the soil tube. This bung also carries a capillary tube open to the air. The reciprocator on filling causes an inflow of air to the reservoir and liquid is forced up into the lift tube until the liquid level falls below the lower end of the lift tube. The liquid in the wide part of the tube falls back but the liquid column in the narrow upper

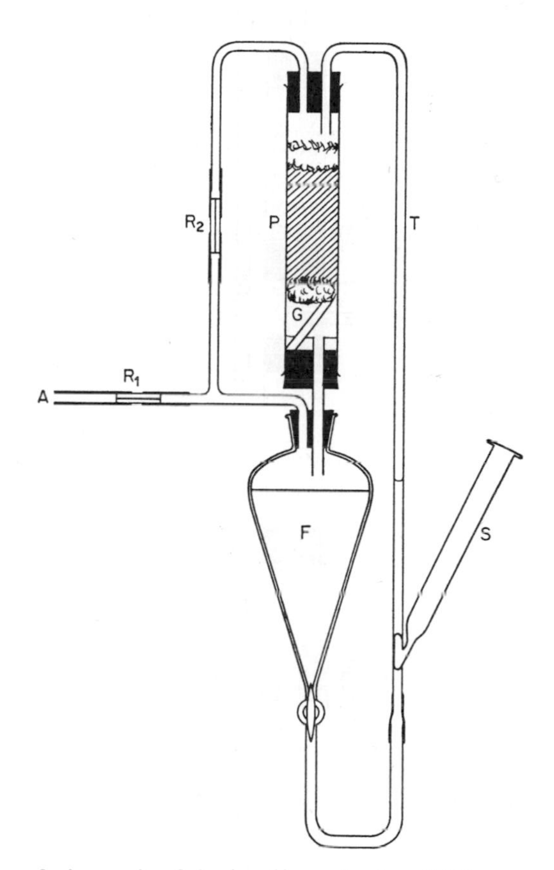

FIG. 2. Soil perfusion unit of Audus (from Audus, 1946).

part of the lift tube is carried upwards onto the soil and, as soon as the lift tube is free of fluid, air escapes to the atmosphere slowly via the capillary tube and some air is forced through to the lower end of the soil column to escape to the atmosphere thus clearing the soil of water and aerating it. When the reciprocation unit empties, a suction is applied to the perfusion unit flask and air is drawn into the reservoir via the immersed tube which mixes and aerates the reservoir fluid.

The soil perfusion technique was simplified in the apparatus of Audus (1946), shown in Fig. 2. Constant small suction is applied at A which is transmitted through the lengths of capillary tubing R_1 and R_2 to the top

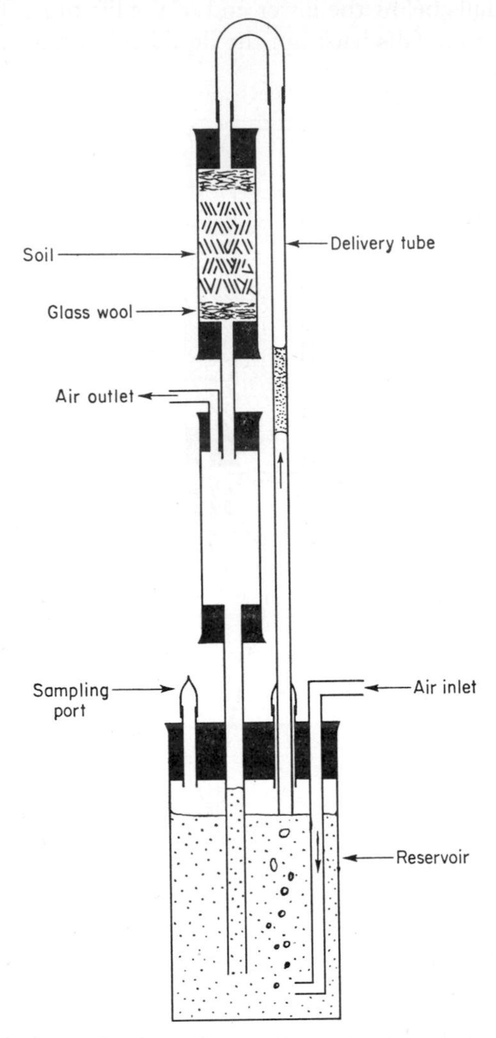

FIG. 3. Soil perfusion unit of Kaufmann (from Kaufmann, 1966).

of the soil column P and thus to the perfusion solution in the tube T. This causes air to be drawn in at the base of the side arm S and liquid is detached from the main bulk, and passes to the top of the soil column. On release of the tension in tube T by this discharge liquid again rises in the tube above the base of tube S until it reaches the level of the solution in F and the process is repeated. This apparatus has advantages in simplicity, greater control over aeration and the wide rates of perfusion obtainable.

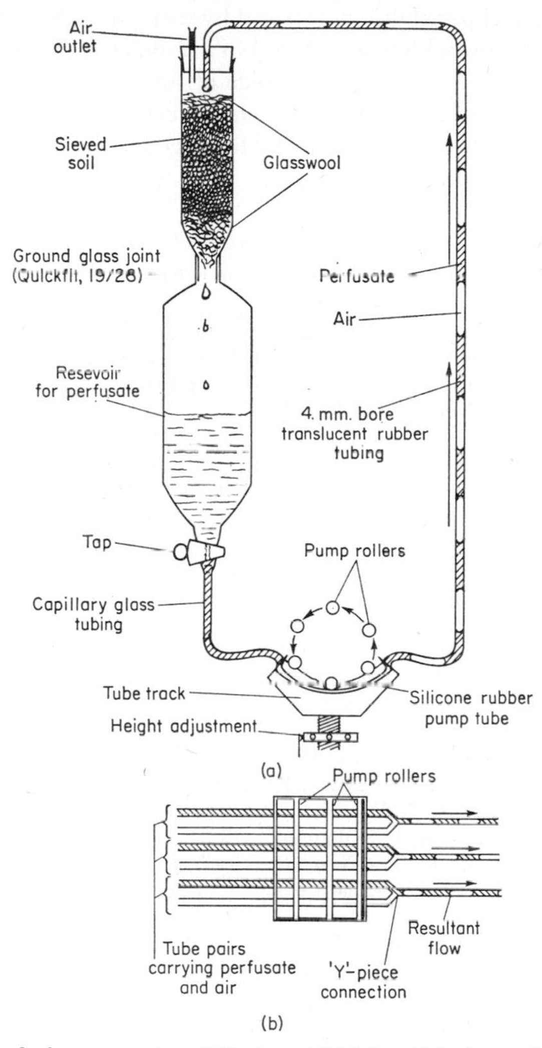

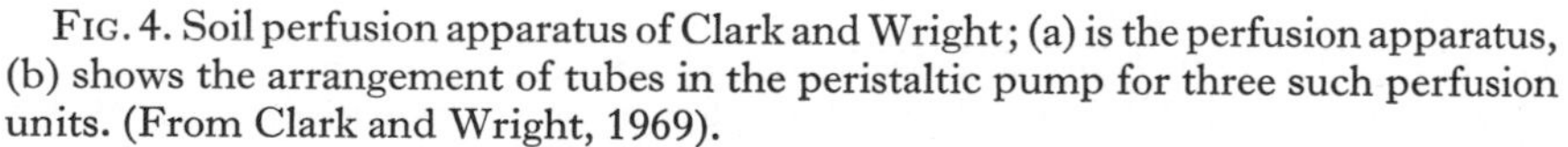

FIG. 4. Soil perfusion apparatus of Clark and Wright; (a) is the perfusion apparatus, (b) shows the arrangement of tubes in the peristaltic pump for three such perfusion units. (From Clark and Wright, 1969).

Kaufmann (1966) introduced a modified perfusion unit for studying degradation of pesticides in soil. This unit, shown in Fig. 3, enables the amount and rate of liquid recycled and the degree of aeration to be controlled by the depth of immersion of the end of the delivery tube in the reservoir and the rate and amount of air introduced into the system. Clark and Wright (1968) have further controlled the perfusion and aeration rate by incorporating a multichannel peristaltic pump and by using one channel in the pump to control liquid flow which is admixed in a Y-piece with air from a second channel on the pump. By suitable choice of tube sizes of the two tubes in the pump the ratio of air to perfusing liquid can be varied and the overall flow rate of the mixture can be varied by the speed of the pump. This apparatus is shown diagramatically in Fig. 4.

REFERENCES

Asao, T., Büchi, G., Chang, S. B., Abdel-Kader, M. M., Wick, E. L., and Wogan, G. N. (1963). *J. Am. chem. Soc.*, **85**, 1706.

Asao, Y., Kikuchi, T., Nobuhara, A., Sasaki, M., and Yokotsuka, T., (1967). *In* "Biochemistry of some food-borne microbial toxins" (Eds. R. I. Mateles, and G. W. Wogan) pp. 131–152. M.I.T. Press, Cambridge, Mass. & London.

Audus, L. J. (1946). *Nature, Lond.*, **158**, 419.

Beadle, G. W., Ryan, F. J., and Tatum, E. L., (1943). *Am. J. Bot.*, **30**, 784–799.

Berliner, M. D., and Neurath, P. W. (1965). *J. cell comp. Physiol.*, **65**, 183–193.

Berliner, M. D., and Neurath, P. W. (1966). *Can. J. Microbiol.*, **12**, 1068–1070.

Bertrand, H., McDougall, K. J., and Pittenger, T. H., (1968). *J. gen. Microbiol.*, **50**, 337–350.

Bindal, A. N., and Sreenivasaya, M. (1945). *J. scient. ind. Res.*, **3**, 368.

Bissett, K. A. (1963). "Bacteria", 3rd Edition, p. 49 E. S. Livingstone, Edinburgh.

Board, R. G. (1967). *Lab. Pract.*, **16**, 479.

Brancato, F. P., and Golding, N. S. (1953). *Mycologia*, **45**, 848–864.

Brock, T. D. (1961). "Milestones in microbiology", Prentice Hall, London.

Bulloch, W. (1938). "The history of bacteriology", Oxford University Press.

Butler, G. M. (1966). *In* "The fungi". Vol. 2. (Eds. G. C. Ainsworth and A. S. Sussman) pp. 83–112. Academic Press, London.

Campbell, R. (1968). *Trans. Br. mycol. Soc.*, **51**, 83–87.

Chadwick, P. (1963). *Can. J. Microbiol.*, **9**, 734–737.

Christensen, L. M., (1944). U.S. Patent 2, 352,168.

Codner, R. C., Sergeant, K., and Yeo, R., (1963). *Biotechnol. Bioengng*, **5**, 185–192.

Comstock, R. F., Gloyna, E. F., and Renn, C. E. (1952). *Sewage ind. Wastes.*, **24**, 1355–1357.

Clark, C. G., and Wright, S. J. L., (1969). *Weed Res.*, **9**, 66–69.

Dawson, P. S. S. (1963). *Can. J. Microbiol*, **9**, 671–687.

Dean, A. C. R., and Hinselwood, C. (1957). *Proc. R. Soc.*, B., 147, 10–20.

De Iongh, H., Beerthuis, R. O., Vles, R. O., Barrett, C. B., and Ord, W. O. (1962). *Biochim. biophys. Acta.*, **65**, 548.

Duguid, J. P. (1959). *J. gen. Microbiol.*, **21**, 271–286.

Duguid, J. P., and Wilkinson, J. F. (1961). *Symp. Soc. gen. Microbiol.*, **11**, 69–99.

Eastermann, E. F., and McLaren, A. D. (1959). *J. Soil Sci.*, **10**, 65–78.

Ebner, E. Enenkel, A. C., and Pohl, K. (1967). *Biotechnol. Bioengng.*, **9**, 357–364.

Fawcett, H. S. (1925). *Ann. appl. Biol.*, **12**, 191–198.

Fowell, R. R. (1965). *J. appl. Bact.*, **28**, 373–383.

Finkelstein, R. A., and Punyashthiti, K. (1967). *J. Bact.*, **93**, 1897–1905.

Fleming, A. Hughes, W. H., Kramer, I. H., and Voureka, A. (1950). *J. gen. Microbiol.*, **4**, 257–269.
Glasstone, G. (1962). "Textbook of physical chemistry", 2nd Edition. MacMillan and Co. London.
Graham-Smith, G. S. (1910). *Parasitology*, **3**, 17–53.
Gregory, P. H. (1961). "Microbiology of the atmosphere". Leonard Hill, London.
Grogan, R. G., and Purdy, L. H. (1954). Phytopathology, **44**, 36–39.
Hao, L. C., Fulmer, E. I., and Unkerkofler, L. A. (1943). *Ind. Engng. Chem.*, **35**, 814–818.
Hesseltine, C. W. (1965). *Mycologia*, **57**, 149–197.
Hill, H. W. (1902). *J. med. Res.*, **7**, 202–212.
Hoffman, H. (1964). *Ann. Rev. Microbiol.*, **18**, 111–130.
Hoffman, H., and Frank, M. E. (1961). *J. gen. Microbiol.*, **25**, 353–364.
Hoogerheide, J. C. (1954). *In* "Industrial fermentations", Vol. II. (Eds. L. A. Underkofler and R. J. Hickey) pp. 122–154. Chemical Publishing Co. New York.
Ingleman, B., and Laurell, H. (1947). *J. Bact.*, **53**, 364–365.
Ingram, M. (1955). "An introduction to the biology of yeasts". Pitman & Sons, London.
Jefferys, G. A. (1948). *Fd. Inds*, **20**, 688–690, 825, 826.
Kaufmann, D. D. (1966). *Weeds*, **14**, 90–91.
Kingsbury, J. M., and Baghoorn, E. S. (1954). *Appl. Microbiol.*, **2**, 5.
Kulka, D., Preston, J. M., and Walker, T. K. (1951). *J. gen. Microbiol.*, **5**, 18–21.
Lees, H., and Quastel, J. H. (1946). *Biochem. J.*, **40**, 803–815.
Lindegren, C. C., and Hamilton, E. (1944). *Bot. Gaz.*, **105**, 316.
Lodder, J., and Kreger van Rij, N. J. W. (1952). "The yeasts". North-Holland Publishing Co.
Mahadevan, P. R., and Tatum, E. L. (1965). *J. Bact.*, **90**. 1073–1081.
Maier, W. J. (1968). *Appl. Microbiol.*, **16**, 1095–97.
Meyer, G. W. (1966). *Bull. Torrey bot. Club.*, **93**, 201.
Meyrath, J. (1965). *J. Sci. Fd Agric.*, **16**, 14.
Mitichison, D. A., and Selkon, J. B. (1957). *J. gen. Microbiol.*, **16**, 229–235.
Moore, H. N. (1940). *J. Bact.*, **40**, 409–413.
Morris, E. O. (1966). *In* "The fungi", Vol. 2. (Eds. G. C. Ainsworth and A. S. Sussman) pp. 63–82. Academic Press, London.
Morris, E. O., and Hough, J. S. (1956). *J. Inst. Brew.*, **62**, 466.
Morton, A. G. (1961). *Proc. R. Soc.*, B, **153**, 548–569.
Mrak, E. M., and Phaff, H. J. (1949). *Wallerstein Labs Comm.*, **12**, 29–44.
Murray, R. G. E., and Elder, R. H. (1949). *J. Bact.*, **58**, 351–359.
Noble, G. T. (1966). *In* "Effluent water treatment manual" 3rd Edn. Thunderbird Enterprises Ltd., London.
Nordin, J. S., Tsuchiya, H. M., and Fredrikson, A. G. (1967). *Biotechnol Bioengng*, **9**, 545–558.
Nugteren, D. H., and Berends, W. (1957). *Recl Trav. Chim. Pays-Bas Belg.*, **76**, 13–27.
Old, D. C., Corneil, I., Gibson, L. F., Thomson, A D., and Duguid, J. A. (1968). *J. gen. Microbiol.*, **51**, 1–16.
Pearce, T. W., and Powell, E. O. (1951). *J. gen. Microbiol.*, **5**, 91.
Pirt, S. J., (1967). *J. gen. Microbiol.*, **47**, 181–197.
Plomley, N. J. B. (1959). *Aust. J. biol. Sci.*, **12**, 53–64.
Poindexter, J. S. (1964). *Bact. Rev.*, **28**, 231–295.

Powell, E. O. (1956). *Jl. R. Microsc. Soc.*, Ser. III, **75**, 235–243.
Pramer, D. (1957). *Appl. Microbiol.*, **5**, 392–395.
Purchase, I. F. H., and Nel, W. (1967). *In* "Biochemistry of some foodborne microbial toxins" (Eds. R. I. Mateles, and G. N. Wogan) pp. 153–156. M.I.T. Press, Cambridge, Mass. and London.
Raper, K. B., and Fennell, D. I., (1965). "The Genus Aspergillus". Williams and Wilkins, Baltimore.
Raper, K. B., and Thom, C. (1949). "Manual of the penicilla". Williams and Wilkins, Baltimore.
Rem, L. T., and Stotzky, G. (1966). *Can. J. Microbiol.*, **12**, 547–563.
Roberts, M. S., Laufer, S., Saletan, L. T., and Stewart, D. (1944). *Ind. Engng Chem.*, **36**, 811–812.
Sergeant, K. Sheridan, A., O'Kelly, J., and Carnaghan, R. B. A. (1961). *Nature, Lond.*, **192**. 1096.
Smith, J. H. (1923). *Ann. Bot.*, **37**, 341–343.
Society of American Bacteriologists (1957). "Manual of microbiological methods". McGraw-Hill Book Co.
Stotzsky, G. (1966a). *Can. J. Microbiol.*, **12**, 831–848.
Stotzsky, G. (1966b). *Can. J. Microbiol.*, **12**, 1235–1246.
Sussman, A. S. Lowery, R. J., and Durkee, T. (1964). *Am. J. Bot.*, **51**, 243–252.
Taylor, C. B. (1950). *J. gen. Microbiol.*, **4**, 235–237.
de Terra, N., and Tatum, E. L. (1963). *Am. J. Bot.*, **50**, 669–677.
U.S. Patent No. 3,244,946 (1965). Socony Mobil Oil Co.
Underkofler, L. A. (1942). U.S. Patent No. 2, 291,009.
Underkofler, L. A., Fulmer, E. I., and Schoene, L. (1934). *Ind. Engng. Chem.*, **31**, 734–738.
Underkofler, L. A., Severson, G. M., Goering, K. J., and Christensen, L. M. (1947). *Cereal Chem.*, **24**, 1–22.
Vaughn, R. H. (1954). *In* "Industrial fermentations" (Eds. L. A. Underkofler and R. J. Hickey) pp. 498–535. Chemical Publishing Co. New York.
Van Veen, A. G. (1967). *In* "Biochemistry of some foodborne microbial toxins" (Eds. G. W. Wogan and R. I. Mateles) pp. 43–50. M.I.T. Press. Cambridge, Mass. and London.
Waksman, S. A., and Carey, C. (1926). *J. Bot.*, **12**, 87–95.
Wilson, G. S., and Miles, A. A. (1962). "Topley and Wilson's Principles of bacteriology and immunity", Vol. 1, pp. 486–488. Arnold, London.
Winogradsky, S. (1891). *Annls. Inst. Pasteur, Paris*, **5**, 93–100.
Yuill, E., and Yuill, J. L. (1954). *Nature, Lond.*, **173**, 643.
ZoBell, C. E. (1943). *J. Bact.*, **46**, 39–56.

CHAPTER XII

The Isolation and Cultivation of Single Organisms

K. I. JOHNSTONE

Department of Bacteriology, The School of Medicine, Leeds, England

I. INTRODUCTION

The cultivation of bacteria from single organisms is especially valuable in the purification of strains, since the only absolute criterion of the purity of a culture is the certainty that it has been derived from the progeny of a single organism. Failure to apply this criterion may lead to much waste of effort in research, especially in regard to the clostridia, apparently pure strains of which may carry latent contaminants for long periods. All strains upon which research is to be based should therefore be rigorously purified at the outset.

The study of the properties of individual organisms, as distinct from those of bacterial populations, is a little-explored field, requiring a rapid and accurate method of isolation to enable batches of isolates to be tested, e.g., the heat resistance of bacterial spores.

Because of the current belief that single-organism culture is necessarily tedious and involves the use of costly and complex apparatus, the technique is often shunned. However, the methods available range from those requiring only the simplest apparatus readily improvised in the laboratory, to those requiring elaborate micromanipulators and ancillary apparatus. The former are only suitable when a few isolations are to be made at rare intervals. When many isolations are required, more complex equipment is necessary to minimize fatigue of the operator. In all cases, it is essential that the method be absolutely reliable and that adequate controls are used to establish the validity of the isolations.

For multiple and rapid bacterial isolations under sterile conditions, it is essential that the microinstrument should be readily removable from the sterile chamber to facilitate interchange of culture materials and should be as readily replaceable. Many of the types of micromanipulator now available are unnecessarily complex for this work and the agar gel dissection methods here described in detail do not require a conventional instrument. A simple type of microforge is, however, a necessity for the satisfactory production of microinstruments.

II. A REVIEW OF THE CHIEF METHODS DEVISED FOR SINGLE-ORGANISM CULTURE

A. Selection of an isolated organism on a randomly inoculated nutrient agar gel

This method, employing a block cut from a lightly-inoculated nutrient agar plate (Ørskov, 1922) and improved by the use of a cast agar block (Gardner, 1925), requires only the simplest apparatus, which can be assembled at very slight cost in the laboratory. The disadvantages are that (a) the method of location of the isolate by drawing a pattern of scratches on the glass slide is tedious, (b) observations on the growing microcolony may extend over many hours, (c) the subculture of the microcolony is a blind operation and (d) aerial contamination of the exposed gel may occur.

B. Formation of droplets with a micropipette

1. *At an air–glass interface* (*Malone, 1918*)

Using a second microscope as manipulator, microdroplets are deposited on the lower surface of a sterile coverglass and are searched for those containing only single organisms. Such droplets are each drawn into separate sterile micropipettes, the tips of which are broken off in nutrient broth tubes. The advantage lies in the use of the $\times 100$ oil-immersion objective, which is not possible in methods operating above the air–gel interface. The disadvantages are (a) the difficulty in excluding the presence of a second organism at the margin of each droplet, (b) the rapid drying of the minute droplets and (c) the sacrifice of a micropipette for each organism isolated.

2. *In an oil chamber* (*de Fonbrune, 1949*)

The coverglass forms the roof of a chamber filled with sterile liquid paraffin and droplets are formed between the glass and the paraffin by means of a micropipette, being examined at intervals until multiplication of a single organism is observed. Such a droplet is removed using a fresh sterile pipette and is cultured. The advantages are (a) the use of the $\times 100$ oil-immersion objective, (b) the complete elimination of drying of the droplets and (c) the ease with which the successive cell divisions may be observed. The disadvantages are that (a) the charged pipette must traverse the paraffin seal during formation of the droplets and the possibility exists of contamination of the oil and subsequent contamination of the subculturing pipette, (b) a clear fluid medium alone can be used and (c) the conditions are anaerobic, and strictly aerobic organisms may not grow. The method is more suitable for observations on the process of multiplication *in situ*, without recovery of the resulting microcultures.

3. *On a mechanically propelled Cellophane strip* (*Reyniers and Trexler, 1943*)

Microdroplets are formed mechanically on the lower surface of a strip of Cellophane, which then passes through the microscope and is taken up on a spool. When a droplet is recognized as containing a single organism, it is cut out on a disc of Cellophane by a micro-fly cutter and falls direct into a culture tube. The method attains the highest degree of mechanization, but relies on the recognition of droplets containing single organisms and requires specialized and complex apparatus.

C. Isolation by micromanipulation on an agar gel surface

1. *Below the air–gel interface* (*Dickinson, 1926*)

Using the lower surface of a coverglass coated with a thin layer of nutrient agar, enclosed to reduce drying of the gel, selected organisms are carried in turn across the agar surface in the water column formed by contact of a vertical needle tip with the gel. The advantages are that (a) the isolates are taken to sterile portions of the gel remote from the inoculum, (b) the organisms can be cultured *in situ*, or portions of the agar may be cut out carrying isolates and cultured separately, (c) the $\times 100$ oil-immersion objective can be used and (d) aerial contamination is excluded. The disadvantages are that (a) there is no rapid and positive method of location of the isolates, their positions being marked with ink on the upper surface of the coverglass, (b) manipulations below the air–gel interface are difficult to carry out and (c) the necessarily thin layer of agar renders dissection of the gel tedious, especially if many isolations are required.

2. *Above the air–gel interface*

(a) *Technique of Koblmüller and Vierthaler* (*1933*). The technique involves the use of a metal needle on a nutrient gel, and cultivation of the isolates *in situ*.

Selected organisms on the upper surface of a nutrient agar gel are drawn in turn to isolated sites on the sterile surface by means of a metallic needle inclined at an angle to the surface. The selected organism floats in the water exuded from the gel around the needle tip. The isolates are cultivated on the agar surface. The advantage lies in the greater ease of manipulation on the upper surface of the gel. The disadvantages are that (a) the method of location of the isolates by ink marks is tedious, (b) a metal microneedle is less easy to prepare than one of glass and (c) a dry objective must be used.

(b) *Technique of Johnstone* (*1943*, *1953*). The technique involves the use of a glass needle and positive location of the isolates on a non-nutrient gel, which is subsequently dissected.

Selected organisms are carried across the upper surface of a cast block of clarified non-nutrient agar gel, with an angulated glass microneedle. The location of each isolate is defined by melting two pits in the gel surface, the organism lying between them. The isolates are cultivated by dissection of the gel, as shown by the pairs of pits, and transfer of portions, each carrying a pair of pits and therefore an organism, to separate tubes of medium. The advantages are that (a) a glass microneedle is used which is readily prepared, (b) manipulations are carried out on the plane upper surface of the gel and are readily observed, (c) the method of location is positive and can be made rapid by mechanical means and (d) any type of culture medium may be used for cultivation of the isolates. The disadvantage, in common with other methods operating above the air–gel interface, is that immersion objectives cannot be used.

(c) *Technique of Holdom and Johnstone* (*1967*). The technique involves the use of a glass microloop on a non-nutrient gel with pre-marked isolation sites and subsequent dissection of the gel.

The cast block of agar gel is pre-marked with 24 isolation sites in a sterile punch. With a glass microloop, prepared in a high-power microforge, selected organisms are carried rapidly above the gel to the isolation sites and are there ejected from the loop on to the agar surface. Dissection of the gel and cultivation of the isolates follow. The advantages are that (a) the isolations are carried out rapidly and (b) when many isolations of the same strain are required, the microloop can be charged with 6–12 selected organisms which are deposited singly on the isolation sites, with great saving of time. The disadvantage is that the microloop must be prepared in a high-power microforge, but such a loop will serve for several thousand isolations.

D. Destruction of all organisms except the one selected

This method (Topley *et al.*, 1921) is unique, in that the selected organism, located in a gelatin film under a quartz coverglass, is protected by a globule of mercury on the coverglass from ultraviolet radiation, which kills all other organisms in the film. After incubation, the resulting microcolony is subcultured. The advantage lies in the simple apparatus required. The disadvantages are that (a) careful controls are necessary to establish the validity of each experiment, especially in the presence of spores, (b) incubation is at 25°C and (c) the method is not suitable if many isolations are required.

III. THE AGAR BLOCK DISSECTION TECHNIQUES

A. The agar gel

The gel must be free from both living and dead bacteria, the latter being abundant in most samples of agar and causing confusion during micromanipulation. They are removed by clarification of a fluid 2·0% (w/v) aqueous solution of New Zealand agar with 2·0% (w/v) of Hyflo Super Cel diatomaceous earth (Koch-Light Laboratories Ltd, Colnbrook, Buckinghamshire) at 60°C for 4 days, with gentle inversion of the bottle twice daily (Feinberg, 1956). This is followed by filtration through Hyflo Super Cel sandwiched between layers of paper pulp and yields a gel free from bacteria, when autoclaved. The reaction should be pH 7·0–7·4. No nutrients are added, since multiplication of the organisms is not desired during manipulations, and a batch of agar can be stored in sealed ampoules for several years, being used for manipulation of a great variety of bacteria. After trial with the microinstrument, the concentration of agar is lowered, by adding water, to the optimum for the method of isolation used.

B. Casting of blocks of gel

A clean 7·6 × 2·5 cm glass slide is separated from a clean 7·6 × 3·8 cm slide by two glass strips 2·0 mm thick to form a casting cell (Fig. 1), the slides being held in apposition to the strips by wire paper clips and the whole sterilized by dry heat in a Petri dish. Diamond-ruled lines on the outer surfaces of the cell serve as guides in sectioning the agar slab.

Clarified agar at 100°C is pipetted into the casting cell at 60°C and, when set, the excess of gel beyond the margin of the smaller slide is removed with a sterile scalpel. The slides are drawn apart with a sliding motion and the gel, which remains adherent to one slide, is cut into blocks 2·5 × 1·3 cm and 2·0 mm thick. Each block is lifted with the back of the scalpel blade and is

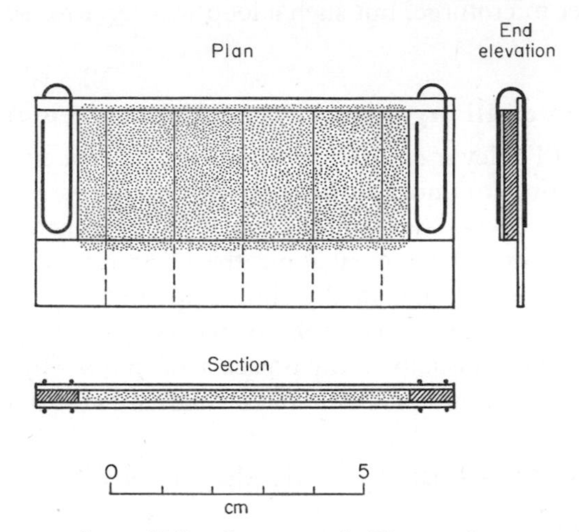

FIG. 1. A glass casting cell for the agar gel. The portion occupied by the gel is stippled.

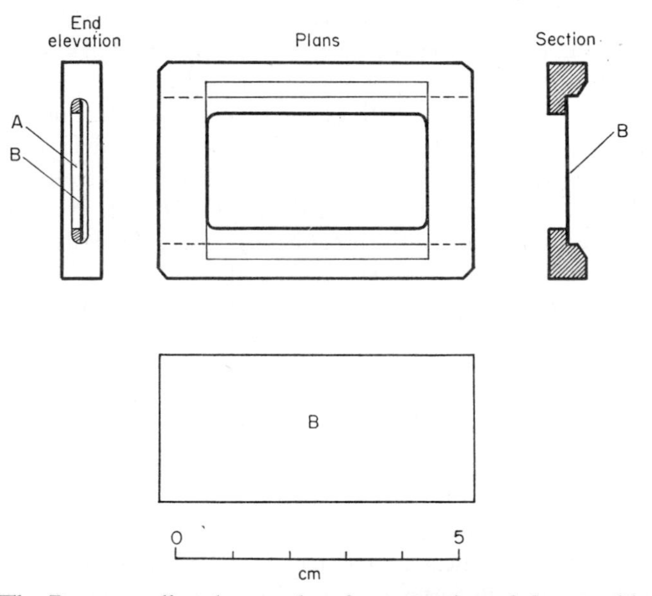

FIG. 2. The Perspex cell and coverglass for protection of the agar block during manipulations. A, the access slot for the microinstrument, or marking needle; B, the coverglass.

deposited on a separate sterile 7·6 × 3·8 cm slide, enclosed in a Petri dish with a Duralumin humidifying trough containing glass wool moistened with distilled water. *At all stages it is essential that the gel be protected from desiccation, as otherwise the plane upper surface will be deformed and the concentration of agar will rise, producing a surface unsuitable for micromanipulation.*

Protection of the agar block from airborne contamination and from desiccation during manipulation is provided by a sterile Perspex cell (Fig. 2), which rests on the supporting slide around the block and carries a 5·5 × 2·5 cm No. 0 coverglass. For optical system 1, the lower surface of the coverglass is 0·9–1·0 mm above the gel surface, but for system 2, this height can be greatly increased, thus allowing more working space for the needle tip. A slot, A, (Fig. 2) gives access to the shaft of the microneedle, a second slot at the opposite end of the cell giving access to the platinum marking needle.

C. Optical equipment

Phase-contrast microscopy renders earlier methods, including dark-ground illumination, obsolete. Positive phase contrast shows bacteria clearly on the gel surface as dark objects against a green background, except for the mature spore, which appears bright before germination. The gel surface appears faintly mottled. Two systems may be used.

1. *A refracting phase-contrast objective* × *40*

The numerical aperture (N.A.) must be at least 0·65, and the working distance 1·0 mm, excluding the coverglass thickness. This has the mechanical disadvantage of a very confined space for operation of the microneedle and also of condensation on the lower surface of the coverglass, requiring a warm-air jet playing on the upper surface of the glass to disperse the water droplets. With a high-absorption phase-plate, the Parachromatic objective × 40 (Watson) gives good resolution.

2. *A Dyson long-working-distance phase-contrast objective* × *40* (*Vickers Instruments*)

Incorporating both reflecting and refracting units, this overcomes both mechanical and thermal disadvantages with a working distance of 12·8 mm, but the N.A. is reduced to 0·57 and the light intensity must be greatly increased to compensate for losses by reflection, with possible lethal action on sensitive organisms.

The phase-contrast condenser must also have a long working distance, to operate through the slide and a 2 mm thickness of agar gel. The illuminant should be of high intensity with a cooling trough, followed by a Chance's ON 20 heat-absorbing filter, and a yellow–green filter to improve optical performance.

D. Use of a second microscope as micromanipulator

A second microscope on a smooth and rigid bench surface contributes the sensitive vertical movement to the microinstrument through its fine adjustment. A nosepiece attachment (Fig. 3) is screwed firmly into the objective thread and the bar A is clamped by the screw B so as to project towards the axis of the observing microscope. This bar is bored to receive the microneedle holder, or the platinum marking needle, as required. Centration of the microneedle tip in the field of the observing microscope is effected by

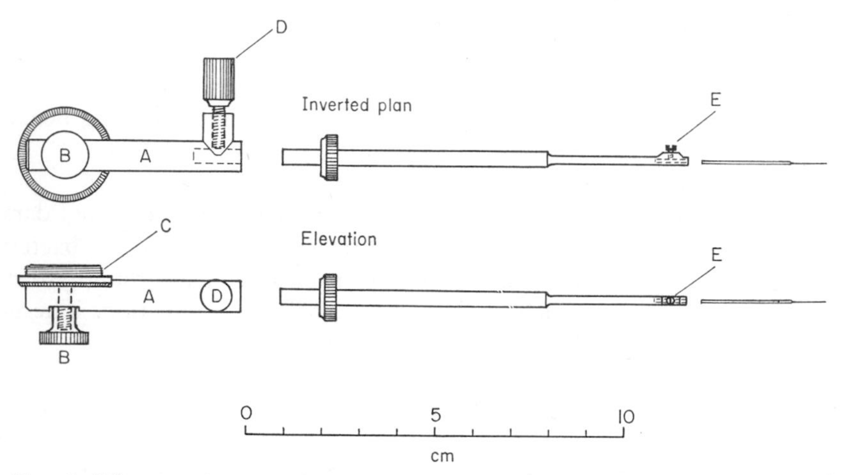

FIG. 3. The nosepiece attachment (left), the microneedle holder (centre) and the microneedle handle (right). C, the r.m.s. objective thread; D, the clamping screw for the microneedle holder; E, the clamping screw for the microneedle.

sliding the manipulating microscope on the bench by finger pressure from both hands applied to the foot of the instrument, while movement of the tip is observed in the field.

Movement in the horizontal plane is obtained by means of the mechanical stage of the observing microscope, which carries the agar block on the sterile slide and therefore gives motion in two dimensions to the gel surface relative to the needle tip, which remains centred in the field.

E. Isolation with a simple angulated microneedle and location of the isolates

The selected organism is carried across the sterile gel surface in the minute pool of water that exudes from the gel around the glass microneedle where it is in contact with the gel. The isolate is under constant observation during its passage and the presence of a second organism of comparable size cannot be overlooked. This method is especially suitable for pure strain isolations, when only a limited number is required.

1. *The angulated microneedle*

Soft-glass rod, 6 mm diameter, is drawn out by hand to 0·9–1·0 mm diameter thus forming the handle of the microneedle. The thin rod is again drawn out after being heated in a minute coal-gas flame burning at the tip of a glass capillary tube, to form the needle shaft 3·8 cm long and 0·15 mm diameter at the free end (Fig. 4a). The tip (Fig. 4b) is formed in a low-power

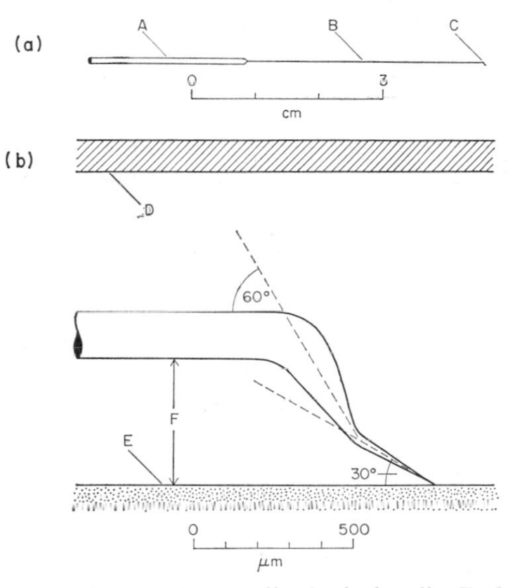

FIG. 4. (a). An angulated glass microneedle. A, the handle; B, the shaft; C, the tip. (b). The tip of the microneedle enlarged, showing its spatial relation to the cover-glass, D, and the agar gel surface, E. The distance F is the clearance of the shaft above the gel.

microforge (Johnstone, 1953) and consists of a portion tapering rapidly from 150 μm to 20 μm at an angle of 60° to the axis of the shaft, and a point inclined at 30° to the same axis, tapering rapidly to 1·5 μm at its tip. The clearance of approximately 400 μm is essential to avoid contact between the horizontal shaft and the gel surface, to which it would adhere owing to surface tension. The shaft must be straight and in the axis of the handle.

2. *The platinum marking needle*

This forms the locating pits in the gel surface on each side of an isolated organism, by means of which the site can readily be seen during dissection of the block. The V shaped heating element (Fig. 5) is made from two 2·7 cm lengths of 40 s.w.g. platinum wire, welded to 24 s.w.g. platinum wires as supports. At the apex of the V, one thin wire is hooked around the second wire 2 mm from its free end and the joint is welded, the terminal 0·5 mm

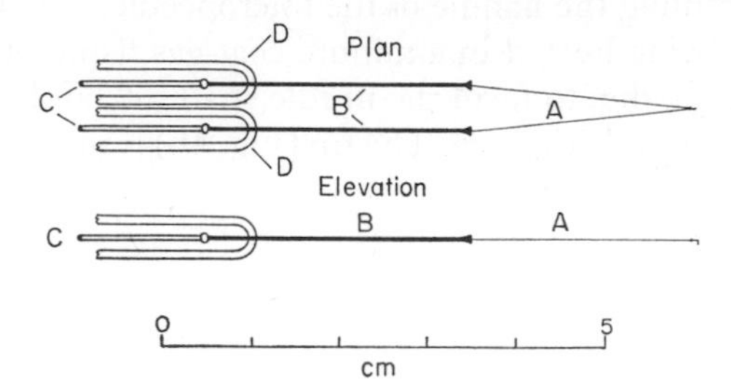

FIG. 5. The platinum marking needle: A, the heating element; B, the supporting platinum wires; C, the copper leads; D, the glass tubes.

of the single wire being bent vertically downwards to form the marking tip. The platinum needle can be mounted in one of two ways—

(a) A simple mount, suitable for occasional use consists of two soft-glass tubes, 4 mm external diameter, bent at right angles. The thick supporting wires of the platinum element are first silver-soldered to 24 s.w.g. copper leads and the platinum wires are then sealed into the ends of the glass tubes (Fig. 6). The vertical ends of the tubes pass through a cork, E, pressed firmly into the nosepiece of a manipulating microscope, whose focusing movements are used to form the pits in the gel surface. Connection is made from the copper leads to a 3 V a.c. supply through push switches S_1 and S_2 (Fig. 7), with variable

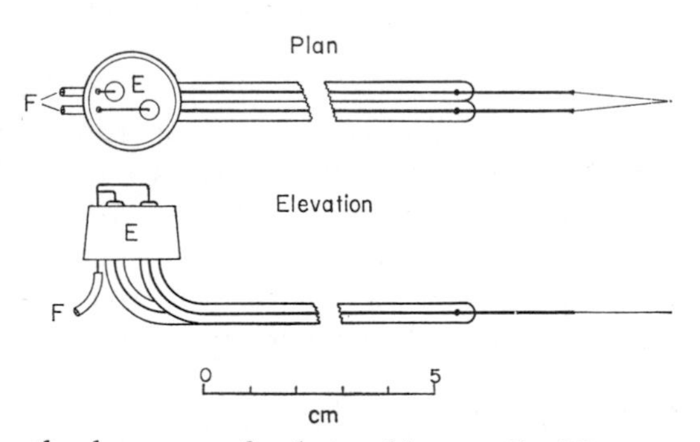

FIG. 6. A simple glass mount for the marking needle. The cork, E, fits into the objective thread of a microscope nosepiece. The electrical leads, F, pass through the cork.

resistors R_1 and R_2, respectively, in series. R_1 is adjusted to bring the element to red heat on closing S_1, the tip being sterilized by conduction. R_2 is adjusted to warm the element sufficiently to melt a pit in the gel surface when S_2 is closed, without causing the gel to boil.

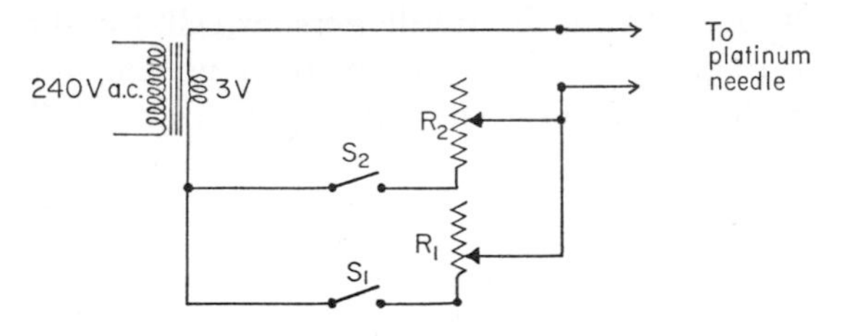

FIG. 7. Heating circuits for the platinum marking needle.

(b) A permanent mount interchangeable with the microneedle holder in the nosepiece attachment is shown in Fig. 8. The steel rods, A and B, carrying the platinum element, are electrically insulated from the stem C, which enters the nosepiece attachment. This mount can with advantage be fitted with adjustable pneumatic dipping and traversing movements and can itself be mounted on a rigid geometric slide replacing the second microscope. The unit may then be withdrawn from the optic axis and rapidly replaced in centre. Also, the locating pits can be formed, as described below, on each side of the isolate while the glass microneedle remains *in situ* and with the $\times 40$ objective in position, thus greatly reducing the time taken.

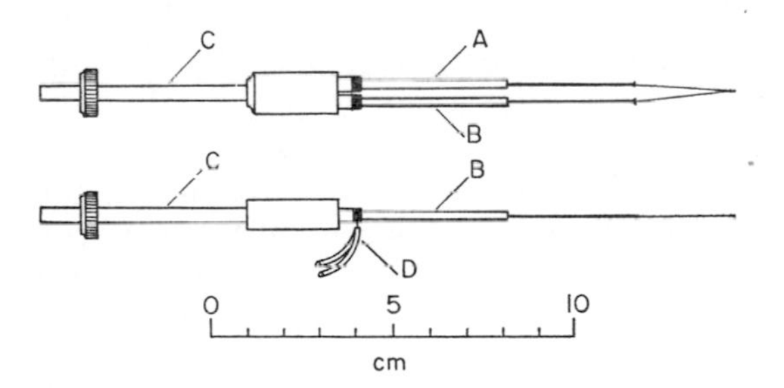

FIG. 8 A permanent mount for the platinum marking needle interchangeable with the microneedle holder in the nosepiece attachment. Electrical connections are made to the rods A and B at D.

3. *Isolation procedure*

During inoculation, the sterile surface of the agar block is protected from bacterial aerosols, created by the inoculating loop, by a sterile metal shield covering all but a strip 2 mm wide at one end. With a 1 mm loop, an inoculum from a just visibly turbid suspension of the organisms is streaked across the upper surface of the gel at the exposed end. The fluid is rapidly absorbed, leaving the organisms on the gel surface and the shield is then removed.

The block is covered with the sterile Perspex cell, the slide is placed on the mechanical stage of the observing microscope, and the margin of the inoculum is located by dark-ground illumination at ×150 magnification

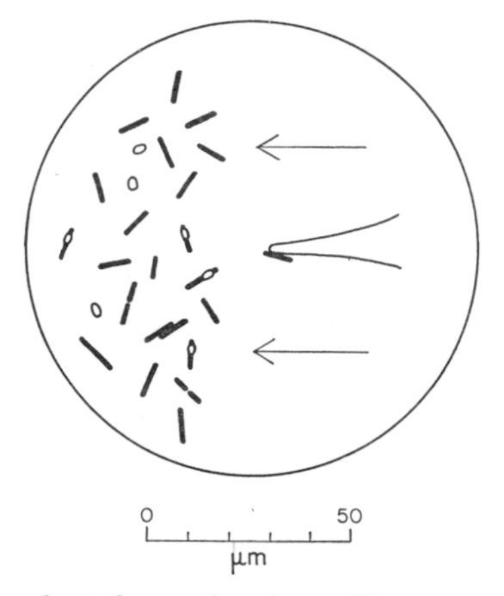

FIG. 9. Isolation of a selected organism by trailing across the agar gel surface with an angulated glass microneedle. The arrows indicate the direction of movement of the gel.

(obtained by the use of a ×10 objective with the ×40 phase-contrast condenser annulus). The microneedle, mounted in the carrier on the nosepiece of the manipulating microscope, is carefully directed into the access slot of the Perspex cell by sliding the manipulating microscope on the bench, taking care to clear the coverglass above and the gel below. The tip is located as it enters the field and is centred first with the ×10 and then with the ×40 objective, while it is poised above the gel surface. By operation of the coarse-focusing movement, the needle tip is lowered until it is just above the gel surface.

A selected, well-isolated organism is brought by means of the stage controls to lie below the needle tip and the latter is then lowered, with the fine-focusing movement, to touch the gel, when the organism floats in the exuded water (Fig. 9). By means of the stage controls, the organism is then carried rapidly over the gel surface, while it and the needle tip remain centred in the field. If the organism escapes from the needle, it is located by following in reverse the needle track, which is visible as a bright line on the gel, until the organism is found. The needle tip must be raised above the gel during reverse movement.

When the isolate has been carried at least 5 mm from the margin of the inoculum, its position is defined by melting two shallow pits in the gel surface 1 mm apart, one above and one below the organism, with the sterile platinum marking needle. The $\times 10$ objective is used unless the pneumatic device for controlling the platinum needle is available, when the $\times 40$ phase-contrast objective can be used and the marking carried out in the presence of the glass microneedle, with great saving of time. Further isolations, up to a total of six, are then made on the same block and should be staggered (Fig. 10a) to facilitate dissection of the gel.

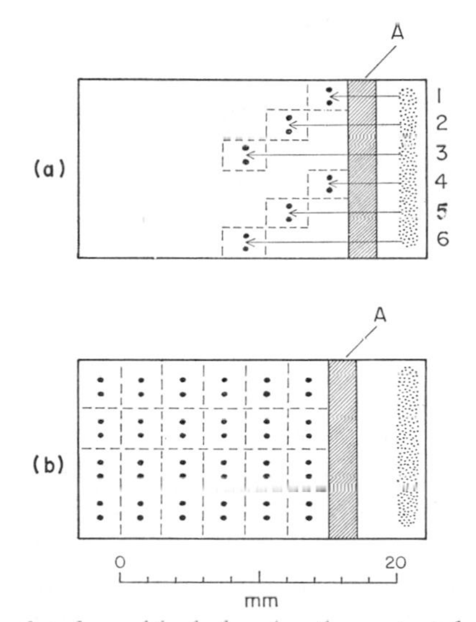

FIG. 10. (a). A completed agar block showing the routes taken by six isolates trailed in turn with an angulated microneedle from the inoculum. (b). An agar block pre-marked with 24 isolation sites for use with a microloop. In both parts of this Fig., the shaded areas are those tested for sterility, the stippled areas carry the inoculum and the broken lines indicate the dissection of the gel.

The concentration of agar must be adjusted, after experiment, for each batch prepared. If too high, the organism will not follow the microneedle: if too low, the excess water exuded around the needle tip will cause displacement of adjacent organisms.

4. *Dissection of the block and culture of the organisms*

The completed block, on the supporting slide, is transferred to the stage of a stereoscopic dissecting microscope of magnification $\times 5$. Aerial contamination is excluded (1) by a cabinet enclosing the stage with a roof of plate glass immediately below the objectives, and preferably also containing a low-pressure mercury-vapour lamp to sterilize the contained air before use, and (2) by a sterile plastic shield resting on the stage around the slide, with an access slot for the dissecting knives and a central aperture, sealed by a coverglass, below the objectives.

The agar block is dissected with sterile, sharp, triangular stainless-steel knives made from safety razor blades soldered into brass handles (Fig. 11).

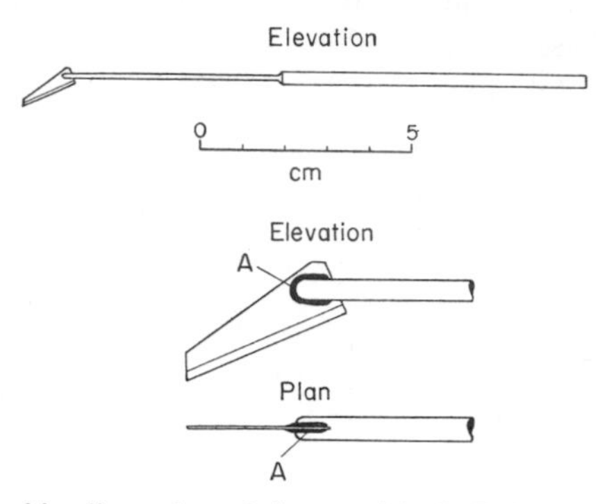

FIG. 11. A knife used in dissection of the agar block. Below, the tip enlarged. At A, a portion of a stainless-steel safety razor blade is soft-soldered into the brass rod.

The site of the inoculum is first separated and is removed with a platinum scoop. The adjacent section of the gel, A (Fig. 10a), is then isolated and transferred to the optimal recovery medium. Unless this is shown to be sterile, the isolations are not valid. The remainder of the block is then systematically dissected and small portions of the gel, each with a pair of locating pits, and, therefore, an organism, are transferred in turn to separate tubes of the medium for incubation. *Single organisms require an optimal medium and optimal conditions of reaction, oxidation–reduction potential and temperature, if a high proportion of the isolates are to prove viable.*

An advantage of this method is that the medium can be fluid or solid, clear or opaque, depending on the nature of the organism. If solid, the small block of agar is deposited with the platinum scoop on to the medium, pits downwards, the isolate being trapped between the gel surfaces and growth appearing around the margins of the block if the organism is viable.

F. Isolation with a microloop on pre-marked sites

This method has the advantage of great speed and is of especial value when many isolates of the same strain are required. The microloop (Fig. 12) is formed in a high-power microforge at the tip of a needle shaft prepared as for an angulated microneedle. The loop is made from a terminal glass filament 1·0–1·5μm diameter and measures internally from $3 \times 1{\cdot}5\mu$m to $14 \times 8\mu$m, depending on the size and the number of organisms to be carried. The loop, which is elliptical in shape, must lie in the plane of the gel surface.

For maximum speed, the sites for 24 isolations are punched in the surface of an agar block by a group of 48 steel needle points, mounted in an enclosed sterile punch. The sites, at 3 mm intervals (Fig. 10b), are in

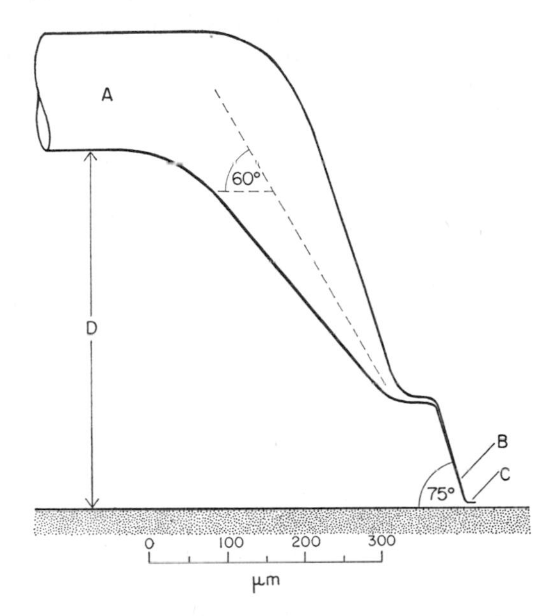

FIG. 12. The tip of a glass microloop in relation to the agar gel surface. A, the shaft; B, the terminal glass filament; C, the loop lying in the plane of the gel surface, which is stippled; D, the clearance between the shaft and the gel.

register with the readings of the mechanical stage of the observing microscope. After inoculation of one end of the upper surface of the block, selected organisms are picked up with the microloop, to the interior of which they adhere firmly by surface tension when raised above the gel. For the isolation of pure strains, organisms are carried singly in the smallest practicable loop to their respective isolation sites, and are there discharged by slight vibration imparted to the bench *while the loop is in contact with the water exuded from the gel surface.*

For multiple isolations from an already pure strain, 6 or 12 organisms are picked up successively by a larger loop, which is then lowered on to the first site. By slight vibration of the bench, one or more organisms are ejected from the loop and remain on the gel surface as the loop is raised. If more than one has been discharged, the excess organisms are again picked up, leaving a single isolate between the pits. The remaining sites are treated in the same way, the whole block of 24 isolations being completed in less than 8 min with a suitable microloop.

Dissection of the block and transfer to the culture medium is carried out as for the angulated microneedle technique.

In the case of large spores, such as fungal spores, a large microloop will readily pick up the spore, but difficulty may be experienced in releasing it. This can be overcome by using a loop with a strong supporting filament and sinking the loop below the surface of the gel. As the agar rises through and

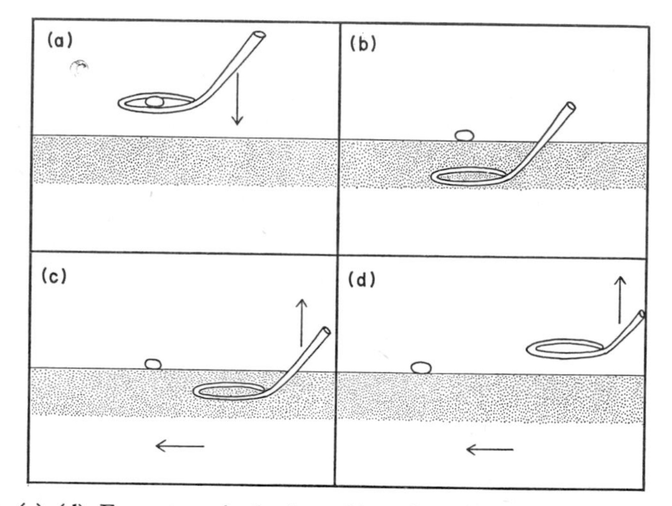

FIG. 13. (a)–(d). Four stages in the deposition of a fungal spore on the agar surface. The vertical arrows indicate movement of the microloop: the horizontal arrows indicate simultaneous movement of the gel. Diagrammatic—not to scale.

around the loop, the spore is carried on the surface of the gel. The microloop is then extracted from the gel by an oblique motion obtained by simultaneous operation of the fine adjustment of the manipulating microscope and of the mechanical stage controls of the observing microscope (Fig. 13).

Sterilization of microinstruments cannot be effected by heat. Immersion in chromic–sulphuric acids cleaning solution, followed by washing in sterile water and in sterile ethanol, yields a clean, sterile instrument.

G. Evidence for the reliability of the methods

The possibility of spreading of the organisms from the inoculation site along the surfaces of the block of agar is excluded by the demonstration of the sterility of the section of the block between the inoculum and the isolation sites. Such a spread of the organisms never occurs on correctly prepared blocks, except when due to carelessness in assembly.

The possibility of displacement of isolates from their marked sites by water exuding from the agar during dissection of the gel does not occur provided that sharp knives are used for the dissection. This can be proved in two ways—

(a) *Microscopical evidence.* After dissection of a block with 24 isolates, followed by transfer of each portion to a second slide, each isolate can be found microscopically between its locating pits.

(b) *Cultural evidence.* Frequently it is found that each of 24 bacterial spores isolated on the same agar block gives rise to growth on transfer to a culture medium. This can only occur if each spore has been transferred to the medium on its allotted portion of the gel. This test is only valid in cases in which all the isolates are viable.

REFERENCES

Dickinson, S. (1926). *Ann. Bot.*, **40**, 273–274.

Feinberg, J. G. (1956). *Nature, Lond.*, **178**, 1106.

de Fonbrune, P. (1949). "Technique de micromanipulation," p. 130. Masson, Paris.

Gardner, A. D. (1925). *J. Path. Bact.*, **28**, 189–194.

Holdom, R. S., and Johnstone, K. I. (1967). *J. gen. Microbiol.*, **46**, 315–319.

Johnstone, K. I. (1943). *J. Path. Bact.*, **55**, 159–163.

Johnstone, K. I. (1953). *J. gen. Microbiol.*, **9**, 293–304.

Koblmüller, L. O., and Vierthaler, R. W. (1933). *Zentbl. Bakt. Parasitkde, Abt I, Orig.*, **129**, 438–446.

Malone, R. H. (1918). *J. Path. Bact.*, **22**, 222–223.

Ørskov, J. (1922). *J. Bact.*, **7**, 537–549.

Reyniers, J. A., and Trexler, P. C. (1943). *In* "Micrurgical and Germ-free Methods" (Ed. J. A. Reyniers), pp. 8–10. C. C. Thomas, Springfield, Illinois.

Topley, W. W. C., Barnard, J. E., and Wilson, G. S. (1921). *J. Hyg., Camb.*, **20**, 221–226.

CHAPTER XIII

Design of Laboratory Fermenters

N. BLAKEBROUGH

Chemical Engineering Department, Indian Institute of Technology, Delhi, India

I. INTRODUCTION

The practice of fermentation has a history stretching over many centuries, but the effort put into the design of fermenters over the last quarter-century far exceeds all that of the preceding years. Hundreds of papers have described fermentation equipment of varying degrees of complexity. It will be the purpose of this Chapter to examine the factors which influence the choice of equipment, especially on the laboratory scale. Some types of fermenter are described by other contributors to this work. Materials of construction, details of fittings, the sterilization of air, equipment and media, and the control of operating parameters are also described elsewhere and will, therefore, be dealt with here only insofar as they relate to fermenter design in general and the degree of sophistication required for any particular purpose. It will be assumed throughout that a basic requirement is that the equipment shall be suitable for the cultivation or utilization of a pure mono-culture, or defined mixture of species, with a high degree of protection against chance contamination.

II. OBJECTIVES

Within the limitations of the basic requirement just stated, there is considerable scope for the use of many types of equipment. In practice the choice has become limited by considerations of convenience and economy, and by the frequent requirement that the apparatus should be suitable for various microbial systems. The most widely used types of equipment are the conical or Erlenmeyer flask and the stirred vessel, usually in the form of a vertical cylinder. The latter, in particular, shows many variations on the original theme, but variations also appear in other types. Whilst many factors may control the final choice it is useful to consider design in relation to process requirements, that is to relate the choice to the objective sought in any particular case. These objectives can be considered generally under four heads.

A. Provision of cells

Here the purpose may be simply to produce a given mass of cells in a reasonable time without heavy investment in equipment. Sometimes a high degree of sophistication may be necessary to produce cells in a particular physiological state, and carefully controlled continuous culture equipment may be required.

B. Provision of products

When the objective is not solely or mainly to encourage the growth of the organisms, the choice of equipment and operating conditions may be related to the formation of a desired product. In such cases simple apparatus might have to be unduly large to produce the desired amount of material. More important it may not be possible to achieve the desired concentration of material without some degree of elaboration. For instance, the concentration of many antibiotics and some vitamins may be increased tenfold or even a hundred-fold in agitated systems as compared with static cultures. Furthermore, the capacity of the equipment may be largely determined by the purpose for which the product is required; whether for analysis, or for field or clinical trials, or for the evolution of processes for the isolation and purification of the product on a commercial scale.

C. Study of growth and metabolism

Provision should be made for the aseptic withdrawal of samples (gas and liquid), in both batch and continuous cultures, and for the measurement of some variables (pH, dissolved oxygen, dissolved carbon dioxide, redox potential, cell concentration, composition of inlet and exhaust gases, etc.) *in situ*. It may also be necessary to provide for the addition of reagents to control pH, antifoam agents, precursors or other metabolites or inhibitors. For continuous microbial systems, some means for varying residence time by adjusting the volume of culture or the rate of medium flow or both is required. It may also be necessary to provide for multistage cultures, with arrangements for transfer from vessel to vessel, and possibly for recycling of a proportion of the effluent stream or of the discharged micro-organisms.

D. Parameters required for plant design

The criteria discussed so far are relevant to the design and selection of laboratory equipment, whether the laboratory operation is to be an end in itself or whether it is to provide information from which larger units can be designed for commercial production. In the latter case, additional criteria may be relevant and may impose restrictions on the designs employed in the laboratory. Ideally, scale-up should be carried out on the basis of biological parameters, or at least of physical parameters such as oxygen absorption rate and mixing time which may be of direct biological significance. Unfortunately, it is frequently difficult to relate such parameters directly to the physical design of equipment in the present state of knowledge. As a result, the uncertainties associated with scaling-up increase unless the equipment used on the various scales of operation maintains the principle of geometrical similarity at least in respect of certain essential features. Criteria which may be of importance to the engineer in achieving an economic design which will meet the requirements of the biologist, include the power absorbed by the system, air-flow rates, effects of pressure, the significance of mixing rate, shear and turbulence, the rheological characteristics throughout the growth cycle and the rates of heat transfer during sterilization and cooling and during the course of the fermentation. Heat transfer capacities of the equipment may assume importance where the performance of the microbial system is significantly affected by changes taking place in the medium during sterilization, and also in the control of temperature during the actual process. For a given design, the ratio of volume to external surface area increases in proportion to the diameter of the vessel as the scale increases. The rate of heat evolution or absorption per unit of external surface will increase approximately in the same order, until the rate of heat transfer through the shell to or from an external temperature-controlled medium may be insufficient to maintain the desired temperature. It will then be necessary to introduce additional surface, usually in the form of internal coils. If strict geometrical similarity is to be maintained on scale-up, dummy coils should be fitted to the laboratory fermenters. It is doubtful, however, whether this complication is worthwhile, at any rate in vessels which are baffled and are used under fully turbulent conditions.

III. OPERATING CONSIDERATIONS

So far we have considered the effects of process requirements on the general nature of the equipment to be employed, and the way these requirements may influence the determination both of equipment capacity and the degree of sophistication required. Before considering actual designs, it will be useful to review the operational factors which affect design.

A. Freedom from contamination

This is a fundamental requirement which affects not only the design and standards of construction of the fermenter itself, and the choice of ancillary equipment such as valves, pumps and sampling arrangements, but also the operational procedures which will be used for cleaning, charging, sterilizing, inoculating, sampling, transferring and harvesting.

The apparatus should be as simple as possible, having clean lines, and being free from crevices and pockets which make cleaning difficult and may impede the access of heat or other sterilizing agents. This same principle should apply to the selection of valves and fittings and to the way in which these items are connected to the vessel. Particular attention should be paid to the design and maintenance of any component which provides a potential conduit between the interior of the vessel and the external environment. This applies particularly to shaft seals on agitators and pumps, sampling points and any temporary connections to other equipment, e.g., inoculators. The number of such components should be kept to a minimum, and maintenance and operating procedures should be designed to minimize the risks of contamination. It is desirable that all surfaces in contact with the vessel contents should be impermeable, resistant to chemical attack by the contents of the vessel (including chemical sterilizing agents, if used) and have a smooth finish. In most cases the materials should not be adversely affected by steam sterilization. Frequently provision has to be made for addition to or abstraction from the vessel during the progress of the biological process, and such operations should not result in contamination. Often large quantities of air (occasionally other gases) must be supplied, and an effective and economical means must be provided to reduce the risks of entry of foreign micro-organisms from this source to an acceptable level. The ingress of contaminants through joints or as a result of developing leaks is usually reduced by maintaining the pressure in the vessel above that of the surrounding atmosphere. This may be unacceptable when cultivating pathogenic organisms, when the vessels may be operated at sub-atmospheric pressure, to provide protection to the operators. In these cases the avoidance of contamination requires extremely high standards of construction and maintenance, or the use of specially designed equipment.

B. Aseptic removal of samples

There are two aspects to this problem. One is to avoid contaminating the withdrawn sample, especially important if the sample is to be used to check the purity of the culture. The other is to ensure that the sampling procedure does not involve the risk of contaminants entering the fermentation vessel.

Sampling devices may be of a detached kind, for example a sterile hypodermic syringe introduced through a self sealing diaphragm and withdrawn after abstracting the sample. The principal disadvantage of this type is that effective post-withdrawal sealing is most readily achieved using fine probes, whilst rapid sampling or the removal of a sample containing large cellular aggregates or other coarse suspended matter requires wide tubes. Alternatively the sampling device may be an integral part of the equipment, through which a sample may be drawn either by suction or by using the pressure in the vessel. The main risk in using such devices arises from the possibility of leaving, after sampling, a continuous column of stagnant nutrient medium interrupted only by a valve. Additional precautions must, therefore, be taken to prevent the possibility of contamination by "growing-back" from the atmosphere. This is usually achieved either by capping or by steam-sealing.

Similar problems arise when transferring inoculum between one vessel and another.

C. Aseptic additions

These may include antifoam agents, nutrients, inhibitors, water, air and inoculum ("aseptic" in this latter case being interpreted as "uncontaminated"). Some reference has been made to these points above, but some further amplification is required. Such additions may be made from permanently attached reservoirs initially holding sufficient material for a complete experiment. Alternatively, additions may be made from a hypodermic syringe as required or from some other form of device which is attached temporarily and withdrawn when the addition has been effected. The careful design of fittings for such attachments reduces the risk of contamination both directly and also by reducing the chances of manipulative errors during attachment and removal. If a permanently attached vessel is to be recharged during the course of an experiment, it may be necessary to provide that this vessel, with or without contents, can be sterilized without interfering with the operation of the fermenter.

D. Sterilization

The vast majority of sterilization operations are carried out using steam. Equipment may be sterilized empty, to be charged aseptically with medium sterilized separately, or the vessel and medium may be sterilized together. Small and simple items of equipment are most conveniently sterilized by autoclaving or heating in an air oven. Sterilization *in situ* is virtually mandatory for equipment with a working capacity above 50 litres. For smaller vessels sterilization *in situ* or in an autoclave is a matter of convenience. If an autoclave is used, as much ancillary equipment as possible should be attached before sterilization to reduce the risks of contamination during subsequent assembly. A consequence of autoclave sterilization of vessels containing media is that the latter cannot be agitated during sterilization, so that heat penetration is slow, and cycles applicable to larger units cannot be simulated. This may be a major disadvantage if the medium and subsequent biological performance are sensitive to the effects of heat degradation. In such cases it may be necessary to sterilize some ingredients separately or to sterilize the medium in a continuous sterilizer from which it can flow into a pre-sterilized fermenter. Care should be taken to ensure that non-sterile air is not sucked into sterilized vessels during the cooling cycle. Particular attention should be paid to the design and assembly of fermenters with glass bodies to avoid the formation of gaps or the introduction of undue stress on the glass arising from differential expansion and contraction of the glass and metal components.

Sterilization *in situ* allows the equipment to be completely assembled and connected to the air supply and agitator drive before sterilization. On the other hand it complicates the service connections in a way that may be troublesome with small fermenters. Again problems may arise with glass bodied fermenters because of limitations on the pressure (and thus steam temperature) that they will safely withstand.

If the vessel is to be sterilized empty an antiseptic solution may be used, though it is difficult to ensure that all surfaces will be treated, and some components, such as air filters, will have to be sterilized separately by heating. A more satisfactory solution uses a volatile fumigating agent such as ethylene oxide or β-propionolactone, though the former requires careful handling because of its toxicity to humans. If chemical sterilizing agents are employed it is essential to ensure that they do not attack the equipment (especially sealing and gasket compounds) and that they can be completely removed subsequently without contaminating the apparatus.

Ancillary equipment, such as pumps, that will be in contact with sterile fluids must be so constructed that the contact parts are accessible for cleaning and can withstand sterilization.

E. Control and measurement

Measurement and/or control of numerous parameters may be desired. These include cell concentration, growth rate, product concentration, pH, respiration rate, heat evolution, flocculation or other forms of aggregation, and sporulation, which are characteristic of the response of the organism to its history and current environment. Other factors such as temperature, pressure, air flow-rate, intensity of agitation, medium feed rate and composition, may be considered as external and independent of the culture. Control of the biological process is exercised by manipulating one or more of these external variables to maintain an environment favourable to the desired biological activity. Each external variable may be fixed or may be varied according to a pre-determined pattern based on experience of the process. Alternatively, an intrinsic biological parameter may be measured and its value used to adjust the external parameters to achieve the desired result. The number and nature of measurements to be made and the nature of the measuring techniques and control systems may affect not only the basic design of the fermentation equipment but also its size. Where, for instance, measurement is made on a withdrawn sample, the size and frequency of sampling must not significantly affect the volume of culture in the vessel. If measurements are to be made *in situ*, provision has to be made for incorporating the appropriate sensing devices. Measuring optical density or operating photo-synthetic processes requires a provision for the passage of light. Where sensors are to be in direct contact with the biological system (for instance, electrodes for determining pH, redox potential, dissolved oxygen, level detectors, foam detectors) these must not act as foci for contamination and must be able to withstand the conditions obtaining in the fermenter. The risk of leakage and subsequent contamination will be minimized by locating the entry glands for such fittings well above the liquid level.

In some cases, the number of factors to be measured and/or controlled may determine the minimum size of equipment required to accommodate the necessary sensors and addition devices.

F. Miscellaneous considerations

In addition to the factors considered in some detail above, robustness, visibility of the contents, equipment handling and access to the interior for cleaning and modification may be important. It is not possible to provide effective illumination and viewing ports in the cover of any vessel with a diameter less than about 12 in. If viewing of the contents is considered important, small vessels must be glass-bodied or incorporate viewing panels in the body. Glass bodies are less robust than metal ones and viewing panels increase the risk of leakage and contamination. This can be overcome to some extent by using heavy walled tubing, such as industrial pipe-line, though this may make it more difficult to achieve good temperature control unless the heating and cooling systems do not depend on transfer across the wall.

If the vessels are not to be cleaned, charged and sterilized *in situ*, it may be desirable to limit the capacity so that the vessels can be transferred manually between cleaning, charging, sterilizing and operating points without the need for mechanical aids. Except for simple all-glass equipment, such as aspirators, this limits the working capacity to about 5 litres.

All glass equipment, such as aspirators and flasks, can usually be cleaned by brushing, supplemented by using detergents and chemical cleaners and, in appropriate cases, by partial demounting. The simplicity of such an arrangement is partially offset by lack of robustness and by the limitation which is set to variations of internal geometry such as the size and type of stirrer and the fitting of baffles.

There are advantages, in relation to cleaning and modification, in designs which permit ready access to the interior of the vessel. This can be achieved with both glass-bodied and metal-bodied units by fitting a metal cover secured by a bolted flange. The cover is made sufficiently rigid to carry the agitator bearing assembly, air inlet and exhaust pipes, and fittings for ancillary equipment such as addition and sampling points, pH electrode and foam probe. This arrangement inevitably leads to the presence of a joint between the body and the cover. This can be avoided, in metal vessels with a top entering agitator with a working capacity greater than about 1000 litres, by fitting a manhole in the cover. A manhole can be fitted into the top of smaller vessels when bottom entry agitator shafts are used. However, the use of bottom entry agitator shafts involves more problems with the design and maintenance of the stuffing box for the agitator shaft.

The basic requirements in fermenter design are similar whether operation is to be batch-wise or continuous. However, continuous operation leads to requirements for a continuous controlled supply of nutrient, withdrawal of product and regulation of the volume of culture in the vessel. If multistage operation is desired, the withdrawal device must act also as a delivery device to the next vessel in the series. The simplest form of withdrawal device is a weir. This has the advantage that it also acts as a self-regulating level controller, but difficulty can occur when handling organisms which form large aggregates since these may be retained on the weir, particularly at low volumetric flow rates. Transfer may also be effected by pumping, though difficulties may arise from sedimentation in the transfer lines at low rates of flow. In both types of device the problems associated with low flowrates can be overcome using intermittent transfer at intervals sufficiently frequent as not to represent a significant departure from continuous operation.

A decision in favour of continuous or batch operation can be a major factor in determining equipment size. Continuous processes with a short mean generation time (say 20–120 min.) will require much smaller equipment for a given rate of output than batch equipment, in which a large proportion of total process time may be spent in turning round the equipment between batches and in the lag between inoculation and maximum rate of biological activity. As the residence time increases the disparity of sizes for a given rate of output will diminish.

TABLE I

Effect of operating variables on relative capacity and output of batch and continuous fermenters

$A(h)$	$-B(h^{-1})$	X	$\log_2 X$	$T(h)$	O_C/O_B	R
6	0·5	64	6	100	12·2	8·2
6	1	64	6	100	8·0	12·5
6	10	64	6	100	3·3	30·3
6	10	64	6	500	4·2	23·8
6	0·5	512	9	500	14·4	6·9
6	1	512	9	500	9·6	10·5
6	10	512	9	500	5·8	17·2
6	10	512	9	100	4·0	25·0
12	0·5	32	5	200	20·0	5·0
12	1	32	5	200	11·6	8·6
12	10	32	5	200	3·7	27·0
12	30	32	5	200	2·6	38·5
12	0·5	512	9	500	22·7	4·4
12	1	512	9	500	14·4	6·9
12	10	512	9	500	6·2	16·1
24	0·5	512	9	500	39·1	2·6
24	1	512	9	500	22·5	4·4
24	10	512	9	500	6·3	15·9

O_C/O_B shows the relative outputs from continuous and batch operation in a vessel of a given capacity. R shows the relative capacity of a continuous culture vessel as a percentage of the volume of the batch culture vessel to give the same output for specified values of A, B, X and T.

Similar tables can be derived for other conditions (e.g., when θ is not equal to B) or for the output of a fermentation product if relationships can be written for the time rate of formation.

The influence of several factors on the relative sizes and outputs of batch and continuous fermenters is illustrated in Table I, based on a somewhat simplified picture of the production of cell materials showing exponential growth. It applies also to the formation of growth-associated products, but not to products related to the concentration of cells rather than to their reproduction.

The values for relative fermenter size and output are derived as follows—

Let A = turn-round time (i.e., time from harvesting of one fermentation to initiation of exponential growth in subsequent fermentation).

B = doubling time of organism in the absence of nutrient limitation or product inhibition.

N_B = Cell concentration at harvest in batch culture

N_C = Steady-state cell concentration in continuous culture.

X_B, X_C = Ratio of final cell concentration: initial cell concentration in batch and continuous culture, respectively.

Then, for batch fermentation, number of doublings

$$n = \log_2 X_B \tag{1}$$

$$\begin{aligned}\text{Time of batch culture} &= A+Bn \\ &= A+B\log_2 X_B\end{aligned} \tag{2}$$

If V = fermenter volume, total output per batch = N_BV. Hourly output over full cycle

$$= \frac{N_BV}{A+B\log_2 X_B} \tag{3}$$

For continuous culture,

$$\begin{aligned}\text{total time} &= A+Bn+T \\ &= A+B\log_2 X_C+T\end{aligned} \tag{4}$$

Where T = time at steady state.

$$\text{Hourly output at steady state} = N_Cv = \frac{N_Cv}{\theta} \tag{5}$$

where v = hourly flowrate

θ = residence time.

Output over continuous fermentation cycle

$$= \frac{N_CVT}{\theta} + N_CV \tag{6}$$

$$\text{Hourly output over whole cycle} = \frac{N_CV(T+\theta)}{\theta(A+B\log_2 X_C+T)} \tag{7}$$

Then

$$\begin{aligned}\frac{O_C}{O_B} &= \frac{N_CV(T+\theta)}{\theta(A+B\log_2 X_C+T)} \cdot \frac{(A+B\log_2 X_B)}{N_BV} \\ &= \frac{N_C}{N_B} \cdot \frac{(T+\theta)}{\theta} \cdot \frac{(A+B\log_2 X_B)}{(A+B\log_2 X_C+T)}\end{aligned} \tag{8}$$

where O_C = hourly output in continuous culture;

O_B = hourly output in batch culture.

If $N_C = N_B$ and the initial cell concentrations are the same for batch and continuous culture, then $X_C = X_B = X$. If also, T is large compared with A, equation (8) becomes—

$$\frac{O_C}{O_B} = \frac{(T+\theta)}{\theta} \cdot \frac{(A+B\log_2 X)}{(T+B\log_2 X)} \tag{9}$$

If the continuous culture system is operated at maximum dilution rate $\theta = B/0{\cdot}693$, and equation (9) further reduces to—

$$\frac{O_C}{O_B} = \frac{(0{\cdot}693T+B)}{B} \cdot \frac{(A+B\log_2 X)}{(T+B\log_2 X)} \tag{10}$$

Table I shows the values of O_C/O_B and of the relative volumes of batch and continuous fermenters for a given output, computed from equation (10). This tends to exaggerate the advantages of continuous culture, since, in practice, operation at the maximum dilution rate is not possible, and the cell concentration in continuous culture may be somewhat less than that economically attainable in batch culture. Nevertheless, the relative output rates and comparative volumes are of the right order of magnitude and the table gives a valid indication of the effects produced by variations in turn round time, growth rate and the time of operation at steady state. For any particular case, more accurate values can be calculated from equation (8) when the values of N_C, N_B and θ have been determined.

There are advantages in continuous operation other than a saving in equipment size, particularly as an investigational tool, but this aspect is dealt with elsewhere in this work (Tempest, this Series, Vol. 2). It must also be borne in mind that continuous operation involves investment in ancillary equipment, such as pumps, which is not required for batch-wise operation, and this must be taken into account in relation to initial and operating costs. Continuous equipment is by nature more complex than batch equipment. It is necessary to consider for any particular case whether this additional complexity is justified by any advantages which may arise.

IV. SELECTION OF EQUIPMENT

Having considered the factors that must be taken into account in selecting equipment, it is proposed to discuss the characteristics of equipment that is commercially available or that can be assembled from readily available components or fabricated by a glass-blower. No attempt will be made to deal with the details of mechanical design. Attention will be given first to the two most commonly used laboratory designs—the flask and the stirred fermenter including variants of the latter suitable for continuous operation. Later, consideration will be given to other less common types which may have special virtues such as cheapness or special characteristics but which have not found wide acceptance.

A. Flask fermenters

By far the most common type in this category is the conical or Erlenmeyer flask, though round-bottomed, flat-bottomed and bolt-head flasks may also be used, together with special designs such as the Pasteur flask used for yeast culture. If the rate of mass transfer between the culture medium and the overlying gas is unimportant, stationary cultures may be employed and the selection of flask type and the ratio of flask to culture volume will be generally unimportant also. Such cultures may be readily developed in laboratory incubators or on shelves in a room maintained at the appropriate temperature.

However, some organisms grow in the form of a surface film, mat or pellicle in stationary culture. Agitation will usually result in a more dispersed form of growth such as single cells, flocs, pellets or filaments. This may have profound biochemical effects and the choice of stationary or agitated culture may rest on the significance of these effects.

Where gas/culture mass transfer is an important factor or where surface growth is to be avoided, the choice is more limited, and the conical flask has found almost universal acceptance. It has a stable shape, is easy to clean and for a given nominal capacity and culture volume offers a larger gas-liquid interfacial area than any other pattern. Moreover, it can be easily secured in clips or other simple holding devices to a platform which can provide a gyratory or oscillating movement. This further increases the interfacial area and also induces movement both within the liquid and at the interface, thus improving the rates of mass transfer. These effects can be further enhanced, especially on gyratory (rotary) shakers by forming indentations around the skirt, though this increases the cost and makes cleaning rather more difficult. The great virtues of the shaken flask over other types of laboratory equipment lie in its cheapness, simplicity, ease of storage and preparation and low space requirement. For preliminary multifactor screening it is unsurpassed, and is widely used for comparing strains or mutants, and for the comparison of media. By varying the volume of medium in each flask, and the amplitude and frequency of shaking, valuable information can often be obtained on the effects of aeration and agitation. Although such results are not easy to interpret quantitatively in relation to the levels of aeration and agitation required in fermenters of more sophisticated design, they may reduce significantly the amount of experimentation required on a larger scale. The necks of the flasks are normally plugged with cotton wool or porous plastics. This arrangement prevents contamination but permits an exchange of gas between the interior of the flask and the surrounding atmosphere. The rate of exchange is dependent on the porosity of the plug, which may vary appreciably from flask to flask, especially when formed manually from cotton wool. If a controlled rate of flow of air or other gas is required, the flask can be fitted with a rubber stopper carrying inlet and outlet tubes flexibly connected to suitable manifolds. Whilst this may be necessary for special purposes, it represents a significant departure from the essential simplicity of the apparatus, and should only be adopted after careful consideration.

The maintenance of a high surface/volume ratio, and avoidance of plug-wetting place a limitation on the volume of medium in each flask (typically 100 ml. in a 250 ml. flask, 500 ml. in a 2 litre flask).

If large numbers of flasks are to be incubated at the same temperature, the shaking machines may be installed in a temperature-controlled room. When the machines are to be installed in rooms operating at elevated temperatures (above 35°C) special attention needs to be given to the selection of electric motors, wiring and lubrication of bearings. For smaller numbers of flasks or for cases where temperature may be a significant experimental variable, self-contained units may be a more economical solution; but these units are not usually designed to operate below ambient temperature. Equipment of this kind is available from Gallenkamp in the United Kingdom and from the New Brunswick Scientific Company in the U.S.A., among others.

B. Stirred vessels

The only fermenter to rival the flask in popularity and versatility is the stirred vessel in the form of a vertical cylinder with a flat or (more usually) dished base and a centrally placed agitator. Many such vessels have been described, most differences being in details of construction or the incorporation of modifications for some special purpose. These designs can, however, be further subdivided into two main types—(a) the fully-baffled vessel with sparger aeration, (b) the vortex fermenter, which is unbaffled and provides aeration by entraining air from the headspace into the vortex produced by the action of the impeller. Interchangeability between the two types can be achieved by the incorporation of removable sparger and baffles. These features can be incorporated readily in most equipment suitable for laboratory operations. They are less readily incorporated in production fermenters, but this limitation is not significant in practice since the choice between the two types will usually have been made before this stage is reached.

The sparged, baffled vessel is superior to the vortex-aerated vessel in respect of mass transfer and mixing of the contents. It can also be scaled up with a greater degree of confidence particularly in systems in which the viscosity varies widely over the course of the fermentation. The principle advantage claimed for the vortex fermenter, apart from its greater simplicity, is its smaller requirement of antifoam agent in cases where foaming is a problem. This advantage arises from the re-entrainment of the foam into the vortex, but is achieved at the price of reduced mass transfer rates and reduced effective fermenter capacity, arising from greater gas hold-up (i.e., the volume of undissolved gas entrained in unit volume of liquid) and the volume occupied by the vortex. The system may commend itself in cases where the use of antifoam agents is to be rigorously avoided.

For batch operations fermenters with an operating volume of 3 to 5 litres (total volume 5–10 litres) are probably most popular, giving a satisfactory balance between economy and the needs of sampling; and permitting the inclusion of most features of larger vessels whilst maintaining the advantages of portability and general ease of handling. For continuous fermentation, especially when the residence time in the system is low, smaller vessels are often more appropriate for laboratory operation. Considerable modification of the basic design may then be necessary to incorporate all the measurement and control features which may be desired. This point will be reverted to later.

Most laboratory fermenters include a stainless steel head, carrying the agitator, sparger, sampling and exhaust tubes, thermometer pocket and so on, surmounting a glass or stainless-steel body. Stainless-steel bodies are usually fabricated in one piece with a dished base. Glass bodies may be in the form of a mould-blown glass cylinder with integral flat or hemispherical base, or they may consist of a tube secured between the head-plate and a stainless-steel base-plate, with the incorporation of suitable gaskets. Where the latter arrangement is used some of the fittings referred to above may be located in the base-plate, a useful feature particularly in the smaller sizes, though this will usually preclude standing the vessel subsequently in a water-bath, a very convenient system of temperature control which is widely used. The use of a body made from tube open at both ends increases the risk of contamination, but provides a convenient way of locating a withdrawal or transfer device in continuous fermenters.

Fig. 1 shows two fermenters with a working capacity of five litres, one with a stainless-steel body, the other with a body formed from a glass tube secured between stainless-steel head and base-plates, and complete with drive unit, removable baffle assembly, air inlet and exhaust filters, addition vessels and sampling tube. Temperature control is achieved by using a water-bath, and each vessel is fitted with a conventional drive through a stuffing-box.

Fig. 2 shows a glass-bodied vessel with built-in heating and cooling elements. Power is transmitted to the agitator through a magnetic coupling, eliminating the need for a conventional stuffing-box. The removable baffle assembly, and various connectors in the cover can be clearly seen.

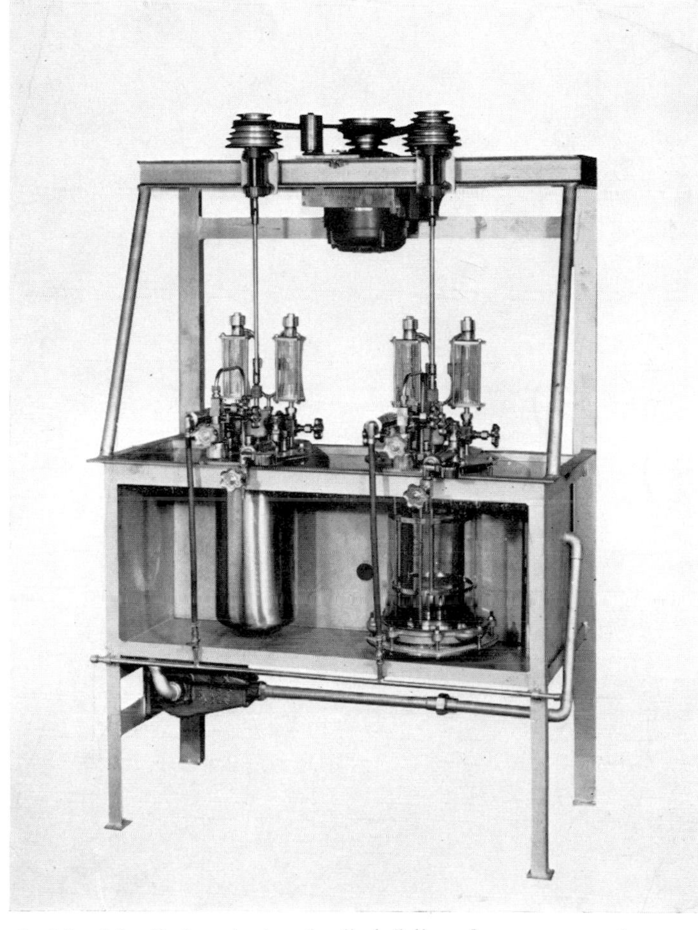

FIG. 1. Metal-bodied and glass-bodied 5-litre fermenters and accessories.

The units as illustrated are suitable for batch fermentations. Both designs of glass-bodied vessels could be adapted for continuous culture by installing a withdrawal device. In many cases this can be simply an overflow tube inserted through the base plate with a suitable seal to permit adjustment of the tube height, thus varying the volume and residence time in the culture vessel. Alternatively, the culture volume can be controlled by an adjustable level detector working in conjunction with a pump or solenoid-actuated valve. The latter arrangement involves the complication of a pump or valve and possible failure due to malfunction of the level detector. In small vessels

FIG. 2. Glass-bodied 5-litre fermenter with magnetic drive and integral heating and cooling.

this usually has to take the form of a conducting probe insulated except at the tip. When the latter is in contact with the culture fluid a circuit is completed through the wall or head of the vessel to a detector relay operating the pump or valve. Malfunction arises most commonly through

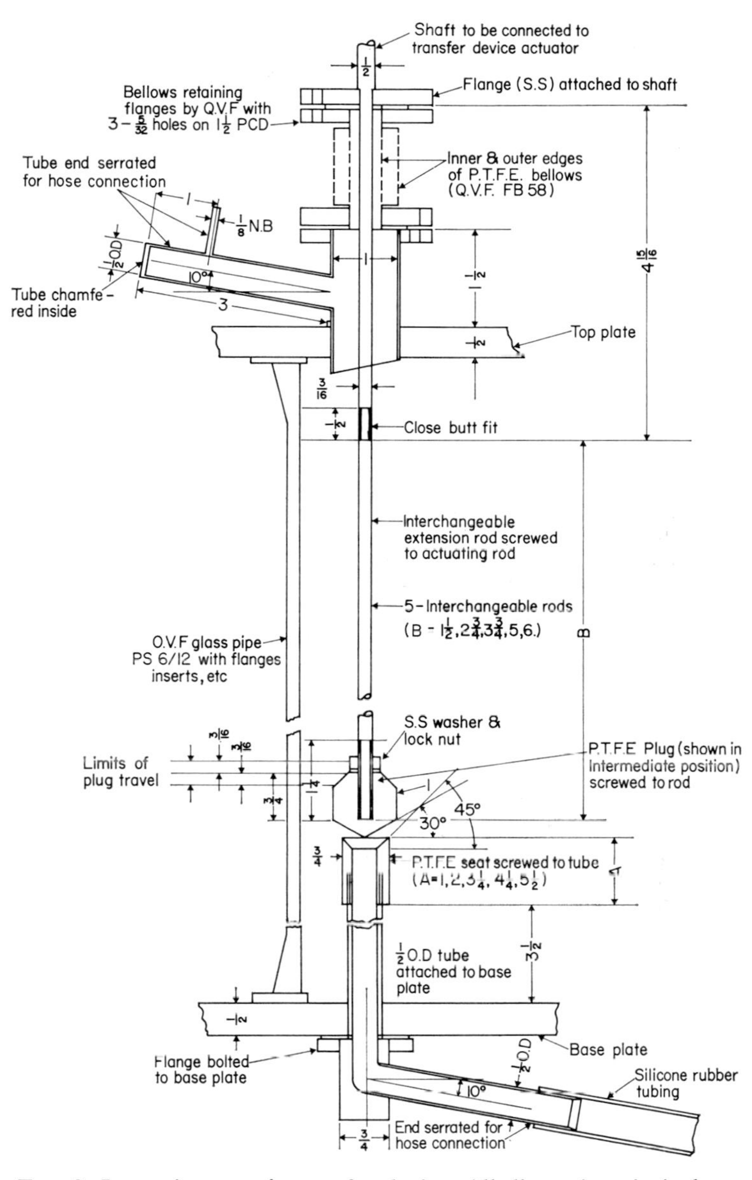

FIG. 3. Intermittent weir transfer device. All dimensions in inches.

splashing or growth of the organism on the electrode sheath causing shortcircuiting.

It has been stated above that the weir, typified by the overflow tube, is highly desirable because of its simplicity and inherent self-regulating characteristics. Its disadvantages are two-fold—(1) When the rate of flow over the weir is low, suspended solids may be retained on the weir, (2) Adjustment of the weir height by raising or lowering the tube may lead to difficulties in multivessel systems, because of changes in the spatial relationships of the vessels.

Both these difficulties can be overcome by the use of a device similar to that illustrated in Fig. 3. In this arrangement the weir is closed by a plug. As fresh feed enters the vessel the liquid level rises. The plug is raised at intervals and the accumulated excess fluid is discharged at flow-rates much higher than the average rate of flow into the vessel, thus eliminating the problem of solids hold-up. Flow and transfer are not truly continuous but the departure from continuity is generally so small as to be of no significance. For example a 15 cm diameter vessel has a culture depth of 17 cm for 3 litres working volume. A submergence of 0·5 cm at the time the plug is raised will be sufficient for all except the most difficult cutures. This represents a maximum variation in culture volume and nominal residence time of 3%. For cultures in which solids build-up is not a problem the plug can be held in the raised position permanently. Intermittent operation is achieved by using a timer actuating a solenoid or pneumatic cylinder to raise the piston. The interval between successive openings is adjusted to give the desired submergence of the weir, and is not critical since volume regulation depends on the open phase not the closed phase. It is usually sufficient that the weir should remain open for 20 to 30 seconds on each cycle. Volume variation is achieved by the use of interchangeable plugs and seats. A disadvantage of this arrangement is that the culture volume cannot be changed during the course of a run, so that residence time variation depends on varying the volumetric rate of feed. This is not generally a serious drawback and the arrangement has the advantage for multi-stage systems that adjustment of the weir height does not affect the location of the connections between one vessel and another. Although the system could be adapted to vessels arranged on a single level it is preferred that each succeeding vessel should be lower than its predecessor. In this way there is no hold-up in the transfer line and the possibility of oxygen depletion is eliminated. Even if pumps or solenoid valves are used in conjunction with a level controller, the multilevel arrangement is to be preferred to eliminate this problem. If a single-level arrangement has to be adopted, interconnecting pipes should be as short as possible to reduce the inter-stage residence time.

Fig. 4 shows the arrangements of head and base-plates suitable for a

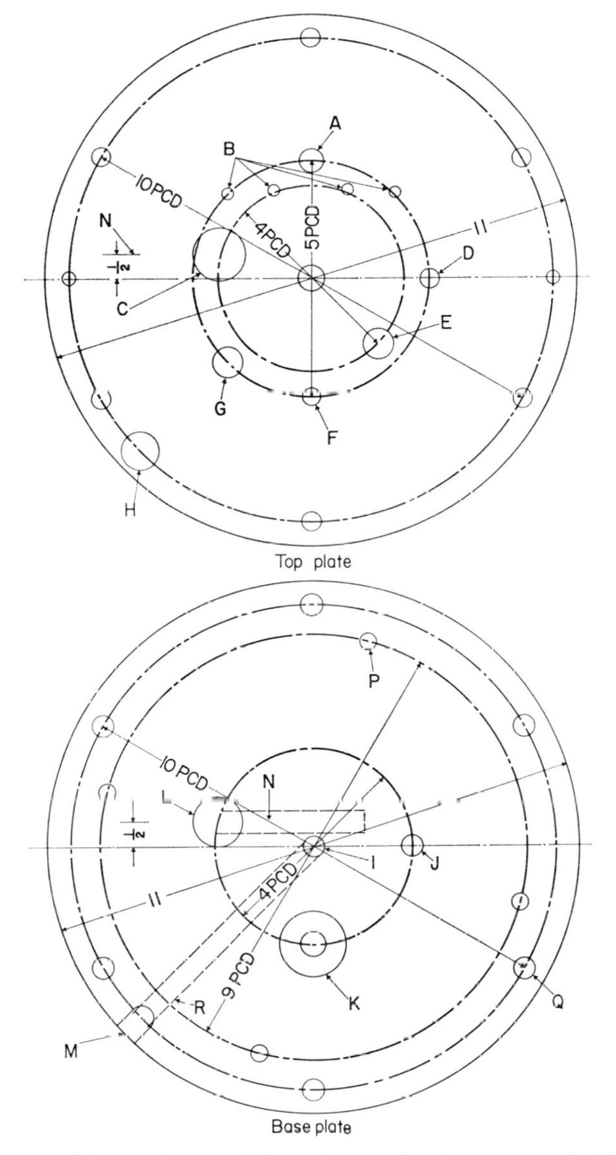

FIG. 4. Head and base-plate for 3-litre glass-bodied fermenter. A, Mounting for foam probe; B, 4 hose nipples $\frac{3}{16}$ bore; C, entry for transfer inlet device; D, mounting for air outlet and filter; E, cold finger; F, pocket fitted with resistance thermometer (10 in. long below the plate); G, mounting for pH electrode holder type 761-02 (W. G. Pye Co. Ltd.) and screw cap; H, mounting for barrel nipple to carry air inlet pipe and filter; I, entry for air pipe; J, mounting for s.s. pocket fitted with bayonet heater (app. 100 W); K, sample outlet; L, mounting for transfer outlet; M, outlet for air inlet pipe; N, alignment of transfer pipe; O, 2 $\frac{5}{16}$ holes diametrically opposite for mounting purposes; P, 4 $\frac{11}{32}$ holes for leg bolts; Q, 6 $\frac{7}{16}$ holes. All dimensions in inches.

15 cm diameter fermenter. These include provision for the transfer device, agitator, cold finger and heater for temperature regulation, sampling device, probes for pH and antifoam electrodes, air inlet and exhaust, and hose nipples to permit the introduction of medium, antifoam agent or other additives. Conversion to vortex aeration is made simply by detaching the air sparger pipe below the head plate and removing the baffle assembly.

Vessels of this kind can be made with working volumes from about 1 litre upwards. In the smaller sizes it becomes increasingly difficult to incorporate all the features, and problems associated with slow rates of flow become

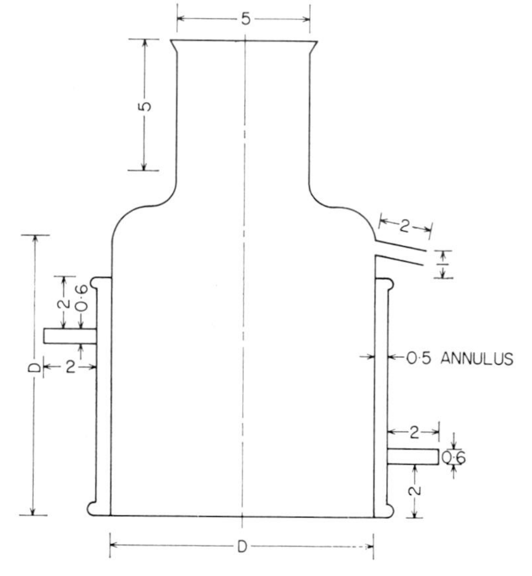

FIG. 5. All-glass fermenter with integral attemperating jacket. All tubulures are riffled for hose connection. All dimensions are in cm.

accentuated. Correspondingly, these problems are eased as capacity increases allowing the attachment of more fixed ancillary items and the provision of truly continuous flow. Units can vary from the very simplest with provision only for the control of air flow-rate, temperature and agitator speed, to elaborate arrangements with the control of many variables, a high degree of instrumentation and analysis, and automatic control of the sterilization cycle.

An alternative design is illustrated in Fig. 5. This is based on a design evolved at the Microbiological Research Establishment, Porton Down, and combines the virtues of low cost, simplicity and small capacity. It is particularly suitable for systems where required mass transfer rates are low to moderate, and where residence times in continuous systems are low. It has proved to be of great utility in biochemical studies of microbial systems. The unit is fabricated in borosilicate glass and has an integral jacket, through which water is circulated from a thermostatically controlled bath to maintain the desired temperature. The neck can be formed from glass tube, or a standard industrial pipe-line flanged end can be used. The top can be closed with a rubber bung or a stainless-steel flange, carrying tubes for air supply, medium feed and other additives, and for the withdrawal of samples if required. For robustness the tubes should normally be made of stainless steel. Agitation is provided by means of a magnetic stirrer, preferably using a follower coated with polytetrafluoroethylene (P.T.F.E.) to eliminate metallic contamination. Satisfactory stirring over a wide range of speeds is best achieved when the base of the vessel is as flat as possible. In the absence of baffles some vortexing occurs, though this is reduced by the presence of the various tubes. The degree of vortexing and thus the volume of culture retained in the vessel depends, in any given system, on the stirrer speed. In experiments in which residence time is a significant factor, this effect must be taken into account. Nominal capacities vary from approximately 100–800 ml for body diameters between 5 and 10 cm. Using a standard neck size for all capacities provides a degree of interchangeability. This size is sufficient to provide for a variety of fittings, including, if desired, an electrode for pH or dissolved oxygen determination. If more fittings, including baffles, are desired in the larger sizes, wider necks can be readily fitted. The overflow should lead to a sterile receiver to prevent contamination. If the vessel is to be used for batchwise culture the overflow tube should be closed with a bung or sealed tube. Samples may be withdrawn either by hypodermic syringe or by the use of a Heatley sampler. The latter consists of a hooded tube, the hood being screwed internally to accommodate a standard screw-neck sample bottle (McCartney or Universal pattern) and being fitted with a side-arm (see Evans *et al.*, this Series, Vol. 2). The latter is fitted with an air-filter to which suction can be applied, orally or by means of a vacuum pump, to draw a sample into the bottle. The bottle with sample is then removed and immediately replaced by a fresh sterile bottle.

The Waldhof fermenter incorporates some of the features of both sparged/baffled and vortex aerated vessels. A tube surrounding the impeller shaft encourages entrainment of headspace air and foam, whilst a baffle reduces the tendency to swirl and improves mixing. Aeration can be achieved from the headspace only or from a submerged sparger or through hollow sparger shaft and agitator blades. These fermenters have been widely used, especially in Germany, for yeast propagation, but have found little favour for other microbial systems.

Another type of all-glass agitated vessel with sparger aeration is shown in Fig. 6, which requires little description. The vessel as illustrated is set up for batch operation, though it includes provision for addition of nutrients

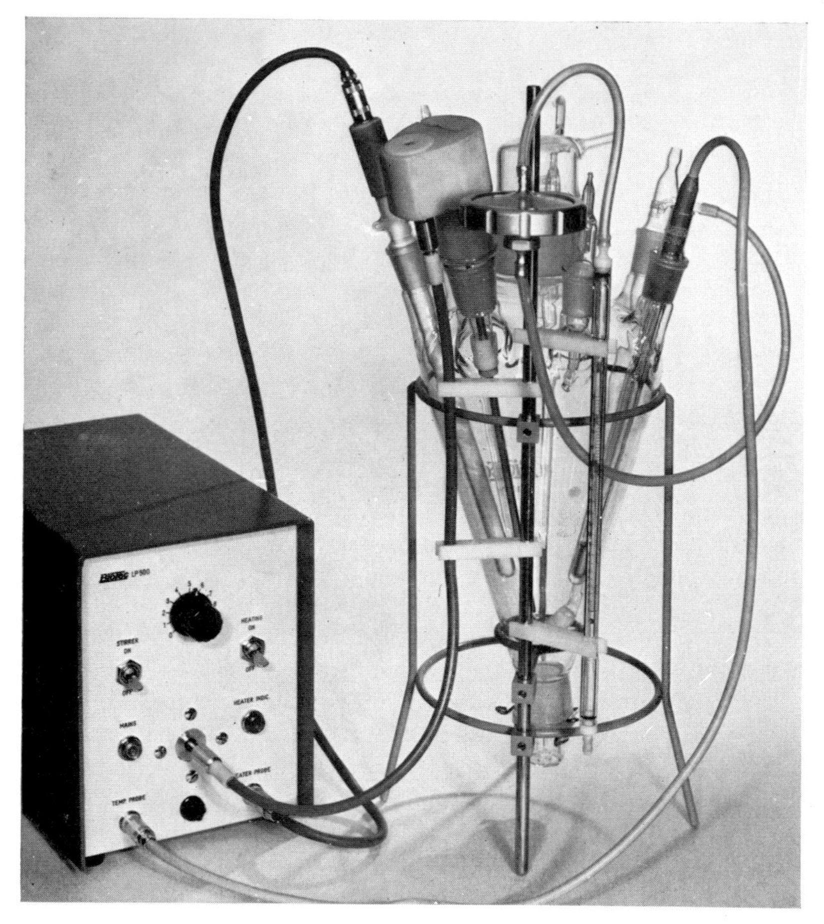

FIG. 6. All-glass fermenter suitable for batch or continuous culture, complete with control unit.

or other materials. Conversion to continuous operation is achieved by replacing the base plug by an overflow tube of appropriate length for the required culture volume. The overflow adapter is so arranged that it can fit directly into the centre neck of a similar vessel. In this way a vertical multivessel system can be built up without interconnecting pipework. It will be seen that the incorporation of a substantial number of control and measuring elements has been achieved by a substantial departure from cylindrical geometry.

It should be noted that neither this unit nor the glass Porton unit previously described here (Fig. 5) is well suited for scaling-up beyond the laboratory scale.

C. Other fermentation and culture vessels

In the two foregoing Sections attention has been focussed on the shaken flask and stirred vessel.

A search of the literature will reveal that many other devices or variants have been used. For instance, a cheap fermenter with a capacity of about 50 litres can be obtained by converting a discarded domestic washing machine,

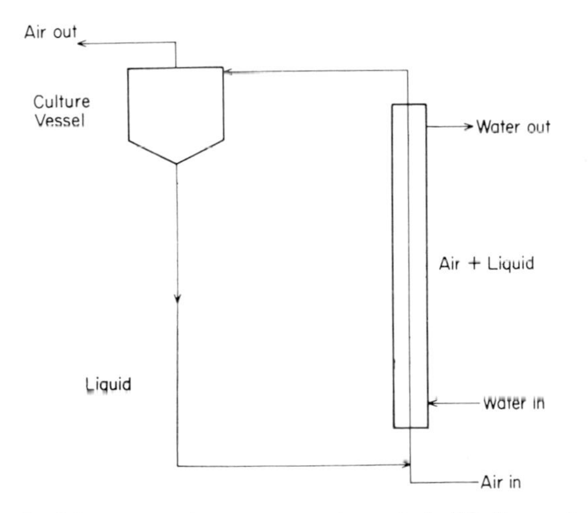

FIG. 7. Diagrammatic representation of air-lift fermenter.

preferably with a stainless steel tub. Domestic preserving jars and ordinary laboratory aspirators have also been used. Such devices may offer a quick and inexpensive solution when all that is required is an apparatus in which micro-organisms can be grown, but they have serious limitations if a fermentation is to be investigated systematically and are incapable of being scaled up.

Among items described in the literature as having been used on the laboratory and production scales, is the rotary horizontal drum fermenter, consisting of a cylinder fitted with baffles and mounted between trunnions. Aeration is achieved by liquid spilling from the baffles through the headspace into the bulk of the liquid, but oxygen uptake rates are well below those attainable in vessels of the type previously described. Moreover the culture volume is usually only about one-third of the total capacity, and there are formidable mechanical problems in making large fermenters of this type.

One design which may not have received due attention is the air-lift fermenter. This appears to be widely used in Eastern Europe for the production of food yeast and it has been claimed to give rates of output comparable with those of sparged, agitated vessels. Little attention has been paid to its laboratory potential or to systematic design and assessment on the industrial scale. Fig. 7 shows schematically a laboratory arrangement that can be

FIG. 8. Multiplate vessel suitable for surface culture of tissue cells.

readily constructed in glass. Air entering the riser serves to provide oxygen and to circulate the culture suspension, whilst temperature control is achieved by circulating water through a jacket surrounding the riser. No mechanical agitator is required and the circulation rate can be varied at a given aeration rate by adjusting the height of liquid in the culture vessel. By restricting the circulation rate an appreciable gradient of oxygen concentration can be established in the culture between the top of the lift and the point where the returning fluid re-enters the aeration zone. This phenomenon could be used to investigate the effect of fluctuations in oxygen concentration on the behaviour of micro-organisms. The results could be useful in improving the performance of conventional fermenters in which uneven distribution of oxygen occurs.

The fermenters which have been described so far are suitable for the cultivation of aerobic micro-organisms. By disconnecting the air supply or by substituting inert gas for air, most designs can also be used for the cultivation of anaerobic micro-organisms. In recent years there has been increasing interest in the cultivation of animal tissue cells, often under microaerophilic conditions. Whilst this can be achieved in equipment of the kind already described, the morphology of cells grown in submerged culture may be quite different from that of cells grown on a solid surface. Conventional techniques for such surface culture are laborious, especially when large quantities of cells are required. In the culture vessel illustrated in Fig. 8, a series of flat plates rotates only partially submerged in the culture medium. The depth of immersion and speed of rotation of the discs can be adjusted to suit the cultural requirements, thus offering a compact and convenient alternative to the labour- and space-consuming method of growing the culture on the inside walls of slowly rotating bottles.

In the production of beer by continuous fermentation an alternative to the stirred vessel is the tower fermenter. Wort is admitted together with a small amount of air to the base of a vertical tube. By selecting a suitably flocculating yeast, and an appropriate velocity for the fluid flow up the tube, the yeast becomes fluidized and a high yeast concentration can be maintained in the tube with a low yeast content in the overflow. As a result of the high yeast concentration in the tube, fermentation times can be reduced from 1 or 5 days to a few hours. This system does not appear to have been widely applied to other fermentations or to cell production, but its essential simplicity merits further investigation.

V. CONCLUDING REMARKS

Many descriptions can be found in the literature, particularly in the past two decades, of apparatus suitable for the cultivation of micro-organisms and tissue cells or for carrying out fermentations on the laboratory or pilot plant scale. Some of these refer to equipment devised for special purposes or to make use of generally available laboratory equipment, whilst others are suitable for a wide variety of purposes. There is no ideal laboratory fermenter, suitable for all situations.

By far the most widely used are the shaken flask and the stirred vessel. Little need be said of the former except to stress again its cheapness, usefulness and versatility as an investigational tool, which can provide small amounts of cells or fermentation products.

The stirred fermenter can be a very simple piece of equipment or it can be highly sophisticated and fitted with arrangements for manual or automatic measurement and control of a wide variety of fermentation variables.

In selecting a suitable piece of equipment, it is first necessary to clarify the experimental requirements. If the purpose is merely to provide a small

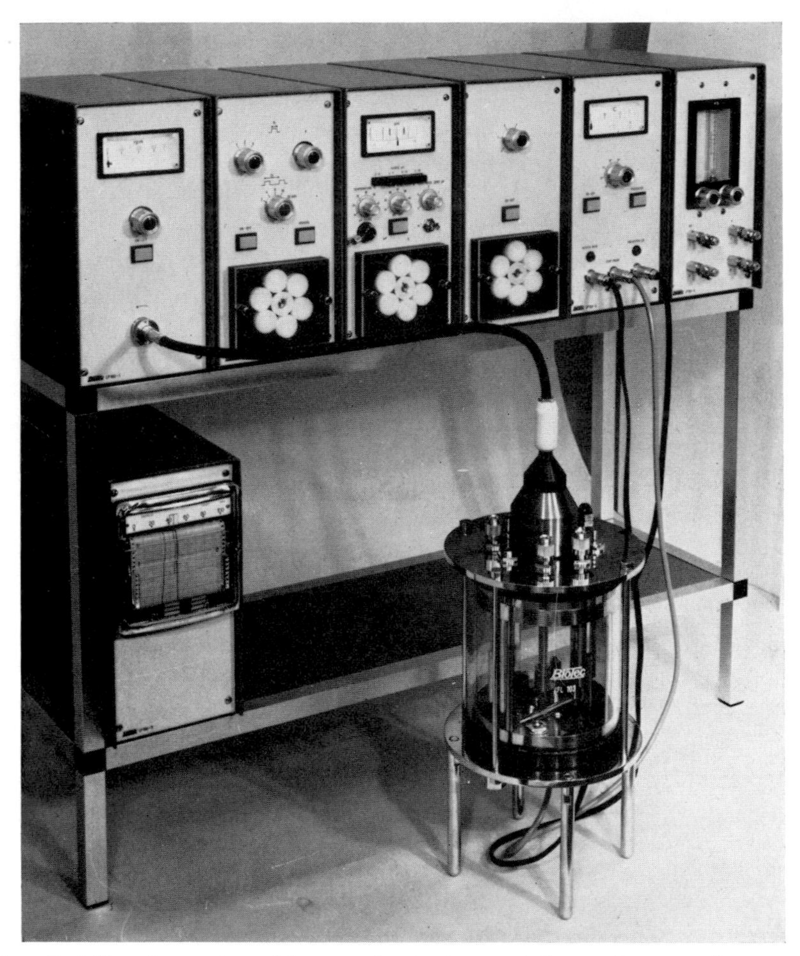

FIG. 9. 5-litre fermenter with extensive instrumentation and automatic controls.

quantity of cells or a fermentation product and the operation is not likely to be repeated, there is a lot to be said for adopting a low investment design, even though yields may be less than the best attainable. If the culture or fermentation equipment is to be an important feature of a continuing programme, a balance must be struck taking account of first cost, maintenance, operating costs including labour, convenience, reliability, required output and the information required. A number of standard designs is available from manufacturers, generally featuring a basic unit which can be built up to any desired degree of sophistication by adding one or more accessories. Fig. 9 shows a 5-litre fermenter with modular units (reading from the left) for speed control, antifoam control, pH control, medium flow control,

FIG. 10. Pilot plant fermenter and seed vessel with accessories.

temperature indication and control, and gas flow indication and control, together with a multipoint recorder. The antifoam, pH and medium control units include peristaltic pumps.

Fig. 10 shows a larger unit comprising seed and pilot fermenters. The unit is extensively equipped with accessories including facilities for manual or automatic operation of the sterilization cycle.

Fig. 11 shows a 150-litre fermenter equipped for automatic control of pH, temperature and foam, together with a range of glass vessels for inoculum and other additives. In each case the complete unit has been built up around a simple basic unit.

If an assessment of engineering parameters is desired, additional features may include arrangements to modify sparging, stirrer and baffle systems and means for measuring power uptake, gas hold-up and mixing effectiveness. However, a detailed discussion of these topics is inappropriate here.

Yields in microbiological processes are discussed elsewhere in this work, but it may be useful to remark here that yields, both in terms of cell and product concentrations and in terms of rate of production, may be greatly

FIG. 11. 150-litre fermenter and controls with auxiliary glass culture and medium vessels.

affected by the nature of the apparatus and the operating conditions chosen, as well as by the strain of organism and the medium. Under suitable conditions cell concentrations in excess of 20 g/litre (dry weight basis) can be obtained, and the requirements of even the most rapidly growing organisms can be met.

Whilst it can be expected that further improvements in fermenter design will be achieved, especially in relation to scaling-up and economy of operation, there now exist designs capable of meeting most of the needs of the microbiologist on the laboratory and pilot plant scales at reasonable cost and with a high degree of reliability.

The costs of culture equipment vary enormously, not only from one type to another but for various models of the same capacity and similar basic design. Some cost difference between similar designs from different manufacturers represents differences in exterior finish rather than in the fitness of the equipment for its purpose. Some is undoubtedly due to differences in robustness and standards of construction, and some to the adaptability of the equipment. Some designs may include, as standard, features that a particular user may not require and that other manufacturers may offer as optional extras. Where the equipment has been designed to allow ready conversion, say from one type of aeration system to another, or to permit the addition of accessories, the cost of the basic equipment may be higher than that of a similar less adaptable design. If a less adaptable design will meet a particular user's needs, this can be a decisive factor in making a choice. It is essential, therefore, in making such a choice to be as definite as possible about immediate and probable future requirements, and to consider manufacturer's price specifications in this light. Solomons (1967) has made a useful collection of data on the cost of stirred laboratory fermenters and has shown that instrumentation and ancillary equipment costs may be three to four times the cost of the basic fermenter.

A simple reciprocating shaking incubator (Gallenkamp, U.K.) complete with temperature controller, variable amplitude and shaking speed, interchangeable trays to take a variety of conical flasks, beakers and test tubes, heater, cooling coil and thermometer will cost £110–£120 depending on the number of trays chosen.

A rotary shaking machine to take forty-nine 250 ml flasks, with facilities for varying amplitude and rotational speed, suitable for installation in a thermostatically controlled room will cost around £200 (Apex Construction Co. and L. H. Engineering Co., U.K.). A similar machine to take about twenty-four 250 ml conical flasks, complete with interchangeable trays, enclosed in a thermostatically controlled cabinet with air-circulation and provision for fresh air intake costs about £400 (Gallenkamp) or $1500 (New Brunswick Scientific Co., U.S.A.).

The fermenter vessel illustrated in Fig. 5 could be constructed to order for under £5. A magnetic stirrer (Gallenkamp) costs around £15, and provision has to be made for a thermostatically-controlled source of attemperating water for the jacket, and for inlet and outlet connections for air and medium, etc.

A simple stirred vessel as illustrated in Fig. 1, with water-bath for temperature control, heater, thermostat, recirculating pump, two addition vessels, variable speed drive and working volume of 5 litres will cost from about £600. A similar six-vessel unit with adjacent pairs operating at the same speed costs about £2200 (Taylor Rustless Fittings Co., U.K.). With comprehensive instrumentation, and automatic control of foam and pH the cost of the six-vessel unit rises to £8500. The New Brunswick Scientific Co. offers similar equipment at comparable cost, though operating in sets of three instead of two.

A basic stainless steel 100-litre fermenter, with variable-speed agitator costs £1000–£1200 (John Dore & Co. Taylor Rustless Fitting Co., T. Guisti & Sons, U.K.). On the other hand a 50-litre fermenter can be constructed by conversion of a domestic washing machine for a figure of the order of £100.

A series of four fermenters in cascade, suitable for batch or continuous operation, incorporating the transfer device and head and base-plates illustrated in Figs 3 and 4, costs approximately £3000, including individual temperature indicator-controllers, independent stirrer speed control for each vessel, transfer timers, pumps and reservoirs and foam-control. (Taylor Rustless Fittings Co., John Dore & Co., U.K.).

ACKNOWLEDGEMENTS

The illustrations for Figs. 1 and 11 were supplied by the Taylor Rustless Fittings Co. Ltd., Leeds, England, and those for Figs. 2, 6, 8, 9 and 10 by Biotec A.B., Stockholm, Sweden through the courtesy of Dr. C. G. Hedén of Karolinska Institutet, Stockholm.

BIBLIOGRAPHY

The following bibliography has been selected as representative of the literature. References have been collected under separate headings, but many contain information on aspects other than those under which they have been classified. References marked with an asterisk include reviews or comparisons of fermenter designs.

Shaken flasks

Auro, M. A., Hodge, H. M., and Roth, N. G. (1957). *Ind. Engng Chem.*, **49**, 1237.
Shu, P. (1953). *J. agric. Fd Chem.*, **1**, 1119.
Smith, C. G., and Johnson, M. J. (1954). *J. Bact.*, **68**, 346.

General purpose fermenters

Bartholomew, W. H., Karow, E. O., and Sfat, M. R. (1950). *Ind. Engng Chem.*, **42**, 1827.
Brown, W. E., and Peterson, W. H. (1950). *Ind. Engng Chem.*, **42**, 1769.
Chain, E. B., *et al.* (1954). *Rc. Ist. sup. Sanità*, **17**, 61.
Clarke, W. C. (1956). *Chem. Process., Lond.*, **19**, 146.
Fortune, W. B. *et al.*, (1950). *Ind. Engng Chem.*, **42**, 191.
Friedland, W. C., Peterson, M. H., and Sylvester, J. C. (1956). *Ind. Engng Chem.*, **48**, 2180.
Gordon, J. J., *et al.* (1947). *J. gen. Microbiol.*, **1**, 187.
van Hemert, P. (1964). *Biotechnol. Bioengng*, **6**, 381.*
Kroll, C. L., *et al.* (1956). *Ind. Engng. Chem.*, **48**, 2190.

Lumb, M., and Fawcett, R. (1951). *J. appl. Chem., Lond.*, **1**, S94.
Nelson, H. A., Maxon, W. D., and Elferdink, T. H. (1956). *Ind. Engng Chem.*, **48**, 2183.
Paladino, S., Ugolini, F., and Chain, E. B. (1954). *Rc. Ist. sup. Sanità*, **17**, 87.
Rivett, R. W., Johnson, M. J., and Peterson, W. H. (1950). *Ind. Engng Chem.*, **42**, 188.
Solomons, G. L. (1967). *Process Biochem.*, **2**, 7.*

Waldhof fermenter

Brown, W. E., and Peterson, W. H. (1950). *Ind. Engng Chem.*, **42**, 1823.
Saeman, J. F. (1947). *Analyt. Chem.*, **19**, 913.

Rotary drum fermenters

Gastrock, E. A., *et al.* (1938). *Ind. Engng Chem.*, **30**, 782.
Herrick, H. T., Hellbach, R., and May, O. E. (1935). *Ind. Engng Chem.*, **27**, 681.
Stubbs, J. J., *et al.* (1940). *Ind. Engng Chem.*, **32**, 1626.
Ward, G. B., *et al.* (1938). *Ind. Engng Chem.*, **30**, 1233.
Wells, P. A., *et al.* (1937). *Ind. Engng Chem.*, **29**, 653.
Wells, P. A., *et al.* (1939). *Ind. Engng Chem.*, **31**, 1518.

Air-lift fermenters

Blakebrough, N., Shepherd, P. G., and Nimmons, I. (1967). *Biotechnol. Bioengng*, **9**, 77.
Kretzschmar, G. (1962). *Zellstoff Pap. Berl.*, **11**, 14.*
Lundgren, D. G., and Russell, R. T. (1956). *Appl. Microbiol.*, **4**, 31.

Cyclone fermenter

Dawson, P. S. S. (1963). *Can. J. Microbiol.*, **9**, 671.

Flask and jar fermenters

Calam, C. T., Driver, N., and Bowers, R. H. (1951). *J. appl. Chem., Lond.*, **1**, 209.
May, O. E., *et al.* (1934). *Ind. Engng Chem.*, **26**, 575.

Extended solid surface culture systems

Porter, F. E., and Nash, H. (1960). *J. biochem. microbiol. Technol. Engng*, **2**, 177.

Photosynthetic apparatus

Gafford, R. D., and Richardson, D. E. (1960) *J. biochem. microbiol. Technol. Engng*, **2**, 299.
Gaucher, T. A., Benoit, R. J., and Bialeski, A. (1960). *J. biochem. microbiol. Technol. Engng*, **2**, 339.

Fastidious systems, tissue cells and pathogens

Cherry, W. R., and Hull, R. N. (1960). *J. biochem. microbiol. Technol. Engng*, **2**, 267.
Merchant, D. J., Kuchler, R. J., and Munyon, W. H. (1960). *J. biochem. microbiol. Technol. Engng*, **2**, 253.
Nickell, L. G., and Tulecke, W. (1960). *J. biochem. microbiol. Technol. Engng*, **2**, 287.

Rightsie, W. A., McCalpin, H., and McLean, I. W. (1960). *J. biochem. microbiol. Technol. Engng*, **2**, 313.*
Achorn, G. B., Jr. (1956). Bioengineering Symposium, Rose Polytechnic Institute, Terre Haute, Indiana.*
Achorn, G. B., *et al.* (1959). *J. biochem. microbiol. Technol. Engng*, **1**, 27.
Holmström, B., and Hedén, C.-G. (1964). *Biotechnol. Bioengng*, **6**, 419.
Ubertini, B., *et al.* (1960). *J. biochem. microbiol. Technol. Engng*, **2**, 327.

Engineering aspects

Blakebrough, N., and Hamer, G. (1963). *Biotechnol. Bioengng*, **5**, 59.
Blakebrough, N., and Sambamurthy, K. (1964). *J. appl. Chem., Lond.*, **14**, 413.
Blakebrough, N., and Sambamurthy, K. (1966). *Biotechnol. Bioengng*, **8**, 25.
Calderbank, P. H. (1958). *Trans. Inst. chem. Engrs*, **36**, 442.
Calderbank, P. H. (1959). *Trans. Inst. chem. Engrs*, **37**, 173.
Calderbank, P. H., and Jones, S. J. R. (1961). *Trans. Inst. chem. Engrs*, **39**, 363.
Mack, D. E., and Kroll, A. E. (1948). *Chem. Engng Prog.*, **44**, 189.
Hamer, G., and Blakebrough, N. (1900). *J. appl. Chem., Lond.*, **13**, 517.
Midler, M., Jr., and Finn, R. K. (1966). *Biotechnol. Bioengng*, **8**, 71.
Oldshue, J. Y. (1966). *Biotechnol. Bioengng*, **8**, 3.
Taguchi, H., and Miyamoto (1966). *Biotechnol. Bioengng*, **8**, 43.
Wilhelm, R. H., *et al.* (1966). *Biotechnol. Bioengng*, **8**, 55.

Miscellaneous

Brandl, E., Schmid, A., and Steiner, H. (1966). *Biotechnol. Bioengng*, **8**, 297.*
Perret, C. J. (1957). *J. gen. Microbiol.*, **16**, 250.
Special Articles on Continuous Culture (1965). *Lab. Pract.*, **14**, 1140, 1145, 1151, 1162, 1169.*

CHAPTER XIV

The Deep Culture of Bacteriophage

K. Sargeant

Microbiological Research Establishment, Porton, Salisbury, Wilts., England

I. GENERAL

A. Methods available for producing bacteriophage

Bacteriophages are bacterial viruses and require to be grown in living cells. It follows therefore that any consideration of how to produce bacteriophage is primarily concerned with how to produce the bacterial host cells.

Bacterial cultures grown on agar surfaces have been used to make small quantities of high-titre bacteriophage (Hershey *et al.*, 1943; Swanstrom and Adams, 1951) by washing off the layer of lysed bacteria with a small volume of liquid. The method suffers from the disadvantage that scale-up involves the use of many small containers and is therefore tedious.

Bacteria grown in suspension using shake flasks have also been used for producing bacteriophage. This method has the advantages that growth conditions can be made more nearly uniform throughout the culture, that materials may be added to the growing culture, and that representative samples may be taken at any stage. Final bacteriophage yields/ml are limited by the relatively low bacterial cell densities that can be achieved readily in shake flasks. Nevertheless, high final bacteriophage concentrations can be achieved in those few cases in which unlysed bacterial cells, replete with bacteriophage, may be collected and lysed separately in a small volume. Scale-up in flasks to volume greater than a few litres has not been achieved.

For producing large masses of bacteria, "deep culture" in stirred culture vessels may well become the standard method. It has the advantage not only that larger volumes of culture can be grown, but also that, as a result of having a more controlled environment, the cell densities attained are often greatly in excess of those achieved in shake flasks. Recently it has been shown that bacteria grown to high cell densities in bacterial culture vessels are suitable for producing three different bacteriophages, $\mu 2$, MS2, and T7 (see later). The higher bacterial cell densities successfully employed have led to improved bacteriophage yields/ml of culture. Culture volumes of up to 300 litres have been used. It is likely that the method will prove generally applicable for producing bacteriophage in bacterial culture vessels of all sizes.

B. The need for large quantities of bacteriophage

Despite early optimism, bacteriophages have not found a place in the treatment of disease, and as no other commercial outlet exists industry has taken no interest in their manufacture. Only very small quantities are demanded for biological studies, and these are readily available by laboratory methods. The largest requirement is for the gram quantities needed in chemical work on the structure and function of these particles.

Interest in the chemical structure and function of bacteriophages at first centred on the "T even" coliphages, which are relatively easy to produce in gram quantities. It is only recently, since the smaller bacteriophages, such as $\mu 2$ (Maccacaro, 1961) and MS2 (Davis *et al.*, 1961), have been discovered that some difficulty has been experienced in easily producing the amounts required.

C. Calculation of bacteriophage yield

The enumeration of bacteriophage particles is normally carried out by the plaque assay method. This method indicates the minimum number of particles present in the assay sample.

If the particle mass, or molecular weight, M, of a bacteriophage is known, the mass of bacteriophage, m (in grams), in a sample is calculable from the plaque assay by using equation (1)—

$$m = \frac{Mx}{Ap} \tag{1}$$

where x is the number of plaque-forming units (p.f.u.) in the sample, A is Avogadro's number ($6{\cdot}02 \times 10^{23}$) and p is the absolute efficiency of plating of the plaque assay method, i.e., the ratio of the plaque count under the assay conditions to the total number of virus particles in the sample.

If p for the plaque assay method is unknown, the minimum mass of bacteriophage in a sample is calculable from equation (1) by assuming $p = 1$, i.e., by assuming that every virus particle present has produced a plaque. The value of p can be determined by the method of Luria *et al.* (1951).

II. BACTERIOPHAGE PRODUCTION UNDER CONDITIONS OF RELATIVELY POOR AERATION

A. The importance of aeration

When the most commonly used bacteriophage host organism, *Escherichia coli* is grown in suspension in a liquid medium, growth is usually exponential until lack of an essential nutrient causes a reduction in growth rate. Frazer (1951) showed that 1 litre cultures of *E. coli* grown in flasks into which air was bubbled did not maintain exponential growth to as high a cell density as did 10 ml experiments run in test tubes. He concluded that the growth-limiting nutrient was oxygen, and went on to design culture apparatus, based on a "cyclone" type foam separator, which enabled 2 and 6 litre cultures to be grown exponentially to a density of about $1{\cdot}8 \times 10^{9}$ cells/ml.

With the aid of the 2 litre version of this apparatus, which could be operated under conditions of low and high aeration, Frazer showed that the maximum yield of bacteriophage T3 obtainable when cultures were inoculated with bacteriophage at a bacterial cell density of $1{\cdot}8 \times 10^{9}$/ml was about 7×10^{11} p.f.u./ml of culture lysate. To obtain this yield, a result comparable with that obtained on a small scale, it was necessary to use the apparatus under conditions of maximum aeration.

Apparatus of the Frazer type is little used today because of the advent of stirred bacterial-culture vessels (Section IIIA), which give even better aeration.

A study of bacteriophage production under conditions of relatively low aeration was made by Sargeant and Yeo (1966) using a 3 litre bacterial culture vessel to grow bacteriophage $\mu 2$ on a strain of *E. coli*. Conditions of aeration were such that the growth rate of cultures of the host organism became limited by lack of oxygen at a cell density of about 5×10^{8}/ml. From this stage, the pH value began to fall, and bacterial growth ceased at about 20 h, when the bacterial cell density was about 3×10^{9}/ml and the pH value was 5·0. When similar cultures were grown with pH control at 7·0–7·1 afforded by the automatic addition of sodium hydroxide or phosphoric acid as required, the maximum cell density obtained was about 10^{10}/ml at 36 h.

TABLE I

Three-litre experiments on the production of bacteriophage $\mu 2$ under conditions of low aeration

Run	pH regime	Viable bacteria at infection $\times 10^8$/ml	Number of phage particles/ bacterium at infection	Age of culture when phage was added, h	Maximum phage count after chloroform treatment $\times 10^{12}$ p.f.u./ml	Age of culture at maximum phage count, h	Number of phage particles produced/ bacterium
1	Non controlled, phage added at 6·0	29·2	3·4	9	0·042	27	14·5
2	Controlled, 7·0–7·1	15·2	16·2	7·5	1·57	13·5	1030
3	Controlled, 7·0–7·1	24·7	2·6	11	1·34	21	540
4	Controlled, 7·0–7·1	29·0	3·2	11	2·37	21	820
5	Controlled, 7·0–7·1	40·2	1·6	19	0·85	41	200

All cultures were grown to an age of at least 28 h. After phage infection, samples were taken at intervals, shaken with chloroform and assayed for phage content. The maximum phage count so obtained from samples of whole culture lysate are recorded.

The results given in Table I show the bacteriophage yields obtained when cultures grown under these conditions of low aeration were infected with bacteriophage $\mu 2$ at different stages. In run 1, which was carried out without pH control, a culture infected at 9 h, when the cell density was $2{\cdot}92 \times 10^{9}$/ml gave only 14·5 p.f.u./cell at 27 h. Runs 3 and 4, which were pH controlled, were infected at 11 h, when the cell densities were $2{\cdot}47 \times 10^{9}$/ml and $2{\cdot}90 \times 10^{9}$/ml respectively. The peak $\mu 2$ production occurred in both cases at 21 h and corresponded to 540 p.f.u./cell and 820 p.f.u./cell, respectively. These results show that it is not so much the low aeration rate itself as the fall in pH value associated with dissimilative bacterial growth under conditions of low aeration that causes the reduction in bacteriophage yield. This is not however the whole story, because under conditions of high aeration (see later) average $\mu 2$ yields of about 3000 p.f.u./cell were obtained.

B. Shake-flask methods for bacteriophage production

Although conditions of relatively low aeration obtain in flask cultures, even when they are vigorously shaken, it is often possible to prepare useful amounts of bacteriophage very easily by using them. The shake-flask method of bacteriophage production is therefore recommended as the primary method to be tried in all cases. Only when it has been shown that the yields so obtainable are too low should the more elaborate stirred culture vessel method be adopted. Even then the shake-flask method still has a place in investigations of the optimum medium composition, or pH value for bacteriophage production, because it is possible to carry out many experiments simultaneously at little expense.

Since automatic pH control for shake-flask cultures is technically difficult to achieve, it should be accepted that there is an upper limit, depending on the aeration capacity of the system, to the cell density at which bacterial cultures can be infected with bacteriophage and lysed to give maximal yields of progeny bacteriophage/cell. It is not possible to predict the optimum conditions, which must be determined for each case.

Many workers have carried out the infection at cell densities in the range 2×10^{8} 1×10^{9} cells/ml, using about 5 p.f.u./cell of bacteriophage to ensure that most cells become infected at about the same time. Culture lysis is usually complete after 2 h or less, and is usually accompanied by foam production, but in some cases agents, such as chloroform or lysozyme, must be added to release the progeny bacteriophage. Again no generalization is possible. Conditions must be determined for the individual case.

C. Summary of results obtained with culture flasks

The results given in Table II summarize the work of several investigators. In each case the minimum yield of bacteriophage produced has been estimated from the plaque count and the molecular weight of the bacteriophage

TABLE II

Bacteriophage production in culture flasks

Bacteriophage	Authors	Culture volume, litres	Viable bacteria at infection $\times 10^8$/ml	Bacteriophage count in culture lysate $\times 10^{12}$, p.f.u./ml	Particle size of bacteriophage $\times 10^6$, daltons	Minimum mass of bacteriophage/litre of lysate, mg†
$\mu 2$	Isenberg (1963)	2	2	1	3·6	6
R17	Paranchych & Graham (1962)	0·09	2	2	4·19‡	14
MS2	Strauss & Sinsheimer (1963)	7	4	2–40	3·6	12–240
$\phi \times 174$	Sinsheimer (1959)	2	10	0·2–0·8	6·2	2–8
T2	Siegel & Singer (1953)	18	5	up to 0·6	200§	up to 200
T2, T4, T6,	Wyatt & Cohen (1953)	0·35	30	up to 1·0	200§	up to 330
T3	Frazer (1951)	6	18	0·7	49¶	57

† Calculated in each case from the particle mass.
‡ Enger *et al.* (1963).
§ Stent (1963).
¶ Swarby (1959); Bendet *et al.* (1963).

from equation (1), Section IC. The nature of the culture flasks used was not always stated, but in most cases, either the aeration was relatively poor, or the full capabilities of the system for bacteriophage production were not utilized. Wyatt and Cohen (1953) did achieve reasonably good aeration by growing 350 ml cultures in vigorously shaken 2 litre flasks. The effectiveness of their method is reflected in the high yields of bacteriophages T2, 4 and 6 obtained. Frazer (1951) used the special culture apparatus discussed earlier (Section IIA) for T3 production.

III. BACTERIOPHAGE PRODUCTION USING STIRRED BACTERIAL CULTURE VESSELS

A. Stirred bacterial culture vessels and aeration

It has been shown that lack of adequate aeration is the most important factor limiting the production of bacteriophage in shake flasks. Stirred bacterial culture vessels were designed to enable bacteria to be grown to a high cell density, and one of the most important factors involved was found to be the provision of adequate aeration. Such vessels are now in common use in many laboratories. For descriptions the reader is referred to Elsworth *et al.* (1958), van Hemert (1964) and Solomons (1967). The chemostat described by Evans *et al.*, in this Series, Volume 2 when operated batchwise, falls into this category.

B. The production of bacteriophage μ2

1. *Growth of the host organism*

This bacteriophage was produced in 3, 20 and 150 litre stirred bacterial culture vessels by Sargeant and Yeo (1966), using as host *E. coli* K12 58–161F^+ Fimσ^+.

Seeds were made by transferring the contents of a freeze-dried ampoule of the host organism to Robertson's meat broth (10 ml) in a 1 oz. bottle. After incubation for 4–18 h, the culture was used at the rate of 1 ml per 8 oz medical flat bottle (agar surface area, 60 sq. cm) and 5 ml per Roux bottle (agar surface area, 200 sq. cm) to initiate cultures on tryptic meat agar. Twelve hours later, the agar surfaces were washed with distilled water (8 ml per 8 oz. bottle, 50 ml per Roux bottle) and the resulting suspension, which contained 1×10^{10}–2×10^{10} viable organisms/ml was used to start 3, 20 and 150 litre cultures at between 10^6 and 10^7 viable organisms/ml. Such suspensions could be kept at +4°C for 2 weeks without appreciable loss of viability.

The medium employed in the culture vessels was—

Medium

Casein hydrolysate (Oxoid L.41 grade, Oxo Ltd, London)	30 g
Glycerol (AnalaR grade, BDH Ltd, Poole)	20 g
Yeatex (light grade, English Grains Co. Ltd, Burton-on-Trent)	1 g
KH_2PO_4	5 g
$MgSO_4.7H_2O$	1 g
Demineralized water to make	1 litre

pH 7·0

KH_2PO_4 and $MgSO_4.7H_2O$ were each dissolved and sterilized separately and added to the main bulk later. The pH was adjusted with 10 M NaOH after sterilization.

Under normal operating conditions air was supplied through the bottom of the culture vessel immediately under the impeller at a rate of one culture volume/min. The 3, 20 and 150 litre vessels dissolved about 220, 150 and 80 mmoles of O_2/litre/h, as measured by the sulphite oxidation method of Cooper *et al.* (1944).

Cultures of the host organism were grown in each vessel and the CO_2 content of the culture effluent gas was measured. In each case a peak value was reached, after which the CO_2 content continued at a high level for some hours. The peak value was lower for the larger vessels because of their lower aeration capacity (the higher stirrer speeds necessary to attain aeration rates in the larger vessels comparable with that in the 3 litre vessel would be accompanied by unacceptably rapid wear). In subsequent work the CO_2 content of the culture effluent gas was used as an immediate guide to the bacterial cell density, which was determined later by plating-out culture samples.

2. *Special precautions taken in producing bacteriophage μ2 in stirred culture vessels*

These were all aimed at one thing—the containment of the bacteriophage.

(a) *Sterilization of the culture vessel following bacteriophage production.* The 3 litre glass vessel was satisfactorily sterilized by adding 10 ml of 40% formalin solution and passing steam through at atmospheric pressure for 4 h. The larger vessels were sterilized by steam at 15 p.s.i. for 30 min.

(b) *The preparation of bacterial seed cultures.* Great care was taken to use bacteriophage-free bacterial seed cultures. A supply of freeze-dried, sealed ampoules of the host organism was prepared before the bacteriophage was used. Seed cultures were produced from these ampoules in a separate laboratory into which bacteriophage was never taken.

(c) *Containment of the culture vessels.* The 3 litre culture vessel was operated in the normal manner, but was enclosed in a cabinet, ventilated by means of an electrically driven fan, which expelled air through an exhaust filter to prevent the escape of aerosols. The provision of such a cabinet was probably unnecessary, because the larger vessels were not so enclosed. Both had leak-proof seals of the type described by Elsworth *et al* (1958). They were also tested for, and found to be free from, leaks before each experiment. It is still possible that bacteriophage aerosols escaped during sampling, which was *via* a hypodermic needle into evacuated bottles. However, it was found possible to grow cultures of $\mu2$-susceptible bacteria in adjacent culture vessels of similar design without special precautions, and it was therefore concluded that the normal sterile techniques used were sufficient to control the spread of bacteriophage.

(d) *Sterilization of the KCl bridge to the calomel electrode.* Standard technique for the growth of bacteria in culture vessels fitted for pH measurement does not require that the KCl bridge be dismantled and sterilized between experiments. Steam-sterilization of the vessel is sufficient because the glass sinter that connects the KCl bridge to the culture vessel is impervious to bacteria. Unfortunately bacteriophage particles can pass through the glass sinter, contaminate the KCl solution and therein survive normal sterilization procedures. In order to prevent the premature bacteriophage contamination of the next culture it was therefore necessary to dismantle, empty, heat-sterilize and re-fill the KCl bridge system after each bacteriophage culture. The calomel electrode (RS.23, Electronic Instruments Ltd, Richmond, Surrey) was satisfactorily sterilized by treatment with steam at 100°C for 15 min.

(e) *Sterilization of the inlet and outlet air.* It is important that both the inlet air to, and the outlet air from, the culture vessel be filtered to prevent uncontrolled ingress or egress of bacteriophage. A suitable filter material is Owens–Corning FM 003 resin-bonded fibre glass (Chemicals Trading Co. Ltd, 25 Berkeley Square, London W.1). For packing details see Elsworth (this Volume, page 123).

(f) *Prevention of foaming.* During the bacterial lysis, which accompanies the release of bacteriophage, foaming tends to occur. This is particularly so in stirred culture vessels of the type used by Sargeant and Yeo (1966), in which air is forced into the bottom of the culture. These authors found that MS antifoam emulsion RD (Hopkin & Williams Ltd, Chadwell Heath, Essex) gave good foam control in the production of $\mu2$. [Rushizky *et al.*

(1965) used Dow–Corning antifoam AF to obtain satisfactory foam control in the production of MS2.]

3. *The influence of bacterial cell density at infection on the yield of μ2 in a 3 litre stirred culture vessel.*

A series of experiments was carried out in which cultures were infected with μ2 at an input ratio always greater than 1 p.f.u./cell. Cultural conditions were maintained for a further 3 h. Chloroform (15–20 ml/ litre of culture) was added, and stirring, without aeration, was continued for another hour. The concentration of infective particles in the crude lysate was determined.

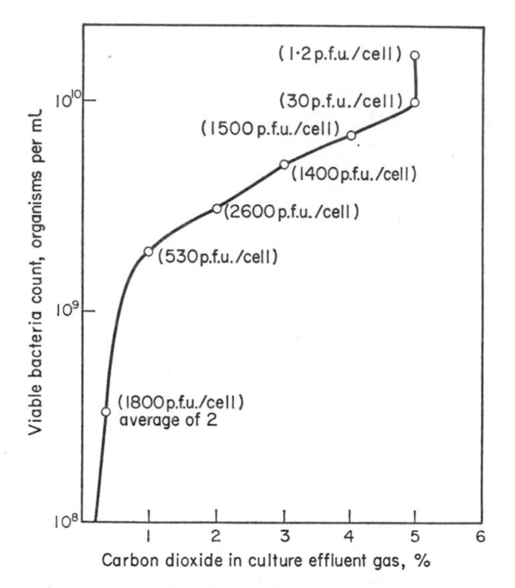

FIG. 1. Viable bacteria count at the time of infection with μ2 (input ratio > 1 p.f.u./cell) versus carbon dioxide content of the culture effluent gas for 3-litre cultures of the host, *E. coli* K12 58–161 F^+ Fimσ^+. The figures in brackets are the μ2 yields, expressed as p.f.u./cell, found 4 h after infection, following chloroform treatment.

In six experiments, cultures infected at viable bacterial cell densities varying from $1{\cdot}3 \times 10^8$ to $6{\cdot}9 \times 10^9$/ml showed no systematic variation in the bacteriophage yield/cell, which averaged 1625 p.f.u. However in two experiments, infected at viable bacterial cell densities of $9{\cdot}9 \times 10^9$ and $1{\cdot}67 \times 10^{10}$/ml, only 30 and 1·2 p.f.u./cell were produced, respectively. The results are expressed graphically in Fig. 1.

Clearly the best time for bacteriophage infection was the latest at which the yield of bacteriophage/cell was maintained at its peak level. As a result of these experiments conducted on the 3 litre scale, it was concluded that this was when the culture effluent gas first attained a CO_2 content of 4%.

However, in view of the immense penalty of delaying infection beyond this stage, and to allow for small culture–culture variations in the CO_2 evolution pattern, it was decided to carry out production experiments by infecting cultures when the CO_2 content of the effluent gas first reached 3%.

The reason for the abrupt fall in bacteriophage yield from the denser cultures is not that the cultures became O_2 limited, because quite high yields of $\mu 2$ can be obtained from less dense cultures grown under conditions of O_2 limitation (see Section IIA). The most probable reason is that in the 3 h of growth allowed following infection the culture medium became exhausted of some essential nutrient or nutrients. This view is supported by the observation that in the densest culture infected, which gave only 1·2 p.f.u./cell, the CO_2 concentration in the effluent gas had collapsed to only 0·2% 3 h after bacteriophage infection.

4. *The production of* $\mu 2$ *in a 3 litre stirred culture vessel*

A series of experiments was carried out in which the cultures were infected with bacteriophage when the CO_2 content of the culture effluent gas first reached 3%. In every case the CO_2 content of the culture effluent gas continued to rise, reached a peak short of the peak that would have been attained in the absence of bacteriophage, and fell rapidly to a low level 3 h after infection. The concentration of infective particles in the crude lysate averaged about 3000 p.f.u./cell, or $2{\cdot}1 \times 10^{13}$ p.f.u./ml, from cultures infected at an average viable bacterial cell density of $7{\cdot}1 \times 10^{9}$/ml (see Table III). This is at least 124 mg of bacteriophage/litre of culture lysate (see Section IC for basis of calculation). The bacteriophage yield did not vary significantly with bacteriophage–bacterial cell input ratios in the range 1·8–410.

TABLE III

Production of bacteriophage $\mu 2$ by different methods

Culture apparatus	Volume of culture, litres	Viable bacteria at infection $\times 10^{8}$/ml	Bacteriophage count in culture lysate $\times 10^{12}$, p.f.u./ml	Minimum mass of bacteriophage/ litre of culture lysate,† mg
Frazer apparatus	2	2	1	6
Stirred culture vessel	3	71	20·7	124
Stirred culture vessel	20	63	16·8	100
Stirred culture vessel	150	43	12·6	75

† Calculated from particle mass of $3{\cdot}6 \times 10^{6}$ daltons (Section IC).

5. *The production of* $\mu 2$ *in* 20 *litre and* 150 *litre stirred culture vessels*

This was carried out by infecting cultures of the host organism with a bacteriophage–bacterial cell input ratio greater than unity, when the CO_2 concentration in the effluent gas had reached 2% (20 litre cultures) and 1·2% (150 litre cultures). Thereafter, cultural conditions were maintained, and lysis was assisted as described for the 3 litre cultures. In both cases the CO_2 content of the culture effluent gas changed in a manner similar to that in the 3 litre experiments.

For the 20 litre culture, the average yield of infective particles was 2650 p.f.u./cell, or $1{\cdot}68 \times 10^{13}$ p.f.u./ml, from an average viable cell density of $6{\cdot}3 \times 10^{9}$/ml (see Table III). For the 150 litre cultures an average of 3550 p.f.u./cell, or $1{\cdot}26 \times 10^{13}$ p.f.u./ml, was obtained from an average viable cell density of $4{\cdot}3 \times 10^{9}$/ml (see Table III). These yields amount to at least 100 and 75 mg/litre of culture lysate, respectively (see Section IC for basis of calculation).

Yields in the larger vessels were lower only because the average cell density at which cultures were infected with bacteriophage was lower in the larger cultures. This was in turn a consequence of the lower aeration capacity attainable in the larger vessels. Nevertheless, the yields are much higher than the 10^{12} p.f.u./ml attained from 2×10^{8} viable bacteria/ml by Isenberg (1963) with a Frazer apparatus.

6. *The purification of bacteriophage* $\mu 2$

The culture lysate was separated from the bulk of insoluble debris by slow-speed centrifugation, and the supernatant was made 2 M with respect to $(NH_4)_2SO_4$ to precipitate the bacteriophage, which was purified by CsCl equilibrium density gradient ultracentrifugation (H.R. Matthews, and H. Isenberg, personal communication).

The crude precipitate was suspended in 1mM EDTA, pH 7·0 (approximately 1 ml/g of wet precipitate) and dialysed against the same solvent (250 ml/g) for 24 h with constant agitation to disaggregate and re-suspend the bacteriophage. The dialysate was clarified by centrifugation for 20 min at 18,000 rev/min in the 8 × 50 ml rotor of the MSE 18 centrifuge. The supernatant was diluted to 108 ml with 1 mM EDTA, pH 7·0, and 68·4 g of CsCl were added and dissolved with continuous stirring. The mixture was centrifuged for 18 h at 37,000 rev/min in the 40 rotor of the Spinco Model L ultracentrifuge. The rotor was allowed to stop without using the brake, and the tubes were removed.

The bacteriophage appeared as a fine opalescent band near the middle of the tubes. The band was recovered by removing the plug from the centrifuge tube cap, puncturing the bottom of the tube, and allowing the contents of the tube to flow out at the bottom while controlling the flow rate by means of a finger on the plug hole. The purified bacteriophage was further concentrated and purified by adding 55% CsCl to the separated fractions and repeating the ultracentrifugation once or more.

The purified bacteriophage was transferred from CsCl to any desired solvent by dialysis. One batch of 15 litres of lysate, containing $4{\cdot}2 \times 10^{17}$ p.f.u. gave 15g of pure $\mu 2$, containing $4{\cdot}8 \times 10^{17}$ p.f.u. Thus for this sample the yield of pure $\mu 2$ was 1g/litre, and the absolute plating efficiency was 19%.

C. The production of other bacteriophages

1. *Bacteriophage MS2*

The study of Sargeant and Yeo (1966) on $\mu 2$ is the only systematic investigation reported for bacteriophage production in culture vessels. However Rushizky *et al.* (1965) used a stirred culture vessel for producing another small RNA bacteriophage, MS2 [mass = $3{\cdot}6 \times 10^6$ daltons (Strauss and Sinsheimer, 1963)], on a 300 litre scale.

The host organism was *E. coli* C.3000, grown in the following medium—

Medium

Tryptone	10 g
Yeast extract	1 g
NaCl	8 g
Na_2HPO_4	3 g
KH_2PO_4	1·5 g
Dow–Corning medical antifoam AF	0·33 g
Cerelose	6·7 g
$CaCl_2$	0·29 g
Thiamine hydrochloride	0·01 g
Distilled water to make	1 litre

All but the last three components were dissolved in water and held under pressure 125°C for 45 min. Aqueous solutions of the last three were autoclaved separately and added to the main bulk, which had been cooled to 37°C.

The bacterial seed consisted of 12 litres of an overnight growth (at 23°C). The authors did not state the sulphite oxidation rate attainable in their vessel, but the 300 litre culture was grown to a density of 4×10^9 cells/ml, determined optically, using a stirrer rate of 120 rev/min and a bottom air flow rate of 276 litres/min. The culture, which had a pH value of 6·53, was then inoculated with MS2 at an input ratio of 10 p.f.u./cell, and aeration was increased by increasing the stirrer speed to 210 rev/min and the air flow rate to 342 litres/min. Foaming was controlled by the automatic addition of antifoam diluted 1 : 50 and sterilized. Despite the increase in aeration, the pH value of the culture fell rapidly following bacteriophage infection, and the fall was only checked by the addition of 453g of Na_2HPO_4. Four hours after infection, when the aeration was stopped, the culture contained $2{\cdot}95 \times 10^{18}$ p.f.u. of MS2 (approximately 1×10^{13} p.f.u./ml). This was equivalent to 2500 p.f.u./cell, a figure close to that obtained by Sargeant and Yeo (1966) for the related bacteriophage $\mu 2$.

The culture was treated with chloroform (2 litres) and EDTA (500 g, free acid) and stirred for 5 min. The solution, pH 5·3, was treated with $(NH_4)_2SO_4$ (300g/litre) at 4°–6°C in glass-lined vessels. All subsequent operations were carried out at below 10°C. The precipitate was recovered with a Sharples centrifuge type AS–16P at 15,000 rev/min and a flow rate of 1000–1600 ml/min. The pellet (2973g) was suspended in water (40 litres). The cell debris was spun off, re-extracted with 10 litres of water, and the combined supernatants were adjusted to contain 300g of $(NH_4)_2SO_4$/litre. After 1 h the suspension was centrifuged for 10 min at 2000 rev/min and the supernatant solution discarded. The precipitate was extracted twice with 3 litre portions of 0·001 M EDTA (monosodium salt; pH 6·1) and then suspended in 1 litre of buffer and dialysed twice against 12 litres of the same buffer for 16 h each. The insoluble material was removed, and the combined supernatants and extracts were centrifuged for 4 h at 21,000 rev/min in the No. 21 rotor of the Spinco Model L preparative centrifuge at 0°C. The pellets were soaked in water overnight; the solution was centrifuged for 10 min at 2000 rev/min and the pale yellow supernatant stored over chloroform at 4°C. The yield was 19·2 g of MS2 (Absorbance at 260 nm = 8·0 for 1 mg/ml). The final yellow supernant contained $2{\cdot}1 \times 10^{18}$ p.f.u., which represents a 71% recovery from the lysed culture. The absolute efficiency of plating for the final product was 66% (calculated from equation (1), Section 1C).

Strauss and Sinsheimer (1963) also grew MS2 by using the same host organism at 4×10^8/ml, and a similar medium "aerated vigorously" on a 7 litre scale. They claimed culture lysate titres of from 2×10^{12} to 4×10^{13}/ml, corresponding to between 5000 and 100,000 p.f.u./cell. These authors, the first to purify MS2, obtained 40 mg of pure bacteriophage/litre of lysate, with an absolute efficiency of plating of 15–20%.

A step-by-step comparison of the two methods is not possible on the data available, but it is clear that the 300 litre process devised by Rushizky *et al.* (1965) represents an improved method for producing MS2, because of the improved recovery of bacteriophage/unit culture volume, and because of the larger scale of operation. It is tempting to speculate that a further improvement in yield of MS2 might be obtained by using a vessel with a very high aeration efficiency.

2. *Bacteriophage T7*

Bacteriophage T7, which has a particle weight of 38×10^6 daltons (Davison and Freifelder, 1962), was prepared by Lunan and Sinsheimer (1956) using *E. coli* B, grown in a glycerol–casamino-acids medium as the host organism. One litre batches, grown with "vigorous aeration", were infected with T7 at an input ratio of 1 : 10, when the cell density was 10^9/ml. After lysis, the T7 titre was 1×10^{11}–3×10^{11} p.f.u./ml, with an average of $2{\cdot}43 \times 10^{11}$ p.f.u./ml, or 243 p.f.u./cell for 14 batches.

Recently K. Sargeant and R. G. Yeo (unpublished work) used a 3 litre culture vessel, having an oxygen dissolution rate of 220 mmoles/litre/h to prepare T7. The medium was similar to that of Lunan and Sinsheimer (1956), and the same as medium 2 of Sargeant and Yeo (1966). Because of the high aeration, it was possible to infect cultures at a higher cell density than that used by Lunan and Sinsheimer, and still obtain satisfactory lysis. In 3 cultures infected with T7 at an average input ratio of 1 : 2 when the average cell density was $1{\cdot}9 \times 10^{10}$ cells/ml, the average culture lysate titre was $1{\cdot}45 \times 10^{12}$ p.f.u./ml, or 76 p.f.u./cell. This was a 6-fold increase in bacteriophage production/unit culture volume, as compared with the earlier method, though the yield/cell was less.

It is already clear that the optimum conditions for T7 production in the system under study involve the infection of denser bacterial cultures than in the case of $\mu 2$ production.

D. Other possible uses of stirred culture vessels in bacteriophage production

It is probable that the use of stirred culture vessels in bacteriophage production will increase. The high yields obtainable already make this the method of choice for cases in which yields are normally low.

The method should prove applicable to the manufacture of temperate bacteriophages in improved yields, because no insuperable difficulty is envisaged in carrying out induction on a large scale.

Finally, the method should enable a more detailed study of the mechanisms of bacteriophage replication to be carried out by making available larger quantities of bacteriophage-infected cells part way through the replication cycle. A procedure for the manufacture, on a 50 litre scale, of *E. coli*, infected with bacteriophages $T2r^+$, $T4r^+$ or $T6r^+$ has been described (Zimmerman, 1966). The yield was only 2·0–2·5 g wet wt/litre. A similar procedure for the growth, on a 150 litre scale, of MS2-infected *E. Coli* gave between 2·3 and 2·7 g wet wt/litre (Billeter and Weissmann, 1966). It is felt that these two methods could be improved and give much higher yields by fully using the aeration capacity of a culture vessel. For example, to assist in a study of the replication of $\mu 2$ in the system described earlier, it proved possible to arrest the growth cycle by cooling 30 litres of a 150 litre culture from 37°C to 7°C, between 27 and 34 min after multiple infection with bacteriophage when the cell density was 4×10^9/ml. This was done with the aid of an A.P.V. Paraflow cooler, type H.T.A. (The A.P.V. Company Ltd, Wandsworth Park, London S.W.18) and gave 300 g of wet infected cells equivalent to about 75 g dry wt (2·5 g dry wt/litre). Under the experimental conditions employed, phage release would have been complete in 90–120 min.

REFERENCES

Bendet, I., Schachter, E., and Lauffer, M. A. (1962). *J. molec. Biol.*, **5**, 76–79.
Billeter, M. A., and Weissmann, C. (1966). *In* "Procedures in Nucleic Acid Research" (Ed. G. L. Cantoni, and D. R. Davies), pp. 501–502. Harper and Row, New York and London.
Cooper, C. M., Fernstrom, G. A., and Miller, S. A. (1944). *Ind. Engng. Chem.*, **36**, 504–509.
Davison, P. F., and Freifelder, D. (1962). *J. molec. Biol.*, **5**, 635–642.
Davis, J. E., Strauss, J. H., and Sinsheimer, R. L. (1961). *Science, N.Y.*, **134**, 1427.
Elsworth, R., Capell, G. H., and Telling, R. C. (1958). *J. appl. Chem.*, **21**, 80–85.
Enger, M. D., Stubbs, E. A., Mitra, S., and Kaesberg, P. (1963). *Proc. natn. Acad. Sci. U.S.A.*, **49**, 857–860.
Frazer, D. (1951). *J. Bact.* **61**, 115–119.
Hershey, A. D., Kalmanson, G., and Bronfenbrenner, J. (1943). *J. Immun.*, **46**, 267–279.
Isenberg, H. (1963). M.Sc. Thesis, University of Birmingham.
Lunan, K. D., and Sinsheimer, R. L. (1956). *Virology*, **2**, 455–462.
Luria, S. E., Williams, R. C., and Backus, R. C. (1951). *J. Bact.*, **61**, 179–188.
Maccacaro, G. A. (1961). *Annali. Microbiol. Enzimol.*, **11**, 169–172.
Paranchych, W., and Graham, A. F. (1962). *J. cell. comp. Physiol.*, **60**, 199–208.
Rushizky, G. W., Greco, A. E., and Rogerson, D. L., Jr. (1965). *Biochem. biophys. Acta.*, **108**, 142–143.
Sargeant, K., and Yeo, R. G. (1966). *Biotech. Bioengng*, **8**, 195–215.
Siegel, A., and Singer, S. J. (1953). *Biochem. biophys. Acta.*, **10**, 311–319.
Sinsheimer, R. L. (1959). *J. molec. Biol.*, **1**, 37–42.
Solomons, G. L. (1967). *Process Biochem.*, **2**, 7–12.
Stent, G. S. (1963). *In* "Molecular Biology of Bacterial Viruses", p. 46. Freeman, London.
Strauss, J. H., Jr., and Sinsheimer, R. L. (1963). *J. molec. Biol.*, **7**, 43–54.
Swanstrom, M., and Adams, M. H. (1951). *Proc. Soc. exp. Biol. Med.*, **78**, 372–375.
Swarby, L. G. (1959). Ph.D. Dissertation, University of Pittsburgh, U.S.A.
van Hemert, P. (1964). *Biotech. Bioengn*, **6**, 381–401.
Wyatt, G. R., and Cohen, S. S. (1953). *Biochem. J.*, **55**, 774–782.
Zimmermann, S. B. (1966). *In* "Procedures in Nucleic Acid Research" (Ed. G. L. Cantoni and D. R. Davies), pp. 309–311. Harper and Row, New York and London.

CHAPTER XV

Evaluation of Growth by Physical and Chemical Means

M. F. MALLETTE

Department of Biochemistry, Pennsylvania State University, University Park, Pennsylvania 16802

I. INTRODUCTION

Most physiological studies in microbiology are at least semi-quantitative. Therefore, some measure of the amount of biological material is necessary. To the extent that such measures reveal an increase, *growth* is indicated.

Those biologists concerned with larger plants and animals always differentiate clearly between the growth of individuals and increases in populations. However, because in the past measurement of changes of individual micro-organisms has been tedious, microbiologists have often used the term growth interchangeably with increase in population, in size of an individual cell or in the total amount of cellular material.

A. Growth and multiplication

As is well known (Hershey, 1939; Hinshelwood, 1946; Levy *et al.*, 1949; Herbert, 1961; Wilson & Miles, 1964; Lamanna and Mallette, 1965), cell size and number may change independently or simultaneously, depending

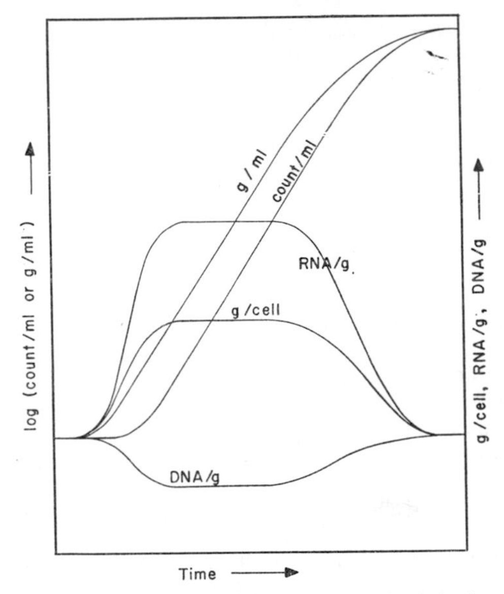

FIG. 1. Changes in population (count/ml), dry weight (g/ml), cell weight (g/cell), RNA/g of cells and DNA/g of cells in an idealized bacterial culture following inoculation with cells in lag phase. Under such conditions, cells enlarge before dividing. During the exponential phase of growth, the average cell size and composition remain constant and change back to the initial values when the stationary phase is reached. The length of the lag period following inoculation varies with the measure taken for "growth." The various parameters have been normalized with respect to the inoculum. (Modified from Herbert, 1961).

upon the interaction of various biological and environmental factors. Stevenson (1962) studied growth of *Arthrobacter globiformis* as a function of cell size, population and levels of macromolecules. Upon transfer of cells from a stationary culture to fresh medium, RNA increased twelvefold, DNA sixfold and protein sixfold (with RNA increasing earliest) before the first cell division occurred. Thus, individual cells increased markedly in size and remained in pleomorphic form until the medium became limiting, and RNA, DNA and protein synthesis stopped. Thereafter, the cells divided for three or four generations yielding smaller, stationary, coccoid forms. Figure 1 shows such changing relationships in idealized form and is based on a similar drawing by Herbert (1961).

In the extreme case, special media may induce growth of bacteria as long filamentous forms. For example, Webb (1949) reported that Gram-positive rods formed filaments or abnormally long chains in magnesium-deficient, complex media. Somewhat similar morphologies were observed at high concentrations of magnesium ion. Weinbaum (1966) described filamentous and branched forms of *Escherichia coli* B grown in a complex medium containing glucose, NaCl, lysine, nutrient broth and casein hydrolysate. *E. coli* E–26, on the other hand, grew more normally with only a slightly increased cell size when compared to standard synthetic medium. Thus, under certain conditions cell size is an important variable.

It is obvious that losses of material from cells during endogenous processes occurring in the absence of exogenous substrates must result in reduction in the dry mass of cells. Other measures of the amount of cellular material may also be affected, e.g., decrease in turbidity (McGrew and Mallette, 1962), loss of NH_3 and CO_2 (Dawes and Ribbons, 1965), decline in carbohydrate content and O_2 uptake (Ribbons and Dawes, 1963), utilization of endogenous poly-β-hydroxybutyrate (Dawes and Ribbons, 1964) or of RNA (Hou *et al.*, 1966; Clifton, 1966).

On the other hand, when only a single required nutrient is missing, the total weight or the turbidity of a cell suspension may increase for a time, reflecting net increase in at least some cellular components. Such observations have been recorded for *E. coli* in media deficient in sulphur (Bohinski and Mallette, 1967), phosphorus (Mallette, *et al.*, 1964) and nitrogen (S. B. McGrew and M. F. Mallette, unpublished work). In the phosphorus-deficient system, cell count increased more rapidly than turbidity, suggesting that daughter cells were reduced in size and again demonstrating the microbiological problem of using the word growth. Hou *et al.* (1966) noted similar behaviour for *Pseudomonas aeruginosa* in phosphorus-deficient systems, and Schmid (1958) reported that cells of *Saccharomyces cerevisiae* incubated without exogenous glucose enlarged, but did not divide until glucose was added subsequently.

Less physiologically normal factors also disturb cell division and alter the relationship between size and number. Radiation inhibited cross septation of cells and led to long multinucleate filaments in *E. coli* according to Adler and Hardigree (1965). Subsequently, cytokinesis was initiated by exposure at 42°C or to pantoyl lactone. Likewise, *p*-fluorophenylalanine affected cell division and mass increase differently when added to this same species, again producing filamentous forms (Previc, 1964). Among others, Allison *et al.* (1962) have reported unbalanced growth and cell division of *E. coli* in the presence of chloramphenicol.

Recent revival of interest in the cycle of bacterial duplication (Kuempel and Pardee, 1963; Marr *et al.*, 1966) has been due largely to development of techniques for measurement of macromolecular components and particularly for measurement of cell volume or length. Toennies *et al.* (1961) applied the Coulter counter to measurement of the distribution of sizes as well as numbers in bacterial suspensions. Since this counter records pulses when bacteria in suspension pass through a small orifice, pulse height is taken as a measure of cell size. The data are distorted to some extent by coincidences in the passage of cells. Harvey and Marr (1966) have modified the circuitry, permitting differentiation of large pulses due to coincidence from those due to the larger cells. Their equipment counts and records the distribution of volumes for latex spheres and bacteria over the range 0·25–20 μm. This method should be especially useful in the study of the division cycle, since it can be applied to viable cells and presumably to synchronously dividing cultures. A more detailed account is given by Kubitschek (this Volume, p. 593).

Primarily on the basis of direct microscopic observation, it has been generally believed that bacteria dividing by transverse binary fission increased largely or entirely in length (Schaechter *et al.*, 1962; Collins and Richmond, 1962). At the same time it has been believed that division yields equal halves, each daughter acquiring half of the original protoplasm. Using carefully controlled electron microscopy, Marr *et al.* (1966) established the validity of both assumptions for *E. coli.* Other rod-like bacteria probably behave in similar fashion.

These observations are supplemented by direct microscopic and photomicrographic data (Schaechter *et al.*, 1962; Collins and Richmond, 1962; Errington *et al.*, 1965) showing that individual bacterial cells increase exponentially in length before division. Furthermore, no hesitation in growth could be detected before or after division. On the other hand, not all cells grow at the same rate, and a 20% variation occurred in the interdivision times of different cells. Sister cells resembled each other closely in size and time of division, but differed from other sets of sisters at other positions in the family tree. The correlation of mother with daughter cells is not yet clear, possibly because of environmental fluctuations in some experiments.

B. Scope of treatment

This Chapter considers the evaluation of growth by physical and chemical means. All chemical and many physical techniques estimate the total amount of biological material. These methods can be converted into population estimates only by means of suitable calibrations, which, as will be seen, are accurate only for particular conditions. Biological, including plate and dilution (most probable number), methods are covered in detail in the Chapter by Postgate (this Volume, p. 611). Since plate counting is widely used in standardizing the methods described in this Chapter, general reference will be made to it.

Electronic counting and sizing by the Coulter counter are described in the Chapter by Kubitschek. A Vickers instrument has been tested (Pegg and Antcliff, 1965) for counting erythrocytes and leucocytes and should apply to enumeration of at least larger species of bacteria and yeasts. Deppe (1965) and Deppe and Norpoth (1965) presented a somewhat different instrumental approach based on conductometric impulses and evaluated the system with *Mycobacterium phlei*. Again reference to such techniques will be made in comparison with data of other types.

Cross-references will be made to still other Chapters when appropriate. These subjects include turbidostats, various chemical analyses, automated microbial assays and growth yield constants.

Of necessity, treatment will be in the form of a selective summary, since the large number of original references precludes inclusion of them all. In addition, much that is relatively elementary and obvious will be presented in the belief that many workers outside of or just beginning in microbiology will profit. Specification of the general method employed in measuring the microbial material is sufficient to inform the knowledgeable reader of the meaning employed. Nevertheless, it is necessary to consider the use to which the data will be put in selecting a meaningful measure of growth. Attention must be paid to such need.

C. General problem of accuracy versus precision

These two terms are variously defined in several more or less related ways. Accuracy is often referred to qualitatively as the difference between the meaning of a set of measurements and the true or actual value of the parameter under investigation. Stated slightly differently, accuracy is an index of the systematic error in a set of measurements. On the other hand, precision refers to the magnitude of the differences between the repeated measurements themselves without reference to the "true" value. Thus, precision is an index of the random error in a set of measurements and is unrelated to accuracy, which may be independently either high or low.

Accuracy in the measurement of bacterial growth is subject to the same kinds of theoretical and practical problems as other types of analytical operations. Thus, assessment of the accuracy of any technique is not possible in an absolute sense. Therefore, it is customary to compare results from different methods using as the base one method that is considered to be the most reliable. The choice usually is made on rather subjective grounds, including estimates of precision, reproducibility in the hands of different investigators and opinions resulting from actual experience in application of the procedures available.

Although precision is susceptible to detailed statistical analysis, the resulting conclusions are not all one would like, since only random errors can be so treated. The potentially much more serious systematic errors so important in accuracy cannot be evaluated. This problem is especially acute in the present context both because of differences in the definitions of growth and because major systematic errors are known or suspected in all available techniques.

A typical comparison of growth measurements is that of Swanton *et al.* (1962) for plate, microscopic and Coulter counts of *Staphylococcus aureus*, *E. coli*, *Serratia marcescens* and latex spheres. Microscopic counts of dried, stained samples, in a haemacytometer or in hanging drops were variably reproducible. The use of hanging drops was thought to be the best. When compared to plate counts, there was some difference at different stages of growth. Pulses in the Coulter counter varied with species and viability. Moreover, responses for some live bacteria were not independent of current flow.

Zellner *et al.* (1963) reported that the Coulter counter was excellent for cell counts of suspensions of *Rhodotorula glutinis*. They did not recommend direct counting in the Petroff-Hausser chamber except at very high populations ($> 10^9$ cells). Drop and spread plate methods were reliable, except during the period of maximum budding. Norris and Powell (1961) made a systematic study of the chamber counting technique and succeeded in eliminating one major source of systematic error, that of error in thickness of the filled chamber. This and other features of the method are discussed later.

On the other hand, Mountney and O'Malley (1966) compared plate and electronic counts, finding greater precision in the electronic data, but better overall growth curves with plate counts. They also commented that absolute accuracy was not determinable.

In cases where sampling errors and uniformity of distribution are in question, e.g., in bacterial counts in sputum and faeces, statistical treatment of data becomes more helpful. Faxen (1960) presents a standard error formula and illustrates numerically its application.

The importance of the foregoing considerations is obvious. One example may be cited to illustrate. Davey *et al.* (1966) proposed that the growth (in this case population after a fixed period of incubation) of micro-organisms was dependent on the microphysical structure of liquid water. They reported population minima as determined by direct cell counts for *Pseudomonas fragi* at ~15°C, *Streptococcus faecalis* at ~30°C, *Bacillus coagulans* at ~45°C and *Bacillus stearothermophilus* at ~60°C. At these same four temperatures, changes occur in the structure of water. Although, such a study does not depend for validity on absolute accuracy in cell counting, it does depend on reproducibility of counts at individual temperatures and stability of systematic errors as a function of temperature. Intuitively one feels that the second condition is probably met, but the possibility of such error is not easily eliminated. In this study and its predecessors, reproducibility seems to be satisfactory, although additional data on the point would have been helpful. In addition, it would have been of interest to employ other types of growth measures. Perhaps the physiological effect is on cell division with no inhibition of increase in total protoplasmic mass. Measurement of dry weight, turbidity, or nitrogen should have settled this point when the original studies were made.

D. Physical and chemical methods

Most of the techniques in these groups include all cells alive or dead, and, except for microscopic procedures, they usually do not distinguish between cells and lysis debris. This inability is a major disadvantage of physical and chemical methods. However, it is to some extent unavoidable in quantitative metabolic studies, which often require a method measuring the total amount of biological material. None of the methods available separately determines mass and number in a single operation, although some may be modified at least in principle so that sequential measurements of number and mass are possible or can be derived from counts and length measurements. Such considerations again emphasize the importance of fitting the approaches to the goals sought.

II. TURBIDIMETRY–NEPHELOMETRY

As the basis of the most widely used methods of estimating total microbiological material in suspension, turbidimetry will be discussed first. Furthermore, since the term turbidity is often used in reference to the scattered light from a suspension as well as the attenuation of the transmitted beam, turbidimetry and nephelometry are treated together.

A. Basis

The various features of the phenomenon of light scattering have been reviewed widely, for example by Oster (1955), van de Hulst (1957) and Hochgesang (1964). They may be summarized briefly for the present purpose as follows.

When light passes through matter, it is scattered in part from its original path by inhomogeneities present. If the inhomogeneities of interest are particles considerably larger than small molecules, scattering becomes relatively intense. Moreover, it is variously dependent upon the concentration, size and shape of the particles, the relative refractive indexes of particle and medium, and the wavelength of the incident light.

Because bacteria lie beyond the size range covered by the Rayleigh expression and the formulations based on it, theoretical derivations have not yet proved generally applicable in microbiology (Lamanna and Mallette, 1965). However, empirical equations of the Beer–Lambert type are important. For suspensions so dilute that the scattering centres are essentially independent—

$$I = I_0 \, e^{-\tau l} \tag{1}$$

where I is the intensity of the beam after attenuation of the incident parallel beam of initial intensity I_0. The length of the path through the suspension is l, e is the natural base and τ the turbidity. In these dilute systems, turbidity is proportional to particle concentration, and we may write—

$$\tau = c\tau' \tag{2}$$

with c for concentration and τ' for turbidity coefficient. Equation (1) may then be written as –

$$\log_{10} \frac{I_0}{I} = \frac{\tau'}{2 \cdot 303} \, cl \tag{3}$$

Many types of photoelectric colorimeters serve microbiologists as turbidimeters and possess scales reading directly in $\log_{10} I_0/I$, usually called absorbance.†

The cuvettes of some instruments provide light paths of 1 cm. In many other cases measurements are made using tubes of a given diameter. When the dimensions of such cylindrical tubes are not definitely known but are kept constant, this factor is combined with the turbidity coefficient forming a single proportionality constant. The working equation for absorbance then becomes—

$$\log_{10} I_0/I = kc \tag{4}$$

† The expression $\log_{10} I_0/I$ is variously called optical density, absorbancy and absorbance: the latter term has been adopted throughout this series.

and k may be evaluated by calibrations as discussed below. To avoid confusion with absorbance, "turbidity" may be defined as $\log_{10}I_0/I$ and represented by τ.

Because of the formal resemblance of equation (3) to that for light absorption by coloured solutions, there is a temptation to equate τ with the absorption coefficient. However, a suspension of particles large with respect to the wavelength of light scatters light primarily in the forward direction (van de Hulst, 1957), i.e., at small angles with respect to the emergent beam. According to Black and Hannah (1963), standard suspensions of fuller's earth scatter 80 to 400 times more light at 15° from the incident beam than they do at 90°. Koch (1961) noted that bacteria scatter primarily in the forward direction. As a result, in most commonly used equipment designed for colorimetry, much of the scattered light is collected by the detector system. Therefore, the mathematical equivalence of the two coefficients is not valid in practice. In fact turbidity ($\log_{10} I_0/I$) is markedly dependent on the geometry of the system, varying widely between instruments according to the characteristics of the cuvettes, the apertures of light-sensing devices and the distance between cuvettes and sensors. Obviously, constancy is required and each set of factors needs its own calibration.

In contrast to the difference measurements during attenuation of a beam (equation 4), it is possible to measure the scattered light itself. Because light is scattered in all directions and the intensity varies with direction, measurement of the total light scattered is impractical. However, the amount of light scattered in a given solid angle is directly proportional to the incident intensity and the concentration. Thus—

$$I_s/I_0 = Kc \tag{5}$$

where I_s is the quantity of light scattered in the given angle, I_0 is the quantity of light in the incident beam and K is the proportionality constant. Note that the quantity of light is proportional to intensity, but that instruments, including those for optical density data, ordinarily register quantity directly rather than intensity. K varies with geometry and all of the factors listed earlier in this Section. These factors may be related mathematically under certain conditions permitting size and shape estimation. The theory does not generally apply to microbial suspensions. Measurement of scattered light is referred to either as turbidimetry or nephelometry.

Because there is angular dependence of intensity on the shape of suspended particles, lengths of bacterial rods have been studied by turbidimetry–nephelometry. Metz and Mayr (1966) described equipment suitable for estimating both cell number and size. Recalling that for *E. coli* at least (Marr *et al.*, 1966) individual cells increased only in length during growth, size in this case can be transformed into data on length. Starka (1962) observed no dependence at 45°C of scattered intensity on cell volume under certain conditions and within certain volume limits. His procedure in part resembles that of Metz and Mayr (1966), but differs from most in yielding data converted directly to cell number in ordinary growing cultures.

A study by Powell and Stoward (1962) was designed to emphasize dependence on cell shape and thus to follow changes in length during growth of synchronized cultures. This work employed flow for partial orientation of bacterial rods during nephelometric observation. As expected, mean cell length increased during the growth cycle up to the time of fission, then dropped.

Without employing the above special techniques *c* of equations (4) and (5) generally is directly proportional to dry weight (Monod, 1942; Spaun, 1962; Powell, 1963). In other words, ordinary turbidity data reflect changes in both cell size and number, and are especially valuable in providing estimates of the total amount of material. Koch (1961) presented a detailed analysis of factors affecting light scattered by bacteria and based it on an examination of approximate solutions of the general theoretical equations. He concluded that the absorbance measurements routinely employed in microbiology are more nearly related to total bacterial mass than to numbers. Data from equations (4) and (5) do not permit separation of the effects of growth of individuals and increases in population. Calibrations based on other techniques may do so, providing the environmental and physiological conditions are controlled carefully because of the differences in the mean sizes of bacteria from old and young cultures, for example.

B. Calibration

For self-consistent turbidity data from a single instrument of fixed geometry, calibration of the instrument itself is seldom necessary. However, when data from different instruments are to be compared, calibration becomes essential. For many purposes a series of cell suspensions may be read in the instruments concerned, providing conversion factors over the appropriate range of turbidities. Since it is not always possible to repeat what can be a time-consuming comparison, relatively stable suspensions of other materials may be used as standards in making regular checks. Such systems should resemble their biological counterparts as closely as possible in average particle size, range of size distribution, particle shape, refractive index relative to the suspending medium, and turbidity. Identity of the experimental material with the non-biological standard is not possible in all of these respects, perhaps accounting for the variety of systems proposed. Some seem to be quite suitable for microbiological work.

Fikhman (1961) polymerized methyl methacrylate with benzoyl peroxide and heat into blocks reported to have turbidities stable for at least 2 years.

Comparison with bacterial suspensions at different wavelengths revealed similar turbidity properties, prompting the conclusion that such blocks will be useful in calibrating instruments for bacteriological work. Recently, Roessler and Brewer (1967) reported permanent turbidity standards also of a stable plastic type. They obtained TiO_2 finally ground for pigment and cosmetic uses to the range 0·18–0·32 μm mean diameter from Glidden Co. (Baltimore, Md.), American Cyanamide Co. (Bound Brook, N. J.) and E. I. duPont de Nemours and Co. (Wilmington, Del). A few milligrams of this pigment was stirred into 100–200 g of Santolite MHP resin (Monsanto Chemical Co., St. Louis, Mo.) that had been heated to the softening point (60°–90°C). This mixture was poured into selected test tubes to a depth of 4–5 cm and allowed to cool slowly. Such solids were stable for 12 years in use as nephelometric standards having turbidities comparable to those of bacterial cultures. Resin 276–VP (Dow Chemical Co., Midland, Mich.) was also satisfactory as a suspending phase. Holders for the standards were described for both nephelometric and transmission readings.

Fikhman (1958) also has recommended, as suitable for bacteriology, a turbidity standard of pulverized glass prepared as follows. Fill a heat-resistant glass flask to one eighth its volume with fragments of the same type of glass approximately 0·5–1 cm in diameter. Cover the fragments with four volumes of water and shake vigorously for 5 h. Discard the liquid, wash the fragments once with distilled water and sterilize for 2 h at 160°C. Again, cover the fragments with four volumes of sterile water containing 1 : 10,000 Merthiolate and shake for 6 h on each of 4 successive days. Decant the supernatant liquid and allow to stand for 24 h to permit settling of any coarse particles carried along with the suspension. After decanting from any such material, the standard is ready for use and is stable for 5–10 years.

Similarly Peterson and Tincher (1961) described the use of sized glass beads packed in standard photometric cells. Turbidity was inversely related to bead size, allowing preparation of standards having different turbidities. In the work on bacterial numbers as distinct from mass referred to above, Starka (1962) employed an opal glass plate for reference.

Polystyrene latex has been recommended as a standard for studying water by Black and Hannah (1963) and in the clinically important thymol method for serum β-globulins by Ferro and Ham (1962). Concentrated suspensions of this latex are available from Dow Chemical Co. with an average particle diameter of 0·8 μm. This size approaches that of the smaller bacteria.

Calibration standards of colloidal sulphur (Rose, 1950) and silica (Jullander, 1949) have been recommended. Since both types of systems have been studied widely by turbidimetry, they might be applicable to bacterial suspensions. The World Health Organization has distributed a reference standard for evaluating the turbidities of vaccines and other biological preparations. Kavanagh (1963) reported that calibration curves for suspensions of living and dead *Staph. aureus* were identical. Moreover, he claimed that such curves could be used for other species of bacteria without appreciable error. In this way, calibrations might be performed with bacteria themselves, although extension of this idea to include all sizes and shapes of bacteria in all types of instruments seems rather unlikely.

Most microbiological work is concerned with a single instrument for turbidity measurement, and relative values of absorbance (τ) adequately reveal changes. However, comparison of data from two sources becomes impossible unless calibrations are presented. Therefore, even in purely relative microbiological studies, data should be given to permit conversion to dry weight, amount of nitrogen or cell number, for example. In following the isolation of components or in kinetic studies requiring expression in terms of the amount of biological material present, reference to one of the above standards suitably calibrated becomes yet more necessary. Or, more commonly and to better effect, a calibration curve is prepared. Such calibration curves permit conversion of the primary instrumental data into mass or some related parameter. These calibration curves possess the great advantage of permitting use of data beyond the linear range. This procedure calibrates the method as a whole, not just the instrument itself as was the situation with turbidity reference standards.

Conclusions drawn from data may vary with the basis of the calibration. Hence investigators may calibrate by more than one method or, as did Hershey (1939), measure several related properties, all of which are useful in evaluating physiological changes in bacterial cultures. The necessary reference data are commonly based on dry weight (Monod, 1942; Palmer and Mallette, 1961; Spaun, 1962; Powell, 1963; Previc, 1964; Eriksson, 1965) or on cell counts (Weinbaum and Mallette, 1959; Hinds and Peterson, 1963; Kavanagh, 1963). Since viable and direct microscopic counts may differ somewhat even during exponential growth (Lamanna and Mallette, 1965), the basis of the calibration must be borne in mind, some authors employing one counting procedure and some another.

Kurokawa *et al.* (1962) described improved linearity of calibration curves obtained for two bacterial forms, *Staphylococcus albus* and *Corynebacterium diphtheriae*, in simple photoelectric colorimeters. They used two equations for different ranges of absorbance. One corresponds to equation (4) and was recommended as a linear function of cell count above absorbances of 0·30 under their conditions. From $0{\cdot}5 < T < 1{\cdot}0$ they recommended $0{\cdot}6 + \log_{10}(1-T)$ where T is transmittance. Since —

$$\text{Absorbance} \equiv -\log T$$

Absorbance decreases as T increases; $T = 0{\cdot}50$ corresponds to absorbance $= 0{\cdot}30$.

Such behaviour apparently arose (Kurokawa *et al.*, 1962) by deviation from linear dependence of turbidity on mass (cell numbers with constant mean size and shape) at low levels of mass. Normally one expects the deviations at high rather than at low population densities, e.g., Koch (1961). Some of the behaviour noted by Kurokawa *et al* (1962) might be due to the design of their instruments although at least one of these types was used extensively in an early model by the present author without noticing significant non-linearity with low populations.

When micro-organisms are suspended in coloured media, absorption of light by the medium must be considered. If wavelengths not absorbed by either the medium or the cells are available, their use avoids the problem. If this approach cannot be applied, the instrument may be adjusted to a "zero absorbance" setting with medium in place of the cell suspension. When the colour intensity is constant throughout the experiment, relative data can be obtained in this way. There are some theoretical objections to this procedure, and they can become serious with increasing absorbance owing to the colour. In practice re-calibration of the instrument with the coloured medium is very helpful. When the intensity of the colour changes with time, kinetic relationships might be established, but would probably negate the virtues of turbidimetric methods, namely their speed and convenience.

Problems presented by absorbing colours when the scattered light is measured appear even more difficult. However, Il'ina and Rozenberg (1954) and Nakagaki and Ikeda (1965) suggested that solutions are possible under some circumstances. It is not yet clear whether their approaches will be applicable to microbiology.

C. Instrumentation

Over the years many instruments have been applied to measurement of turbidity, including scattered light, and several are commercially available. Current interest in development of the instrumentation centres largely around recording turbidimeters–nephelometers, especially for use in automated systems, and around turbidostats in the continuous culture of micro-organisms.

Routine measurements of turbidity are conveniently made with any of the numerous, commercial, photoelectric colorimeters and spectrophotometers. These instruments are normally equipped with both linear and logarithmic scales, the latter designed to read $\log_{10} I_0/I$ directly. This arrangement is most useful in applying equations of the type of (4) for either colorimetry or turbidimetry. Data from the linear scales of some older or homemade instruments often may be converted to turbidity–absorbance by means of equation (6).

Because of the wavelength dependence of light scattering, it can be helpful to change wavelengths either to a region of greater scattering for improvement of sensitivity or to a region in which light is not absorbed by coloured materials. Spectrophotometers, even of the simplest types, are convenient in this respect and have become the most widely used class of instruments in microbiological turbidimetry. Koch's (1961) demonstration that scattering by *E. coli* is inversely proportional to $\lambda^{2.28 \pm 0.018}$ (λ = wavelength) from $\lambda = 350$ to 800 nm shows the importance of this flexibility. Various models from Coleman (Weinbaum and Mallette, 1959; Kurokawa *et al.*, 1962; Ferro and Ham, 1966), Beckman (McGrew and Mallette, 1962; Kavanagh, 1963; Previc, 1964), Bausch and Lomb (Hucke and Roche, 1959–1960; Kavanagh, 1963: Haney *et al.*, 1963; Mallette, *et al.*, 1964) have been mentioned as satisfactory for work with bacteria. Colorimeters (photometers) of various types have been used directly or as components of continuous culture and/or recording systems, Hitachi (Kurokawa *et al.*, 1962), Klett–Summerson (Northrup, 1960; Kavanagh, 1963), Leitz (Ferro and Ham, 1966), Lumitron (Kavanagh, 1963).

Because of their electronic and mechanical simplicity, speed of registration and ability to operate without disturbing sterility, turbidimetric systems have been almost invariably incorporated into the population (mass) controlling elements of continuous culture systems as will be seen in the Chapter by Munson (this Series, Volume 2). Suffice it to say here that circuitry controlling input of nutrients and withdrawal of culture is actuated from the galvanometer circuit of the turbidimeter.

Automatic recording of turbidimetric data in studying growth of micro-organisms is currently employed where many samples must be handled routinely Goldfarb *et al.*, 1962; Horn *et al.*, 1964; Ferrari *et al.*, 1965a, b). In carrying this development a step further, recording of turbidity data as a function of time has been described by Chernov *et al.* (1963) for 12 tubes, by Wentink and La Riviere (1962) for 24 tubes and by Lindberg and Reese (1963) for 90 tubes.

Inevitably, recording of data has found its way into microbial assays for nutrients, vitamins, drugs and testing for sterility, for example, in papers by Gerke *et al.* (1960), Hucke and Roche (1959–1960), Jacob and Horn (1964).

As pointed out above, instruments respond differently when reading a given suspension in a given cuvette. Much of this difference is occasioned by collection of varying fractions of the light scattered in the forward direction. Since sensitivity is affected as well as calibration, those instruments with the greatest distance between the cuvette and the phototube tend to be the most sensitive. Note that the dimensions of many phototubes are rather similar. In some instruments the spacing can be changed by switching cuvette housings. For example, sensitivity of the Beckman model DU is readily increased (Bohinski and Mallette, 1967) by replacing the standard housing with one intended for use with cuvettes having a 10 cm light path. This commercially available accessory is already designed to accommodate 1 cm cuvettes at a much increased distance from the phototube. Flexibility of this type is possible only when cuvette and phototube housings are demountable.

Many commercial instruments accept only cuvettes of specific size and shape. Since these cuvettes are not readily adapted to the maintenance of sterility or cultural purity in microbiological work, their use requires frequent sampling, usually accompanied by loss of the biological material in the sample. More convenient models are provided with adapters fitting culture tubes of various sizes and in this way avoiding contamination when series of culture tubes are involved (Konev, 1959). Frequently, somewhat larger volumes of culture are desired. Capped Erlenmeyer flasks of up to perhaps 1 litre capacity can be fitted with sidearms in such a way that part of the culture may be tipped into the sidearm for turbidity reading without loss of material or cultural purity (Eriksson, 1965; Faith and Mallette, 1966).

Eriksson (1965) also has developed a method for obtaining turbidity data from Petri plates. Although not so applied, this apparatus should be useful to bacteriologists. It will be valuable in studying those organisms that grow only on surfaces and that have not been subject to routine turbidimetric assay in the past.

Nephelometers have never become as useful in routine microbiological works as have those turbidimeters relying on equation (4) instead of (5). This restriction seems to have arisen from difficulty in adapting most commercial nephelometers to temperature control in the biological range and particularly to fitting them to accept conventional culture tubes. The situation is somewhat unfortunate because the data obtainable span a much broader range, especially at low populations where turbidity by spectrophotometers is quite inaccurate. For certain research problems, withdrawal of samples from the growing culture and brief exposure to contamination is not serious, and nephelometry is most helpful, as revealed in survival studies by Harrison (1960), growth curves by Forrest and Stephen (1965), measurement of microbial size and number by Metz and Mayr (1966), virus production by Underwood and Doermann (1947) and changes in cell length by Powell and Stoward (1962).

Instruments for light-scattering research are manufactured by Zeiss, Coleman, American Instrument, Phoenix and Evans Electroselenium Ltd. Of these, the first two have been the most adaptable to examination of cell suspensions. Modification of various photometers for light-scattering measurement has been described by Metz and Mayr (1966) and Dezelic (1961). As is to be expected, recording nephelometers have been built, e.g., by Underwood and Doermann (1951), Wilkins *et al.* (1959) and Forrest and Stephen (1965). The last two were used in studying bacterial cultures and would be useful in routine work if commercially available.

The sensitivity of turbidimetric methods varies widely from about 10^7 *E. coli* cells/ml in the Beckman DU at 400 nm and with the elongated cuvette housing (calculated from Bohinski and Mallette, 1967) to well over 10^9 cells/ml at wavelengths above 600 nm in most instruments provided with suitable calibration curves and small cuvettes. Still greater populations yielding optical densities much above unity probably are best read after dilution with medium.

Nephelometers may be applied (Underwood and Doermann, 1947) to the estimation of bacterial populations ranging from about 10^6 to 10^9 cells/ml. As Koch (1961) pointed out, much depends on choice of wavelength, angle at which scattering is measured and the geometry of the light collector relative to the illuminating beam. It has been possible to detect the scattering excess due to 10^4 *E. coli* cells/ml suspended in solutions free from other colloidal matter (Lamanna and Mallette, 1965).

At high concentrations of moderately large particles like bacteria, part of the scattered light is scattered again, some of it re-entering the transmitted beam. In addition, particles are not completely independent and destructive interference between particles becomes a factor (Koch, 1961). Under these conditions the intensity of the transmitted light is higher than would be expected from data taken with dilutions of the concentrated suspension. The overall effect makes it possible to work with higher populations in nephelometric than in optical density experiments.

D. Errors

Several problems arising in turbidimetry have already been mentioned. Among these are light absorption in addition to scattering, non-adherence to the Beer–Lambert law, the need for calibration curves, and collection of light in low scattering angles. Besides these factors, there are other purely instrumental difficulties. Some galvanometer–potentiometer systems do not react rapidly, and the operator may read too quickly. Some instruments, reading by galvanometer displacement, drift with time or are otherwise inherently unstable; others have scales too short to provide the desired level of discrimination (Kavanagh, 1963). Certain optically simple, commercial spectrophotometers provide light beams of very low purity. In these instruments examination by inserting a strip of white paper into the light path with the room darkened reveals colour mixtures and distributions that can only be described as fantastic. Such instruments obviously cannot be used in studying wavelength dependence, but do serve in turbidimetry when properly calibrated.

Buchanan (1955) pointed out that the optical cells in continuous-culture equipment require cleaning devices. These have been designed, e.g., by Northrup (1960). The cuvettes or culture tubes themselves scatter light, rather extensively sometimes, leading to errors of 30% (Heller and Tabibian, 1957) unless precautions are taken, especially since the tubes in a series may differ considerably. Also see Munson (this Series, Volume 2).

Without occasional stirring, many cells tend to settle out of suspension. This behaviour is so obvious that one tends to mix vigorously and read the turbidity immediately. However, the streaming birefringence, used to advantage by Powell and Stoward (1962), can selectively orient unsymmetrical bacteria and change the apparent absorbance. As a result, readings should be made after a short delay for randomization, or stirring should be continuous and be employed in reproducible fashion throughout both calibration and experiment.

Those cells, which tend to remain aggregated as does *Ps. aeruginosa* at moderate to high levels of phosphate, present difficulties to turbidimetry (Hou *et al.*, 1966). In certain cases, such aggregation and absorption at interfaces may be reduced as recommended by Norris and Powell (1961), by addition of anionic detergents to samples withdrawn from cultures for turbidity measurement. Dubos and Middlebrook (1948) dispersed tubercle bacilli in broth media by adding non-ionic detergents of the Tween and Triton series. For those species whose surfaces are relatively hydrophobic in character, detergents should be at least partially effective in maintaining the dispersion of individual cells. However, species with relatively polar surfaces may also form clumps that are not readily dissociated by detergents. For example, added detergents had relatively little effect on clumps of *Ps. aeruginosa* growing in an excess of phosphate. This characteristic introduced random uncertainties into turbidity data (M. F. Mallette, unpublished work).

Unwanted changes in the biological material also may introduce large uncertainties. When the osmotic situation in a cell is disturbed by transfer to another medium or by washing, the cell/medium ratio of refractive indexes may change and change the turbidity without alteration in cell count or total mass (Mager *et al.*, 1956; Kuczynski-Halmann and Avi-Dor, 1958; Abram and Gibbons, 1961; Bernheim, 1963; Rogers, 1963). Fikhman (1963) suggested a microscopic technique for the immersion refractometry of cells held in place in gelatin gels as useful for studying alterations in cell surfaces and the related changes in turbidity. Since Bernheim (1963) noted that not all bacteria respond in the same way, Fikhman's method may prove to have comparative value as in the differentiation of normal *Staph. aureus* from protoplasts (Palkina and Fikhman, 1964). Removal of cell walls should change light-scattering properties by changing the refractive index, so that turbidities of equal populations of cells and protoplasts would be unequal. All manner of environmental changes and physiological factors are potentially important and must be kept constant or evaluated for their effects on turbidity.

Dead cells also scatter light, but not necessarily to as great an extent as live ones (McGrew and Mallette, 1962). Under some conditions, for example, when death is due to phosphate deprivation (Mallette *et al.*, 1964) turbidity may first decline then increase above its original level as cells die and the lysis debris aggregates. The intensity and duration of mixing affect the mean size of the clumps and the resulting turbidity. Under these conditions data are not very reproducible.

When significant numbers of cells in a suspension are dead and a measure of the total (both living and dead) is desired, perhaps the best procedure may be to kill the remaining live cells before measurement. Thereafter, the turbidity responses of all cells should become more nearly alike. In fact, samples may be brought routinely to 0·5% Formalin to kill the cells (Norris and Powell, 1961). This procedure has the additional merit of preventing any further growth during the period of handling. It will be obvious that turbidity measurements following such treatment must be calibrated for the presence of formaldehyde.

For the collection of data that are truly relative, all turbidimetric methods must be recalibrated following any procedural change. Even such care does not ensure accuracy in any absolute sense, but it does aid in the comparison of data from different experiments.

E. Applications

A complete tabulation will not be attempted here. However, some idea of the scope of turbidimetric technique in microbiology is indicated by the representative references cited above in other contexts. The primary function is the determination of total microbiological material as already discussed. However, cell numbers and sizes are also studied by special techniques (Powell and Stoward, 1962; Starka, 1962; Metz and Mayr, 1966).

Particular attention has been paid to study of growth using turbidimetry by Hershey (1939) Sartory *et al.*, (1947), Starka and Koza (1959), Powell and Stoward (1962) and Forrest and Stephen (1965). Enzyme biosynthesis and diauxic growth are often referred to turbidity (Monod, 1942; Weinbaum and Mallette, 1959; Palmer and Mallette, 1961). These techniques have been widely employed in studies of endogenous metabolism (Harrison, 1960; McGrew and Mallette, 1962; Bohinski and Mallette, 1967).

Likewise, there have been investigations of age, size of inoculum, effects of sterilization, osmotic changes, and permeability (Toennies and Gallant, 1948; Abram and Gibbons, 1961; Bernheim, 1963; Rogers, 1963). Turbidity may serve as the growth measure in microbiological assays (Snell, 1950; Tittsler *et al.*, 1952; Ishikura *et al.*, 1964), in routine checks of sterility (Hucke and Roche, 1959–1960), in estimating biodegradability (Prochazka and Wayne, 1965), in evaluating biosynthesis and effects of analogues (Previc, 1964), in drug action on bacteria (Wilkins *et al.*, 1959) and in surface-active agents and clumping of cells (Brown and Richards, 1964).

III. DRY WEIGHT

A. Procedural requirements

The combined weights of the cells harvested from cultures can be determined by direct weighing. Estimates of wet weights are neither precise nor accurate because of difficulty in evaluating the weight contributions of intracellular water and water wetting the cell surfaces (Lamanna and Mallette, 1965). Fikhman's (1965) refractometric technique providing refractive-index data and calculation of weight from cell volume is a possible exception, but data ordinarily are reported on a dry weight basis. The general requirements become simply removal of cells from the medium, drying and weighing.

Dry weight is rather widely measured, especially for reference in isolation and purification work and in the basic calibration of other methods. It provides data proportional to mass, in which rôle it is of pre-eminent importance. No single modification of technique is clearly superior to the others. All are rather precise, and as usual none can be evaluated in terms of absolute accuracy. Systematic errors of several types are anticipated.

Separation of cells from medium follows two patterns. In the first, cultures are filtered, and the dried filter bearing the collected cells is reweighed. Sartory *et al.* (1947) applied such a procedure to a study of the total nitrogen in *E. coli* cells grown in either ammonium or peptone medium. They employed an ultrafiltration membrane prepared from cellulose acetate. (Such membranes are now commercially available in various sizes and porosities, for example, from Gelman Instrument Co., Ann Arbor, Mich.). The membrane is first washed and dried to constant weight, then placed in a suitable funnel or holder. Bacteria are collected by filtering a sample of the culture, and the membrane plus bacteria is dried to constant weight *in vacuo* at 40°C. Sartory *et al.* (1947) worked in the dry weight range of 10–90 mg of cells from 100 ml of culture and populations up to a little more than 2×10^9 cells/ml. Nikitina and Grechushkina (1965) reported that this type of technique was the best available for study of growth in media containing liquid hydrocarbons.

The method might be made more sensitive by using a newer type of polyvinyl chloride membrane that appears to be especially useful for gravimetric work. This membrane contains little material extractable with water and in use does not adsorb much moisture. Therefore, drying of the membrane should be rapid and reproducible. When weighed with a semi-microbalance (to the nearest 0·01 mg), the sensitivity of the method might be extended downward to 1 mg of dried cells. Before drying, the cells and filter may be washed with a small volume of water to remove medium still present in the membrane. Washing is less important with polyvinyl chloride membranes not wet with water than it is with cellulose acetate membranes. In either case, washing should be minimized to reduce the leaching of material from inside the cells.

When filtration is not employed, cells are harvested by centrifugation and the medium removed by decantation, aspiration or pipetting. The cells are washed by re-suspending, usually in water, and re-centrifuging. After one to five washes, depending on the relative volumes of wash solution and pellet of cells, a sample is transferred to a container for drying. Small weighing bottles and planchets are suitable. Thin, hand-blown glass bulbs are even better because they can be made lighter than dishes, bottles or plates.

Drying is continued to constant weight under various conditions. Hadjipetrou *et al.* (1964) dried at 105°C, Eriksson (1965) at 100°C, Sartory *et al.* (1947) at 40°C *in vacuo*, Palmer and Mallette (1961) at 80°C *in vacuo*, Weinbaum (1966) under infrared lamps. A jet of dry air directed through a capillary at the cells markedly facilitates drying. When oxidation may be detrimental, air is replaced by inert gas.

Although bacteria are somewhat hygroscopic, they need not be weighed in closed vessels if modern, high-speed balances are used and if the dried samples are transferred quickly from desiccator to balance. Any gain in weight then is small relative to the sample size and is probably little greater then the changes in weight that follow the handling of closed containers like weighing bottles or pigs. Moreover, such devices are relatively heavy when they are large enough to contain several millilitres of cell suspension for drying to constant weight. This relatively high mass gives rise to appreciable uncertainty because dry weight then becomes the difference between two numbers both large compared to the sample.

Dry weights are small ranging upward from about 10^{-13}g/dry bacterium (3×10^{-13}g for *E. coli*; Palmer and Mallette, 1961). Therefore, sensitive balances (10^{-6}–10^{-5}g) are needed in order to obtain valid data at low populations or with relatively small samples of cultures. In such cases it is especially important that the weight of the filter or container be minimized.

When changes in cellular components with growth stage are to be compared, some relatively fixed point of reference is necessary. Cell counts or dry weight (Tepper, 1965; Polakis and Bartley, 1966) or both (Candeli *et al.*, 1960) are widely used. This requirement is illustrated in the following by Forrest *et al.* (1962) who found that heat evolution by *Aerobacter aerogenes* was directly related to dry weight throughout the growth curve. On the other hand, changes in respiration did not parallel dry weight in synchronized cultures of the protozoan *Astasia longa* (Wilson and James, 1963) or of *S. cerevisiae* (Scopes and Williamson, 1964). Previc (1964) found that *E. coli* fission and increase in dry weight were inhibited separately by added *p*-fluorophenylalanine. Thus, whereas heat evolution could be used as a measure of mass in the first study, respiration and cell counts could not be so used in the last three.

A so-called colorimetric method for dry weight has been suggested by Bailey and Meymandi-Nejad (1961). In this procedure, biological material (amounting to 0·2–2·0 mg dry wt) in not more than 1 ml of H_2O or inorganic buffer is heated with 2 ml of 2% potassium dichromate in sulphuric acid at 100°C for 30 min. The mixture is cooled, diluted with H_2O to 5 ml and the absorbance measured at 580 nm against a reagent blank. Dry weight is read from calibration curves based on samples of known dry weight treated in the same way. Fundamentally, the method is a determination of carbon in the biological material.

B. Errors

Several types of errors arise. The filtration methods tend to overestimate the total mass when cells are not washed, but they underestimate mass because water soluble components are leached out when cells are washed. Centrifugation methods probably underestimate the mass because a few cells may not be deposited from suspension. Still more cells are likely to be re-suspended by turbulence when the centrifuge is stopped and the tubes are handled. These cells are lost when the supernatant medium or washing liquid is removed. Again, washing removes soluble cellular components. Repeated washing with distilled water so stabilizes *E. coli* suspensions electrostatically that the cells cannot be centrifuged into a stable pellet at ordinary fields (S. B. McGrew and M. F. Mallette, unpublished work).

Drying introduces still another set of errors. These arise from decomposition of biological material during the dehydration process. Water and other volatile materials involved in structure are lost to an indeterminate extent. Moreover, such decrease in weight may be compensated by weight gained during partial oxidation of biological material dried in air at elevated temperatures. Also, Niemierko (1947), suggested that SO_2 was formed by reaction of organic material with the acid and contaminated biological samples dried over H_2SO_4. He recommended addition of $K_2Cr_2O_7$ to the H_2SO_4 and compartmental separation of sample and acid with a porous material to keep droplets of spray away from the sample. Degassing of H_2SO_4 *in vacuo* may cause droplets to form.

In spite of these sources of error, dry weight must be regarded as the best reference method for use in work relating other parameters to total amount of biological material. When not used directly, it serves as a useful basis for calibration of other procedures.

IV. CELL PACKING BY CENTRIFUGATION

This method consists of centrifuging a sample of cell suspension in a special tube and calculating the packed cell volume from the height of the column of solid material (Hopkins, 1913). Clinical laboratories routinely run haematocrit values on human blood by centrifuging the formed elements of blood and reporting as volume per cent. The principal component, of course, consists of erythrocytes. Quantities of micro-organisms can be measured in the same way although the number of studies employing this method is relatively small.

Practical restriction in use is due partly to inconvenience when compared with turbidimetry. Readings cannot be taken as quickly. Relatively instaneous data for growth curves cannot be obtained. Moreover, one must withdraw individual samples, which are not easily kept free from contamination. Although special tubes with graduated segments of reduced diameters at the lower ends are available, they are most useful only at population densities beyond the range of maximum interest to microbiologists. Haematocrit readings for blood are normally 40–50% of the total volume, but liquid cultures of micro-organisms rarely yield more than a small fraction of such a volume.

Working with baker's yeast, Conway and Downey (1950) studied permeability problems using heavy-walled capillary tubing of small, uniform bore. Suspensions were drawn into lengths of this tubing by suction, the length conforming to the dimensions of the centrifuge. An insulating air space was provided at one end of the tube, and this end was sealed with paraffin wax. When placed downward, the paraffin effectively plugged the bore for at least 20 min at 3000 rev/min (no details were given to permit an estimate of the centrifugal field).

Centrifugation does not suffer from some of the problems of turbidimetry, such as clumping of cells and light adsorption by coloured materials. There is still no differentiation of live from dead cells, and osmotic factors may alter cell volume as much as they do the scattering of light. Both approaches require calibration, although centrifugation would appear to be intrinsically somewhat simpler in this respect.

For study of problems of growth, nutritional requirements, biological assays, metabolism and the like, centrifugal methods would seem not to have general merit. However, for permeability work on micro-organisms, populations rather than single cells or samples of the permeability membranes must be used. These may be dense populations, and data on packed cell volumes are likely to be necessary in any interpretation. In this connection, Reid and Frank (1966) described an isotopic dilution method in which radioactive, saturated inulin solutions were used for improving sensitivity in estimating packed volumes of microbial cells and spores. This method was applied to populations of spores of *Bacillus subtilis* var. *niger* totalling $1{\cdot}6 \times 10^9$ spores/ml and was especially designed to avoid errors in directly measured, packed volumes. Such errors become large when cells or spores are centrifuged from dilute solutions of NaCl or other electrolytes, including buffers. Apparently they arise from failure of the organisms to pack reproducibly from dilute salt solutions. Although valuable in permeability studies, this method is not practical for routine measurement of growth of micro-organisms, the chief concern of this Chapter.

V. NITROGEN DETERMINATION

A. Chemical basis

Analysis for any major cellular component provides a relatively direct method for estimating microbial growth. The determination of nitrogen is the most widely used method of this class, primarily because of its first importance in the development of agricultural chemistry. Because the earliest procedures required rather large samples, microbiologists seldom analysed for nitrogen until they had general access to microbalances. Even then the conventional volumetric methods work best with samples of a few milligrams and, therefore, apply only to relatively large populations, e.g. 1 mg of *E. coli* approximates to $3{\cdot}2 \times 10^9$ dry cells. Improvements in the sensitivity of ammonia determination have helped.

Available methods for nitrogen depend first on conversion of all nitrogen present to a single form, then determination of this common product. Two types of conversion processes are common. In the Dumas methods the nitrogen of micro-organisms is oxidized when the sample is mixed and heated with cupric oxide in a stream of CO_2. The end-product N_2 is collected in a manometer over KOH which removes the CO_2 carrier gas. Any nitrogen oxidized beyond the elementary form is reduced by hot copper included for this purpose in the combustion train (see Johns, 1941). Since many bacteria contain 10–15% N, the smallest useful sample is equivalent to about 1 mg of dried cells in the manual Dumas methods. Hozumi and Kirsten (1962) described an ultramicromanometric method capable of determining N below the microgram range. Although of potential importance for special problems, this technique is too time consuming or specialized for general use.

Bacterial nitrogen may be converted instead to ammonia by a sulphuric acid digestion. Takagi and Honda (1962) suggested a mixture of iodic and phosphoric acids. Though this procedure has not been widely investigated, it might become useful by shortening the time required for digestion. Galanos and Kapoulas (1966) recommended perchloric acid digestion of biological materials as rapid and quantitative. It is doubtful that all heterocyclic nitrogen compounds and oxidized forms of nitrogen would be reduced to NH_3 in the presence of this acid. In micro-Kjeldahl methods (Johns, 1941; Horwitz, 1965; Bradstreet, 1965), samples are heated at the boiling point of concentrated sulphuric acid. Additives are included to catalyse and complete (Beet, 1957) the digestion, and in some cases, they reduce any oxidized nitrogen to NH_3. After digestion is complete, the mixture is made alkaline with NaOH, and NH_3 is distilled into acid and determined by difference or direct titration depending on the modification employed. Once again, 1 mg is about the smallest sample that can be used in common manual methods.

However, if NH_3 is measured colorimetrically, sensitivity is increased. Nessler methods have been widely employed, but generally apply only for more than 50 μg of NH_3 (Wilczok, 1959; Rusanov and Balevska, 1964). Galanos and Kapoulas (1966) used a version applied to as little as 5 μg, and Lang (1958) extended the range to 1 μg of NH_3. Mathies *et al.* (1962) carried sensitivity into the submicrogram range by determining the Nessler complex by X-ray spectroscopy. Lack of speed and convenience would limit this last technique to special problems.

Ammonia may be determined in other ways. Fels and Veatch (1959) and Kanchukh (1961) used the colour based on reaction with ninhydrin, although sensitivity is not yet comparable to the best Nessler methods. Techniques employing Conway microdiffusion chambers appear to be better. For example Borsook (1935) and Parker (1961) digested protein-containing material by Kjeldahl processes and determined NH_3 with a range of 0·5–25 μg, Parker applying his technique to bacterial suspensions. After distillation of the NH_3 in Conway dishes, the amount of NH_3 may be determined by a modification (Borsook, 1935) of the Berthelot colour reaction based on addition of alkaline phenol and hypochlorite solutions. Vardanyan (1960) modified the Conway vessel for improved sensitivity, and Searey *et al.* (1965) studied the specificity of the Berthelot reaction. They found that addition of a phenol–nitroprusside reagent to the NH_3 solution followed by alkaline hypochlorite gave the most reliable results and avoided losses of ammonia when an alkaline solution was added first.

The above techniques are all somewhat time-consuming for individual samples. However, several of them, including Kjeldahl, Nessler and Conway methods, can be applied to series of samples withdrawn from biological systems. In this way a rather large number of samples can be run per day. Obviously the samples are destroyed in the process. For details of these procedures see Herbert *et al.* (this Series, Volume 5).

B. Instrumentation

All large laboratory supply houses stock standard equipment for Dumas and Kjeldahl microanalyses and Conway diffusion dishes. No special apparatus is used in Nessler techniques in which colour intensity is measured spectrophotometrically. As would be expected for widely used analytical procedures, however, equipment has been designed for automation.

Earlier semi-automatic nitrogen analysers include one described by Charlton (1957) and the Coleman models operating on the Dumas principle. Flokstra and Nadolski (1962) and Mazoyer (1964) evaluated the Colman instrument in detail and ran five or six samples/h with it. Monar (1965) automated the reading of the azotometer (manometer) and expanded the analysis to include carbon, hydrogen and oxygen, as well as nitrogen all completed within 13 min. Blumenthal and Cander (1966) described a linear nitrogen analyser stable to 0·005% nitrogen and used it in conjunction with an analogue computer.

Other highly automated instruments for C, H and N have been described by Walisch (1964) for samples to 0·1–0·5 mg and completing five analyses/h and by Clerc *et al.* (1963) and Condon (1966) for 0·5–2 mg. samples in less than 13 min each. Whitehead (1961) automated the Kjeldahl procedure with an automatic sample changer feeding into the digester, and Hofstader (1966) measured the coloured indophenol complex for N after a Kjeldahl digestion, achieving a capacity of 30 samples/h. Using a continuous digestion system from Ferrari (1960), Ferrari *et al.* (1965a, b) analysed solid and solution samples, including bacterial pastes and media.

Availability of these new instruments will make routine nitrogen analyses practical in microbiology. They remove a major barrier to common usage by markedly reducing the time needed for analysis of many individual samples. They ultimately offer the possibility for continuous sampling and automatic analysis for the cellular nitrogen of growing cultures.

C. Errors

Cells must be removed from nitrogen-containing media and washed free from the medium. Harvesting and washing involve losses of cells, and the latter leaches out some soluble nitrogenous material just as in dry weight determinations. On the other hand, problems of standardization of the methods of nitrogen analysis are not so serious as in turbidimetry, for example. Nevertheless, known compounds should be run periodically as checks, especially on the conversion of organic nitrogen to N_2 in Dumas methods or to NH_3 in Kjeldahl digestions. Some stable nitrogen heterocycles are notoriously difficult to degrade. Among these are quinoline and pyridine derivatives (Johns, 1941) and pyrazines (Mallette *et al.*, 1947), whose biological relatives include calcium dipicolinate and the folate coenzymes. Errors from the latter are likely to be small, but dipicolinate is abundant in the coats of bacterial spores and contributes significantly to the total nitrogen. Standards for calibration purposes include highly purified amino-acids and proteins. Tris(hydroxymethyl)aminomethane (Tris or Tham) is probably even better (Rodkey, 1964), since it is available in primary standard grade.

Many workers have discussed the precision of their techniques in various statistical terms. Among these are Kanchukh (1961) for a Kjeldahl–ninhydrin method and Flokstra and Nadolski (1962), Clerc *et al.* (1963), Walisch (1964), Condon (1966) and Blumenthal and Cander (1966) for automatic C, H and N methods. To illustrate the somewhat misleading sense of security which can result, Leibetseder (1961) compared relatively precise macro- and micro-Kjeldahl methods and obtained significantly lower data on the microscale. He attributed this difference to errors in pipetting small volumes. On the other hand, most analysts report close correspondence between such methods, as would be expected if analyses of standards are near theoretical values.

The general reliability of nitrogen analyses should be adequate for practically all microbiological work, except perhaps when applied at the lowest limits of sensitivity. However, one must always keep in mind that nitrogen content may vary considerably with the nutritional and physiological states of micro-organisms, and this variation will affect the conclusions drawn from such measurements.

Candeli *et al.* (1960) reported that the weight of N/cell and the weight of N/g of dry wt increased during growth of *E. coli* from the end of the lag phase through the remainder of the growth phase. Virtanen and De Ley (1948) cultured *E. coli* in media of normal and low (less than 40 mg of N/litre) nitrogen contents and reduced the cellular nitrogen from 13 to 6·5% of the dry weight. In studying the survival of *E. coli* in nitrogen-free medium, S. B. McGrew and M. F. Mallette (unpublished work), observed cell division and increased turbidity. Presumably the cells utilized nitrogenous reserves, a possibility suggest by the work of Virtanen and De Ley. Tepper (1965) grew *M. phlei* in several media finding the highest nitrogen content in complex media. When glycerol was present, the dry weight exceeded that of cells from glucose medium although the nitrogen content/cell was the same in both cases. Glycerol increased the storage of lipid and carbohydrate. Because of such variation, there must be a physiological calibration when nitrogen content is used as a measure of growth, and the calibration must be checked whenever the biological situation changes.

VI. ANALYSES FOR OTHER CELLULAR COMPONENTS

As with determination of microbial nitrogen, measurement of other cellular components can be used as direct indices of growth. Those analyses that have been and are likely to be most useful for this purpose are reviewed in the Chapter by Herbert *et al.* (this Series, Volume 5), to which reference is made for practical details. A limited selection of references and some comments pertinent to the present subject follow. Chemical and spectrophotometric methods for carbon and hydrogen, phosphorus, protein, DNA and RNA are included.

Methods for carbon, hydrogen and phosphorus are believed to be quite reliable and have been automated for routine application. McDonald (1963) reviewed automatic methods for carbon and hydrogen. Interference appears to be unlikely by elements and compounds normally encountered in biological materials. Improvements in instrumentation described by Monar (1965) have reduced the time required per sample to about 10 min, including nitrogen and oxygen as well as carbon and hydrogen. Reproducibility also is becoming satisfactory according to the study by Prezioso (1966) for carbon, hydrogen and nitrogen analyses using a new automatic analyser.

Bennett and Williams (1957) reported that the determination of total phosphorus is a reliable measure of bacterial mass during growth of *E. coli* and *Micrococcus pyogenes*. Since then, problems in the methods available have been re-examined, and the importance of the concentrations of molybdate and acid emphasized by Yuasa (1959), Pujic *et al.* (1961) and Meshcheryahov (1965). Pujic *et al.* (1961) also studied stability of the colour, showing that changes occur. Thus, spectrophotometric analysis should follow a standard time interval. Several different reducing agents for colour development have been recommended, but Duval (1963) believed that ascorbic acid was best, following a comparison in which the minimizing of blank colour and the stabilities of reagents were considered. An ultramicroanalysis for phosphorus has been described by Koscianek (1961), and automated procedures by Weinstein *et al.* (1964) and Hofstader (1966), the latter running 30 samples/h.

Elementary analyses for carbon, hydrogen, nitrogen, oxygen or phosphorus can be correlated with mass only when residual medium components containing the element in question have been removed. Therefore, in all of these cases the micro-organisms must be washed free from the medium before analysis. More than any other, this factor retards widespread application of the newer automated procedures because collection and washing of cells cannot be short cut as yet.

To some extent assays for protein are subject to this same disadvantage because several bacteria produce extracellular protein in considerable amounts. Cultures of those species that do not yield extracellular protein might be analysed directly for cellular protein, thus saving time and the mechanical losses associated with harvesting and washing. However, many methods for protein record other nitrogen-containing components and therefore require separation of the protein, usually by precipitation (Pande *et al.*, 1961; Parvin *et al.*, 1965). In addition, the cells usually must be ruptured before the analysis. Thereafter, Folin–phenol (Miller, 1950; Lowry *et al.*, 1951), biuret (Siltanen and Kekki, 1960; Itzhaki and Gill, 1964) and spectrophotometric (at 210 nm; Tombs *et al.*, 1959) methods may be applied. Van de Loo (1962) compared biuret, absorption at 280 nm and refractometric determinations and outlined their most favourable applications. Leavitt and Umbarger (1960) applied the Lowry method to the estimation of growth in cup assays. Finally, as would be expected, automatic analyses are available (Gerke, 1965). These have been reviewed by Marsh (1961).

As for proteins, analytical techniques for nucleic acids probably number in the hundreds. Mitchell (1950) studied the problems of nucleic acid estimation in intact bacterial cells, and Munro and Fleck (1966) reviewed developments in measurement of nucleic acids. Some of the most recent improvements include those of Giles and Myers (1965), Hirschman and Felsenfeld (1966) and Byvoet (1966). Gerke (1965) has automated the determination of RNA and DNA in *E. coli* simultaneously with turbidity and protein. This system should be potentially valuable in work on all kinds of physiological kinetics. It has been so used by Watson (1965) and Ferrari *et al.* (1965a, b).

Methods for nucleic acids and proteins are relatively precise, but are subject to systematic errors, including particularly the need to rupture the cells without destruction of the macromolecules. Moreover, cells contain nucleotides and peptides much smaller than nucleic acids and proteins but giving the same colour reactions and having interfering absorption spectra. Fractionations have been devised to reduce the interference, but they are not completely successful. Some treatments are harsh enough to cause a little degradation. This factor may partially compensate for incomplete removal of normally occurring fragments.

Specific cellular compounds should correlate with growth in particular circumstances and receive wider attention than they have. Two cases will be considered here. In the first, el-Shazly and Hungate (1966) used diaminopimelic acid as a measure of bacterial growth in the rumen. Microbial activity was stopped by addition of an equal volume of 0·1 M HCl, the suspension centrifuged for 20 min at 3500 *g*, the residue dried and ground. A 2 g sample was hydrolysed in a sealed system with 100 ml of 6 M HCl at 105°C for 24 h. The hydrolysate was filtered while hot, washed and concentrated to 5 ml. Diaminopimelic acid was separated from other amino-acids on a Dowex 50 column and analysed by measurement of the yellow colour produced on treatment with ninhydrin at low pH. Reaction at low pH was chosen to avoid interference by other amino-acids that might contaminate the samples. This choice of parameter permitted differentiation from the effects of protozoa, which would not have been possible using a measurement for an element or a broadly distributed major component, such as RNA. Along a somewhat different line, Namekata (1960) showed that catalase activity paralleled growth of mycobacteria during much of the growth phase. Any enzyme, such as catalase, by remaining in constant ratio to mass during specified conditions of growth should be a useful growth parameter under those conditions.

Variation of composition with nutritional and physiological factors again requires biological calibration before any of these parameters can be taken as measures of growth. Among many studies of these effects, the reports by Stevenson (1962) on *A. globiformis* and Polakis and Bartley (1966) on yeast show that the DNA, RNA and protein contents do vary but in cyclic fashion with the stages of growth. RNA particularly seems to fluctuate with growth phase, and many observations of the interrelationship have been published (Neidhart and Magasanik, 1960; Postgate and Hunter, 1962; Rudner *et al.*, 1964; Rosset *et al.*, 1966).

VII. ANALYSIS FOR SUBSTRATES AND EXTRACELLULAR END PRODUCTS

At least under some physiological conditions, growth is determined by the nature and amount of carbon and/or energy source utilized. This relationship is developed in the Chapter by Stouthamer (this Volume, p. 629) on the determination of molar growth yields. Two aspects of this dependence provide useful indexes for the measurement of microbial growth.

First, a correlation between the total mass of cells in a suspension and the amount of molecular oxygen utilized by aerobes would be anticipated. The relationship could vary with temperature, species, strain, partial pressure of oxygen at low pressures, and growth phase; but it should be reproducible for a given set of conditions. Mandels and Siu (1950) and Young and Clark

(1965) used manometric methods to follow growth of fungi and bacteria. Scopes and Williamson (1964) compared O_2 consumption measured polarographically with dry weight in synchronously dividing cultures of *S. cerevisiae* and found that dry weight increased continuously, whereas O_2 consumption increased in steps. Non-synchronized cultures do not show this type of behaviour, and the rate of increase correlates well with the rate of growth.

Oxygen is taken up by living micro-organisms without growth, often at rather constant rates. Therefore, increases in rates may indicate growth, providing major physiological changes like adaptation to a new energy source are not occurring. The necessary differential measurements are more readily obtained by polarography than by manometry according to Strickland *et al.* (1961). Nevertheless, manometric techniques are valuable, especially when recording manometers like those described by Reineke (1961) and Arthur (1965) are employed. Manometric methods are discussed in detail in the Chapter by Beechey and Ribbons (this Series, Volume 6) who also describe polarographic methods in another Chapter (this Series, Volume 6). Evaluations of growth by respirometric techniques are described by Elsworth (this Volume, p. 123).

Besides the relationship to O_2 consumption, growth depends even more universally upon the carbon and/or energy source. Therefore, a record of the utilization of a major substrate of this type reflects growth of the culture. Again there are variations with conditions, but probably not to the same extent as for oxygen consumption. Coupling of growth with carbon and energy source requires conversion of a constant proportion of the substrate into cellular material. Stouthamer discusses the required determination and utilization of yield coefficients in his Chapter (this Volume, p. 629), and Ecker and Lockhart (1961) and Wright and Lockhart (1965) studied some effects of limiting substrate on growth rate. One would expect growth yield/mole of substrate to decline at very low levels where the proportion of the carbon and energy source required solely for maintenance becomes significant (see Tempest, this Series, Volume 2). Since the maintenance level appears to be quite low (Mallette, 1963; Marr *et al.*, 1963), this possibility should not affect ordinary microbial cultures.

Many measurements of growth from the point of view of substrate utilized might involve glucose as the carbon and energy source for two reasons. This substrate is a common medium component, and its analysis has been studied extensively. Ferrari *et al.* (1965a, b) applied an automated version of the ferricyanide oxidation to glucose determination in bacterial systems. The more recent glucose oxidase methods (Washko and Rice, 1961; Pazur *et al.*, 1962) are of even greater interest because of the greater inherent sensitivity and specificity. Automated modifications are also available for this assay (Hill and Kessler, 1961; Getchell *et al.*, 1964; see also Dawes, this Series, Volume 6).

A constant yield coefficient during growth, as shown by Elsden (1965) for five different micro-organisms and by Fujimoto (1963) for three, requires conversion of a constant proportion of the free energy of the substrate into cellular material. As suggested above, many major constituents or substrates might serve as measures of growth. At the same time the remaining free energy of the carbon and energy source must also be constant. It is transferred to the metabolic by-products or to the environment as heat. In either case additional growth parameters are provided.

Battley (1960) showed a close relationship between entropy, free energy and enthalpy and the growth of baker's yeast. Therefore, calorimetric data for ΔH should relate well to growth. This possibility has been confirmed by Senez and Belaich (1965) for *Pseudomonas lindneri* and *E. coli* and by Forrest *et al* (1962) for *Aer. aerogenes* who also resolved major experimental and instrument design difficulties. Calorimeters are referred to by the foregoing authors and by Calvet (1953), Forrest (1961) and Forrest (Volume 6).

Other extracellular reaction products than heat can be taken as measures of growth. The two most common are CO_2 and acid. Many standard manometric techniques are applicable to the former according to Beechey and Ribbons (this Series, Vol. 6). Goksoyr (1962) incorporated a small Geiger tube and stirrer into Warburg flasks and recorded respiratory $^{14}CO_2$ continuously. This special approach would indicate growth during the steady-state utilization of radioactive carbon-energy sources (see also Wang, this Series, Vol. 6). Fatt (1964) adapted a CO_2 measuring electrode to use in the gas phase, and Pagano *et al.* (1962), Gerke *et al* (1962) and Haney *et al.* (1963) automated measurement of the CO_2 produced by *Candida tropicalis* and used it in assays of antibiotics, for example. As was the case for O_2 consumption, correlation of the rate of growth with CO_2 evolution would be via the rate of increase in the rate. Also once again, the rate of evolution of CO_2 would be directly proportional to the mass of cells present under standard conditions. Reproducibility of conditions may pose the greatest problem in using CO_2 data for this purpose; small changes in the pH value markedly affect the amount of CO_2 evolved.

Microbiological assays have always depended on measurement of growth either directly as measured by turbidity, dry weight, cellular nitrogen or cells counts, or as measured indirectly, from data on CO_2 or lactic acid production (Snell, 1950). In a Chapter in this Series, Vol. 7, Marten discusses automated microbiological assays, the obvious solution to the routine microbiological assay of large numbers of materials. van der Linden (1942) and Dawbarn *et al.* (1961) discussed several of the problems met in typical assays and proposed solutions for them. Wood (1947) and Smith (1965) proposed mathematical procedures for relating response, acid production and growth in standardizing assays. Tittsler *et al.* (1952) reviewed the turbidimetric and acidity titration methods, and Dawbarn *et al.* (1961) have shown that these two measures of growth can be comparable. Cell counts and titrable acidity are also comparable according to Smith (1965).

VIII. MICROSCOPIC DETERMINATION OF CELL NUMBERS

A. Basis

The related methods of this category are mostly dependent upon direct microscopic observation. They may be used for routine estimation of cell numbers as well as for the standardization of other techniques, e.g., turbidity. The general basis for light microscopy lies in resolution of small objects, which depends in turn upon contrast in the image (Lamanna and Mallette, 1965). Two major factors contribute to contrast, colour difference and refractive index difference.

Since micro-organisms seldom show enough natural colour contrast with their environments for effective resolution, colour contrast is enhanced by staining. Many biological stains may be applied for total counts, usually after fixing for attachment of the cells to a slide and for killing to make them permeable to stain. Staining techniques are especially important in making counts on dry materials and tissues in which situations they are probably the most important methods.

For total counts on liquid cultures, staining is usually unnecessary and is omitted in order to avoid mechanical losses of cells associated with fixing, staining and washing. However, when an estimate of the ratio of live to dead cells is desired, staining again becomes important, since the two states often take up dyes differently. Kadlekova (1965) reported that 2, 3, 4-triphenyltetrazolium chloride permitted this differentiation and gave more accurate counts of viable cells than did plate and membrane filter methods (see Postgate, this Volume, p. 611). The relative advantage claimed might be expected when counting microbial strains forming chains of cells because the average number of cells per chain could be estimated. However, data in the absence of chains or with few and small chains ought to be rather comparable to those from other techniques. Detailed discussion of this problem is given by Postgate (this Volume, p. 611).

Colour contrast in micro-organisms also may be increased by means of fluorescence. Derkacheva *et al.* (1959). de Repentigny (1961) and de Repentigny and Sonea (1961) observed that several species of bacteria and *Candida albicans* display characteristic fluorescent colours and intensities. In addition to taxonomic use, such behaviour should aid in counting the more fluorescent species with the fluorescence microscope. While treatment with fluorescent antibody would raise the intensity sufficiently for easy counting, this modification would seem to be more trouble than it is worth for any but special situations. On the other hand, Derkacheva *et al.* (1959) recommended addition of pseudoisocyanin chloride, which fluoresces strongly when adsorbed by the cells present but which fluoresces only slightly or not at all when in dilute solution. The optimal concentration of pseudoisocyanin chloride was found to be in the range 0·1–0·4 mg/ml. Most species of bacteria adsorbed the dye with a λ_{max} of 547 nm for the fluorescence. Bacterial populations of from 10^6 to 5×10^8 cells/ml were determined with a fluorometer, making the method rather comparable in sensitivity to nephelometric techniques.

Improvements in image contrast depending on manipulation of the refractive index of cells relative to the medium may be achieved by placing the cells in another medium. Thus, Fikhman (1959, 1963) used an anoptral microscope as a microrefractometer to differentiate live from dead bacteria suspended in gelatin. The technique should be useful in studying turbidity changes associated with other types of changes in cell surfaces.

One of the more easily applicable instruments is the phase microscope as employed for cell counts by Weibull (1960), who immobilized the cells in agar, by Razumov and Korsh (1962) for counts on membrane filters without staining, and by Cook and Lund (1962) with chambers having different depths. Schaechter *et al.* (1962) and Hoffman and Frank (1965) carried the development a step farther with either dark-field or phase-contrast microscopes and time-lapse photography in studying cell elongation as a function of time. This is one of the few ways by which microscopy might be adapted to the estimation of mass increase rather than simply increase in cell numbers (see also Quesnel, this Series, Volume 5).

B. Counting of slide cultures

Postgate *et al.* (1961) and Meisel *et al.* (1961) devised an effective method of discriminating live from dead micro-organisms by a brief incubation for the development of microcolonies following division of the live cells. These small-scale slide cultures are then examined by phase microscopy. In his Chapter on viability determination (this Volume page 611), Postgate discusses this procedure in detail. Parker (1965) modified slide cultures for enumerating obligate anaerobes by diluting an aliquot into a suitable agar medium, drawing this suspension into 1 ml pipettes and incubating. Colonies were counted with a lower-power microscope. The method appears to be a simple and effective one where aggregates or chains of cells are not significant. Although not truly a slide culture, Niemela (1965) applied a similar idea to counting bacterial colonies on membrane filters by placing the filter on medium, incubating briefly and counting the microcolonies. Not all bacteria could be counted in this way since 15 out of 24 species did not form point colonies at the original loci.

C. Proportional counts with internal standards

Microscopic counting of micro-organisms in cultures requires that all the cells be counted in a sample of known volume. Alternatively, cells are counted in a specific and reproducible volume whose actual size is unknown, but which includes a counting standard. Since the very small volumes examined microscopically are relatively difficult to measure, this latter system presents some advantages. Basically, the procedure involves mixing a known volume of the standard suspension of known particle concentration with a known volume of cell suspension of unknown concentration. After thorough mixing, both cells and standard particles are counted in the same volume of suspension, and the concentration of the former calculated from —

No. of cells/ml of culture

$$= \frac{\text{(Average No. of cells/field) (No. of particles/ml)}}{\text{(Average No. of standard particles/field)}} \times \frac{\text{Volume of particle suspension, ml}}{\text{Volume of cell culture, ml}}$$

For this purpose, standards should be rather uniform in size, and they must be easily distinguished from the cells to be counted. Klapper (1962) advocated the use of fungal spores as standard particles. She found that natural field-grown, fruiting bodies of the giant puffball *Calvatia gigantea* could be used as a source of small, uniform particles for proportional counting. Air-dried specimens were suspended by shaking vigorously 0·5 g samples in 100 ml of 0·85% NaCl–1·5% formalin contained in 250 ml Erlenmeyer flasks. Debris was removed by suction-filtering the spore suspension through a pad of glass wool 10 mm thick. Clumps of spores were broken up by addition of commercial detergent and mixing for 1–3 min at high speed in a small Waring Blendor cup. Although a few clumps still remained, more vigorous treatment ruptured individual spores and was avoided. Other fungi that can be grown in the laboratory were examined. Spore suspensions prepared in the same way were found to be satisfactory. As rather typical examples, spores of *C. gigantea* had an average size of 4·0 μm with a standard deviation of 0·27, and spores of *Penicillium nigricans* were 2·8 μm in dia, standard deviation 0·36.

Commercial polystyrene latex would be perhaps more convenient when available in more uniform and large enough sizes. It was employed in enumerating *Mycoplasma laidlawii* by Anderson *et al.* (1965) who used the electron microscope for which the usual small latex particles (0·264 μm in dia from Dow Chemical Co., Midland, Mich.) are quite good. Special counting chambers are not needed when applying the proportional method. Either entire fields are counted or zones delineated by a grid included in the field of view.

D. Chambers for microscopic counting

Except when the proportional method is used, cells must be counted in samples of known volume. Obviously such volumes must be small both for reasonable counts/sample and so that the depth of the chamber does not much exceed the depth of focus of the microscope. Commercial counting chambers have been widely used, for example by G. Weinbaum and M. F. Mallette (unpublished work), whose data correlated with plate counts within 10%; by Zellner *et al.* (1963) who found the Petroff-Hausser chamber was reliable only at populations exceeding 10^9 *R. glutinis*/ml; by Cook and Lund (1962), who reported that a chamber 0·02 mm deep was better than one 0·10 mm deep for counting bacterial spores; by Norris and Powell (1961) who designed a chamber for accurate measurement of the depth when filled for counting; by Swanton *et al.* (1962), who found that bacterial counts made in haemacytometers were unreliable. Fikhman (1962) solved the problem of changes due to motility by adding 4% polyvinyl alcohol to the cell suspension before filling the chamber and thus immobilizing the bacteria. He claimed that the alcohol caused no morphological damage and no aggregation of cells.

Substitutes for precisely manufactured chambers include capillaries filled by the capillary rise of cell suspension and measured for length and diameter by an ocular micrometer. Gabe (1957) mounted his capillaries horizontally and counted with oil immersion. He claimed better results than with commercial chambers. Swanton *et al.* (1962) spread 0·01 ml samples over a 1 sq. cm area on slides, dried, fixed and stained. Some problems were encountered at particular growth phases. They also described a hanging-drop method as more successful, using drops of known volume and having selected diameters for their upper surfaces. In a somewhat similar way, Borzani (1960) placed a 0·01 ml sample of yeast suspension on a coverglass, added another coverglass, mounted on a slide, sealed the edges with Vaseline and counted the cells in several fields. After measuring the field diameter with a calibrated slide, the number of cells/ml was given by $(\overline{N}/0{\cdot}01)\,(S/S_0)$ where $\overline{N}$ is the average number of cells/field, S is the area of coverglass and S_0 is the area of the microscopic field. The membrane-filter technique of Korsh (1961) is similar in principle, but designed for suspensions containing relatively few cells, which are collected by suction filtration and counted directly on the filter. MacLeod *et al* (1966) adopted the same approach, but labelled the cells with $^{32}PO_4H^{2-}$ and estimated the number of cells collected on the filter by radioactivity measurements. Since the level of radioactivity per cell would vary markedly with various factors, general utility of this method is doubtful.

Weibull (1960) suspended bacteria and a little indian ink in agar, placed the sample on a warm slide, covered, cooled and counted the cells in several fields. Agar depth varied, but was measured by focusing on carbon particles at the top and bottom of the layer. With measurement of the field diameter, cell concentration was calculated and greater accuracy claimed than with conventional chambers.

At least in principle, cell and colony counting can be automated by flying-spot scanning. A microscopic unit has been described by Buck (1958 that avoids double counting and classifies according to size. Glaser and Wattenburg (1966) developed a scanner suitable for the repeated examination of colony growth on 30 Petri plates. Techniques of this type probably are not yet generally useful, and the equipment may not be available commercially for some time.

E. Counting errors

Most methods of counting stained preparations appear to suffer from mechanical losses of cells. Even when a measured sample of cell suspension is dried on the counting slide and the cells fixed without washing in any way, removal of excess stain is almost certain to take off some cells. Adding a small volume of a dilute solution of stain and drying in place should be better, but improvement in colour contrast would be slight unless the cells took up the stain very intensely. In addition, aggregates of precipitated stain might mask cells, cause them to clump or even (but less likely) be mistaken for cells. Addition of stain or materials fluorescent after adsorption to samples of cell suspension and direct counting of the resulting liquid mixture would seem to be a better approach, but one not generally used perhaps because live cells often do not take up stains effectively.

Koch and Kaplan (1964) studied a membrane-filter method for viable counts. They obtained results that were a little lower than expected, but well within the 10% uncertainty often assigned to plating methods.

The statistics of direct microscopic counts of both dried films and suspensions in chambers have been outlined by Cassell (1965) who presented graphical procedures designed to reduce computational time and to minimize counting time in accord with the precision desired. This method is based upon plots prepared for confidence levels of 90, 95 and 99% for means and illustrated as Figs. 2–4, respectively. These Figures are used as follows.

First, the desired level of confidence in the result is chosen, for example 90% in Fig. 2. Confidence intervals in this Figure are expressed in two ways: as percentage variation of the average count /field ($\pm\% \bar{X}$), designated by

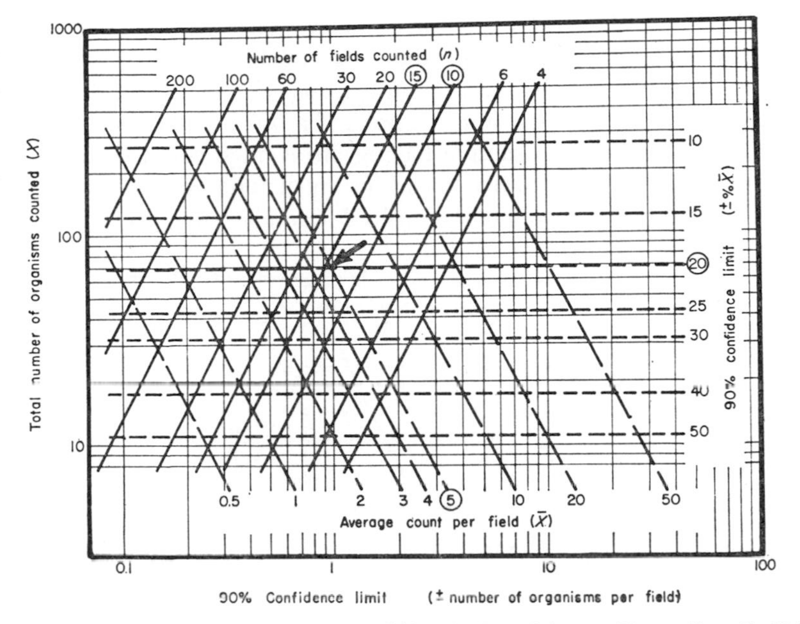

FIG. 2. Schedule of precision at the 90% level of confidence. (From Cassell, 1965.)

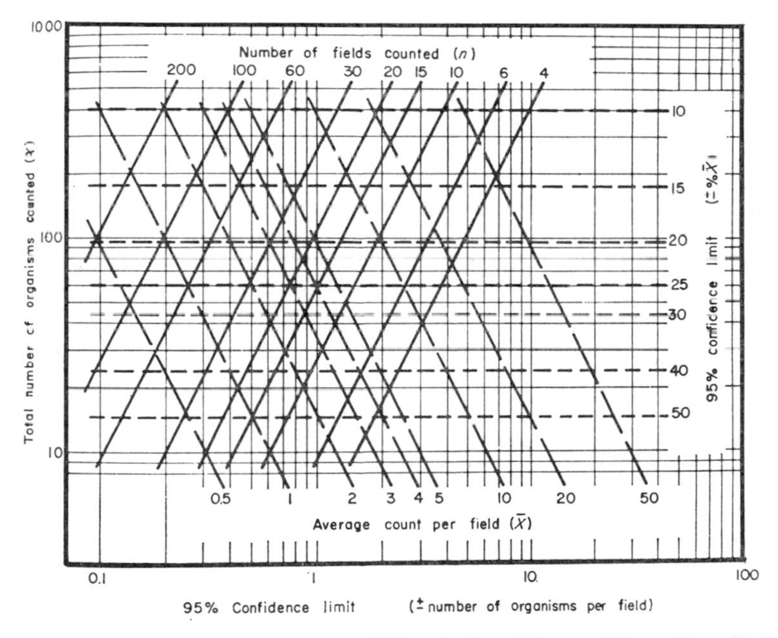

FIG. 3. Schedule of precision at the 95% level of confidence. (From Cassell, 1965.)

the horizontal dashed lines and the right ordinates and as variation ($\pm$) in the number of organisms/field, indicated by the bottom abscissa. Secondly, select the percentage of the average count/field ($\pm \% \bar{X}$) that the desired confidence interval represents. For example, if a 90% confidence of interval $\pm 20\% \bar{X}$ is chosen for a sample having an average count/field of 5, the probability is 0·90 that the true mean lies in the chosen range 4–6 (± 1 of $\bar{X} = 5$). If the confidence interval desired is smaller, i.e., greater precision

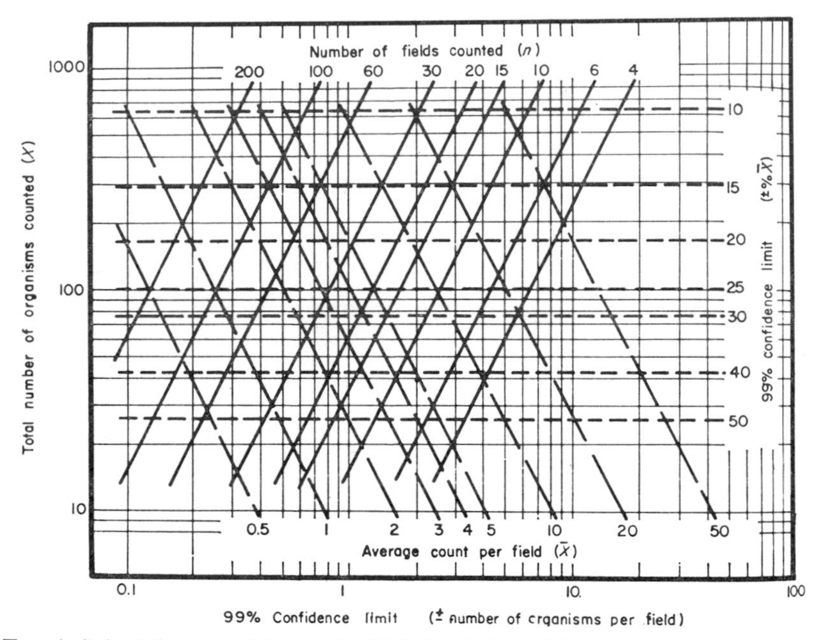

FIG. 4. Schedule of precision at the 99% level of confidence. (From Cassel, 1965.)

is wished, more fields must be counted to ensure that the mean count $\bar{X}$ is 90% certain to lie within the chosen confidence interval of the true mean. Next, count about four fields and obtain a rough average count/field. This value corresponds to one of the sloping dashed lines that are identified by the next-to-bottom abscissa. The intersection of this line (here taken as $\bar{X} = 5$) with that of the 90% confidence interval (here taken as $\pm 20\% \bar{X}$) is denoted in Fig. 2 by a heavy arrow. This point reveals (by interpolating vertically to the bottom abscissa) that the 90% confidence interval also is ± 1 organism/field. On interpolating horizontally to the left ordinate, one sees that 69 organisms must be counted, and, on interpolating between the sloping solid lines, one sees that between 10 and 15 (approximately 14 fields must be counted. These Figures are employed in this way to specify the counting which must be done to establish a given level of precision. It should be emphasized that these Figures are based on assumed Poisson distributions of the counts/field. This assumption is probably valid when the organisms are non-colonial and do not aggregate.

Although the above treatment implies that precision is affected only by statistical variables, Norris and Powell (1961) investigated the problem in detail and found a dependence of precision on the investigator, the counting chamber, the organism being counted, and methods of filling the chamber, handling of the cell suspension and cleaning of the glassware. They recommended the following procedure, which substantially reduced the variance of replicates.

The surface volume ratios of suspensions to be counted should be minimized by using containers small enough to keep the samples in compact shapes. Suspensions should not be spread thinly over surfaces, since micro-organisms tend to concentrate at liquid–glass interfaces, thus reducing the populations in the aliquots withdrawn for counting. Secondly, a small amount of an anionic detergent should be added to reduce aggregation and reduce adsorption at interfaces. The recommended level is indicated by free frothing of the suspension. Cationic detergents are likely to cause cell lysis and should be avoided. Glassware cleaned by conventional washing procedures is preferred over chemically clean, grease-free glassware. The foregoing conditions permit one to store cell suspensions for 24 h without significant change in count. Then for filling the counting chamber, a wire loop should be used, not a pipette or rod, and the sample should be placed directly over the ruled area.

Norris and Powell (1961) suggested the following stock diluent as effective in precise counting work. Normal saline containing the anionic detergent is made 0·5% in formalin, and Na_2HPO_4 is added to make the pH 7·5. The formaldehyde prevents growth, including that of contaminants, and the phosphate avoids aggregation due to the acid sometimes occurring in saline solutions.

Perhaps even more important, Norris and Powell (1961) found that accuracy in employment of counting chambers was subject to a considerable systematic error of as much as 50% in the actual thickness of the layer of cell suspension. Carefully filled chambers usually held liquid at depths greatly exceeding the depth of the empty chamber. However, the presence of granular material in the cell suspension led occasionally to a sub-normal depth, to changes in the depth in the chamber with time and, after passage of a longer time, to a sudden increase in depth. Because of these problems, Norris and Powell urged that the depths of filled chambers be measured just before counting as a means of avoiding possible large systematic errors. They described an aluminizing technique and microinterferometer for rapid measurement of the depths of filled counting chambers. When this equipment becomes commercially available, it should prove valuable in routine use.

Some of these technical aspects of the use of counting chambers probably are responsible for the concern expressed by Cook and Lund (1962), Zellner *et al.* (1963) and others over the comparative merits of this method under certain conditions. Although many investigators comparing plate and chamber counts have not encounted errors of the magnitude reported by Norris and Powell (1961), there is no doubt that the last authors have made a useful contribution. Much of the agreement between methods reported in other studies may arise by compensation of errors.

Finally, it must always be borne in mind that cell numbers and the total mass of cells are not always related in the same way. All factors influencing cell size at division affect the relationship and must be evaluated whenever cell counts are taken as the measure of total cellular material.

REFERENCES

Abram, D., and Gibbons, N. E. (1961). *Can. J. Microbiol.*, **7**, 741–750.

Adler, H. I., and Hardigree, A. A. (1965). *J. Bact.*, **90**, 223–226.

Allison, J. L., Hartman, R. E., Hartman, R. S., Wolfe, A. D., Ciak, J., and Hahn, F. E. (1962). *J. Bact.*, **83**, 609–615.

Anderson, D. L., Pollock, M. E., and Brower, L. F. (1965). *J. Bact.*, **90**, 1764–1767.

Arthur, R. M. (1965). *Appl. Microbiol.*, **13**, 125–127.

Bailey, J. M., and Meymandi-Nejad, A. (1961). *J. Lab. clin. Med.*, **58**, 667–672.

Battley, E. H. (1960). *Physiologia Pl.*, **13**, 628–640.

Beet, A. E. (1957). *Nature, Lond.*, **175**, 513–514.

Bennett, E. O., and Williams, R. P. (1957). *Appl. Microbiol*, **5**, 14–16.

Bernheim, F. (1963). *J. gen. Microbiol.*, **30**, 53–58.

Black, A. P., and Hannah, S. A. (1963). *Am chem. Soc., Div. Waste Water Chem.* Preprints pp. 1–5.

Blumenthal, W. S., and Cander, L. (1966). *J. appl. Physiol.*, **21**, 1099–1102.

Bohinski, R. C. and Mallette, M. F. (1967). *J. Bact.*, **93**, 1316–1326.

Borsook, H. (1935). *J. biol. Chem.*, **110**, 481–493.

Borzani, W. (1960). *Stain Technol.*, **35**, 49.

Bradstreet, R. B. (1965). "The Kjeldahl Method for Organic Nitrogen". Academic Press, New York.

Brown, M. R. W., and Richards, R. M. E. (1964). *J. Pharm. Pharmac.*, **16**, (Suppl.) 41T–45T.

Buchanan, R. E. (1955). *A. Rev. Microbiol.*, **9**, 1–20.

Buck, W. A. (1958). U.S. Atomic Energy Commission Report TID–7568 (Pt 3), pp. 82–97.

Byvoet, P. (1966). *Analyt. Biochem.* **15**, 31–39.

Calvet, E. (1953). *Meml. Servs. chim. Etat*, **38**, 209–246; *Chem. Abstr.*, **49**, 6731.

Candeli, A., Mariotti, F., and Antonioni, A. M. (1960). *Boll. Ist. sieroter. milan.*, **39**, 450–457; *Biol. Abstr.*, **36**, 20782.

Cassell, E. A. (1965). *Appl. Microbiol.*, **13**, 293–296.

Charlton, F. E. (1957). *Analyst, Lond.*, **82**, 643–648.
Chernov, V. N., Bereznikov, V. M., Drevush, V. P., and Kolbasov, A. N. (1963) *Vest. Adad. Nauk SSSR*, **33**, 77–79; *Biol. Abstr.*, **45**, 38128.
Clerc, J. T., Dohner, R., Sauter, W., and Simon, W. (1963). *Helv. chim. Acta*, **46**, 2369–2388.
Clifton, C. E. (1966). *J. Bact.*, **92**, 905–912.
Collins, J. F., and Richmond, M. H. (1962). *J. gen. Microbiol.*, **28**, 15–34.
Condon, R. D. (1966). *Microchem. J.*, **10**, 408–426.
Conway, E. J., and Downey, M. (1950). *Biochem. J.*, **47**, 347–355.
Cook, A. M., and Lund, B. M. (1962). *J. gen. Microbiol.*, **29**, 97–104.
Davey, C. B., Miller, R. J., and Nelson, L. A. (1966). *J. Bact.*, **91**, 1827–1830.
Dawbarn, M. D., Forsyth, H., and Kilpatrick, D. (1961). *Aust. J. exp. Biol. med. Sci.*, **39**, 305–322.
Dawes, E. A., and Ribbons, D. W. (1964). *Bact. Rev.*, **28**, 126–149.
Dawes, E. A., and Ribbons, D. W. (1965). *Biochem. J.*, **95**, 332–343.
Deppe, H. D. (1965). *Arch. Hyg. Bakt.*, **149**, 51–68.
Deppe, H. D., and Norpoth, K. (1965). *Arch. Hyg. Bakt.*, **149**, 161–172.
de Repentigny, J. (1961). *Trans. R. Soc. Can.*, **55**, Section V, 5–14.
de Repentigny, J., and Sonea, S. (1961). *Bact. Proc.*, **61**, 73.
Derkacheva, L. D., Zhebandrov, N. D., and Khan-Magnetova, S. D. (1959). *Biofizika* (*Transl.*), **4**, No. 1, 119–122.
Dezelic, G. (1961). *Croat. chem. Acta*, **33**, 51–54.
Dubos, R., and Middlebrook, G. (1948). *J. exp. Med.*, **88**, 81–88.
Duval, L. (1963). *Chim. analyt.*, **45**, 237–250.
Ecker, R. E., and Lockhart, W. R. (1961). *J. Bact.*, **82**, 80–84.
Elsden, S. R. (1965). *Colloques int. Cent. natn. Rech. scient.*, **124**, 305–306.
el-Shazly, K., and Hungate, R. E. (1966). *Appl. Microbiol.*, **14**, 27–30.
Eriksson, R. (1965). *Physiologia Pl.*, **18**, 976–993.
Errington, F. P., Powell, E. O., and Thompson, N. (1965). *J. gen. Microbiol.*, **39**, 109–123.
Faith, W. T., and Mallette, M. F. (1966). *Archs Biochem. Biophys.*, **117**, 75–83.
Fatt, I. (1964). *J. appl. Physiol.*, **19**, 550–553.
Faxen, N. (1960). *Acta path. microbiol. scand.*, **49**, 271–272.
Fels, G., and Veatch, R. (1959). *Analyt. Chem.*, **31**, 451–452.
Ferrari, A. (1960). *Ann. N. T. Acad. Sci.*, **87**, 792–800.
Ferrari, A., Catanzaro, E., and Russo-Alesi, F. (1965a). *Ann. N. Y. Acad. Sci.*, **130**, 602–620.
Ferrari, A., Gerke, J. R., Watson, R. W., and Umbreit, W. W. (1965b). *Ann. N. Y. Acad. Sci.*, **130**, 704–721.
Ferro, P. V., and Ham, A. B. (1966). *Am. J. clin. path.*, **45**, 166–171.
Fikhman, B. A. (1958). *Lab. Delo*, 53–55; *Chem. Abstr.*, **53**, 21388.
Fikhman, B. A. (1959). *Zh. Mikrobiol. Epidemiol. Immunobiol.* (*Transl.*) **30**, 121–125.
Fikhman, B. A. (1961). *Lab. Delo*, 52–54; *Biol. Abstr.*, **38**, 1917.
Fikhman, B. A. (1962). *Lab. Delo.* 55–56; *Biol. Abstr.*, **39**, 6749.
Fikhman, B. A. (1963). *Zh. Mikrobiol Épidem. Immunobiol.*, **40**, 60–65.
Fikhman, B. A. (1965). *Lab. Delo*, 45–49; *Biol. Abstr.*, **47**, 8423.
Flokstra, J. H., and Nadolski, E. B. (1962). *Ann. N. Y. Acad. Sci.*, **102**, 76–82.
Forrest, W. W. (1961). *J. Scient. Instrum.*, **38**, 143–145.
Forrest, W. W., and Stephen, V. A. (1965). *J. Scient. Instrum.*, **42**, 664–665.

Forrest, W. W., Walker, D. J., and Stoward, P. J. (1962). *Nature, Lond.*, **196**, 990–991.
Fujimoto, Y. (1963). *J. theoret. Biol.*, **5**, 171–191.
Gabe, D. R. (1957). *Mikrobiologiya*, **26**, 115–124.
Galanos, D. S., and Kapoulas, V. M. (1966). *Analytica chim. Acta*, **34**, 360–366.
Gerke, J. R. (1965). *Ann. N. Y. Acad. Sci.*, **130**, 722–732.
Gerke, J. R., Haney, T. A., and Pagano, J. F. (1960). *Ann. N. Y. Acad. Sci.*, **87**, 782–791.
Gerke, J. R., Haney, T. A., and Pagano, J. F. (1962). *Ann. N. Y. Acad. Sci.*, **93**, 640–643.
Getchell, G., Kingsley, G. R. and Schaffert, R. R. (1964). *Clin. Chem.*, **10**, 540–548.
Giles, K. W., and Myers, A. (1965). *Nature, Lond.*, **206**, 93.
Glaser, D. A., and Wattenburg, W. H. (1966). *Ann. N. Y. Acad. Sci.*, **139**, 243–257.
Goksoyr, J. (1962). *Analyt. Biochem.*, **3**, 439–447.
Goldfarb, M. L., Goldfarb, D. M., Prozorovskii, S. V., and Sobolev, V. M. (1962). *Lab. Delo*, 41–45; *Chem. Abstr.*, **43**, 6659.
Hadjipetrou, L. P., Ligeri, P., Gerrits, J. P., Teulings, F. A. G., and Stauthamer, A. H. (1964). *J. gen. Microbiol.*, **36**, 139–150.
Haney, T. A., Gerke, J. R., and Pagano, J. F. (1963). *In* "Analytical Microbiology" (ed. F. Kavanagh). Academic Press, New York.
Harrison, A. P. (1960). *Proc. R. Soc.*, **B152**, 418–428.
Harvey, R. J., and Marr, A. G. (1966). *J. Bact.* **92**, 805–811.
Heller, W., and Tabibian, R. M. (1957). *J. Colloid Sci.*, **12**, 25–39.
Herbert, D. (1961). *Symp. Soc. gen. Microbiol.*, **11**, 391–416.
Hershey, A. D. (1939). *J. Bact.*, **37**, 285–299.
Hill, J. B., and Kessler, G. (1961). *J. Lab. clin. Med.*, **57**, 970–980.
Hinds, A. E., and Peterson, G. X. (1963). *J. Bact.*, **86**, 168.
Hinshelwood, C. N. (1946). "The Chemical Kinetics of the Bacterial Cell". Oxford University Press, London.
Hirschman, S. Z., and Felsenfeld, G. (1966). *J. molec. Biol.*, **16**, 347–358.
Hochgesang, F. P. (1964). *In* "Treatise on Analytical Chemistry" (Ed. I. M. Kolthoff and P. J. Elving), Part I, 5 Section D–3, pp. 3289–3328. Interscience, New York.
Hoffman, H., and Frank, M. E. (1965). *J. Bact.*, **89**, 212–216.
Hofstader, R. A. (1966). *Microchem J.*, **10**, 444–445.
Horn, G., Jacob, H. E., and Bockel, W. (1964). *Z. allg. Mikrobiol.*, **4**, 1–12.
Hopkins, J. G. (1913). *J. Am. med. Ass.*, **60**, 1615–1617.
Horwitz, W. (1965). *In* "Official Methods of Analysis of the Association of Official Agricultural Chemists". Tenth edition. Association of Official Agricultural Chemists, Washington.
Hou, C. I., Gronlund, A. F., and Campbell, J. J. R. (1966). *J. Bact.*, **92**, 851–855.
Hozumi, K., and Kirsten, W. J. (1962). *Analyt. Chem.*, **34**, 434–435.
Hucke, D. M. and Roche, C. H. (1959–1960). *Antibiotics A.*, **7**, 556–562.
Il'ina, A. A. and Rozenberg, G. V. (1954). *Dokl. Akad. Nauk SSSR*, **98**, 365–368; *Chem. Abstr.*, **50**, 27.
Ishikura, T., Sakamoto, T., Kawasaki, I., Tsunoda, T., and Narui, K. (1964). *Agric. biol. Chem.*, **28**, 700–709.
Itzhaki, R. F., and Gill, D. M. (1964). *Analyt. Biochem.*, **9**, 401–410.
Jacob, H. E., and Horn, G. (1964). *Abh. dt. Akad. Wiss. Berl. Kl. Chem., Geol. Biol.*, 320–326.

Johns, I. B. (1941). "Laboratory Manual of Microchemistry". Burgess Publishing Company, Minneapolis.
Jullander, I. (1949). *Acta chem. scand.*, **3**, 1309–1317.
Kadlekova, O. (1965). *Biológia, Bratisl.*, **20**, 575–580; *Biol. Abstr.*, **47**, 118062.
Kanchukh, A. A. (1961). *Biokhimiya*, **26**, 393–398.
Kavanagh, F. (1963). *In* "Analytical Microbiology" (Ed. F. Kavanagh), pp. 141–217. Acadamic Press, New York.
Klapper, B. F. (1962). *Appl. Microbiol.*, **10**, 487–491.
Koch, A. L. (1961). *Biochim. biophys. Acta*, **51**, 429.
Koch, W. and Kaplan, D. (1964). *Nature, Lond.*, **203**, 896–897.
Konev, I. E. (1959). *Lab. Delo*, 49–52; *Biol. Abstr.*, **36**, 41465.
Korsh, L. E. (1961). *J. Hyg. Epidem. Microbiol. Immun.*, **5**, 349–356.
Koscianek, H. (1961). *Bull. Acad. pol. Sci. Cl II Sér. Biol.*, **9**, 285–286.
Kuczynski-Hallmann, M., and Avi-Dor, Y. (1958). *J. gen. Microbiol.*, **18**, 364–368.
Kuempel, P. L., and Pardee, A. B. (1963). *J. cell. comp. Physiol.*, **62**, Pt. II, Suppl. I, 15–30.
Kurokawa, M., Hatano, M., Kashiwazi, N., Saito, T., Ishida, S., and Homma, R. (1962). *J. Bact.*, **83**, 14–19.
Lamanna, C., and Mallette, M. F. (1965). "Basic Bacteriology". Williams and Wilkins Company, Baltimore.
Lang, C. A. (1958). *Analyt. Chem.*, **30**, 1692–1694.
Leavitt, R., and Umbarger, H. E. (1960). *J. Bact.*, **80**, 18–20.
Leibetseder, J. (1961). *Z. Tierphysiol., Tierernähr. Futtermittelk.*, **17**, 129–131.
Levy, H. B., Skutch, E. T., and Schade, A. L. (1949). *Archs Biochem.*, **24**, 199–205.
Lindberg, D. A. B., and Reese, G. (1963). *In* "Biomedical Sciences Instrumentation" (Ed. F. Alt), Vol. 1, pp. 11–20. Plenum Press, New York.
Lowry, O. H., Rosebrough, N. J., Farr, A. L., and Randall, R. J. (1951). *J. biol. Chem.*, **193**, 265–275.
MacLeod, R. A., Light, M., White, L. A., and Currie, J. F. (1966). *Appl. Microbiol.*, **14**, 979–984.
Mager, J., Kuczynski, M., Schatzberg, G., and Avi-Dor, Y. (1956). *J. gen. Microbiol.*, **14**, 69–75.
Mallette, M. F. (1963). *Ann. N. Y. Acad. Sci.*, **102**, 521–535.
Mallette, M. F., Taylor, E. C., and Cain, C. K. (1947). *J. Am. chem. Soc.*, **69**, 1814–1816.
Mallette, M. F., Cowan, C. I., and Campbell, J. J. R. (1964). *J. Bact.*, **87**, 779–785.
Mandels, G. R., and Siu, R. G. H. (1950). *J. Bact.*, **60**, 249–262.
Marr, A. G., Nilson, E. H., and Clark, D. J. (1963). *Ann. N. Y. Acad. Sci.*, **102**, 536–548.
Marr, A. G., Harvey, R. J., and Trentini, W. C. (1966). *J. Bact.*, **91**, 2388–2389.
Marsh, W. H. (1961). *Protides biol. Fluids*, **9**, 11–26.
Mathies, J. C. Lund, P. K., and Eide, W. (1962). *Analyt. Biochem.*, **3**, 408–414.
Mazoyer, R. (1964). *Bull. Ass. fr. Étude Sol*, 282–287.
McDonald, A. M. G. (1963). *Ind. Chemist*, **39**, 265–267.
McGrew, S. B., and Mallette, M. F. (1962). *J. Bact.*, **83**, 844–850.
Meisel, M. N., Medvedeva, G. A., and Aleekseva, V. M. (1961). *Mikrobiologiga* (*Transl.*) **30**, 669–704.
Meshcheryakov, A. M. (1965). *Pochvovedenie*, 75–80; *Chem. Abstr.*, **62**, 15420.
Metz, H., and Mayr, A. (1966). *Zentbl. Bakt. ParasitKde Abt. I Orig.*, **200**, 110–112.
Miller, P. (1950). *J. gen. Microbiol.*, **4**, 399–409.

Mitchell, P. (1950). *J. gen. Microbiol.*, **4**, 399–409.
Monar, I. (1965). *Mikrochim Acta*, 208–250.
Monod, J. (1942). "Recherches sur la Croissance des Cultures Bacteriennes". Hermann, Paris.
Mountney, G. J., and O'Malley, J. (1966). *Appl. Microbiol.*, **14**, 845–849.
Munro, H. N., and Fleck, A. (1966). *Analyst, Lond.*, **91**, 78–88.
Nakagaki, M., and Ideda, K. (1965). *J. pharm. Soc. Japan*, **85**, 693–699.
Namekata, H. (1960). *Jap. J. Bact.*, **15**, 478–482.
Neidhardt, F. C., and Magasanik, B. (1960). *Biochim. biophys. Acta*, **42**, 99–116.
Niemela, S. (1965). *Ann. Acad. Sci. Fenn. Ser Suomal. Tiedeakat. Toim. Sarja A IV Biol.*, **90**, 1–63.
Niemierko, W. (1947). *Acta Biol. Exptl Warsaw*, **14**, 195–197; *Chem. Abstr.*, **42**, 8863.
Nikitina, K. A., and Grechushkina, N. N. (1965). *Vest. mosk. gos. Univ. Ser. VI. Biol. Pochevoved*, **20**, 45–49; *Biol. Abstr.*, **47**, 88299.
Norris, K. P. and Powell, E. O. (1961). *Jl R. microsc. Soc.*, **80**, 106.
Northrup, J. H. (1960). *Jl gen. Physiol.*, **43**, 551–554.
Oster, G. (1955). *In* "Physical Techniques in Biological Research" (Ed. G. Oster and A. W. Pollister), Vol. I, pp. 51–71. Academic Press, New York.
Pagano, J. F., Haney, T. A., and Gerke, J. R. (1962). *Ann. N. Y. Acad. Sci.*, **93**, 644–648.
Palkina, N. A., and Fikhman, B. A. (1964). *Lab. Delo*, 491–494; *Biol. Abstr.*, **46**, 57844.
Palmer, I. S., and Mallette, M. F. (1961). *J. gen. Physiol.*, **45**, 229–241.
Pande, S. V., Tewari, K. K., and Krishnan, P. S. (1961). *Arch. Mikrobiol.*, **39**, 343–350.
Parker, C. A. (1961). *Aust. J. exp. Biol. med. Sci.*, **39**, 515–520.
Parker, R. B. (1965). *Appl. Microbiol.*, **13**, 1042.
Parvin, R., Pande, S. V., and Venkitas, T. A. (1965). *Analyt. Biochem.*, **12**, 219–229.
Pazur, J. H., Shadaksharaswany, M., and Meidell, G. E. (1962). *Archs Biochem. Biophys.*, **99**, 78–85.
Pegg, D. E., and Antcliff, A. C. (1965). *J. clin. Pathol.*, **18**, 472–478.
Peterson, E. C., and Tincher, A. H. (1961). *J. Colloid Sci.*, **16**, 87–89.
Polakis, E. S., and Bartley, W. (1966). *Biochem. J.*, **98**, 883–887.
Postgate, J. R., and Hunter, J. R. (1962). *J. gen. Microbiol.*, **29**, 233–262.
Postgate, J. R., Crumpton, J. E., and Hunter, J. R. (1961). *J. gen. Microbiol.*, **24**, 15–24.
Powell, E. O. (1963). *J. Sci. Fd Agric.*, **14**, 1–8.
Powell, E. O., and Stoward, P. J. (1962). *J. gen. Microbiol.*, **27**, 489–500.
Previc, E. P. (1964). *Biochim. biophys. Acta*, **87**, 277–290.
Prezioso, A. N. (1966). *Microchem. J.*, **10**, 516–521.
Prochazka, G. J., and Wayne, W. J. (1965). *Appl. Microbiol.*, **13**, 702–705.
Pujic, Z., Odavic, R., Sabovljev, A., and Melicevic, V. (1961). *Glasn. Drust. Hemicara Tehnol. Nr Bosne Hercegovine*, **10**, 37–43; *Chem. Abstr.*, **61**, 7693.
Razumov, A. S., and Korsh, L. E. (1962). *Mikrobiologya*, **31**, 288–291.
Reid, D. F., and Frank, H. A. (1966). *J. Bact.*, **92**, 639–644.
Reineke, E. P. (1961). *J. appl. Physiol.*, **16**, 944–946.
Ribbons, D. W., and Dawes, E. A. (1963). *Ann. N. Y. Acad. Sci.*, **102**, 564–586.
Rodkey, F. L. (1964). *Clin. Chem.*, **10**, 606–610.
Roessler, W. G., and Brewer, C. R. (1967). *Appl. Microbiol.*, **15**, 1114–1121.
Rogers, D. (1963). *J. Bact.*, **85**, 1141–1149.

Rose, H. E. (1950). *J. Inst. Wat. Engrs*, **5**, 310–320.
Rosset, R., Julien, J., and Monier, R. (1966). *J. molec. Biol.*, **18**, 308–320.
Rudner, R., Prokop-Schneider, B., and Chargaff, E. (1964). *Nature, Lond.*, **203**, 479–483.
Rusanov, E., and Balevska, P. (1964). *Izv. Inst. po. Fisiol. Akad. Nauk*, **7**, 181–188; 189–197, 199–204; *Chem. Abstr.*, **62**, 11131–11185, 8105.
Sartory, A., Meyer, J., and Buri, K. (1947). *Bull. Soc. Chim. biol.*, **29**, 168–170, 171–177, 178–183.
Schaechter, M., Williamson, J. P., Hood, J. R., and Koch, A. L. (1962). *J. gen. Microbiol.*, **29**, 421–434.
Schmid, W. (1958). *Biochem. Z.*, **329**, 560–567.
Scopes, A. W., and Williamson, D. H. (1964). *Expl. Cell. Res.*, **35**, 361–371.
Searey, R. L., Simms, N. M., Foreman, J. A., and Bergquist, L. M. (1965). *Clin. chim. Acta*, **12**, 170 175.
Senez, J. C., and Belaich, J. P. (1965). *Colloques int. Cent. natn. Rech. scient.*, **124**, 357–369.
Siltanen, P., and Kekki, M. (1960). *Scand. J. clin. Lab. Invest.*, **12**, 228–234.
Smith, K. L. (1965). *J. Dairy Sci.*, **48**, 741–742.
Snell, E. E. (1950). *In* "Vitamin Methods", (Ed. P. Gyorgy), Vol. I, pp. 327–421. Academic Press, New York.
Spaun, J. (1962). *Bull. Wld Hlth Org.*, **26**, 219–225.
Starka, J. (1962). *J. gen. Microbiol.*, **29**, 83–90.
Starka, J., and Loza, J. (1959). *Biochim. biophys. Acta*, **32**, 261–262.
Stevenson, I. L. (1962). *Can. J. Microbiol.*, **8**, 655–661.
Strickland, E. H., Ziegler, F. D., and Anthony, A. (1961). *Nature, Lond.*, **191**, 969–970.
Swanton, E. M., Curby, W. A. and Lind, H. E. (1962). *Appl. Microbiol.*, **10**, 480–485.
Takagi, T., and Honda, T. (1962). *Japan. Analyst*, **11**, 286–289; *Chem. Abstr.*, **57**, 14103.
Tepper, B. S. (1965). *Am. Rev. resp. Dis.*, **92**, 75–82.
Tittsler, R. P., Pederson, C. S., Snell, E., Hendlin, D., and Niven, C. F. (1952). *Bact. Rev.*, **16**, 227–260.
Toennies, G., and Gallant, D. L. (1948). *J. biol. Chem.*, **174**, 451–463.
Toennies, G., Iszard, L., Rogers, N. B., and Shockman, G. D. (1961). *J. Bact.*, **82**, 857–866.
Tombs, M. P., Souter, F., and MacLagan, N. F. (1959). *Biochem. J.*, **73**, 167–171.
Underwood, N., and Doermann, A. H. (1947). *Rev. scient. Instrum.*, **18**, 665–669.
Underwood, N., and Doermann, A. H. (1951). *Phys. Rev.*, **83**, 489.
van de Hulst, H. C. (1957). "Light Scattering by Small Particles". Wiley, New York.
Van de Loo, J. (1962). *Protides biol. Fluids*, **10**, 335–338.
van der Linden, A. C. (1948). *Analytica chim. Acta*, **2**, 805–812.
Vardanyan, V. A. (1960). *Izv. Akad. Nauk armyan. SSR, Biol. Nauki*, **13**, 81–84; *Chem. Abstr.*, **55**, 21963.
Virtanen, A. I., and De Ley, J. (1948). *Archs Biochem.*, **16**, 169–176.
Walisch, W. (1964). *Lab. Pract.*, **13**, 132–133: Belgian Patent 641,850; *Chem. Abstr.*, **63**, 15565.
Washko, M. E., and Rice, E. W. (1961). *Clin. Chem.*, **7**, 542–545.
Watson, R. W. (1965). *Ann. N. Y. Acad. Sci*, **130**, 733–744.
Webb, M. (1949). *J. gen. Microbiol.*, **3**, 410–417.

Weibull. G. (1960). *J. Bact.*, **79**, 155.
Weinbaum, G. (1966). *J. gen. Microbiol.*, **42**, 83–92.
Weinbaum, G., and Mallette, M. F. (1959). *J. gen Physiol.*, **42**, 1207–1218.
Weinstein, L. H., Bozart, R. F., Porter, C. A., Mandl, R. H., and Tweedy, B. G. (1964). *Contr. Boyce Thompson Inst. Pl. Res.*, **22**, 389–397.
Wentink, P. and La Riviere, J. W. M. (1962). *Antonie van Leeuwenhoek*, **28**, 85–90.
Whitehead, E. C. (1961). *Protides biol. Fluids*, **9**, 70–73.
Wilczok, T. (1959). *Chemia analit.*, **4**, 981–998; *Chem. Abstr.*, **54**, 15505.
Wilkins, J. R., Allen, W. W., and Alway, C. W. (1959). *Appl. Microbiol.*, **7**, 173–176.
Wilson, B. W., and James, T. W. (1963). *Expl Cell Res.*, **32** 305–319.
Wilson, G. S., and Miles, A. A. (1964). "Topley and Wilson's Principles of Bacteriology and Immunity", Vol. I, pp. 100–126. Williams and Wilkins Co., Baltimore.
Wood, E. C. (1947). *Analyst, Lond.*, **72**, 84–90.
Wright, D. N., and Lockhart, W. R. (1965). *J. Bact.*, **89**, 1082–1085.
Young, J. C. and Clark, J. W. (1965). *Wat. Sewage Wks*, **112**, 251–255.
Yuasa, T. (1959). *J. chem. Soc. Japan, pure chem. Sect.*, **80**, 1201–1202; *Chem. Abstr.*, **55**, 4244.
Zellner, S. R., Gustin, D. F., Buck, J. D., and Meyers, S. P. (1963). *Antonie van Leeuwenhoek*, **29**, 203–210.

CHAPTER XVI

The Evaluation of Mycelial Growth

C. T. CALAM

Imperial Chemical Industries Ltd, Pharmaceuticals Division, Alderley Park, Macclesfield, Cheshire, England

I. INTRODUCTION

Difficulties in the evaluation of the growth of mycelial organisms, such as fungi and actinomycetes, arise from the nature of the mycelium itself. Yeasts and bacteria have a rapid rate of reproduction. At the end of growth, the cells usually form a fairly stable suspension, though sometimes they undergo autolysis. Fungi and actinomycetes grow much more slowly, the threads extending and changing in composition as they age. At any given time, parts of the culture are growing while other parts are ageing and dying. The composition of the mycelium is continually changing. The mycelium may also store reserve substances and fats, so that a weight increase can occur without the formation of new cells or without consumption of nitrogen. Growth may be evaluated by measuring increases in weight or by the quantity of nitrogen being removed from the medium. Both methods give useful information, but they are not necessarily strictly related nor do they mean the same thing. In view of this, some workers, such as Borrow *et al.* (1961), have distinguished between "growth" and "proliferation", the former implying true growth of cells with the synthesis of new protein materials, the latter meaning weight increases due to the formation of reserve substances.

This brief consideration of the situation is sufficient to indicate the main difficulties of the evaluation of mycelial growth. It might well be concluded that a really reliable evaluation is impossible; that it is only possible to express it by giving several sets of figures. This view is largely correct. At least it is always necessary to consider carefully why growth is being measured, and to make sure that the methods used give the desired information. For practical reasons, however, it is frequently necessary to use a single method to express growth: the main thing is to make sure that when a single index is used its limitations are realized.

Another point of importance is that mycelial cultures are often grown on complex media containing chalk as a buffer. Such media contain a good deal of insoluble material that interferes with the determination of mycelial weight. This is often the case with media used for industrial fermentations, and attempts must be made to correct for this source of error in growth determinations.

It is proposed to begin by summarizing the reasons for which mycelial growth is measured, and suggest methods suitable for different purposes. After this, the different methods will be described. The approach is strictly practical. Problems of sampling and assay will be discussed and some practical results are given to illustrate the type of data obtained in ordinary experimental work.

II. REASONS FOR MEASURING MYCELIAL GROWTH AND CHOICE OF METHOD

The need to measure growth arises for various reasons, and in each case the most suitable method must be chosen. It is the object of this Section to discuss typical cases that illustrate the way in which different analytical methods are used.

A. Studies of growth and metabolism

In this type of work an accurate measure of growth is frequently essential. Metabolic behaviour is related to growth, and it is necessary to make a balance sheet comparing nutrients consumed and products obtained. The work of Raistrick and Rintoul (1931) is a classic example of work depending on the accurate measurement of growth. Not only is the determination of cell weight important, but also the avoidance of interference from extraneous substances that might make interpretation difficult. Direct weighing of the mycelium, followed by analysis, is the most suitable approach. Time is not usually a limitation, and nothing is gained by the use of rapid indirect methods.

B. Control of fermentations by quantity or morphology of mycelium

Rate and extent of growth are useful criteria for assessing the development of the organism, and in the operation of various processes, measurement of growth is valuable for control purposes. Control may imply the obtaining of a record or it can also mean the use of growth measurements in the operation of the fermentation process. For instance feeds may be started or additions made when a certain degree of growth is achieved or a certain morphological change occurs in the mycelium. For process control, speed and ease of working are desirable, and determinations of packed cell volume or wet mycelial weight are particularly useful. Indirect methods may also be used such as measurements of ammonia utilization or a change in pH.

C. Control of growth of inoculum

In most fermentations with fungi and actinomycetes, inoculation is carried out using a young mycelial culture. For measuring growth, cell volume or wet weight may be used, but it is often found that the culture is ready for use when a sample, on standing, shows no separation and the liquid remains full of cells. Rapid growth is usually essential, and it is necessary to specify the time in which an adequate degree of growth must be achieved. As the culture passes beyond the "full of cells" stage, it will probably become less suitable for use as an inoculum.

D. Estimation of insoluble products held in the mycelium

Fermentation products are sometimes insoluble in water and are filtered off with the mycelium, being subsequently recovered by extraction. Extraction may take place with the mycelium in the wet or dry state. If the culture is filtered in several lots, or on a rotary filter with a pre-coat, the composition of the filtered mycelium will vary considerably, and it can be very difficult to get a satisfactory estimate of the total mycelial weight and the quantity of product in the mycelium. Under these circumstances very careful sampling and assaying are necessary if even an approximate result is to be obtained, and expert advice is needed to do this in the most economical way.

III. PROBLEMS OF ASSESSMENT OF GROWTH ARISING FROM THE USE OF COMPLEX MEDIA AND FROM THE AGEING OF CELLS

When media containing only sugar, a nitrogen source, phosphates and trace elements are used, a direct determination of mycelial weight is easy. In stirred cultures using complex media conditions are very different. If the medium contains a complex nitrogen source together with chalk and a vegetable oil, the mycelium weight obtained by filtering a sample and drying the insoluble portion will be far too high. Some correction is possible if the fat is extracted with petroleum ether and, after re-weighing, the residue is washed so that allowance can be made for insoluble calcium compounds. The situation can be even worse when an insoluble organic material is present that is filtered off with the mycelium and makes an accurate analysis impossible. With fermentations of this sort it is customary to use approximate estimations, for instance by measuring the packed volume of cells after centrifugation, and converting this to g/litre by some factor obtained from special experiments. If a more accurate cell determination is needed, an attempt can be made to work out a special method of analysis, but it is better to use a medium with which the problem is not too difficult.

Another problem is that mycelial cultures usually contain a considerable number of small air bubbles so that the density of the culture is reduced. Falls in sp. gr. to 0·8–0·9 are common, while with some highly agitated systems even lower densities may occur. With mycelial cultures, volume measurements are also very difficult and unreliable. For these reasons samples for determinations of mycelial weight should always be weighed rather than measured volumetrically.

Serious difficulties sometimes arise because of slow filtration. This is usually because ageing of the culture produces fine particles in suspension, which block the filter, or because starch or proteins in the medium make filtration difficult. Under these conditions it may take an hour or more to filter 25 ml of culture and washing is almost impossible. Concern may well be felt that under these conditions further deterioration of the culture might occur, though it is of course possible to conduct the filtration at a low temperature.

Methods used to deal with these problems are: (a) careful choice of porosity of the filter medium, grade 2 porosity sintered-glass filters being often the most suitable—they may pass some of the finer particles; (b) often a filter will pass a certain amount of liquid before it blocks and careful choice of filter size and volume of sample will prove helpful; (c) the culture may be treated before filtration, e.g., by heating or by adding a coagulant, such as $CaCl_2$ or $Al_2(SO_4)_3$ (1–3 g/litre may be used, but experimentation is necessary), so as to make it easier to filter; (d) an amylolytic or proteolytic enzyme may be added to digest and render starch or protein soluble; (e) filter aid (e.g., Hyflo-Supercel, Johns-Mandeville & Co., London) may be used in carefully weighed amounts. Of these the first two are best, since with the others filtrate or mycelium may be changed by the treatments used.

The latter comment may also be borne in mind when considering washing and other treatments of the mycelium. These may include treatment with a solvent to remove fat or with acid to remove chalk, or boiling the culture to coagulate it and make it easier to filter. Washing procedures lasting a long time may also be harmful. When it is only a matter of obtaining a value for the dry weight of the cells these treatments may make little difference, as hyphae are rather resistant, but if it is intended to make analytical studies, it may be found that important constituents may have been lost or dispersed. If it is desired to measure hyphal walls, the dried mycelium should be ground in a mortar, than suspended in water and frozen and thawed; the suspension should then be treated overnight at 25°C with trypsin to render the protein soluble, and finally washed several times on the centrifuge until the walls are seen, under the microscope, to be free from contamination. They may then be filtered off and weighed (cf. Applegarth, 1967).

During the course of a fermentation, the mycelium of a micro-organism undergoes various changes. In the early stages fungal hyphae are young and filled with rapidly growing protoplasm. Later the protoplasm deteriorates and vacuolation and other changes occur. Eventually the cells lose their contents and the cell walls may autolyse. It is not clear how important these changes are when the weight of cells present is assessed. The process of vacuolation may have no effect on cell weight and degenerative changes do not necessarily alter biosynthetic activity. In some cases at least ageing is accompanied by a reduction in the proportion of nitrogen in the cells, whereas in other cases fat and starch accumulate. The occurrence of senescence and of a fall in mycelial weight at the end of a fermentation are, of course, well known.

Fragmentation of mycelium is not uncommonly observed. This is particularly marked in the case of certain actinomycetes, such as *Streptomyces rimosus*, in the oxytetracycline fermentation, where vigorous initial growth is followed by fragmentation and the culture continues to grow as short instead of long threads. Fragmentation is sometimes attributed to shearing due to the vigorous agitation used in stirred culture, but it seems more likely that fragmentation is associated with some phase of growth in the cells. Changes in the morphology of the culture have been used by some workers for fermentation control.

The above considerations, which will be dealt with in more detail in the next Section, indicate some of the problems involved in mycelial cultures. There is no means whereby they can be readily solved. This means that it is essential to consider each fermentation separately, and to determine the precise object of the analyses that are to be made, and the best way to acquire whatever information is desired. It has to be admitted that when the mycelium undergoes very extensive changes in morphology it may be very difficult to obtain meaningful results when attempting to measure growth (for instance if sporulation sets in). It may be necessary to use some rather arbitrary method for assessing growth, or, if it is possible, to alter the experimental conditions in some way so as to minimize the problem.

IV. METHODS OF ASSESSMENT OF MYCELIAL GROWTH

In describing the methods used, it is only proposed to give an outline of the procedure. This is partly because full accounts of the general procedures are given in other Chapters, but also because each situation has its own particular needs and it is better to leave the people doing the work to settle the details as seems most suitable. An important factor here is the degree of accuracy and reproducibility required in the results obtained. This may vary from the highest possible, time and effort being available to meet the most stringent standards, to rather rough and ready results when large numbers of samples have to be dealt with as quickly as possible and with a minimum of effort.

It is assumed that, since mycelial organisms are being considered, ample quantities of culture are available (litres) so that large samples (20–30 g) can be taken. Mycelial cultures are so thick that small samples may not be representative. With flask cultures (50–100 ml of culture) it is better to take the whole flask contents, mix them thoroughly and remove a representative sample for analysis. The whole culture should be used if only a small quantity is available. With laboratory cultures, care should be taken to note growth deposited on the walls, as this can sometimes make up a large part of the mycelium produced.

A. Direct methods of estimation

In these, the method of estimation depends on measuring the mycelium itself. The most obvious way is the weighing of a sample of culture that is then filtered, the mycelium being weighed wet or dry. As mentioned above, to avoid errors in sampling it is advisable to use culture samples of 20–50 g in weight. The sample may be weighed to the nearest 0·1 or 0·01 g, according to the degree of accuracy required. The dried mycelium may be expected to weigh between 1·0 and 0·05 g, and should be weighed to the nearest 0·001 or 0·0001 g in order to provide a corresponding degree of accuracy.

Samples may be filtered using Buchner funnels with filter papers, sintered glass funnels or Gooch crucibles. Filter papers 7 cm in dia are usually convenient and Whatman's Grade No. 1 is suitable. With sintered-glass filters a disc of Grade 2 porosity, 30 mm in dia is suitable. Borrow *et al.* (1961) have described a convenient filter apparatus in which the sample can be weighed and filtered without a transfer.

Washing is usually satisfactorily accomplished by briefly washing with two or three changes of water. It is preferable to stop filtration and stir the mycelium with the water and then re-start filtration. This procedure may not entirely free the mycelium from traces of medium. If more complete washing is necessary, a system should be devised based on practical tests with the mycelium concerned.

1. *Wet weight of mycelium*

A 20–50 g sample of culture, in the sample tube or in a beaker, is weighed and then filtered through double filter papers and washed with water. The container is re-weighed empty and the weight of the sample noted. The filtered mycelium is squeezed dry by pressing firmly between sheets of blotting paper or filter paper until no more moisture can be extracted. A small pair of rubber rollers may be used. Using a similar damp filter paper as a counterpoise, the mycelium is now weighed. In many cases the mycelium can be removed from the paper and weighed directly. The mycelium can be again squeezed between absorbent paper and re weighed. The mycelial pad can finally be dried to give the dry weight, as described below.

The wet mycelial weight is useful in providing a rapid estimate of growth. It also gives a measure of the actual amount of mycelium in the suspension in its original physical form. This is of particular interest in studies in agitation or in investigations of the viscosity of cultures. The wet weight is usually 2·5–5 times the dry weight. As thick cultures reach dry weights of 20–40 g/litre of culture; the amount of wet mycelium will be about 50–200 g/litre.

2. *Dry weight of mycelium*

A sample of culture is weighed and filtered and washed in the same way as with the wet weight. Filter papers (7 cm dia) vary in weight by ± 10 mg and sometimes more, so that for accurate results the papers should be weighed before use after drying in the same way as it is intended to dry the mycelium. After drying the mycelium, the weight of the paper can be deducted or, for rough results, another paper can be used as a counterpoise. Filter funnels or Gooch crucibles should be well washed to remove medium that may remain on the walls.

The mycelium may be dried at room temperature in a high vacuum over P_2O_5, or at 80°C in a vacuum oven, or at 110°C in air or *in vacuo*. Mycelium can also be dried at higher temperatures if desired to obtain faster drying. Drying usually takes a few hours, and trials should be made to ensure that a constant weight is achieved. Drying at room temperature under high-vacuum conditions normally produces a pale-coloured friable product suitable for further investigations. As the temperature is increased, the mycelium tends to become harder and more intractable; this, however, is not important if only the weight is required. For rapid spot determinations of dry weight, the mycelium and filter paper may be dried in 5–10 min by exposure to an infrared lamp at a distance of 5–8 cm.

When using a filter paper for separating the mycelium, we have found it convenient to wrap the filter paper and mycelium in a larger piece of filter paper during drying. This is a convenient method that avoids the loss of bits of mycelium during handling.

3. *Packed cell volume*

A 10–25 ml sample of culture is measured into a calibrated centrifuge tube and centrifuged under moderate conditions so that the cells pack down to minimum volume. The total volume and that of the cells are read off and the volume of cells is calculated as a percentage (v/v). Although it is convenient to use calibrated centrifuge tubes, plain tubes can be used and the height of liquid and cells measured with a ruler. When working with a large number of samples and only rough results are needed, this might prove more economical.

This method has obvious advantages when the cultures are difficult to filter. The apparatus required and the procedure are very simple.

Culture volume may also be measured in young cultures by allowing the culture to stand for 10–30 min and noting the height reached by the culture after settling in relation to the total liquor depth. This method is convenient for obtaining a quick estimate of growth, especially with young cultures, but after dry mycelium weights have reached about 10 g/litre, the culture remains full of mycelium even after standing for a considerable time.

4. *Treatments to remove contaminating substances*

When a synthetic medium containing no suspended solids is used, direct filtration and weighing of the mycelium is sufficient to give a reliable estimate of mycelial growth. With many types of media, however, a considerable amount of insoluble matter is present, such as chalk added to control pH, which may be converted to calcium sulphate or phosphate. Proteinaceous materials may be precipitated from corn-steep liquor. Insoluble nutrients are also used, such as soya flour or coarsely ground meals. The addition of vegetable oil is a common practice, and the usual range of pH, 6·0–7·5, and the presence of phosphates leads to precipitation of inorganic salts. Hydrocarbon oils are sometimes used, but fermentations based on these substances have not yet been developed with mycelial organisms. Although insoluble materials are inconvenient in use, they are valuable because they provide a medium that gives a continuous supply of nutrients by slow solution into the medium. In some cases, interference can be avoided, e.g., by adding EDTA to render cations soluble, but as a rule the problem remains a serious one. With typical media, undissolved solids may range from 5–15 g/litre or more.

When mycelial weight is only being used as an index of growth and an absolute value is not needed, a correction can be made based on experience, or the crude dry weight can be used. If packed cell volume is estimated by centrifuging, the layers of different materials can be distinguished from the mycelium and due allowance made.

When a quantitative estimate of mycelial weight is required, attempts can be made to reduce interference from insoluble substances by washing with acid to remove chalk and with solvent to remove fat. These treatments may be applied before or after filtration. The writer has found it convenient to extract the dried mycelium with petroleum ether in a Soxhlet apparatus and, after re-weighing, to ash the residue. In this way fat and ash can be allowed for. Data obtained in this way are given in Tables I, II and III. It should be noted that the data in Table II refer to mycelium that had been treated with acetone and dilute hydrochloric acid before drying. Protein is more difficult to deal with, but often it is digested after a few hours' cultivation, and can thereafter be ignored.

In one case, when a medium containing soya and other flours was used, an uninoculated fermenter was set up to which protease and amylase were added, and the loss in weight of the insoluble material was observed. This was used to give a rough correction in mycelial weight determinations (Table IV).

B. Indirect methods

1. *Absorbance, reflectance and viscosity*

With bacteria, growth is frequently measured in terms of cell concentration. As the cells usually form even suspensions of fine particles, with a reasonable degree of transparency even up to 10^{10} cells/ml or more, absorbance serves as a convenient measure of cell concentration. For mycelial cultures, this method is not usually appropriate. The culture soon becomes too thick for satisfactory light transmission, while clumping of the hyphae into aggregates leads to unevenness. If the culture is diluted, the cells settle out giving an uneven layer. Various dilution procedures are possible, but there is no advantage over more reliable methods such as packed cell volume.

A possible advantage of the light-absorption methods used with bacteria is the possibility of continuous monitoring of growth using flow-through absorption cells. This would only be possible with mycelial organisms if special cells were developed. The writer is not aware of such equipment being available. Equipment is now being developed, for instance at the Water Pollution Research Laboratory, Stevenage, Hertfordshire, (Ministry of Technology) for the measurement of the opacity of sewage suspensions, using the difference of light intensity measured by two photocells at different distances from a light source. So far apparatus of this type has only been used up to densities of 10 g/litre, but an apparatus of this type might be developed to deal with mycelial cultures.

Calderbank (1967) has described the use of reflectance for measuring the opacity of suspensions in stirred vessels. In work on culture selection, Growich and Deduck (1964) used reflectance to detect particular types of culture. Reflectance measurements are fairly simple and could be taken through the wall of a glass fermenter. Unfortunately considerable interference would arise from suspended solids and air bubbles in the medium, and it is unlikely that reflectance could be readily used as a measure of growth.

Viscosity of cultures can be measured quite easily with a viscometer of the rotating cylinder type (e.g., Ferranti viscometer). The viscosity usually rises from the low value of the original medium (about 1 cP) up to values equivalent to 500–1500 cP. Cultures are non-Newtonian in character and it is difficult to assign any specific meaning to the readings obtained. The viscosity arises from several factors, such as concentration of cells present, the type and age of the mycelium and the viscosity of the suspending liquid. It would only be possible to express viscosity as a scale reading under specified operating conditions. Viscosity readings could therefore have a certain utility for evaluating growth, but in this case also it is doubtful whether viscosity would be of more use than one of the rapid methods for cell determination.

2. *By analysis of nitrogen, nucleic acids etc.*

The possibility of turning to a chemical analysis such as that of nitrogen, instead of mycelium weight, has been used for many years as an indication of mycelial growth. When a synthetic medium with a single source of nitrogen is used, measurement of the disappearance of nitrogen gives a good index of the amount of mycelium formed, especially if the percentage of nitrogen in the mycelium is precisely known. Difficulties arise if the proportion of nitrogen in the mycelium varies, or if nitrogen is absorbed by the mycelium and then returned to the medium. Hosler and Johnson (1953) estimated nitrogen in a weighed sample of culture and in the filtrate and took the difference as mycelial nitrogen. A large sample of culture should be digested (e.g., 10 g) so as to avoid sampling errors. This method can be used with complex media, but when insoluble proteins are present (e.g., with soya meal) a method of this type would fail. Many workers feel that the nitrogen present in the cells represents the quantity of enzymes therein and therefore is more representative of "active cells" than is the mycelial weight, which includes active material, reserve substances and also dead cells. Cells usually contain 4·5–8·5% of nitrogen.

Similar arguments arise over the possibility of using nucleic acids as a basis of effective cell mass. As an example of typical values obtained for dry weight, nitrogen uptake and nucleic acid formation see Table V. The paper referred to (Gottlieb and Van Etten, 1964) gives full experimental details, but, briefly, the culture was grown on the surface of 30 ml of medium in 500 ml flasks.

In the writer's view there is considerable value in measuring nitrogen uptake. Especially when ammonia is the sole nitrogen source, rapid analyses can be carried out giving a useful estimate of mycelium formation. Nonetheless if a single index is desired, a direct estimate of mycelial weight is to be preferred. It is not without its difficulties, but it does give a value for the weight of cells present, and this is usually what is required. Determination of nucleic acids is relatively academic, and is better done for its own sake. In any case, the analysis is rather inconvenient and therefore unsatisfactory as a rapid guide to growth.

3. *By consumption of sugar or by CO_2 production or O_2 uptake*

When cells are growing freely, the cell weight is proportional to the quantity of nutrients consumed. In an experiment, five fungi were inoculated into a medium containing glucose (20 g/litre) and corn-steep liquor solids (10 g/litre), diluted to different degrees from full strength down to 1 : 300. After allowing growth for 4 days on a shaker, cell weights were measured. In all cases, a straight-line relationship was observed between dry cell weight and nutrient concentration, the average figures being 1·0 g of cells requiring 2·5 g of nutrient. Righelato (1967) found a value of 2·56 : 1 for the ratio glucose used: dry cells produced for *Penicillium chrysogenum* cultures grown on synthetic media. Although this type of behaviour is common, there are many occasions when sugar is converted to other substances, such as gluconic acid, and large amounts of sugar can disappear without any increase in cell weight. Much of the sugar is oxidized rather than assimilated. Sugar analyses are, however, very easy, either by a copper–

iodimetric method like that of Shaffer and Hartmann (1920) or by an enzymic method (Blood Sugar Biochemical Test Combination, Boehringer Corporation Limited, London), and sugar consumption as an index of growth is useful provided the metabolism of the organism is understood.

CO_2 production is another possible index of growth, although the relation between the two factors is complex and will certainly vary as the fermentation proceeds. CO_2 can be measured in the effluent air from stirred cultures using automatic analysers, such as the infrared equipment described by Telling *et al.*, 1958 which provides a means of monitoring the process. Less conveniently, CO_2 can be measured by an Orsat or Haldane volumetric apparatus. Especially during the early stages, CO_2 output is useful as indicating when rapid growth and vigorous cell activity are occurring, and later the decline in metabolic activity can be observed.

Measurement of O_2 uptake is complementary to CO_2 production and is also a measure of cell activity. Oxygen analysers are available, for example the Servomex (Servomex Controls Ltd., Crowborough, Sussex.). Changes in O_2 concentration due to respiration are relatively small, and it is necessary to make sure that the apparatus has an adequate degree of accuracy and sensitivity. The ratio of CO_2 produced/O_2 used gives further information about the growth process.

V. PROBLEMS OF CHANGING ANALYSIS WITH AGE

The type of change occurring during the growth cycle of mycelial organisms has already been referred to and is described in detail below. An initial growth phase is followed by a period of biochemical activity, during which the weight of cells may increase or decrease owing to accumulation of storage products (carbohydrates or fat) or to vacuolation or autolysis.

Borrow *et al.* (1961) give an extensive discussion of the changes occurring during growth of *Gibberella fujikuroi*. Gottlieb and van Etten (1964) have described the biochemical changes in *Penicillium atrovenetum* (see Table V) covering total nitrogen, cold trichloroacetic acid soluble nitrogen, ribonucleic acid and protein. Although this work was done in surface culture, the results are similar to those obtained in submerged conditions. In a recent paper Fencl and Ricica (1968) have given a summary of current ideas on this point and described the advantages of continuous culture in which cell populations are more homogeneous.

VI. SAMPLING PROBLEMS, ACCURACY OF TESTS, MEASUREMENT OF CULTURE VOLUMES, TIMING

In the case of stirred and shaken cultures thought must be given to the production of satisfactory samples for mycelial weight determinations or for analysis. The culture may be very thick so that samples of uneven quality are obtained, especially if the samples are taken intermittently via a long narrow tube. Difficulty due to settling may also be troublesome if the sample is received from the fermenter in a bottle or jar and then allowed to stand for a time. When chemical analyses of whole cultures are involved, it may be necessary to use a Waring Blender to homogenize the sample, but this is undesirable in mycelial estimations as the cells are damaged.

The precise method of taking a sample from a large fermenter so as to obtain a representative sample will depend on circumstances, but care must be taken to avoid taking samples from a badly stirred part of the fermenter or at the surface where a poorly stirred layer may develop. With shake flasks a ring of culture may adhere to the walls and fall into the culture when shaking stops. Whenever signs of unhomogeneity appear steps must be taken to ensure that representative samples are taken. With small stirred fermenters mycelial weights may vary unexpectedly owing to mycelium being thrown out onto the walls of the fermenter and later falling back again. When samples are taken from steam-sterilized sample points, sample dilution can occur if the steam valve leaks or is not fully closed. The sample line must also be carefully flushed out. Careful training of personnel is necessary if fully reliable samples are to be obtained.

Sampling problems arise with filtered mycelium especially when the mycelium contains a product that is to be recovered. Owing to the presence of material from the fermenter walls, etc., the filter cake may be of uneven composition and removal of water may also be uneven. In order to obtain a representative sample, the mycelium should be carefully mixed, and preferably an appropriate sampling technique applied such as "quartering" (i.e., selecting diagonally opposite quarters of the whole and repeating the process by quartering the selection). When the mycelium is dried, the dry material should be ground in a mill and well mixed before sampling. Sampling problems also occur on the large scale when a fermenter batch is harvested and the mycelium recovered on a rotary filter. Initial samples are likely to contain excessive amounts of filter aid, as are those at the end, while water content will vary throughout. The product will probably be collected in drums or bogies, and it is necessary to sample these on the assumption that there will be considerable variations both within and between containers. For an account of the methods used for sampling, see Strouts *et al.* (1962).

Although it may appear a little naïve, it is felt that attention should be drawn to the need for careful choice of times of sampling and to accuracy of time measurement in the estimation of mycelial growth. When a growth curve is being obtained at least five or six points are necessary and careful choice of sampling times must be made to ensure that these are obtained.

All too often, when it comes to the examination of experimental data it is found that the results would have been more valuable if the sampling times had been more carefully worked out beforehand.

Calculation of growth rates obviously calls for an accurate knowledge of the period of growth. It is easy to quote sampling times as, say, 6 h, 9 h, 12 h, without specifying the accuracy of the sampling times. An error of 10% here is as important as a similar error in mycelial weight; especially in plant work it is not uncommon for a sampling time to be stated, e.g., 8 a.m., whereas in fact the samples are taken from several fermenters one after another over a fairly long period of time. Errors arising from this practice must be prevented if confusion and inaccuracy are to be avoided.

This Section attempts to indicate the problems of accuracy and of sampling that arise in measuring mycelial growth. It is the writer's view that sampling is just as much a source of error as are analytical errors, probably in fact sampling errors are more serious because they are unnoticed. Sampling problems can usually be overcome by taking suitable precautions, but whenever unexplained irregularities in results appear, sampling errors should be suspected and stringent precautions taken to eliminate or reduce them.

VII. MICROSCOPY OF MYCELIUM

The changes that take place during growth also result in changes in the microscopic appearance of the mycelium. There are large differences in the behaviour of different organisms, and growth conditions also have a considerable effect. Numerous investigations of these changes have been described in the literature, and a few of them will be mentioned here as typifying the general approach. In some cases, the mycelium is examined during fermentations, and the observations made are used in process control. One European firm (Chemap AG, Männendorf, Switzerland), makes a microscope that can be fitted to fermenters so that observations can be carried on during fermentations.

Duckworth and Harris (1949) gave an interesting description of the morphological changes in *Pen. chrysogenum* during growth. It was convenient to fix samples of culture by adding an equal volume of 10% Formalin. Slide preparations were made by using lactophenol with cotton blue (phenol, 10 g; lactic acid, s.g. 1·21, 10 g glycerol, 20 g distilled water, 10 g; cotton blue, 0·05 g/100 ml). Slides can be made semi-permanent sealing the coverslip with a mixture of equal parts of beeswax and gum-damar applied in the molten state. Usually a magnification of about × 500 is suitable.

When conidia were germinated on agar, the germ tube developed after about 8 h. Growth was confined to the hyphal tips and proceeded for some hours before the first signs of vacuolation appeared. Mycelium was usually in this condition when production cultures were inoculated, and inoculation was followed by a period of rapid growth. Quite often a somewhat distorted appearance developed, but hyphal tips were characteristically rather narrow. The mycelium was full of protoplasm that stained readily. After this, vacuolation took place, at first rather slight, later large vacuoles formed and "ladder" vacuolation (so called from its appearance) might occur. Eventually the protoplasm disappeared and empty hyphae were seen. A feature of this and other organisms was the appearance of spherical cells either alone or as part of the mycelium; these may be degenerate forms, but often they represented resting bodies from which new generations of mycelium arose. These and other effects are illustrated in two plates in Duckworth and Harris' paper.

Depending on the type and size of inoculum used, and on other factors, cultures grow either as a suspension of threads ("mycelial culture") or in tufts or pellets. Pellets develop from a central tuft of mycelium. Owing to lack of O_2 and nutrients at the centre, the central area dies and is surrounded by a layer of active cells. With some rapidly growing fungi with stiff hyphae, pellet formation can prove a difficulty as the pellets increase in size to form large irregular balls of mycelium. In other cases tiny pellets are formed resembling sand, which are sometimes very active. These cultures show a low viscosity when compared with the mycelial type. Thus in evaluating growth, the appearance of the culture should be noted as well as the mycelial weight.

Another description of *Pen. chrysogenum* mycelium is illustrated in plates in an article by Hockenhull (1963). Probably a higher yielding mutant than that used by Duckworth and Harris was the subject of the photographs. Initially, well grown, straight threads were produced, full of protoplasm. As penicillin production became maximal, there was some vacuolation and some swelling, giving a wavy effect. There were many short branches with pointed tips: about half the hyphae were full of protoplasm. In the next stage, the start of senescence, most of the hyphae were vacuolated, few healthy hyphal tips were visible and more swollen forms. At the end of the fermentation the culture consisted mainly of short, rather swollen vacuolated hyphae. Growth had stopped and the hyphal tips were swollen. This development of numerous short hyphae is a common event. It has been suggested that this fragmentation is due to mechanical shear, but it is equally likely to represent a stage of growth.

In the case of *Pen. chrysogenum* the cells frequently show various particles, and oil droplets can be seen. Nuclei are very small and are not easy to see.

Butler and Harris (1949) developed a useful method of scoring the morphology of *Pen. chrysogenum* that could well be used in other cases. An important feature was the separation of the different features when considering the various factors involved—

(i) The sample was treated with formalin and allowed to stand in a 30 ml McCartney bottle. The volume of culture relative to total volume was noted, as a guide to degree of growth.
(ii) The formalized sample was poured into a Petri dish and examined for form of growth (pellets, sludge, etc.) and colour. The ratio of sludge to pellets was expressed on the scale 0/6 to 6/0 via 5/1, 3/3 etc.
(iii) The sample was then examined microscopically for the following features—

Branching
Vacuolation
Autolysis
Presence of conidiophores
Spherical bodies
Fragmentation

Each factor was evaluated and expressed by a number from zero to six, e.g., all mycelium 0, all large pellets 6; or all healthy cells 0, completely autolysed 6. The value of the method of presentation lies in the separation of the different features that facilitates the description of a culture. The system could be readily modified to suit other organisms. An example of the use of this method is given in Table III.

Describing the growth of *Streptomyces rimosus*, Doskočil *et al.* (1958) observed the occurrence of five stages. These may be summarized as follows—

Stage 1: lag phase, up to 90 min when inoculum was very small.
Stage 2: growth of primary mycelium up to 4–5 g/litre, 10–25 h thick Gram-positive hyphae.
Stage 3: fragmentation of primary mycelium, 10 h.
Stage 4: growth of secondary mycelium, thin Gram-negative filaments, ca. 25 h.
Stage 5: stationary phase, fragmentation of thin filaments—remainder of fermentation.

The very extensive morphological changes, accompanied by different patterns of metabolism, are an interesting feature of the oxytetracycline fermentation. A very interesting description of *Strept. aureofaciens* cultures has been given by Tresner *et al.* (1967).

A good account of the techniques suitable for use in the examination of fungi is given in a book by Smith (1960).

VIII. MEASUREMENT OF SURFACE CULTURES

The measurement of surface cultures grown on liquid medium present no particular problems. The surface felt can be filtered off, rinsed and dried in the same way as is done with submerged cultures. As the surface cultures develop, numerous changes are apparent in colour, sporulation, thickness of growth and the proportions of aerial and submerged mycelium. The mycelium is thus variable in quality, and it is advisable that after drying it should be ground and thoroughly mixed before analysis. As a recent example of work with surface cultures, reference may be made to the paper by Gottlieb and Van Etten (1964) already mentioned and to Table V.

It may be necessary to measure the weight of colonies that have grown on agar. One possibility is to cut out the colony with a minimum of agar, dry it *in vacuo* and weigh, when the dried agar medium will probably have a negligible weight. Alternatively the colony and the agar may be treated with boiling water to dissolve the agar; the colony can then be filtered off, washed and weighed.

A kinetic study of the growth of surface cultures has recently been made by Pirt (1967). Colony size was proportional to the diameter, and during growth the diameter increased linearly with time. Colony sizes were measured using an enlarger with a projection screen (the Shadowmaster, Buck and Hickman, Watford, Herts, England). Colony sizes can also be measured with a ruler or with callipers, but the projection method is considerably more convenient.

IX. ILLUSTRATIVE EXAMPLES

The results given in Table I were obtained from an experiment in which a strain of *Penicillium* was grown on a corn-steep liquor medium containing a small quantity of arachis oil and chalk. No sugar was added, but a feed of sugar was applied as soon as the culture became thick. At intervals, 25 ml samples were weighed (50 ml at 3 h) and the mycelium was filtered off, dried in a vacuum oven at 80°C and weighed to the nearest 0·0002 g; the dried material weighed 0·4–0·8 g.

The first section of Table I gives the crude dry weight obtained in this way, together with the allowances for petrol-extractable oil and ash. The dried mycelium was extracted with petroleum ether (b.p. 60°–80°C) in a Soxhlet extractor for 2–3 h and then dried as before. It was ashed to constant weight by heating it in a silica crucible at red heat. An appreciable amount of oil was present at first, but this disappeared after 13 h, a trace reappearing later on. Percentage of ash was high at first, as would be expected since there were few cells to offset the chalk. As expected, ash fell as the mycelium grew, but then increased as the rise in pH brought about precipitation of phosphates.

In this experiment growth was extremely rapid at first, but owing to changes in metabolism, in this case probably a shortage of rapidly usable nitrogen, growth soon slowed down. The mycelial weight fell, rose and fell again until at 89 h steady growth re-started. The experiment thus illustrates the type of behaviour that makes growth evaluation so difficult. During a short initial phase, up to 13 h, growth was probably exponential: later stages probably involved degeneration of the cells followed eventually by slow growth giving cell-weights higher than at 13 h. Presumably adaptation of the cells must have occurred to allow this to happen. No microscopic studies were made, but they would undoubtedly have shown significant changes due to these factors.

Although the measurements show that an interesting series of changes in growth occurred during the experiment, and give hints as to the timing of the various events, it is clear that they are insufficient to give any information about growth rates. Thus rapid growth occurred between 3 and 13 h, followed by a fall in mycelium weight. To follow this in detail two hourly or better one-hourly measurements would be necessary.

Wet weights were also measured, the mycelium being filtered and pressed several times between filter-papers until no more moisture could be removed. Packed cell volumes were measured by centrifuging about 10 ml of culture in tapered and graduated centrifuge tubes at 2000 rev/min with a radius of 20 cm (490 *g*) for 15 min. Trials showed that this was sufficient to reduce the cell volume to a steady value. The data given show an adequate degree of reproducibility (standard errors of less than 5%).

The last two columns compare the wet weights and packed cell volumes with corrected dry weights. The ratios obtained are reasonably consistent, but there are some changes during the run, probably indicating that a shrinking of the cells has accompanied changes in metabolism.

Table I shows that crude or corrected dry weights, wet weights and packed cell volumes can be used for evaluating growth. All give useful results, but are subject to variations caused by changes in the metabolism of the cells as well as to the presence of oil and ash which arise from the constituents of the medium. If analytical studies are in progress, dry weights are convenient as the dried cells are readily stored and analysed. If the extent of growth is the only object of interest, wet weight and packed cell volume are as good as the corrected dry weight, while the crude dry weight can be misleading especially with very young cultures. With young cultures, wet weights seem a little better than packed cell volumes.

Table II gives data from a penicillin fermentation, using a medium containing lactose, arachis oil, corn-steep liquor and chalk. In this case the

TABLE I

Growth evaluation in a stirred culture of *Penicillium*

Age, h	Mycelial dry weight, g/litre				Wet wt, g/litre	Packed cell volume, %	Ratios of means	
	crude dry wt g/litre	fat, %	ash, %	corrected dry cell wt			wet wt/corrected dry wt	packed cell volume/corrected dry wt
3	7·16	11·7	36·8	3·69	15·0	6		
	7·45	11·9	38·8	3·67	15·0	6	4·2	1·7
	7·06	13·4	40·3	3·27	14·7	..		
13	27·25	2·0	18·0	21·80	92·0	55		
	28·45	1·3	17·7	23·04	97·0	55	4·2	2·4
	27·78	0·3	19·8	22·20	93·8	52		
30	22.11	0	27·6	16·01	68·0	40	4·1	2·5
	21·06	0	28·7	15·02	61·3	41		
	21·70	0	28·5	15·52	63·0	35		
65	24·68	0	16·25	20·67	79·2	53		
	26·89	0	16·6	22·47	87·7	54	3·7	2·4
	25·48	0	17·0	22·15	81·8	51		
89	23·88	1·8	16·2	19·58	65·8	37		
	23·02	1·5	16·4	18·90	63·5	38	3·3	2·0
	22·64	2·0	16·6	18·43	59·0	37		
113	31·42	3·6	15·7	25·39	111·0	53		
	31·74	0·7	16·1	26·41	105·7	61	4·2	2·1
	31·34	0·8	16·3	25·98	108·9	57		
Standard error	0·64	..	..	0·65 g/litre	3·21 g/litre	2·1%	..	..

TABLE II

Penicillin production culture: changes in mycelium analysis and mycelial nitrogen

Age, h	Mycelium weight g/kg: dry wt (uncorrected)	Mycelium weight g/kg: analysis of mycelium oil, %	Mycelium weight g/kg: analysis of mycelium ash (oxide), %	Corrected mycelial wt	Mycelial nitrogen g/kg: whole culture	Mycelial nitrogen g/kg: filtrate	Mycelial nitrogen g/kg: difference (=mycelial N)	Mycelial nitrogen g/kg: N as % of dry wt
14	15·4	53·8	3·7	6·5	3·1	2·3	0·8	12·5
26	27·0	57·7	2·0	10·9	3·1	1·7	1·4	12·8
50	26·3	35·6	3·6	17·1	3·4	1·3	2·1	12·3
74	36·5	22·8	7·4	25·5	4·0	1·6	2·4	9·1
98	41·0	13·6	13·6	29·8	4·4	2·4	2·3	7·7
121	40·7	11·2	13·5	30·6	4·5	2·9	1·6	5·2

mycelium weight is given with and without correction for oil and ash, together with analyses of mycelial nitrogen obtained by comparing the nitrogen content of whole culture and filtrate, determined by the Kjeldahl method on 10 g samples. To remove chalk the mycelium was treated with hot acetic acid. A 20 g sample of culture was diluted to 100 ml with water, 10 ml of glacial acetic acid was added and the mixture heated to nearly boiling. It was then filtered through a Whatman No. 52 filter paper, thoroughly washed with water and dried. It is obvious that although the acid removed much of the ash at first, later it was ineffective. This was probably due to the chalk being changed to sulphate during the fermentation. The level of nitrogen in the mycelium (%) was high at first, but fell to less than half this value at the end of the run. Crude dry and mycelial nitrogen give a misleading picture of the amount of cells present. The nitrogen analyses show up the deterioration of the mycelium, at a time when its activity also declines.

Table III gives the results of another penicillin fermentation which was carried out under different conditions. The quantity (%) of nitrogen, ash, fat and carbohydrate (reserve) are given. This Table includes the results of microscopic examinations of the mycelium made according to the method of Butler and Harris (1949) and illustrates the method of scoring the results. It shows in particular the increase in pellet formation in the middle of the run and the onset of vacuolation, autolysis and fragmentation.

Table IV gives data from a comparative experiment in which the mycelial weight of a *Fusarium moniliforme* was followed when grown on a medium containing soya-flour and other insoluble materials. Alongside the experimental fermenter was set up a control to which were added enzymes which slowly dissolved the insoluble materials. With the aid of the control fermentation 10, containing 5 g of each of Panazyme and Bacterase (both obtainable from Associated British Maltsters Ltd, Woodley, Cheshire) and of Takadiastase (Parke, Davis & Co., Hounslow, London), corrected mycelial weights (fermentation 11, for which the culture was on the same medium as fermentation 10, but without the enzymes) were obtained. Although the corrected mycelial weights give a reasonable graph there is no reason to suppose the figures are necessarily correct, and the experiment is intended mainly to illustrate the difficulties that arise with this type of medium.

Table V is based on data given by Gottlieb and Van Etten (1964) for a surface culture experiment with *Pen. atrovenetum*, grown on the surface of 30 ml of medium in 500 ml flasks. It is of value in showing the changes occurring in the contents of nitrogen and nucleic acids during growth, as well as giving typical results for a surface-culture experiment. At the intervals shown, complete flasks were harvested and the contents analysed.

TABLE III

Metabolism of pilot plant batch 506B—Penicillin

Initial volume 160 gallons, including inoculum

	Mycelial analysis				
Age, h	Weight (uncorrected) g/kg	N, %	Ash, %	Carbo-hydrate, %	Fat, %
0 (sown)	4·7			11·5	20·9
15	14·7	4·9	16·8	..	..
27	15·6	5·3	17·1	20·3	18·1
39	Partly lost	5·6	14·8	..	..
51	20·6	5·6	13·2	..	..
63	21·2			28·0	8·5
75	29·5	5·2	13·8	27·5	13·3
87	41·1	4·6	13·0	26·9	5·1
99	42·0	4·7	13·5	27·6	4·7
111	41·4	..	..	28·5	7·0
123	44·4	4·3	11·7	31·0	7·3

TABLE III.—*Continued*

		Morphology (Harris and Butler's scale)						
Age, h	Volume, % full (in equal vol. of formal.)	Consistency ratio sludge/pellets)	Branching (not very branched)	Vacuolation (6 = all cells vacuolated)	Autolysis (6 = all autolysed; 3 = about ½)	Conidiophores	Spherical Bodies (including all types of swollen tips)	Fragmentation (2 = slight; 4 = moderate; 6 = severe)
0	50	6/0	2	3	0	0		
15	70	5/1	3	2·5	0	1(?)		0
27	90		3	3	2	0		0
39	90	2/4	2	4	1	0		0
51	90							
63	90	3/3	2	5	4	0		0
75	90		2	6	2	0		2
87	90	3/3	2	6	3	0		
99	90							
111	90	4/2	2	6	3	0	+	6
123	90	5/1	2	6	4	0	+	6

Twisted and distorted hyphae were present at 111 and 123 h. Small, living fuzzy pellets were visible at all stages.

TABLE IV

Comparison of enzyme-treated medium with growth of culture

Fermentation 10, enzymes				Fermentation 11, (*Fusarium moniliforme*)	
age h	dry wt, g/litre	total carbohydrate in solution, g/litre	total N in solution g/litre	total dry wt, g/litre	corrected dry wt, g/litre
0	31·8	50·5	1·02	32·5	0·7
18	23·3	60·8	1·48	33·7	10·4
27	24·3	..	1·56	38·5	14·2
44	20·4	61·5	1·72	43·4	23·0
68	17·7	62·4	1·76	49·8	31·0
99	21·5	64·0	1·82	51·5	30·0
115	23·7	61·0	1·80	49·3	25·6

TABLE V

Biochemical changes during growth of *Penicillium atrovenetum* (Gottlieb and Van Etten, 1964)

Age, h	Dry mycelial wt/flask, mg	Composition of mycelium		
		N,%	total DNA, %	total RNA, %
0	Trace	6·3	0·0	3·1
24	5	8·0	0·80	4·7
48	36	7·5	0·60	3·3
60	120	6·6	0·60	3·0
72	210	5·7	0·60	2·9
96	330	4·0	0·60	2·3
120	372	3·6	0·43	1·8
144	378	4·0	0·50	1·8
168	354	..	..	..

REFERENCES

Applegarth, D. A. (1967). *Archs. Biochem. Biophys.*, **120**, 471–478.

Borrow, A., Jefferys, E. G., Kessel, R. H. J., Lloyd, E. C., Lloyd, P. B., and Nixon, I. S. (1961). *Can. J. Microbiol.*, **7**, 227–276.

Butler, M., and Harris, G. C. M. (1949). *J. gen. Microbiol.*, **3**, vi.

Calderbank, P. H. (1967). *In* "Biochemical and biological engineering science" (Ed. N. Blakebrough), p. 106. Academic Press, London and New York.

Doskočil, J., Sikyta, B., Kašparova, J., Doskočilova, D., and Zajiček, J. (1958). *J. gen. Microbiol.*, **18**, 302–314.
Duckworth, R. B., and Harris, G. C. M. (1949). *Trans. Br. mycol. Soc.*, **32**, 224–235.
Fencl, Z., and Ricica, J., (1968). *Process Biochem.*, **3**, No. 1, 41–45.
Gottlieb, D., and Van Etten, J. L. (1964). *J. Bact.*, **88**, 114–121.
Growich, J. A., and Deduck, N. (1964). British Patent 952,820.
Hockenhull, D. J. D. (1963). *In* "Biochemistry of industrial microorganisms" (Ed. C. Rainbow and A. H. Rose), p. 227. Academic Press, London and New York.
Hosler, P., and Johnson, M. J. (1953). *Ind. Engng Chem.*, **45**, 871–874.
Pirt, S. J. (1967). *J. gen. Microbiol.*, **47**, 181–197.
Raistrick, H., and Rintoul, W. (1931). *Trans. R. Soc.*, **B 220**, 1 and succeeding papers.
Righelato, R. C. (1967). Ph.D. Thesis, University of London.
Shaffer, P. A., and Hartmann, A. F. (1920). *J. biol. Chem.*, **45**, 365–390.
Smith, G. (1960). "An introduction to industrial mycology", 5th Ed. Arnold, London.
Strouts, C. R. N., Wilson, H. N., and Parry-Jones, R. T. (Eds.) (1962) "Chemical Analysis", Vol. I, p. 46. Clarendon Press, Oxford.
Tolling, R. C., Elsworth, R., and East, D. N. (1958). *J. appl. Bact.*, **21**, 26–44.
Tresner, H. D., Hayes, J. A., and Backus, E. J., (1967), *Appl. Microbiol.*, **15**, 1185–1191.

CHAPTER XVII

Counting and Sizing Micro-organisms with the Coulter Counter

H. E. Kubitschek

Argonne National Laboratory, Argonne, Illinois, U.S.A.

I. INTRODUCTION

The Coulter counter (Coulter Electronics Inc., Hialeah, Florida, U.S.A.) has become increasingly important in counting and sizing micro-organisms in suspension. The counter, which consists of a transducer (or particle detection system) and an analyser system, is currently used primarily to determine cell concentrations, although its unparalleled potential for determining volumes of microbial cells was demonstrated rather early (Kubitschek, 1958). Routine application to sizing of micro-organisms has been facilitated by the development of large-capacity, transistorized electronic instrumentation. Accurate and rapid measurement of complete cell-volume distributions requires a multichannel analyser (Nuclear–Chicago

Corporation, Des Plaines, Illinois, U.S.A.; Nuclear Data, Inc., Palatine, Illinois, U.S.A.).

With large-capacity analysers, both cell volumes and cell concentrations can be determined far more rapidly and accurately than by other techniques; micro-organisms can be counted at rates of about 50,000 cells/min with some instruments, and complete cell-size distributions can be determined within 10 sec. Thus, both cell numbers and cell volumes can be determined continuously during the course of growth of even the most rapidly growing cultures. These characteristics make Coulter counter–analyser systems especially valuable for studies of synchronous cell cultures.

II. THEORY OF OPERATION

A. Conditions for the proportionality of pulse amplitude to particle volume

To be counted, particles (or cells) must be suspended in a conducting fluid that passes through a minute aperture through which an electric

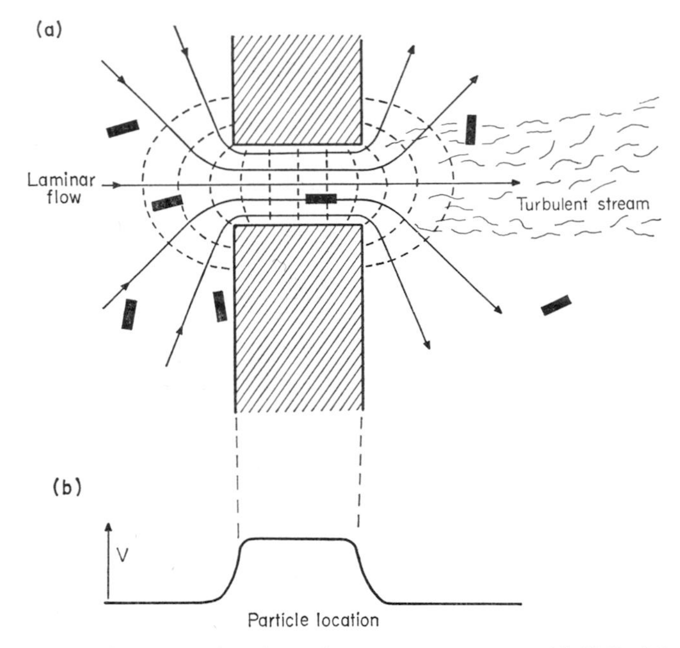

FIG. 1. Flow of a suspension through a counter aperture. (a) Cylindrical particles passing through the aperture. Flow is laminar as particles enter the aperture and are within it, but large flow rates cause the formation of a turbulent jet at the exit. The network of solid and dashed lines represent lines of force and equipotential lines. (b). The relative pulse amplitude (voltage) resulting from the position of a particle as it passed through the aperture.

current also flows (Fig. 1a). During the passage of any relatively non-conducting particle, the electrical resistance of the fluid within the aperture is increased, giving rise to a transiently increased voltage across the aperture (Fig. 1b). The voltage pulse generated by each particle is amplified and registered electronically, giving a count of the number of particles flowing through the aperture. The pulses also are sorted electronically, since under appropriate conditions the amplitude of the voltage pulse is essentially proportional to the volume of the particle; thus the distribution of pulse amplitudes represents the relative distribution of the volumes of the particles counted. The pulse-amplitude distribution usually is displayed on an oscilloscope for immediate reference, and, when desired, can be recorded permanently in graphical or numerical form.

In order to understand the requirements for obtaining valid size distributions, it is necessary to consider more closely the manner in which the aperture resistance changes during the passage of a particle. In the presence of a particle of low conductivity the aperture resistance is increased. If a constant aperture current I is maintained, then the aperture voltage will be increased by an amount $\Delta E = I\,\Delta R$, where ΔR is the increase in resistance. Since the same relationship, $E = IR$, also holds for aperture voltage E and resistance R in the absence of particles, the fractional increases in voltage and resistance are equivalent—

$$\frac{\Delta E}{E} = \frac{\Delta R}{R} \tag{1}$$

In some instruments a constant aperture *voltage* is maintained and a passing particle causes a transient decrease in electric current, which in turn is amplified and measured. For these instruments the fractional *decrease* in electric current, $\Delta I/I$, is equivalent to the fractional increase in aperture resistance.

The change in resistance can be calculated most easily for cylindrical particles, shorter than the aperture, and constrained to pass through it with their axes parallel to the axis of the cylindrical aperture (Fig. 1a), since, in this case, a particle within the aperture and the remaining conductive fluid can be considered as resistances in parallel. Suppose the aperture has length L, cross-sectional area A, and volume V and that the corresponding dimensions of the particle are l, a, and v, and suppose that ρ_0 is the resistivity of the solution and ρ is the resistivity of the particle. Then the resistance of the unoccupied aperture is—

$$R = \frac{\rho_0 L}{A} \tag{2}$$

and the corresponding electrical resistance of any aligned particle within the aperture is $\rho l/a$. The increase in aperture resistance due to the presence of a single particle is—

$$\Delta R = \frac{v\rho_0}{A^2}\left(\frac{\rho}{\rho-\rho_0}-\frac{a}{A}\right)^{-1} \tag{3}$$

From equations (1), (2) and (3)—

$$\frac{\Delta E}{E} = \frac{\Delta R}{R} = \frac{v}{V}\left(\frac{\rho}{\rho-\rho_0}-\frac{a}{A}\right)^{-1} \tag{4}$$

This relationship simplifies for particles with radii that are small compared to that of the aperture; then $a \ll A$, and pulse amplitudes become proportional to particle volumes—

$$\frac{\Delta E}{E} = \frac{v}{V}\frac{\rho-\rho_0}{\rho} \tag{5}$$

Further, it is theoretically possible to obtain counter response effectively independent of particle resistivity, permitting comparative sizing of cells with different resistivities, as might be expected for different kinds of microorganisms, or even cells of the same strain at different times during their cell-division cycle. To obtain this independence, the resistivity of the solution must be much smaller than that of the cells or particles, $\rho_0 \ll \rho$. Then to an excellent approximation—

$$\frac{\Delta E}{E} = \frac{v}{V} \tag{6}$$

B. Effect of particle shape and other factors

This relationship between particle volume and pulse amplitude also holds for spherical particles provided that the same approximations are made again for a and ρ_0 (Kubitschek, 1960). Thus, sizing would appear to be independent of particle shape. However, a more accurate analysis considering electric currents and equipotential lines would show that output pulse amplitudes do depend somewhat upon particle shape. Fortunately, this dependence is usually negligible for cells with smooth surfaces. For example, when compared to a sphere of the same volume, a non-conducting prolate ellipsoid of revolution with an axial ratio of 4 to 1 should give rise to a pulse amplitude only about 3% larger.

In principle, the counter may be used as an absolute instrument for determining volumes of particles with smooth surfaces when equation (6) is satisfied. The major difficulty for such a determination is that of accurately measuring the effective aperture volume V in equation (6). Such a measurement would require a correction for the contribution to the electrical resistance by the electrolyte immediately outside the aperture. Mercer (1966) found, empirically, that the effective aperture volume also depends upon the diameter of the particles flowing through it. A more direct approach would avoid the measurement of aperture volume by using equation (3), or a modified form of this equation. Because of the difficulties of an absolute calibration, it is customary to calibrate Coulter counters by using secondary volume standards. The polystyrene and polyvinyltoluene microspheres, available as monodisperse latexes in several diameters comparable to bacterial dimensions, are suitable (Dow Chemical Co., Midland, Michigan, U.S.A.; Coulter Electronics Inc., Hialeah, Florida, U.S.A.).

It should be noted that the derivation of equation (6) assumed that there are no effects due to surface charges or surface conductivity. In fact, the complexity of pulse shapes generated by bacteria argues for the presence of just such effects, and consequently equation (6) can only be considered an approximation. These and other difficulties will be considered later.

Gregg and Steidley (1965) described an alternative approach to the theory of operation of an instrument designed to count and size mammalian cells, which was tested by Adams *et al.* (1967).

III. SAMPLE PREPARATION AND OPERATIONAL PROCEDURES

A. Counting solutions

Cells may be counted accurately in various electrolytes if background counts due to foreign particles and electronic noise are low. Usually, cells can be counted in the medium in which they are growing, a procedure that is highly advantageous when cell concentrations are small. At low concentrations, counting accuracy requires the removal of foreign particles from the growth medium before inoculation with a culture, and this can be accomplished by filtration through washed bacteriological membrane filters with a pore diameter of about 0·5 μm. These filters must be washed before use because they contain detergents to reduce capillary forces, as well as other substances that can affect cell growth.

At high cell concentrations, accurate counting requires dilution of cell suspensions to reduce the probability of coincident passage of two or more cells through the aperture, since cells in coincidence are counted as a single cell. Saline is frequently used as a diluent. However, dilution into saline could conceivably lead to a change in osmotic pressure sufficient to change cell volume for some microbial cells, thereby distorting cell-size distributions. For cells of *Escherichia coli*, such volume changes occur far more readily with stationary-phase cultures than with exponential phase cultures, which frequently show no discernible effect. Phosphate buffers are sometimes added to prevent volume changes. Alternatively, changes in cell volume may be prevented by the addition of formaldehyde to a final concentration of 0·4–4%. Formaldehyde is also commonly used to prevent further growth when samples are counted in growth medium.

When counting bacterial spores, or micro-organisms with a relatively impermeable mucopolysaccharide coat, or with a rigid cell wall that prevents volume changes, or some bacteria during exponential growth, the use of HCl should be considered. For such cells 0·1 M HCl has several advantages. It is prepared easily, remains sterile, as well as particle-free when stored in plastic containers, and prevents further growth or metabolism of micro-organisms. In addition, the mobility of the H^+ ions, which move more rapidly than other cations, should allow the most rapid electrical response to the passage of a cell. This greater mobility would be expected to reduce inhomogeneity of ion concentration, thereby decreasing the amplitudes of electrical pulses arising from fluctuations of ion concentrations in solution. As a result, accuracy of sizing should be improved, and smaller cells or particles should be detectable.

If bacteria are grown in a medium of lesser conductivity, dilution into saline or HCl will lead to reduced pulse amplitudes in many instruments. This change arises from the decreased resistivity of the counting electrolyte, leading to a decreased voltage difference across the aperture. However, it should be balanced by a corresponding decrease in pulse amplitudes for calibrational microspheres. On the other hand, in those instruments in which particles are detected by changes in electric current, changes in electrolyte conductivity will have little effect upon pulse amplitude when steady-state currents are maintained constant. For example, with microspheres or cells of *E. coli* in the exponential growth phase, when electrolyte conductivity was changed by a six-fold factor the corresponding change in pulse amplitude of one instrument was less than 10%, and smaller conductivity changes, such as that between saline and 0·1 N HCl, gave no detectable change in pulse amplitude (unpublished observations).

B. Small apertures

Counting and sizing spores or small bacteria requires the use of very small apertures, of the order of 10–30 μm in diameter. Although the applied voltage and current values may be relatively modest, say about 10 V and 0·1 mA, the very small dimensions of the aperture itself leads to large values for the voltage gradient and the current density within the aperture, with values greater than 1000 V/cm and 100 A/cm^2. These values are so large that the temperature of the fluid within the aperture may be increased markedly by resistive heating. Maximum permissible aperture currents and voltages are limited by such temperature increases. With increasing voltage there is an increase in electronic noise level as cavitation begins, and at higher voltages boiling occurs and fluid flow is interrupted by a vapour block. Even at operational voltages, the laminar fluid flow through the aperture is so rapid, with an average velocity of the order of 1m/sec, that the exiting stream breaks into a clearly visible turbulent jet (Fig. 1a); and cavitation may occur in this jet near enough to the orifice to contribute to the noise background.

Operational voltages and currents are usually selected to be about as large as possible, consistent with a maximum ratio of signal to electronic noise background. Actual values depend not only upon aperture length and diameter, but also upon the conductivity of the electrolyte. With apertures 12–15 μm in diameter, operational values are about 10–20V and 40–100 μA when the electrolyte is 0·1 N HCl or saline. Corresponding aperture resistances are about $10^5\ \Omega$, roughly ten times greater than the resistances of apertures 100 μm in diameter that are commonly employed in counting erythrocytes.

When operational voltages are exceeded and a vapour block occurs, or when the aperture is blocked by a foreign particle, the interruption of fluid flow is easily eliminated with apertures larger than 25 μm in diameter. Frequently, flow is re-established simply by allowing air to flow through the aperture. Alternatively, foreign particles can usually be removed by rubbing the surface of the aperture disc with a small piece of rubber tubing or with the finger.

As aperture diameters are diminished, it becomes increasingly difficult to maintain uninterrupted flow, and further, these apertures become increasingly difficult to clean. With apertures 10–15 μm in diameter, blocking occurs so frequently that it is necessary to provide some mechanism to reverse the pressure differential across the aperture, permitting transient reversals of fluid flow. Foreign material usually can be expelled from the aperture by a rather small reverse pressure, such as that applied by hand to a rubber bulb. When this procedure fails, the aperture can usually be cleared by passing methanol or acetone through it. If foreign particles are not removed shortly after blocking, they frequently form a more permanent attachment, and are more difficult to dislodge. Much greater pressures may be required, and these must be applied in the proper direction. Even more tightly bound matter may require the application of nitric acid (approximately N) to remove it, or in very extreme cases a very brief exposure to a 1% solution of HF. All else failing, the aperture can sometimes be rescued by heating it and the counting tube to annealing temperatures, 500° to 550°C, in air.

The rather frequent blocking that occurs with the smallest apertures demands continual monitoring of counter operation, and this is best accomplished by observing pulse shapes as they appear upon a pulse-triggered oscilloscope. If the aperture becomes thoroughly blocked, a "noise pattern" of large irregular pulses may be observed or, alternatively, large sinusoidal waves may be induced by laboratory electrical supply lines. More serious is the partial blocking of the aperture, increasing aperture resistance and pulse amplitudes only slightly, with correspondingly small decreases in fluid flow and counting rates. This kind of partial blocking leads to spurious volume distributions as well as to spurious counts. Partial blocking is readily detected as an increase in the pulse widths displayed upon the oscilloscope.

Frequency of blocking can be reduced by a programme of daily maintenance, washing out the aperture tube and refilling with freshly filtered solutions, and cleaning the aperture with acetone and alcohol.

Although counting and sizing micro-organisms is generally more difficult than counting and sizing larger cells, micro-organisms provide the advantage of such small size that they settle out of suspension only very slowly. For this reason, counts of dead bacteria (fixed or in HCl) and of bacterial spores remain constant for periods of an hour or longer without the need for agitation.

IV. CELL COUNTING

Rather different problems are involved in obtaining accurate cell counts at low and at high concentrations. At low concentrations accuracy requires low background counts. With cell volumes greater than 1–2 μm^3, background counts arise predominantly from rare contaminating particles; under these conditions it is possible to measure bacterial concentrations as small as several hundred cells per millilitre. However, as cell volumes are reduced below 1 μm^3, electronic noise levels are relatively larger, and contribute to the background count. Further, the sensitivity of detection must be increased, and previously undetected foreign particles are counted. These elevated background counting rates result in greater errors in counting.

At very high cell concentrations, small concentrations of contaminant particles can be ignored, but accuracy is now limited by coincidence counts that arise when the counting orifice is occupied by more than one cell. Since only a single count is recorded for coincident cells, the observed count underestimates the true count. Coincidence corrections usually are negligible at rates below 10,000 counts/min. They may become rather large, however at higher counting rates, as much as 10% to 30% of the total count at rates from 25,000 to 50,000/min with apertures designed to size accurately. Coincidence losses also increase as aperture volume increases. These losses can be reduced by the use of short apertures customarily supplied by manufacturers. The effect of short apertures upon size distributions is discussed elsewhere.

The proper calculation of the true count from the observed count has been investigated repeatedly (see, for example, Princen and Kwolek, 1965; Mercer, 1966) with results leading to the same first approximation for coincidence correction derived earlier for nuclear particle counters—

$$N_{\text{true}} = N_{\text{obs}} + kN^2_{\text{obs}} \tag{7}$$

where N_{true} is the true count, N_{obs} is the observed count, and k is a constant.

The exact coincidence correction developed by Princen and Kwolek (1965)—

$$N_{\text{obs}} = N_{\text{true}} - cN^2_{\text{true}} \tag{8}$$

where c is a constant, it is recommended for more accurate corrections, which are most precisely determined according to the method described by Princen (1966).

In practice, an empirical coincidence correction curve must be obtained for each new aperture, since aperture volumes differ for each. Such correction factors are obtained by counting a standard, concentrated suspension of cells or particles, and then counting accurately diluted samples of the standard suspension. Alternatively, the correction for coincidence can be avoided if all samples are diluted to a concentration known to involve little or no coincidence correction.

Coincidence frequencies can also be reduced electronically. Harvey and Marr (1966) reduced coincidences by a pulse-shaping technique that decreases pulse widths, thereby reducing the probability that pulses would overlap.

Even when coincidence is negligible, counting errors may be greater than those predicted for sampling cells or particles in suspension. Ideally, individual counts are expected to vary according to a Poisson distribution about the mean count. In a given count of N particles, the expected error is $\sqrt{N}$. When performing a series of short-term counts without interruption of fluid flow or electrical current, errors may approach theoretical values. This is not the case, however, when fluid flow is interrupted between sets of counts. Differences between means of each set can be significant even though great care is taken to mix particles thoroughly. These differences are due, in part, to minor obstructions of the aperture or its entrance, and in part, to the particular distribution, near the aperture, of contaminant particles (or bubbles) too large to enter the aperture. Such large particles often flow toward and away from the aperture repeatedly, diving toward the aperture only to be shunted off to one side by the action of more centrally located and more rapidly moving slipstreams, which cause the particles to rotate rapidly as they travel in loops like eddy-current patterns. These particles may interfere, momentarily, with fluid flow through the aperture.

The counter must be calibrated for absolute particle concentration when counts are taken for a fixed time period under a constant pressure head, rather than for a metered volume. This calibration can be made by measuring the volume flow rate when apertures are large. However, with apertures 10–15 μm in diameter, volume flow rates may be so small, of the order of 10^{-3}ml/sec, that a direct measurement of this kind is difficult. A more common method of calibration is to compare counter values with measurements of bacterial concentrations obtained by plating viable cells. For this kind of calibration it is necessary to grow bacteria under conditions in which essentially all cells are viable. Broth cultures in the exponential growth phase are customarily used.

V. SIZING

A. Conditions necessary for accuracy

Several conditions must be met for accuracy in sizing. First, pulse amplitudes must be at least several times as large as electronic noise levels. If not, signals from cells or particles will be submerged in the noise background with little or no resolution.

Second, extremely large pulses from any source (noise, contaminant particles or very large cells in the distribution) must be relatively infrequent, since pulses of large amplitude momentarily perturb the pulse-height analyser, preventing proper sizing.

Third, accurate sizing requires counting rates low enough to make coincidence rare; coincident passage of cells will give rise to a single pulse having an amplitude proportional to the sum of their volumes. Such pulses severely distort distributions, increasing apparent frequencies at large volumes. The maximum permissible counting rate at which cells may be sized is determined for any aperture in the same manner as that described earlier for establishing coincidence corrections: distributions of successively diluted samples are obtained in order to determine that counting rate for which the shape of the distribution is unaffected by further dilution.

Fourth, aperture lengths, flow rates, and amplifier response must be matched to yield output pulses of the proper amplitude. Since all commercial instruments examined by the author have failed to size accurately because they failed to generate pulses of this kind, this requirement will now be discussed in some detail.

According to equation (6), particles or cells should generate "square waves" as they pass through the aperture (Fig. 1b); i.e., the aperture resistance should increase rapidly as the particle approaches the aperture, remain essentially constant while the particle remains within the aperture, and decrease again as the particle exits. With an amplifier of high fidelity the maximum pulse amplitude should be proportional to cell volume. Amplifiers of the highest fidelity, however, also amplify electronic noise over a very broad band of frequencies, thereby giving an inferior signal-to-noise ratio. For this reason, noise levels are usually reduced by decreasing

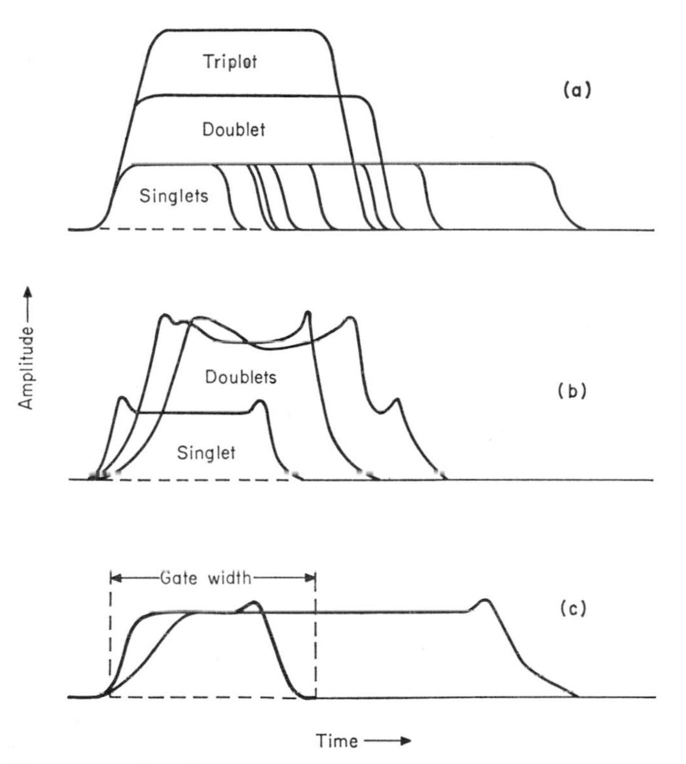

FIG. 2. Observed pulse shapes.

(a). Small microspheres: pulse widths are variable, and depend upon transit times through the aperture. Pulse amplitudes are proportional to the number of particles passing coincidently through the aperture as well as to individual particle volume. (b). Large microspheres: pulses are more complex, with a peak at each end. (c). Pulses with a final peak only: depending upon flow rates and other time constants either the leading or final peak may be unobservable. Two pulses are shown. If the pulse-amplitude detector is designed to respond only within the gate width shown, then the maximum amplitude will be measured only for the shorter pulse.

amplifier bandwidth, setting a "cut-off" frequency beyond which little or no amplification of signal or noise occurs. This attenuation of the high frequency response results in a rounding of the corners of the square-wave pulses, but it need not affect sizing since the proper amplitudes can be reached.

With further decreases in cut-off frequency, pulse shapes become more greatly distorted, and ultimately the proper maximum pulse amplitude cannot be reached before the particle leaves the aperture because the amplifier response has become too slow. This distortion could be tolerable if all particles traversed the aperture in the same time, but they do not. Pulse widths can vary by about a factor of 5 (Fig. 2), corresponding to the same variability in transit times (Kubitschek, 1962). Thus, the amplifier response may be sufficiently rapid to develop the proper maximum pulse amplitude for particles moving slowly near the wall of the aperture and generating long pulses, but those cells travelling down the middle of the aperture may pass too rapidly to permit the proper amplitude to be reached. In extreme cases, particles of the same volume can generate pulse amplitudes differing by a factor of 2 or more. In addition, the dependence of the velocity of a particle upon its trajectory through the aperture leads to difficulties in pulse shaping. Thus reduction of pulse widths by the differentiation–integration method described by Harvey and Marr (1966) may account for the difference between their results and the somewhat increased resolution to be discussed later.

B. Resolution: measurement and improvement

Poor resolution is readily detected with monodisperse latex particles 1–2 μm in diameter. Measured volume distributions of these particles should be unimodal and very nearly symmetrical, as, for example, the singlet peak in Fig. 3. A quick measure of the resolution is given by half the width w of the singlet distribution taken at a height that is half the maximum value of the distribution. With good resolution, values of this half width and half maximum have been about 6% of the distribution mean (Kubitschek, 1958, 1960). Some analysers, especially the older instruments that were designed primarily for counting rather than for sizing, do not provide direct measurements of volume distributions (differential distributions) because they have only a single sizing or "threshold" level. These analysers give an integral distribution of the kind shown in Fig. 3, which must be differentiated graphically or analytically in order to obtain the corresponding volume distribution. Figure 3 also shows the manner in which electronic noise pulses are recorded unless provision is made to discriminate against these pulses.

When much larger, and therefore inferior, values of resolution are observed, good resolution can usually be restored either by using longer apertures (Kubitschek, 1962) or by decreasing fluid flow rates. Often it is necessary to reduce flow rates two- to three-fold below that provided by the manufacturer. (Some apertures are so short that lines of force are curved within the aperture, so proper resolution can never be obtained at any flow

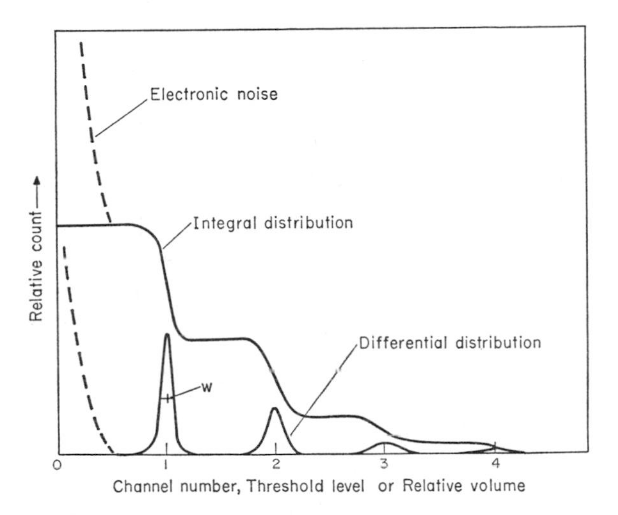

FIG. 3. Characteristic integral and differential distributions for aggregated microspheres. Electronic noise increases sharply at low pulse amplitudes, and therefore it is usually eliminated by appropriate electronic design. The peaks represent the distributions for single particles and aggregates of two, three and four particles. The width of the singlet distribution, w, is measured at half-maximum amplitude.

rate; such apertures should be replaced.) If flow rate is controlled by a fixed mercury manometer, the best way to reduce the flow rate is to sever the manometer at its lowest point, and to connect the two pieces with plastic tubing, well tied. With this flexible connection the horizontal U tube can be raised or lowered with respect to the manometer reservoir to obtain arbitrary pressures. First, flow rate should be greatly decreased so that the best value of the resolution can be obtained and measured. Then, flow rates can be increased again up to the maximum rate at which this resolution is still maintained.

Another method of increasing resolution is that of hydrodynamic focusing (L. Spielman and S. U. Goren, private communication). With this method all particles entering the counter aperture are confined to very similar trajectories by injecting them into the same stream-line. This was accomplished by feeding the suspension through a second aperture only a few millimeters away. The method may lead to substantial improvements in resolution.

C. Pulse shapes with latex microspheres

Polystyrene and polyvinyltoluene microspheres usually yield square wave pulses, or nearly so (Fig. 2a). With very large microspheres, however, pulses may appear to be convex toward the time axis, as in Fig. 2(b); that is, pulses may appear to have initial and terminal peaks. Under some conditions, and depending upon amplifier response, either the initial peak (Fig. 2c) or the terminal peak may be absent. Pulse shapes of these kinds indicate that factors other than particle volume affect pulse amplitude. The question then arises: what part of the pulse amplitude should be measured as the best approximation to particle volume? The answer, or course, depends upon the causes leading to these more intricate pulse shapes.

A possible explanation for the pulse shapes of Fig. 2(b) is that surface conductance occurs on these microspheres following breakdown of the surface dielectric as these particles enter the aperture, where voltage gradients may approach 10,000 V/cm. This conductance would increase the current through the aperture, thereby decreasing pulse amplitude. In this case, the maximum amplitude of the entire pulse should be measured, rather than curtailing measurements at some lower amplitude by using an electronic "gate". Figure 2(c) shows how "gating" to measure only the leading portion of the pulse can lead to apparently different sizes for particles of constant volume; of the two pulses shown, only the shorter one reaches its maximum amplitude within the "gate", or period of time allowed for sizing.

D. Scale calibration with microspheres

Although pulse shapes are complex, there is good evidence that the smaller microspheres, at least, have pulse amplitudes proportional to volume. When such microspheres are allowed to form small aggregates by suspending them in 0·1 N HCl for about a day, they provide a supply of "particles" with volumes in accurate, integer relationship, since the aggregates consist of multiples of individual microspheres. Measurements of the pulse-amplitude distribution generated by these aggregates gives a series of peaks corresponding to single microspheres, doublets, triplets and larger aggregates, and the mean volume of each aggregate is in excellent agreement with the value expected from the measurement of the (singlet) mean microsphere volume (Kubitschek, 1960). Such aggregates may be used to provide an effective calibration of the volume scale (Fig. 3).

As pointed out earlier, the maximum resolution (obtained with microspheres 1–2 μm in diameter) expressed in terms of the half width at half maximum compared to the mean particle diameter, is about 0·06. In terms of the standard deviation, σ, and the mean volume, V, this value corresponds to a value of 0·05 for σ/V, the coefficient of variation. For particle diameters the corresponding values are reduced by a factor of 3. These values for the dispersion obtained by electronic sizing of the monodisperse latex particles are much larger than those listed by the manufacturer, which were determined from electron micrographs. At least part of the increased dispersion with the counter is due to electronic noise.

For a more accurate calibration, several different latexes must be used since, as Princen (1966) has shown, latex *diameters* given by the manufacturer are only accurate to about 2% (and volumes to about 6%), despite the fact that the labelled values are given to 3 or 4 significant digits.

E. Sizing micro-organisms

Pulses generated by bacteria have more complex shapes than those from microspheres. This complexity is best observed for large bacteria (such as those from broth cultures) with reduced flow rates through the aperture. As with microspheres, twin peaks are again observed, but both of these peaks are now located well in toward the mid-region of the pulse and appear to rest upon a plateau established earlier. A possible interpretation for this kind of pulse shape is—

1. The entrance of the cell into the aperture first generates the primary resistive component as expected according to equation (3), and leads to the initial plateau of the pulse amplitude.
2. The subsequent rise to the first peak is due to capacitive charging of the cell.
3. Fall-off from this first peak occurs when voltage gradients across the cell reach values great enough to cause dielectric breakdown.

These steps would be reversed as the cell leaves the aperture, giving rise to the second peak. With smaller cells these secondary peaks are greatly diminished and usually unobservable.

The complexity of the pulses generated by bacteria indicates a clear departure from the simple theory for pulse amplitudes presented earlier. We cannot expect peak amplitudes to be precisely proportional to cell volume for such cells, nor can we expect the shapes of cell volume distributions to be more than approximate. There is, however, empirical evidence for a fair degree of proportionality from measurements of optical density of bacterial cells of different mean volumes. Such cell samples were produced by adding formaldehyde to cultures in the exponential growth phase, and separating cells by size upon a sucrose gradient (Mitchison and Vincent, 1965). For each sample, cell counts, apparent volume distributions and optical density were determined. It was found that the mean optical density per cell is proportional to the measured cell volume, as shown in Fig. 4 (Kubitschek, 1967). Thus, to the extent that optical density is a measure of cell mass, cell-volume distributions represent mass distributions.

Measurements of the kind shown in Fig. 4 only permit an estimate of the resolution obtainable for cell-volume distributions because of the dispersion in cell volumes. For the distributions in Fig. 4 the half width is about 15%

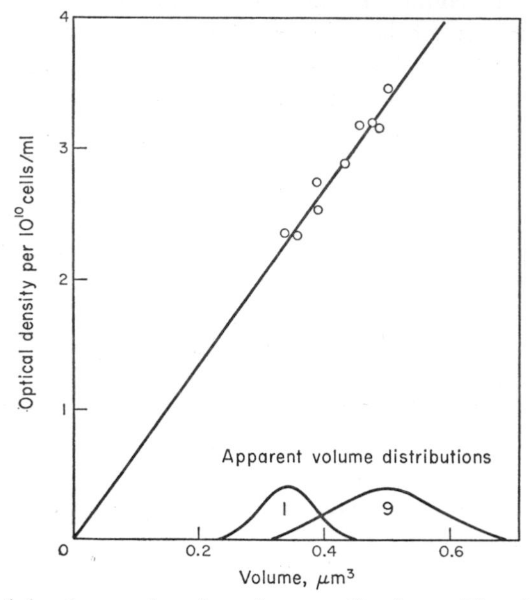

FIG. 4. Optical density as a function of mean cell volume. The observed distributions of relative volume are shown for the first and last point.

of the mean value; spherical cells would have a half width of about 5% for the distribution of their diameters. It also should be emphasized that these are only *relative* distributions of cell volume, and that comparison with standard microsphere volumes may provide an imprecise measurement of the absolute volume of any cell.

A final note of caution is needed when cell-size distributions are examined. In growing populations, some of the larger cells may have cross septa, or have divided more completely but failed to undergo mechanical separation. Such "double" cells will be mistakenly sized as single large cells unless they can be separated, as some of them can, by vigorous agitation. In order to estimate the frequency of such "doubles", microscopic examination is necessary. Similar problems are encountered for cells that aggregate easily and remain tightly bound. To reduce these problems, it is common practice to agitate culture samples briefly.

VI. RANGE OF SIZING

The range of particle sizes determined accurately with a given aperture depends upon both aperture diameter and length. The upper limit of particle size, imposed by the requirement for linearity of instrument response as given by equation (6), is satisfied when particle cross-sectional areas are no more than a percent or two of the cross-sectional area of the aperture. The lower limit of sizing, by any criterion, depends upon electrical noise generated within the aperture, and therefore, upon aperture resistance. Sizing limits are more restrictive than those for counting, since a broader range of particles can be counted than can be measured accurately.

Particles as small as 0·2 μm in diameter have been detected (Kubitschek, 1960), but noise backgrounds were too large to permit their measurement. Apertures 10–14 μm in diameter, permit sizing of the smallest bacteria, such as stationary phase cultures of *E. coli* or *Salmonella typhimurium*, and bacterial spores, such as *Bacillus megaterium* (Kubitschek, 1958). All have volumes of approximately 0·5 μm^3 (effective spherical diameter of about 0·7 μm). The largest animal viruses might be counted with a well designed instrument, but an improvement in sensitivity of several hundred-fold would be needed to count bacteriophage.

The range of size determination with any instrument can be greatly extended by the addition of a second transducer system employing an aperture of different diameter. In the author's laboratory, either of two apertures is used with the same amplifier and multichannel analyser system, and samples of bacteria or red blood cells can be counted and sized, quickly and easily, in any order. The aperture diameters, 13 and 60 μm, permit counting and sizing of cells with diameters from 0·6 to 6 μm (volumes of 0·1 to 100 μm^3).

VII. USE WITH SYNCHRONOUS CELL CULTURES

The Coulter counter provides a unique advantage for studying cell properties as a function of cell size, and for studying synchronized cell cultures, where both counts and size distributions are required simultaneously for maximum information. This advantage is clearly demonstrated by the results shown in Fig. 5 for a culture of *E. coli* B/r, synchronized by the Mitchison–Vincent technique (1965), and growing in a medium of acetate and salts. Cell concentrations increased through two successive doublings (Fig. 5a) and the relative frequencies of divided and undivided cells (for example, at 120 min, Fig. 5b) were in good agreement with that calculated from cell numbers. From data of this kind, cell-division rates and mean volume growth rates can be determined independently during the cycle, as well as the degree of synchrony.

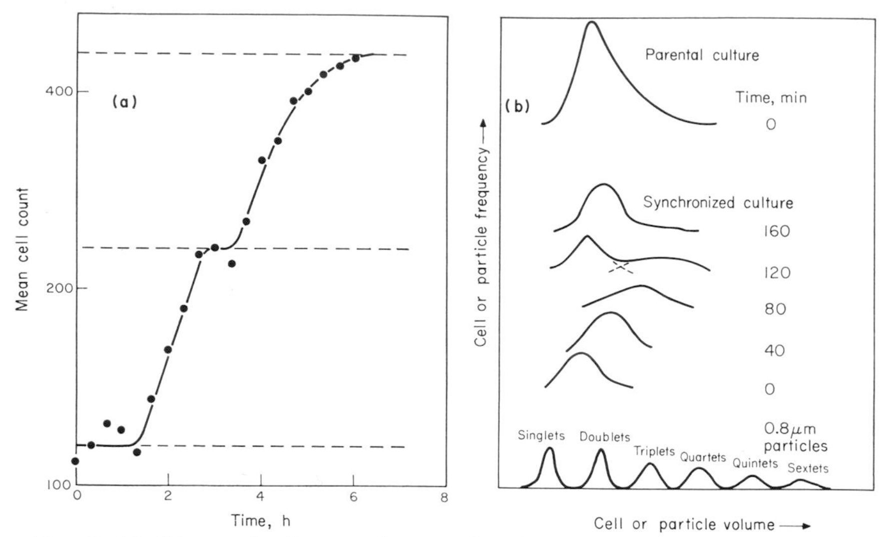

FIG. 5. (a). Observed cell counts in a synchronized culture of *E. coli* B/r. For cell concentrations multiply ordinate by 1000. (b). Relative volume distributions: the synchronized culture is the same as that in (a). The distribution of cells in the parental culture is also shown. Polystyrene microspheres with a nominal diameter of 0·8 μm were used to calibrate the volume scale.

In addition, the size distribution of the parental culture should be determined before synchronization, since the parental distribution can be used to establish a scale for the location of cells within the interdivision cycle as a function of cell volume (Kubitschek *et al.*, 1967). Further, with the establishment of a scale of this kind for steady-state cultures, measurements of the properties of cells of different sizes can be related to their interdivision age without the need for establishing synchrony.

REFERENCES

Adams, R. B., Voelker, W. H., and Gregg, E. C. (1967). *Physics Med. Biol.*, **12**, 79–92.

Gregg, E. C., and Steidley, D. K. (1965). *Biophys. J.*, **5**, 393–405.

Harvey, J. J., and Marr, A. G. (1966). *J. Bact.*, **92**, 805–811.

Kubitschek, H. E. (1958). *Nature, Lond.*, **182**, 234–235.

Kubitschek, H. E. (1960). *Research, Lond.*, **13**, 128–135.

Kubitschek, H. E. (1962). *Rev. scient. Instrum.*, **33**, 576–577.

Kubitschek, H. E. (1967). Argonne National Laboratory Annual Report.

Kubitschek, H. E., Bendigkeit, H. E., and Loken, M. R. (1967). *Proc. natn Acad. Sci., U.S.A.*, **57**, 1611–1617.

Mercer, W. B. (1966). *Rev. scient. Instrum.*, **37**, 1515–1520.

Mitchison, J. M., and Vincent, W. S. (1965). *Nature, Lond.*, **205**, 987–989.

Princen, L. H. (1966). *Rev. scient. Instrum.*, **37**, 1416–1418.

Princen, L. H., and Kwolek, W. F. (1965). *Rev. scient. Instrum.*, **36**, 646–653.

CHAPTER XVIII

Viable Counts and Viability

J. R. POSTGATE

University of Sussex, Falmer, Sussex, England

I. INTRODUCTION

The viable count of a microbial population is the absolute concentration of viable organisms present; the viability is the ratio of that number to the total concentration of microbes, dead or alive. An organism is viable if it is capable of multiplying to form two or more progeny in conditions that are "optimal" for the species and strain of microbe concerned; the question of what is an optimal environment is often a matter of informed opinion rather than of scientific fact. This Chapter will not discuss details of the composition of medium, atmosphere, temperature, pH and so on, that may be optimal for diverse types of microbes, since these are covered elsewhere. Here we shall discuss the assessment of viable numbers and viabilities in general terms, and describe procedures of fairly general validity. It is important, however, to emphasize two points of principle. Firstly, dying or damaged microbial populations may be more exacting with respect to cultural conditions than healthy, unharmed ones; a clear example occurs in the phenomenon of "metabolic injury" among frozen bacteria, in which populations of coliform bacteria obtained from the frozen state may show considerably greater viabilities when tested on rich media than on minimal media. Secondly, media that are selective are rarely optimal for the types selected: well known examples are the early media for *Desulfovibrio*, which were wholly inadequate for enumeration because of their oxidizing E_h value, and MacConkey's medium, which gives a "presumptive" coliform count which is often fairly remote from the true viable count (Childs and Allen, 1953).

II. DETERMINATION OF VIABLE NUMBERS

A. General

1. *Principle*

The microbial population is diluted in a non-toxic diluent and an aliquot of the diluted population is dispersed about a suitable solid medium, such that each viable unit forms, after incubation, a colony. The number of viable individuals or clusters originally present is deduced from the colony count and the dilution. If the population contains both spores and vegetative forms, both a spore count and a cell count will be necessary, since germination of an appreciable proportion of the spores may well not take place unless a heat stress has been applied to the population.

2. *Preparation of samples*

For counting liquid samples, no special precautions beyond thorough mixing are necessary before dilution. Excessive aeration of samples containing anaerobes should be avoided. For non-liquid samples, viable counts are expressed as numbers/g material, and the dispersion of the initial sample such that a representative indication of its microbial population can be obtained presents a problem that must be solved for each individual material. Rocks may be crushed; sewage is homogenized; meat, foods and other biological material may also be homogenized: Pochon (1956) has described a routine procedure for use with soils. Airborne microbes require special techniques for collection.

3. *Dilution procedure*

The best diluent is usually the growth medium, but growth should not take place in it while the dilution is being made; cold shock or dilution shock of susceptible populations must be avoided. Many instruction manuals recommend the operator to prepare serial 1 ml→10 ml dilutions, using test tubes and pipettes, to lower the microbial population to the region of 10^2 to 10^3 viable cells/ml diluent. When the initial population is greater than 10^9 ml, as in a laboratory culture of bacteria, such repeated dilution can introduce appreciable manipulative errors. For this reason, multiple decimal dilutions are not to be recommended, and the micro-pipette method described below is to be preferred. Nevertheless, such procedures are still widely used, so the following description is given, representing a technique that keeps inherent errors to a minimum.

Sterilize a bulk of diluent and distribute 9-ml lots aseptically in $6 \times \frac{5}{8}$ in. test tubes (in no circumstances should measured volumes be prepared non-aseptically and then autoclaved, since evaporation during autoclaving cannot be avoided; this point applies to all dilution techniques described in this Chapter). Take a clean pipette (modern "blow out" pipettes are as reliable as non-blow out pipettes, though this was not true two decades ago), add exactly 1 ml of culture to 9 ml of diluent and discard the pipette. Mix the dilution well, preferably by pressing the tube against a mechanical mixer to sustain a vortex for 20 sec and, taking a fresh pipette, prepare the next dilution similarly. If a mechanical mixer is not available, the traditional bacteriologists' mixing procedure is as follows: with the pipette required for the next dilution, remove 1 ml from the bottom of the first dilution and inject it back at the top; repeat this operation ten times (five times if the decimal dilutions take the form of 0·5 ml culture added to 4·5 ml of diluent) and only then add 1 ml of the mixed dilution to the next portion of diluent. With fragile organisms, such as some of the pneumococci, excessive blowing through the pipette while mixing must be avoided because the organisms, while passing through the pipette orifice, may be subject to sufficient shear to kill a proportion of the population.

Errors in the accuracy of commercial pipettes tend to increase after repeated sterilization, but since the volume changes can be in either direction—either too small or too large—the final error over a series of dilutions is not cumulative. On the other hand, cleanliness is most important: grease or fatty matter on the interior of a 1-ml pipette can lower the volume it discharges up to 15%; surface-active material can increase the volume delivered. Another source of error arises because many bacteria tend to adhere to glass surfaces; the serial decimal dilution procedure repeatedly exposes increasingly dilute suspensions to conditions in which the ratio of glass surface area to liquid volume is fairly large, thus maximizing errors due to adhesion. Coating the interiors of pipettes with baked-on silicone certainly reduces losses due to adhesion, but since commercial pipettes are calibrated for wettable interior surfaces, the volumes discharged by siliconed ones are greater than those marked and re-calibration is necessary.

For these reasons it is desirable to keep the number of diluting operations in a viable count to a minimum and a 2-stage process described below, making use of disposable capillary pipettes and large one-stage dilutions, reduces many of the inherent errors to reasonable proportions.

4. *Choice of dilution*

Aliquots of diluted suspension are spread over replicate dried agar plates (for aerobes) or mixed with molten agar and set (for anaerobes). After incubation the colonies are counted and, since the standard deviation of the count is the square root of the number of colonies counted, an ideal count should lie in the range of 200 to 300 colonies. The gain in percentage accuracy that results from counting more is small unless a very large number indeed is counted. The actual dilution plated should therefore contain a population of around 200 viable cells in a volume of diluent that will be absorbed rapidly on the agar surface being used, or will mix into the agar shake cultures without significantly altering the setting properties of the agar. For conventional Petri dishes or shake tubes it is unwise to exceed 0·2 ml per surface or tube, so, if quintuplicate counts are being made, the dilution should contain 200–300/ml; each plate or shake tube will then have 40 to 60 colonies to be counted.

Various procedures for viable counting are described in detail below.

B. An accurate procedure for colony counts

This description applies to a population of unicellular microbes of about 10^9/ml; for other population densities the dilutions should be altered appropriately.

1. *First dilution* ($1/10^4$)

Sterilize more than 300 ml of diluent. With a clean, sterile measuring cylinder, dispense two lots of 100 ml diluent aseptically into two 250-ml conical flasks. (Note that the error in using measuring cylinders with volumes of about 100 ml is well within the Poissonian error of the average count; as mentioned before, it is not permissible to measure out the volumes of diluent first and then autoclave them.) Take a commercial 10 μl glass capillary and stand it in alcohol for a few minutes supported by its holder. Drummond "microcap" disposable micropipettes (Drummond Scientific Co., Broomall, Pennsylvania, U.S.A.) are suitable. After exposure to alcohol, the pipette is sufficiently aseptic for most work (since the contaminants it may introduce are vastly outnumbered by the organisms being counted), but if sterility is mandatory pipettes may be dry-sterilized in Petri dishes at 160°C for 15 min, before use. Alcohol is removed by touching the capillary to a piece of dry-sterilized filter paper in a Petri dish, as when "spotting" a chromatogram, and in a similar way the pipette is rinsed three times in the culture being counted. (Note that the filter paper is now contaminated and must be sterilized when discarded.) Then take 10 μl of culture in the capillary and rinse them three times into 100 ml diluent in one of the conical flasks at room temperature; use the contents of the flask for rinsing. Less than $1/10^4$ of the microbes remain in the capillary, which can be returned to alcohol for re-use if comparative dilution counts are being made.

2. *Second dilution* ($1/10^2$)

Remove 1 ml of diluent from 100 ml in a conical flask, discard it and replace it by 1 ml of the first dilution. Mix by swirling.

3. *Plating* (*for aerobes*)

Take 5 agar plates, previously dried in air for 15 min at 55°C or 45 min at 37°C; add 0·2 ml portions of the second dilution to each and spread evenly over each surface in turn with a glass spreader: a smooth L shaped glass rod (some bacteriologists favour wire spreaders). Provided the glass spreader is scrupulously clean, less than 1% of ordinary microbes adhere to the spreader in practice; nevertheless, some prefer to spread by spinning the plates and thus avoid errors due to adhesion of cells to the spreader. Incubate in air and count colonies; calculate the original count from the colony count and the dilution.

4. "*Shake*" *culture* (*for anaerobes*)

To five narrow tubes ($6 \times \frac{3}{8}$ in), commercially known as "anaerobic" or "vanilla" tubes, add 0·2 ml of the second dilution, then about 7 ml of molten agar medium that has been held at below 40°C. Mix while still molten by inverting the tubes once (less than 2% of the population will be lost by adhesion to the cotton-wool plug; tidy workers prefer to substitute a sterile bung for the plug before this stage). When set, add about 1·5 cm of sterile molten agar to exclude air. Incubate in air and count colonies in the seeded portion of agar.

This procedure is not suitable for strongly aerogenic anaerobes, because gas formation disrupts the agar; such organisms can often be counted by a modification of the roll-tube procedure (below). Anaerobes that are not exacting may often be counted on ordinary Petri dishes, as described for aerobes, if these are incubated in an atmosphere of nitrogen, hydrogen or argon in an anaerobic jar or desiccator.

5. *Roll-tube methods*

Perform the dilutions as before, but mix the sample with molten agar in a test tube or "bijou" bottle and roll it mechanically while setting so that the agar forms a film. Aerobes may then be incubated in the air as usual; anaerobic counts usually require pyrogallol plugs to be inserted in the tube, or some other device for ensuring an anaerobic atmosphere. (See also Hungate, this Series, Volume 3B).

6. *Pour-plate methods*

Mix the dilution with about 15 ml of molten medium, held at a temperature below that which will kill the organisms by thermal shock, and pour the mixture into a sterile Petri dish. With agar media, a temperature of about 40°C is necessary; with gelatin media lower temperatures may be used. When the medium and dilution have set, incubate the dish; colonies form throughout the agar. The method is unsatisfactory for very exacting anaerobes and is subject to losses in the residue left after pouring; these may be avoided by adding the dilution to Petri dishes before pouring in the agar. Some bacteriologists find the manipulations preferable to those required for conventional surface plating.

C. The drop-plate method

This method, developed by Miles and Misra (1938), is very suitable where large numbers of routine counts on aerobes or non-exacting anaerobes, must be undertaken. It is also useful when the density of the population is not known accurately enough to apply the previous method. It depends on diluting with the use of standard drops and is, therefore, intrinsically less accurate than the method just described, because differences in surface tensions of various diluting fluids influence drop size (milk, for example, gives a 30% larger drop than tap water).

1. *Preparation of pipettes*

With a wire gauge and a glass knife, break several Pasteur pipettes to give an external diameter of 0·036 in. These should deliver 0·02 ml of physiological saline if held vertically and used slowly; check that they do by adding drops to a torsion-balance pan and accepting 20 ± 1 mg as 0·02 ml. Sterilize them in the autoclave.

2. *Dilutions*

Add 9 drops of diluent to each of a sequence of six tubes. Dilute the sample sequentially, using a fresh pipette at each stage, by adding 1 drop to 9; plate out each dilution (below) as the next one is being prepared.

3. *Plating*

Add drops of each dilution to a plate of agar medium, dried as described earlier, so that each drop is well separated radially. Allow 6 drops/plate, let them soak in and incubate. Count colonies in those giving populations of 20–100 viable organisms per drop, and deduce the original count from these figures and the dilution.

D. Most probable number (MPN)

For many routine purposes the viable count need only be known within a factor of about 2. In such cases a method depending on dilution to extinction is adequate (Taylor, 1962). A sequence of 1 ml→10 ml dilutions is made with ordinary pipettes, using a fresh pipette at each stage, and five subcultures of 1 ml of each dilution, into a convenient volume of liquid growth medium, are made. At the levels where the concentration of cells ranges from an average of 10/ml to 0·1/ml, there is a statistical probability that progressively fewer of the subcultures will grow. From the distribution of tubes showing growth at three such levels of dilution, the most probable number of bacteria in the preceding dilution can be obtained from tables (see p. 624). The most probable number in the original sample may be deduced from these. This method has the advantage that manipulation may be undertaken quickly and the method applies equally to aerobes and anaerobes. Its convenience should not be allowed to disguise its relative inaccuracy.

E. Membrane counts

Samples containing less than about 100 microbes/ml, such as urine or clear natural water, often require concentration rather than dilution before counting. This is simply done by filtering a known volume of sample through a sterile membrane of pore size suitable to retain all the required microbes, then transferring the membrane to a nutrient pad and counting the colonies that grow after incubation. Emulsions such as milk must be broken down with a non-toxic detergent before use; viruses and very small organisms such as *Bdellovibrio* will be missed if routine bacteriological membranes are used. For a discussion of membrane filtration techniques see Mulvany (this Volume, page 205).

1. *Preparation of apparatus*

Commercial apparatus for membrane filtration is sterilized by autoclaving; membranes (pore size 0·4–1 μm) are placed between filter papers and autoclaved; the apparatus is assembled aseptically. It is often possible to sterilize the apparatus in the assembled state, but there is an appreciable risk of the membrane cracking. Absorbent pads, cut from cotton wool or washed lint, are autoclaved or sterilized dry in Petri dishes.

2. *Filtration of sample and transfer of membrane*

The filtration apparatus is set up aseptically, a suitable volume of liquid to be examined is filtered with suction and droplets of sample are carefully rinsed from the funnel on to the membrane using sterile culture medium. The apparatus is demounted and the membrane is transferred aseptically to an absorbent pad in a Petri dish to which sufficient culture medium has been added to dampen the pad, but not to flood it. The dish is incubated and the colonies counted. Very translucent colonies may not show up well; if the membrane is transferred to an absorbent pad covered with 0·01% aqueous methylene blue solution, the colonies take up more dye than the membrane and stand out more clearly. With some colonies, the membrane is best exposed to methylene blue for about one minute and then transferred to a pad damped with water.

3. *Use with anaerobes*

With unexacting anaerobes and microaerophiles the filtration is performed as described, but the membrane is laid face downwards on a thin (about 3 mm) layer of nutrient medium set with agar in a Petri dish. A fresh layer of nutrient agar, at 40°C, is poured gently over the membrane, so as not to disturb it, to a depth of 1 cm; addition of a reducing agent (such as sodium ascorbate or thioglycollate) to the agar is often recommended. The plate is incubated and the colony count made, using a colony microscope, while viewing from the underside of the dish. For reasons that are not wholly clear, this procedure is not very successful with very exacting anaerobes, such as certain of the rumen bacteria, the methane bacteria and the sulphate-reducing bacteria.

F. Viable counting of microbes of non-uniform morphology

Moulds, filamentous bacteria, algae and microbes that form zoogleal clusters form populations whose viability may differ from one part of the cluster to another. It is self-evident that no general methods for assessing viable numbers or viabilities in such systems can be prescribed. In practice, cluster or group counts, using appropriate modifications of the methods described, are used to obtain a numerical impression of such populations.

G. Viable counting of viruses

The principle of dilution to extinction, or to a manageable number of viable units per unit volume, applies equally to virus populations, and the problem with these organisms is generally one of choosing a test environment on which viability will be reliably expressed. Bacterial viruses are generally counted as plaque-forming units on plates seeded with sensitive bacteria; viruses of higher organisms are counted as trauma-producing units on whatever host material, plant or animal, will produce an obvious response. Details of appropriate systems for bacteriophages are given in this Series by Billing (Volume 3B) and by Kay (Volume 7).

H. Spore counts

In principle, spores are counted in the same manner as the vegetative microbes described earlier, but the population is first subjected to a heat stress that the spores will survive, but the vegetative organisms will not. It is important that the whole population should reach the appropriate temperature to kill off vegetative organisms; one should remember that, in large liquid volumes contained in thick-walled tubes, heat is transferred only slowly. A typical procedure is as follows.

Maintain a stirred water bath, of at least 2 litres capacity, at 80°C. Homogenize, extract, wash or take other acceptable action to obtain the spore population in homogenous dispersion. Take 1 ml aliquots in thin-walled tubes of a capacity between 20 and 30 ml, heat them in the water bath for 10 min, cool, homogenize again if precipitation has occurred, add 9 ml diluent and treat this as the first decimal dilution of a series of the kind described above.

When possible, spore counts should be accompanied by a spore germination or viability assessment using slide culture (below) since, though heat stress generally provokes germination of microbial spores, the stress of 80°C will not necessarily provoke germination of 100% of the potentially viable units of all species. With highly resistant organisms, a higher stress temperature should be used; correspondingly, with resting forms of only marginal heat resistance, such as the microcysts of Azotobacteriaceae, lower temperatures suitable to the case in question should be adopted.

III. DETERMINATION OF VIABILITY

A. Indirect assessment of viability

Many procedures have been used or proposed as indirect or rapid methods of estimating viability. These include staining with so-called "vital" stains, assessing dehydrogenase activity, measurements of amounts of dye taken up, immersion refractometry, observing changes in the "optical effect" (an optical response to mild osmotic stress), observing leakage of purines, staining selectively for the "fluctuating RNA" whose synthesis precedes cell division. Postgate (1967) discussed these techniques critically and reached the conclusion that they are not of general validity unless the stress leading to loss of viability is one that damages the osmotic integrity of the organisms, so causing refractive-index changes, loss of pool materials and loss of enzymic activity. Such changes do not universally accompany microbial death, so procedures making use of them will not be described further here.

B. Direct assessment of viability

Reliable estimates of viability can be obtained only retrospectively: by observing what proportion of the population multiplied when a sample was incubated in suitable conditions. Since most microbial populations show a lag phase, during which further death of individuals may occur, an unavoidable element of uncertainty appears in the retrospective procedure. Nevertheless, such procedures are generally the methods of choice. They can be divided according to whether the total and viable populations are measured independently or whether their ratio is determined in one operation.

1. *Separate determination of total and viable populations*

Total counts are discussed elsewhere (see Mallette, this Volume, p. 521), and it is, therefore, only necessary here to recapitulate those aspects of the techniques that are particularly relevant to viability measurements.

Procedures making use of shallow glass chambers of standard depth (such as the 20 μm bacterial counting chamber) are unsuitable because they are subject to a systematic error on which the count may be 20–50% high. Such errors are particularly important when the viability is high (>10%); they can be corrected by interferometric checking of the chamber depth each time it is set up, but the apparatus for such measurements is rarely accessible. The systematic error does not operate significantly when the 100 μm chamber is used for counting large microbes such as yeasts or algae.

Methods based on counting a sample area of a standard volume of microbial suspension dried on to a slide suffer from the errors inherent in delivering a small volume accurately on to a glass surface and the tendency of many microbes to cluster on drying. The most reliable apparatus are electronic counters, such as the Coulter counter, which, when behaving satisfactorily, give much more accurate counts. But it is important to realize that, with single bacteria, the Coulter counter is working near the limit of its sensitivity. An orifice of 30 μm diameter or less is essential and, as emphasized elsewhere, a plateau in the plot of threshold setting against count is an essential preliminary to using the instrument, since with such small objects the sensitive range is often close to the noise level of the instrument. Blank counts on the suspending fluid are mandatory. With chain- and cluster-forming microbes (e.g., streptococci or micrococci) the instrument measures microbial units; budding yeasts also score as single microbial units. This property conforms with most laboratory conventions.

Total counts of microbial populations too sparse for chamber counting may be obtained after membrane filtration: the membrane is stained, cleared and standard areas are counted microscopically. Alternatively, a large volume of sample is evaporated so as to yield a microdrop of glycerol, the whole microbial content of which is counted (Collins and Kipling, 1957).

The viability of the population is deduced from the total count, obtained by one of the methods just discussed, and the viable count, obtained by one of the procedures outlined in the first section. It hardly needs saying that the methods chosen should approximately match in accuracy; there is little point in doing an elaborately careful plate count if an uncalibrated chamber is used for the total count; nor should one expect an electronic count to match an MPN determination.

2. *Determination of viability in one step*

Microcultural procedures are readily adapted to viability determination. They give the ratio of viable to dead organisms at once, though they do not give the absolute count of either. Several techniques are discussed by Quesnel (this Volume, p. 365); slide culture, as described below, has been used successfully as a routine procedure over several years with various aerobic organisms; its accuracy and limitations were discussed by Postgate *et al.* (1961).

The best available agar medium for the organism being studied is prepared and membrane-filtered, while hot, to remove the debris (which often includes dead bacteria) frequently to be found in commercial agars. An annular stainless-steel or plastic ring, 22 mm external diameter × 1·2 mm deep, is laid on a clean microscopic slide. This annulus should be capable of supporting a circular coverglass, nominally $\frac{7}{8}$ in, with overlap all round. Slides, annuli and coverglasses are stored in alcohol and burned or wiped dry before use. Molten agar (0·22–0·24 ml) is pipetted into the annulus, sealing it to the slide and, when set, a loop of the microbial population is passed over the agar surface. When the sample has dried in, the agar is covered by a coverglass and sealed by allowing a drop of water to penetrate the interface between the cover glass and annulus (Fig. 1). By this means a microculture of the organisms, plus an air space, is sealed in a well, formed by the annulus on the slide. The population should be diluted (or concentrated by centrifugation) to between 5×10^7 and 10^8 organisms/ml, thus providing 30–50 organisms/field under a 400-fold microscopic magnification. The slide culture is incubated for 3–4 mean doubling times of the population (this period must be determined for each species by inspecting some trial slides; with coliform bacteria, 2–3 h is usually adequate); it is cooled briefly and inspected under phase contrast with the cover removed (cooling prevents fogging of the microscopic objective by vapour from the agar). Microcolonies are counted as representing one viable unit, single organisms as one dead unit. Sufficient whole fields are counted to reach a total exceeding 300 objects, co-incidences are disregarded, and the viability is obtained from the ratio of viable to dead objects.

The technique described above differs in two details from that described by Postgate *et al.* (1961). The brass annuli described originally have come under suspicion because their copper content may influence the growth of some organisms, and the air space is deeper, a detail found important for microbes with intense respiratory activities such as *Azotobacter*.

This procedure is generally successful with coliform and coryneform bacteria, yeasts, bacilli and other simple aerobes; it measures viability with a standard deviation of 3%. With micrococci, streptococci and chain-forming bacteria it suffers from the usual difficulty of deciding what is an initial

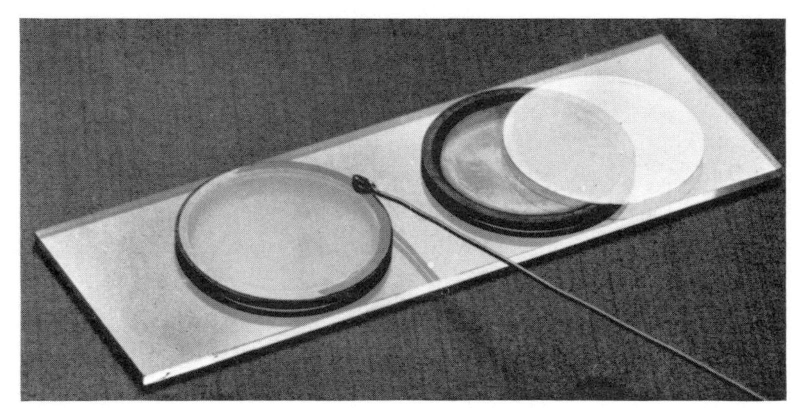

FIG. 1. Slide cultures in preparation. The right hand annulus contains agar, but has not been covered. The left hand annulus has been covered with a coverslip and is in process of being sealed with water from a loop. (From Postgate *et al.*, 1961; Crown copyright.)

viable unit. Filament-forming microbes may obscure the picture if incubated too long, but this problem is less of a nuisance than originally expected, and even such filamentous organisms as *Bacillus megaterium* or *Escherichia coli* strain B can usually be assessed readily. Lysis, too, is less of a problem than might be expected because a recognizable "ghost" is almost always detectable; fragmentation of dead cells can, however, occur with certain large bacteria (e.g., *Spirillum* spp.) and cause difficulties. Table I lists examples of conditions used by the author with various microbes. It must be remembered that exact incubation times must be determined by experiment, usually by setting up two or three slide cultures and judging a suitable period for one from the behaviour of the others. The range is rarely wide: young, log-phase *E. coli* requires 1–1·5 h incubation at 37°C to give clear viabilities: a 2-month-old suspension stored in the refrigerator needs about 4 h. Some stresses increase lags among survivors (e.g., "substrate-accelerated death") and incubation times should be prolonged appropriately.

TABLE I

Conditions for slide culture of various aerobic micro-organisms

Organism	Medium	Temperature, °C	Time, h	Comments
Aerobacter aerogenes	GlYCTMA	37	1·5–4	Depending on age of population.
Escherichia coli	TMA	37	1·5–4	5 strains. American type B forms filaments when old which may be confusing.
Pseudomonas aeruginosa	TMA	37	2–2·5	
Pseudomonas ovalis	TMA	30	2	From 16 h agar slope.
Serratia marcescens	TMA	37	1–2	From 16 h agar slope.
Chromobacterium violaceum	NA	30	4–5	
Alcaligenes metalcaligenes	TMA	RT	17–19	
Pasteurella pestis	TMHA	28	3–6	
Salmonella typhimurium	TMA	37	2·7–3·5	From 3-day-old broth culture.
Vibrio anguillarum	saline NA	30	5–6	
Azotobacter vinelandii	sucrose B	30	16	
Bacillus subtilis	TMA	37	2–3	From 18 agar slope. Filaments obscured slide if incubated longer.
Micrococcus sp.	TMA	37	3·3	From 16 h broth culture. Fairly long incubation needed to distinguish microcolonies from initial clusters.
Bakers' yeast	MYPGA	RT	18	Commercial material as damp block. pH 5.
Candida utilis	GYCTMA	30	4–6	pH 5.

RT signifies room temperature (16–20°C). TMA is tryptic meat agar; H signifies haematin as supplement; Y, yeast extract; C, acid-hydrolysed casein; P, "peptone"; G, glucose; Gl, glycerol; M, maltose. NA is Oxoid nutrient agar; when saline it has 2·5% NaCl (for marine organisms). Sucrose B is Burk's original *Azotobacter* medium. Concentrations are not quoted, since they are conventional ones mentioned elsewhere in this Series.

3. *Viability of viruses and mycoplasmas*

Absolute viabilities of viruses are not easily obtained unless the virus can be clearly observed, unobscured by contaminating material, in electron-microscope preparations. Hence survival studies on viruses are usually related to an initial viable count without regard to what relation this bears to the absolute population of virus particles. Viability determinations on populations of mycoplasmas are discussed by Fallon and Whittlestone (this Series, Volume 3B).

REFERENCES

Childs, E., and Allen, L. A. (1953). *J. Hyg., Camb.*, **51**, 418–424.
Collins, V. G., and Kipling, C. (1957). *J. appl. Bact.*, **20**, 257–264.
Miles, A. A., and Misra, S. S. (1938). *J. Hyg., Camb.*, **38**, 732–742.
Pochon, J. (1956). *Annls Inst. Pasteur, Paris*, **89**, 464–465.
Postgate, J. R. (1967). *Adv. Microbial Physiol*, **1**, 1–23.
Postgate, J. R., Crumpton, J. E., and Hunter, J. R. (1961). *J. gen. Microbiol.*, **24**, 15–24.
Taylor, J. (1962). *J. appl. Bact.*, **25**, 54–61.

APPENDIX: McGRADY'S PROBABILITY TABLES

These Tables indicate the probable number of bacteria of the coliform group present in 100 ml of water, as shown by the various combinations of positive and negative results in the quantities used for test.

TABLE A

Quantity of water put up in each tube	50 ml	10 ml	Probable number of organisms in 100 ml of the original water
No. of tubes used	1	5	
Number of tubes giving positive reaction	0	0	0
	0	1	1
	0	2	2
	0	3	4
	0	4	5
	1	0	2
	1	1	3
	1	2	6
	1	3	9
	1	4	16
	1	5	18+

TABLE B

Quantity of water put up in each tube	50 ml	10 ml	1 ml	Probable number of coliform organisms in 100 ml of the original water
No. of tubes used	1	5	5	
Number of tubes giving positive reaction	0	0	0	0
	0	0	1	1
	0	0	2	2
	0	1	0	1
	0	1	1	2
	0	1	2	3
	0	2	0	2
	0	2	1	3
	0	2	2	4
	0	3	0	3
	0	3	1	5
	0	4	0	5
	1	0	0	1
	1	0	1	3
	1	0	2	4
	1	0	3	6
	1	1	0	3
	1	1	1	5
	1	1	2	7
	1	1	3	9
	1	2	0	5
	1	2	1	7
	1	2	2	10
	1	2	3	12
	1	3	0	8
	1	3	1	11
	1	3	2	14
	1	3	3	18
	1	3	4	20
	1	4	0	13
	1	4	1	17
	1	4	2	20
	1	4	3	30
	1	4	4	35
	1	4	5	40
	1	5	0	25
	1	5	1	35
	1	5	2	50
	1	5	3	90
	1	5	4	160
	1	5	5	180+

Note—The most probable numbers from 0 to 20 are correct to the nearest unit; above 20 are correct to the nearest 5.

TABLE C

Quantity of water put up in each tube	10 ml	1 ml	0·1 ml	Probable number of coliform organisms in 100 ml of the original water
No. of tubes used	5	5	5	
Number of tubes giving positive reaction	0	0	0	0
	0	0	1	2
	0	0	2	4
	0	1	0	2
	0	1	1	4
	0	1	2	6
	0	2	0	4
	0	2	1	6
	0	3	0	6
	1	0	0	2
	1	0	1	4
	1	0	2	6
	1	0	3	8
	1	1	0	4
	1	1	1	6
	1	1	2	8
	1	2	0	6
	1	2	1	8
	1	2	2	10
	1	3	0	8
	1	3	1	10
	1	4	0	11
	2	0	0	5
	2	0	1	7
	2	0	2	9
	2	0	3	12
	2	1	0	7
	2	1	1	9
	2	1	2	12
	2	2	0	9
	2	2	1	12
	2	2	2	14
	2	3	0	12
	2	3	1	14
	2	4	0	14

Note—The most probable numbers from 0 to 20 are correct to the nearest unit; from 20 to 200 are correct to the nearest 5; above 200 are correct to the nearest 50.

TABLE C—*Continued*

Quantity of water put up in each tube	10 ml	1 ml	0·1 ml	Probable number of coliform organisms in 100 ml of the original water
No. of tubes used	5	5	5	
Number of tubes giving positive reaction	3	0	0	8
	3	0	1	11
	3	0	2	14
	3	1	0	11
	3	1	1	14
	3	1	2	17
	3	1	3	20
	3	2	0	14
	3	2	1	17
	3	2	2	20
	3	3	0	17
	3	3	1	20
	3	4	0	20
	3	4	1	25
	3	5	0	25
	4	0	0	13
	4	0	1	17
	4	0	2	20
	4	0	3	25
	4	1	0	17
	4	1	1	20
	4	1	2	25
	4	2	0	20
	4	2	1	25
	4	2	2	30
	4	3	0	25
	4	3	1	30
	4	3	2	40
	4	4	0	35
	4	4	1	40
	4	5	0	40
	4	5	1	50

TABLE C—*Continued*

Quantity of water put up in each tube	10 ml	1 ml	0·1 ml	Probable number of coliform organisms in 100 ml of the original water
No. of tubes used	5	5	5	
Number of tubes giving positve reaction.	5	0	0	25
	5	0	1	30
	5	0	2	40
	5	0	3	60
	5	0	4	75
	5	1	0	35
	5	1	1	45
	5	1	2	60
	5	1	3	85
	5	2	0	50
	5	2	1	70
	5	2	2	95
	5	2	3	120
	5	2	4	150
	5	2	5	175
	5	3	0	80
	5	3	1	110
	5	3	2	140
	5	3	3	175
	5	3	4	200
	5	3	5	250
	5	4	0	130
	5	4	1	170
	5	4	2	250
	5	4	3	300
	5	4	4	350
	5	4	5	450
	5	5	0	250
	5	5	1	350
	5	5	2	600
	5	5	3	900
	5	5	4	1,600
	5	5	5	1,800+

CHAPTER XIX

Determination and Significance of Molar Growth Yields

A. H. STOUTHAMER

Botanical Laboratory, Microbiology Department, Free University, De Boelelaan 1087, Amsterdam, The Netherlands

I. INTRODUCTION

A. Relation between the amount of growth and substrate concentration

Monod (1942) studied the aerobic growth of *Bacillus subtilis*, *Escherichia coli* and *Salmonella typhimurium* in minimal medium with a large number of carbohydrates as carbon and energy source. He concluded that the amount of microbial growth with a limiting amount of energy source is proportional to the amount of carbohydrate added. Thus the amount of growth (G) is related to substrate concentration (C) by—

$$G = KC$$

In this relation K is the yield constant. When the dry weight of bacteria and the substrate concentration are expressed in μg/ml, K gives the amount of dry weight in μg formed during consumption of 1 μg of substrate. Monod (1942) has shown that the value of K for a given substrate can be reproducibly determined and has a fixed value for every substrate. The way Monod expressed growth yields is still widely used. A variant of expressing growth yields, which is only rarely used, is obtained by expressing both the amount of substrate and the bacterial yield in μg of carbon, in which case K gives the amount of carbon in the cells in μg after consumption of 1 μg of substrate carbon (Pirt, 1957).

A more advantageous way of expressing growth yields has been introduced by Bauchop and Elsden (1960). They expressed the amount of dry weight also in μg/ml, but the substrate concentration in μmoles/ml. The yield coefficient defined in this way has the dimensions of μg dry weight formed per μmole substrate. The yield constant expressed in this way is called the molar yield coefficient and is indicated by Y. The utilization of Y enables us to compare directly growth on substrates with different molecular weights and is also necessary if we wish to relate growth yields to the amount of ATP that can be obtained from the substrate. If the ATP yield (moles ATP produced/mole substrate used) for a substrate is known, we can calculate the amount of dry weight produced per mole of ATP formed. Bauchop and Elsden called this ratio Y_{ATP}.

B. Relation between ATP production and molar growth yield

DeMoss *et al.* (1951) studied the growth of three strains of lactic acid bacteria (*Streptococcus faecalis*, *Leuconostoc mesenteroides* and *Lactobacillus delbrueckii*) and found that the growth yield was a linear function of carbohydrate consumption. The molar growth yield (which they measured as increase in optical density on a molar basis) for *St. faecalis* and *Lact. delbrueckii* was about twice the value for *L. mesenteroides* and they concluded that *St. faecalis* and *Lact. delbrueckii* obtained more energy per mole of glucose than did *L. mesenteroides.* Further work (Gunsalus and Gibbs, 1952; Hurwitz, 1958) has shown that *L. mesenteroides,* like all other heterofermentative lactic acid bacteria, ferments glucose by the hexose monophosphate pathway (Fig. 1). Only 1 mole of ATP is formed per mole of

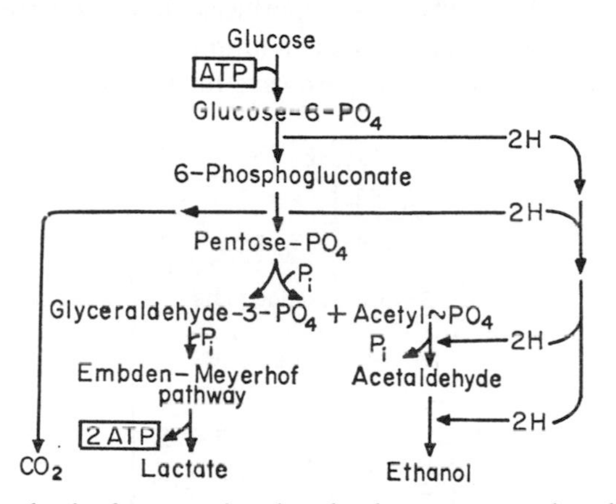

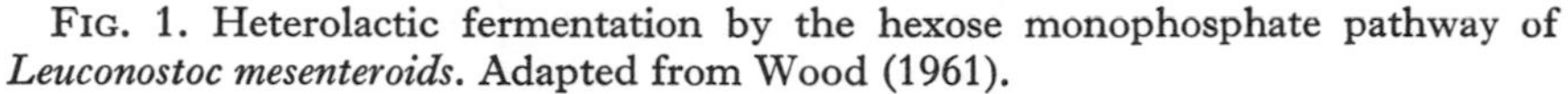

FIG. 1. Heterolactic fermentation by the hexose monophosphate pathway of *Leuconostoc mesenteroids.* Adapted from Wood (1961).

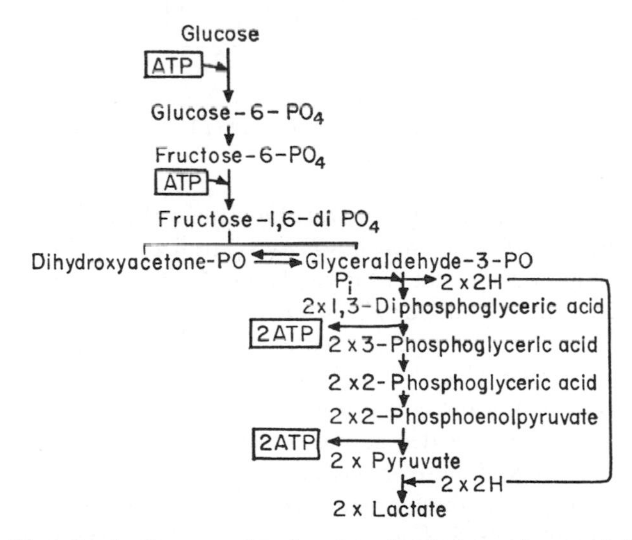

FIG. 2. Homolactic fermentation by the glycolytic pathway of *Streptococcus faecalis* and *Lactobacillus delbrueckii.*

glucose fermented via this pathway, compared to the formation of 2 moles of ATP per mole of glucose when fermented by the glycolytic pathway (Fig. 2) in the homofermentative lactic acid bacteria. Sokatch and Gunsalus (1957) found that *St. faecalis* gave about the same growth yield on glucose and on gluconate. They concluded that the pathways for the degradation of these two substrates produce similar net amounts of coupled energy. This suggested to Bauchop and Elsden (1960) that the amount of growth of micro-organisms is directly related to the amount of ATP produced from the energy source. To verify this hypothesis they measured the growth yields of a number of organisms on a number of energy sources. The organisms chosen utilize different pathways for the degradation of glucose.

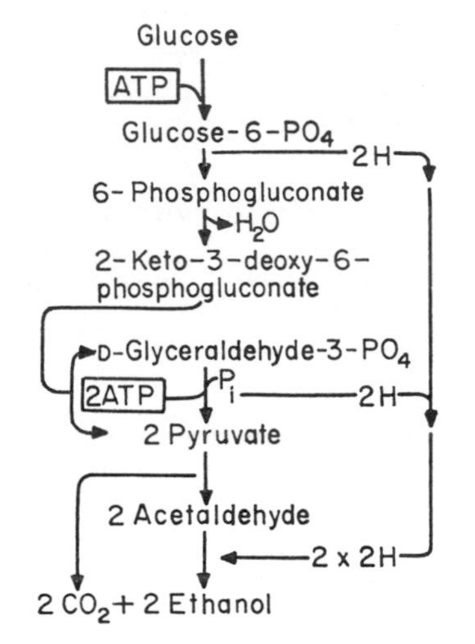

FIG. 3. Ethanolic fermentation by the Entner–Doudoroff pathway of *Zymomonas lindneri*. Adapted from Wood (1961).

St. faecalis and *Saccharomyces cerevisiae* utilize the glycolytic pathway and *Zymomonas mobilis* (*Pseudomonas lindneri*) utilizes the Entner–Doudoroff pathway (Fig. 3). The net ATP yields are, for the first two organisms, 2 moles/mole of glucose and for *Z. mobilis* 1 mole/mole of glucose. Further, they determined the growth yields with several other energy sources, the degradation of which give different ATP yields. For one organism the amount of growth was directly proportional to the amount of ATP that could be obtained from the degradation of the energy source. Further, they found that Y_{ATP} proved more or less a constant among different organisms. Calculation of Y_{ATP} for these organisms gave values ranging from 12·6 to 8·3 with an average of 10·5. Thus Y_{ATP} values could be useful in predicting ATP yields for fermentations of which the ATP yield is unknown. More values for Y_{ATP} have been determined since the work of Bauchop and Elsden. These values are included in Table I.

Further extensions on the theoretical background of the relations between growth yield and ATP production have been given by Gunsalus and Shuster (1961) and Senez (1962).

C. Relation between the amount of growth and oxygen uptake during growth

The previous results all applied to bacteria growing anaerobically in complex media. Under these circumstances very little of the carbon of the energy source is incorporated into cellular material. It is nearly completely used as an energy source (Bauchop and Elsden, 1960; Dawes *et al.* 1966). For bacteria growing anaerobically in complex medium, the ATP yield can easily be calculated, provided that the fermentation pathways utilized by the organisms are known. For anaerobic bacteria, all the ATP produced comes from phosphorylation at the substrate level. During aerobic growth, oxygen is taken up and the ATP yield comes not only from phosphorylation at the substrate level but mostly from oxidative phosphorylation, the efficiency of which is not known. Whitaker and Elsden (1963) suggested that the oxygen uptake during growth is a measure of that part of the glucose that is completely oxidized. They introduced the coefficient Y_{O_2} (g dry weight/mole O_2). Their results showed that Y_{O_2} is a constant for any substrate. Hadjipetrou *et al.* (1964) used the coefficient Y_O (g dry weight/atom O). This last coefficient has the advantage that division of Y_O by Y_{ATP} is identical to the P/O ratio, which is the normal ratio to express the efficiency of oxidative phosphorylation. In chemo-autotrophic micro-organisms growth yields are generally expressed as O_2/CO_2 ratios [= (moles of O_2 consumed)/(moles of CO_2 assimilated)]. This ratio is similar to the Y_O values and in fact from the O_2/CO_2 ratios Y_O values can be calculated and the other way round. An example of this calculation will be given in Section IX.

D. Influence of hydrogen acceptors on anaerobic molar growth yields

Molar growth yields for anaerobic growth are sometimes higher in the presence of hydrogen acceptors (nitrate, fumarate, etc.).

There may be several reasons for this phenomenon—

1. The substrate might be more completely oxidized and consequently no reduced products ought to be formed. The formation of ethanol from acetylphosphate leads for instance to loss of activated phosphate.
2. Oxidative phosphorylation during the reduction of the hydrogen acceptor might occur. The same remarks that have been made on the relationship

between growth yield and oxygen uptake also apply here. Therefore we can express these growth yields by the coefficient—

$Y_{NO_3^-}$ (= g dry weight/mole of nitrate).
$Y_{fumarate}$ (= g dry weight/mole of fumarate).

Most of these coefficients have not yet been measured; further, the significance of these values is still in doubt (see Section X).

II. DETERMINATION OF MOLAR GROWTH YIELDS

A. General considerations

At the start it is necessary to emphasize two points. In the first place determinations of molar growth yields under aerobic and under anaerobic conditions can be performed only under two different conditions—

(a) In a complex medium, where it can be shown that the carbohydrate is completely used as energy source and the assimilation of glucose can be measured by incorporating ^{14}C from uniformly labelled glucose into cell material. Measurements of this kind have been performed by Bauchop and Elsden (1960), Hadjipetrou *et al.* (1964), Belaïch and Senez (1965) and Dawes *et al.* (1966). These experiments have shown that during growth of *St. faecalis* or of *Z. mobilis* on complex media with limiting amounts of glucose assimilation is indeed negligible (1–3% of the glucose added).

(b) In a simple medium, where the carbohydrate serves both as carbon and energy source. In this case the proportion that is assimilated can be determined fairly accurately from the amount of dry weight and the amount of carbon in the cells.

It is thus evident that the choice of the medium must be such that the experiments are performed under either of these two conditions. However, a free choice of the medium is not always possible, as growth in some media is limited by the rate at which some essential nutrients can be synthesized. Under these conditions, uncoupling between growth and energy production may occur (see Section VI B) and this must be considered when the medium is chosen.

Secondly, it is always necessary to determine the molar growth yield at several different substrate concentrations, for several reasons—

(a) The plot of growth yield against substrate concentration should be linear. Only at concentrations which give growth yields on this straight line may we be sure that it is indeed the energy source that is limiting growth.

(b) The plot of growth yield against substrate concentration may yield a curve that does not pass through the origin, but intersects the ordinate at a positive value at zero substrate concentration. This is due to the presence of a small amount of an unknown energy source in the complex medium. This phenomenon has been observed repeatedly, e.g., by Bauchop and Elsden (1961) for *St. faecalis*.

(c) The plot of growth yield against substrate concentration sometimes gives a curve with an inflection point. At both sides of the inflection point a linear relationship exists between growth yield and substrate concentration. At the lowest concentration the slope of the curve is greater than at higher substrate concentrations. This has been found for *St. faecalis* (Forrest and Walker, 1965) and for *Aerobacter aerogenes* (Hadjipetrou and Stouthamer, 1965). It seems that the metabolism of the bacteria is different at the lower and higher concentrations. Forest and Walker (1965) have shown that at low glucose concentrations much volatile acid is formed: at higher concentrations glucose is completely fermented to lactate.

B. Practical execution of molar-growth-yield determinations

1. *In batch culture*

From the previous considerations, it is evident that for the determination of molar growth yields it is necessary to prepare several tubes containing media with different concentrations of the energy source. After inoculation special attention should be paid to the incubation conditions. Anaerobic incubation is performed by gassing with oxygen-free nitrogen, nitrogen+ carbon dioxide, hydrogen or hydrogen+carbon dioxide, or by including an oxygen adsorbent in the culture flask, such as alkaline pyrogallol. The gas mixtures are especially suitable for organisms, which require CO_2 for growth. Aerobic incubation is performed either by placing the cultures on a shaking machine or by forcing compressed air through the cultures. Growth is followed by measuring the optical density of samples taken at intervals from the cultures. When the maximum optical density is reached, the energy source has been completely consumed and the dry weight of the micro-organisms is determined as soon as possible. The following methods of dry-weight determination have been used in studies of molar growth yields.

(a) *Direct dry-weight measurement by centrifugation.* After growth has reached completion the organisms are harvested by centrifugation, washed three times with distilled water and dried at 100°C to constant weight. The direct method for measuring dry weight has been used extensively by Bauchop and Elsden (1960). As an example one of their experiments will be described in detail.

Conical flasks (250 ml) each containing 150 ml medium and a test tube containing 3 ml of 15% (w/v) pyrogallol were sterilized by autoclaving at 15 lb/sq. in. for 15 min. The required amounts of glucose (6, 12 and 18 μmoles/ml for *St. faecalis*) and inoculum were then added. Finally, 3 ml of 10% Na_2CO_3 were added to the test tube containing pyrogallol. The flasks were closed with rubber stoppers, sealed with paraffin wax, and incubated at 37°C. When growth had reached completion the organisms were harvested by centrifugation, washed three times with 100 ml of distilled water, washed into pre-weighed beakers and dried to constant weight.

In several cases the washing with distilled water proves harmful and leads to losses of dry weight. In these cases the washings should be performed with buffer or saline, and a correction should be applied for the contribution of the solutes to the dry weight. This correction may be obtained by drying a volume of the solution in which the bacteria were suspended to constant weight.

(b) *Dry-weight measurements by filtration.* For several organisms the dry weight may easily be obtained by filtration of an aliquot of the culture through a weighed special membrane filter (Sartorius Membranfilter A.G., Göttingen). These cellulose nitrate filters do not contain a softening agent as do normal filters and do not change in weight on drying at 105°C. This last technique is especially suitable for determining the dry weights of yeasts and of bacteria. This method has been used for the determination of molar growth yields of *Bifidobacterium bifidum* (de Vries and Stouthamer, unpublished results). These experiments were performed in the following way.

A 300-ml portion of medium, containing 10 μmoles glucose/ml were inoculated. The flask containing the medium was placed in a McIntosh and Fildes anaerobic jar. Anaerobic conditions were obtained by evacuation and three successive washings with a gas mixture of 95% N_2+5% CO_2. When growth at 37°C was complete the culture was centrifuged. The supernatant was passed through the membrane filter (MF 30), previously dried and weighed. The filtration was performed in a Stefi filter apparatus. Then the sediment was transferred to the filter quantitatively. After a washing with 25 ml distilled water, the filter was dried to constant weight at 105°C, for which about 2–6 h were required.

Turbidimetric methods for determining the dry weight of bacteria could not be used in *Bi. bifidum*. A constant relating optical density to dry weight could not be determined as the bacteria have a strong tendency to aggregate during the washing procedure.

(c) *Turbidimetric method for dry-weight measurement.* In the turbidimetric method, the dry weights are always obtained by reference to a curve relating optical density to dry weight. Standard curves should be prepared from cells at the maximal optical density, as it has been found (Hungate, 1963; Hadjipetrou, *et al.*, 1964; de Vries and Stouthamer, unpublished results) that the light-scattering properties of the cells change after the maximum. With *A. aerogenes* Hadjipetrou *et al.* found, for instance, that the constant relating optical density and dry weight for cells at the moment of maximum optical density was only 91% of the value for cells after this moment. The determination of the constant relating optical density to dry weight seems to be one of the most critical points in the determination of molar growth yields.

The experiments of Hadjipetrou *et al.* (1964) for measuring molar growth yields were performed in the following way.

Nephelometer tubes (2·5 cm dia. × 8·5 cm), containing minimal medium salts (10 ml) plus carbon source were inoculated with 0·1 ml of an overnight culture in minimal medium, with the same substrate as carbon source. For every substrate, five tubes were used with five different concentrations. For glucose, the concentrations used were 0, 0·5, 1·0, 1·5 and 2·0 μmoles/ml for aerobic experiments, and 0, 2·0, 4·0, 6·0 and 8·0 μmoles/ml for anaerobic experiments. Aerobic incubation was performed by placing the tubes on a shaking machine at 37°C. To obtain anaerobic conditions, tubes fitted with rubber stoppers having inlet and outlet tubes were gassed for 30 min with pure nitrogen and then closed. The turbidity of the cultures was measured at regular intervals in a nephelometer until the turbidity reached its maximum value. Then immediately a sample was taken for the more accurate determination of the dry weight by measuring the extinction at 660 mμ in a spectrophotometer.

(d) *Dry-weight determination by measuring total organic matter*. In several cases, turbidimetric determination of dry weight is impossible as the bacteria tend to grow in clumps. In this case cell density can be determined by measuring the total perchloric acid precipitable organic matter by the method of Johnson (1949). This method must also be standardized by determinations on cell suspensions of known dry weight. MacKechnie (1966) has compared the turbidimetric method for the determination of dry weight and a modification of the method of Johnson for *Pseudomonas aeruginosa* and found that both methods gave essentially the same results.

MacKechnie (1966) used the procedure described below for this determination.

A volume of cell suspension, usually 2 ml, was pipetted into a 10 ml hard-glass centrifuge tube and mixed with an equal volume of 0·8M perchloric acid. The mixture was centrifuged hard (14,000 ***g*** for 10 min) and the supernatant carefully decanted. The precipitate was re-suspended in a volume of 0·4M perchloric acid equal to twice the original volume of cells and centrifuged as before. The supernatant was again carefully decanted and the residue suspended in 1 ml of distilled water, a vibrator being used to give a uniform suspension. A 3 ml volume of the oxidizing agent, $K_2Cr_2O_7$ (0·18 (w/v)) in concentrated H_2SO_4 (95% (v/v)), was added and the solution heated in a vigorously boiling-water bath for 20 min. Two reagent blanks containing 1 ml of distilled water and 3 ml of the dichromate solution were set up at the same time. The tubes were closed with pear-drop condensers to exclude extraneous organic matter in the air and the tips of the pipettes used to deliver the oxidizing reagent were wiped with glass filter paper (Whatman Glass Fibre, No. Gf/A), as recommended by Halliwell (1960). The tubes were cooled and the solutions made up quantitatively to 10 ml with distilled water in volumetric flasks. Considerable heat was generated in the solutions in this process and care was taken to make them up to the mark after they had cooled to room temperature (0·2 ml of $Na_2SO_3.7H_2O$ (20%, w/v) was added to one of the reagent blanks to give a totally reduced blank before dilution to 10 ml). All the solutions were then read against the reduced blank at 400 nm in a spectrophotometer. A calibration curve relating extinction at 440 nm to bacterial dry weight over the range 0·1–0·8 mg cells per ml was prepared from a standard suspension of the organism.

2. *Determination of molar growth yields for organisms showing lysis after complete substrate utilization*

The methods outlined above are usually adequate. Occasionally, however, rapid lysis occurs after the energy source has been completely consumed. This has been found for instance for *B. subtilis* and *Selenemonas ruminantium* (Monod, 1942; Hadjipetrou and Stouthamer, 1963; Hobson, 1965). This makes the determination of molar growth yields by the methods given above impossible. Within 5 min of the substrate being completely utilized, this lysis can be demonstrated. In such cases molar growth yields must be determined in another way. A flask with medium containing a certain concentration of energy source is inoculated. At intervals the dry weight is determined turbidimetrically and a sample is centrifuged. In the supernatants the amount of energy source still remaining is determined by conventional methods. In this way a plot of substrate utilization and of increase in bacterial dry weight against time may be obtained. From this the molar growth yield may be obtained. This method has been used successfully for *B. subtilis* by Hadjipetrou (1965).

3. *In continuous culture*

Growth in continuous culture can be used for determination of molar growth yields. In several cases this method offers distinct advantages over the batch methods. The theoretical treatment of continuous culture has been given by Monod (1950), Novick and Szilard (1950) and Herbert *et al.* (1956). The theoretical treatment of Herbert *et al.* allows quantitative prediction of the steady-state concentrations of bacteria and substrate in the culture. In the steady state, the dry weight of bacteria is given by the following relation—

$$X = Y(s_r - s)$$

in which s_r is the concentration of the growth-limiting substrate in the inflowing medium and s is the concentration of the substrate in the culture and thus also in the effluent of the continuous-culture vessel.

In this relationship, s is dependent on the dilution rate (D) which is the ratio of the volume of fresh medium supplied (f) to the growth vessel and the volume (v) of the vessel ($D = f/v$). For the determination of molar growth yields, D must be relatively low. In this case the substrate added is nearly completely consumed and s is negligible in comparison with s_r. This method can be used for determining both aerobic and anaerobic molar growth yields. The determination of the dry weight can be performed by one of the methods given before. If the dry weight is determined turbidimetrically, we should take into account that Herbert (1958), Pirt (1957), Perret (1958) and Button and Garver (1966) have described changes in the size of cells of *E. coli*, *Aerobacter cloacae*, *Bacillus megaterium* and *Torulopsis utilis* when grown at different rates in continuous culture. It is thus evident that the constant that relates optical density to dry weight must be determined with cells growing under identical conditions as in the experiments for the determination of the molar growth yield.

Herbert *et al.* (1956) have found that Y determined in batch culture for growth of *A. cloacae* in minimal medium with glycerol as growth-limiting substrate was the same as that determined in continuous culture at low dilution rates. Y was constant in these experiments at growth rates between 0 and 0·65. In several experiments with rumen micro-organisms, however, Y was found to vary considerably with the growth rate (Hobson, 1965; Hobson and Summers, 1967). The fermentation products also varied with the growth rate, but this was not sufficient to explain the large influence of the growth rate on Y. These experiments will be further discussed in Section VII D.

Bauchop and Elsden (1960) have shown that the growth yield of *St. faecalis* in semi-defined medium with arginine as energy source could not be determined reproducibly in batch culture. In continuous culture, however, such determinations were straightforward and reproducible.

III. DETERMINATION OF Y_O

Several authors (Czerniawski *et al.*, 1965; Johnson, 1967) mention in their papers Y_O values that have been calculated in the following way: the amount of oxygen used in the conversion of substrate into cells is the amount of oxygen necessary to burn the substrate *minus* the amount of oxygen required to burn the product. In this calculation, it is assumed that no carbon compounds other than cells are formed. The experiments of Hadjipetrou *et al.* (1964) with labelled glucose show that this assumption at least in *A. aerogenes* is not correct. Therefore, Y_O values calculated in this way have little significance. The only way to obtain reliable Y_O values is to measure both growth and oxygen uptake.

Determination of Y_O can be performed in batch culture and in continuous culture.

A. Determination of Y_O in batch culture

This determination is performed by growing the bacteria in sterilized Warburg vessels. In this way, oxygen uptake during growth can be measured. Special attention should be paid to the following points.

1. It should be carefully tested whether the growth yield in the Warburg vessels is the same as the growth yield determined otherwise. In several cases lower yields are found in Warburg vessels, which may arise from differences in the shaking and aeration conditions, or from the fact that oxygen uptake during growth is measured in vessels that contain alkali in the centre well, leading to a shortage of CO_2. Hadjipetrou *et al.* (1964) have studied the influence of both factors on the growth yield of *A. aerogenes* growing on minimal medium. They found that the rate of shaking had the largest influence on the molar growth yield. The influence of CO_2 tension was studied by the diethanolamine method of Pardee (1949) for the measurement of respiration in the presence of CO_2. With low substrate concentrations, growth with this method was better than with the normal method (Hadjipetrou and Stouthamer, unpublished results). MacKechnie (1966) has also found that growth of *P. aeruginosa* was much better when growth took place in the presence of CO_2 than in its absence. Thus the diethanolamine method can sometimes have a certain advantage.
2. The moment of maximal growth does not coincide with the moment of maximal oxygen uptake (Hadjipetrou *et al.*, 1964; MacKechnie, 1966). Thus it is necessary to follow both growth and oxygen uptake. Hadjipetrou *et al.* (1964) have demonstrated that acetate accumulates during growth and that it is oxidized after maximal growth. This result explains why maximal O_2 uptake is reached after maximal growth.

As an example of experiments which take into account the factors mentioned above, the experiments of Hadjipetrou *et al.* (1964) with *A. aerogenes* will be described in detail.

Several sterile Warburg vessels containing 3·25 ml of glucose minimal medium were inoculated with 0·15 ml of an overnight culture in minimal medium grown aerobically with glucose as carbon source. The glucose concentrations used were 1·0 and 1·5 μmoles/ml. To the centre wells of the vessels 0·6 ml of the diethanolamine reagent and a piece of filter paper were added. The diethanolamine reagent is prepared by mixing 6 ml diethanolamine, 15 mg thiourea, 3 g of $KHCO_3$, 0·7 ml 6N HCl and water to bring the volume to 15 ml. Before use, the reagent is allowed to stand overnight. The composition of the reagent is chosen to give a concentration of 0·3% CO_2 in the gas phase of the Warburg vessel. Carbon dioxide formed by glucose oxidation is adsorbed by the reagent. Thus the bacteria grow with a constant pressure of CO_2.

The Warburg vessels were incubated at 37°C with good shaking (125 complete 5 cm strokes/min). At intervals, the contents of a Warburg vessel were taken for the measurement of the extinction at 660 nm in a spectrophotometer. In these experiments the rate of oxygen uptake increased exponentially until growth stopped, and thereafter the rate became linear. The extinction reached a maximum at about the same time as the bend in the curve for the O_2 uptake occurred. Y_O values were obtained by dividing the growth yield by the oxygen uptake in atoms at the moment of maximal growth. To obtain a significant value of Y_O by this method, 6–10 replicates of the determination were required.

B. Determination of Y_O in continuous culture

Two methods may be used for these determinations in continuous culture.

In the first method, experiments may be performed in the normal apparatus which is used for aerobic growth in continuous culture. Growth may be limited by using a sub-optimal aeration rate. Pirt (1957) has grown *A. cloacae* with glycerol as carbon and energy source in a continuous culture vessel at different rates of air supply. At insufficient aeration rates, the oxygen supply to the cells is completely determined by the oxygen solution rate in the medium, there being no significant amount of oxygen in solution in the medium. The oxygen adsorption coefficient may be determined by the sulphite method (Cooper *et al.*, 1944). It must be emphasized however, that Pirt and Callow (1958) have shown that the maximum possible oxygen solution rates in bacterial cultures with different aeration conditions are appreciably less (0·5–0·7 times) than those in a sodium sulphite solution under the same aeration conditions. This factor decreases the accuracy of the Y_O values calculated from the results of Pirt (1957), and the second method of determination is preferable.

In this method, the continuous-culture vessel is completely filled with medium (Bulder, 1960; Button and Garver, 1966). Oxygen is supplied only as dissolved oxygen in the fresh medium added to the continuous-culture vessel. The amount of dissolved oxygen in the feed may be regulated by sparging different oxygen-nitrogen mixtures into the feed carboy. This method is only applicable to organisms that do not ferment the energy source, since in the presence of fermentation, the contribution of respiration to energy production may be small (Bulder, 1960, 1963, 1966; Button and Garver, 1966).

IV. DETERMINATION OF $Y_{NO_3^-}$

Hadjipetrou and Stouthamer (1965) found that the anaerobic growth yield of *A. aerogenes* growing on glucose was increased by adding nitrate. In these experiments nitrate is used only as terminal-hydrogen acceptor ("nitrate respiration") and not as nitrogen source, since ammonia is also present in the medium. When nitrate is both hydrogen acceptor and nitrogen source, the growth yield is somewhat lower (about 6%) than in the presence of ammonia (Hadjipetrou and Stouthamer, 1965).

To calculate $Y_{NO_3^-}$ the minimal amount of nitrate was determined that must be added to the medium to increase the molar growth yield on glucose to the value with excess nitrate. In principle, this method can also be used for other values such as $Y_{S_2O_3^{2-}}$. No examples of this kind are available in the literature at the present time.

Direct determination of the amount of nitrate reduced by the method of Middleton (1959) confirmed that the method outlined above gave reliable results. These determinations were performed in the following way.

A flask containing 200 ml of minimal medium, glucose (3 μmoles/ml) and KNO_3 (6 μmoles/ml) was inoculated with 5 ml of an overnight culture grown on the same medium and under the same conditions. Anaerobic conditions were obtained by gassing with pure N_2. Incubation was at 37°C in a water bath. Growth was followed by taking samples at intervals for measurement of the extinction at 660 nm in a spectrophotometer. At the moment the extinction reached its maximum, a sample was taken and centrifuged. The supernatant was used for the determination of the amount of nitrite produced and the amount of nitrate remaining in the medium.

Nitrite was measured by a modification of the method of Nicholas and Nason (1957). To 0·5 ml of a sample, containing maximally 0·1 μmoles of nitrite, were added 0·5 ml of 1% sulphanilamide in 2·5N HCl. After standing at 0°C for 15 min 0·5 ml of 0·02% *N*-(1-naphthyl)ethylenediamine was added. After 30 min at room temperature, 2 ml of distilled water were added and the extinction was measured at 540 nm in a spectrophotometer.

Nitrate was determined after reduction to nitrite by Zn powder by the method of Middleton (1959) and then by determination of total nitrite. Thus in this way the sum of nitrate and nitrite is determined. To 0·5 ml of the sample (containing a maximum of 0·40 μmoles nitrate + nitrite) were added 5 ml 0·55% $Ca(CH_3COO)_2.H_2O$ in 4% (v/v) aqueous ammonia and 0·1 ml of 1% (w/v) $MnSO_4.4H_2O$ in 5% (v/v) acetic acid. After addition of about 0·1 g finely powdered zinc the solution was vigorously shaken for 1 min in a flask shaker and filtered immediately through a dry filter. To 2 ml of the filtrate was added 0·5 ml 1% sulphanilamide in 5N HCl. After 15 min at 0°C 0·5 ml of 0·02% *N*-(1-naphthyl)ethylenediamine was added, and after 30 min at room temperature the extinction was measured at 540 nm. The difference between the nitrite determination after reduction and the direct determination gave the amount of nitrate remaining in the medium. In this way the amount of nitrate reduced during growth was determined. It must be mentioned that nitrate reduction, like oxygen uptake, continues after growth has reached its maximum.

V. DETERMINATION OF Y_{ATP}

For the determination of Y_{ATP} we must know both Y and the amount of ATP produced from the energy source. Y may be determined by any of the procedures in the previous Sections. In most cases the amount of ATP can be determined only for known fermentation pathways. Calculation of ATP production is easiest when the substrate is only used as an energy source and is not incorporated into cellular material. It must be emphasized that it is necessary to determine the fermentation balance for growing cells as differences between fermentation balances for growing and resting cells have been found in several cases (Forrest *et al.*, 1961; Forrest and Walker, 1965). For instance growing cells of *St. faecalis* form large amounts of volatile acids, resting cells ferment glucose to 2 moles of lactate, and the ATP production, calculated from these balances, is quite different.

For bacteria growing in simple media, part of the substrate is used for energy and part is assimilated. For example, for *A. aerogenes* growing anaerobically with glucose in minimal medium, the molar growth yield is 26·1 g/mole (Hadjipetrou *et al.*, 1964). The carbon content of the bacteria is about 40% (Hadjipetrou *et al.*, 1964). We may conclude that $180 - 26{\cdot}1 = 153{\cdot}9$ g of glucose have been fermented. During growth, 0·84 moles of acetate were produced per mole of glucose. The theoretical overall fermentation is 1 glucose →1 acetate + 2 formate + 1 ethanol. As the net gain of ATP per mole of glucose transformed by the glycolytic pathway into pyruvate is 2 moles, and as one extra ATP is formed per mole of acetate produced, the amount of ATP from 153·9 g glucose is $2 \times \frac{153{\cdot}9}{180} + 0{\cdot}84 = 2{\cdot}55$ moles of ATP. From these results we can calculate Y_{ATP} as 10·2.

TABLE I

Molar growth yield, ATP yield and Y_{ATP}, for several organisms growing with glucose

Organism	ATP yield/ mole glucose fermented	Y	Y_{ATP}	Ref.
Streptococcus faecalis	2·0–3·0	20·0–32·0	10·8 ± 0·2(6)	Bauchop & Elsden (1960); Sokatch & Gunsalus (1957); Forrest & Walker (1965); Beck & Shugart (1966)
Zymomonas mobilis	1·0	8·0– 9·3	8·5 ± 0·2(4)	Bauchop & Elsden (1960); Dawes *et al.* (1966); Senez & Belaïch (1965); Belaïch & Senez (1965)
Saccharomyces cerevisiae[1]	2·0	18·8–21·0	10·0 ± 0·2(4)	Bauchop & Elsden (1960); Battley (1960); Bulder (1963)
Saccharomyces rosei[2]	2·0	22·0–24·6	11·6 (2)	Bulder (1963, 1966)
Lactobacillus plantarum	2·0	18·8	9·4	Oxenburgh & Snoswell (1965)
Aerobacter aerogenes[3]	3·0	26·1	10·2	Hadjipetrou *et al.* (1964)
Escherichia coli[4]	3·0	25·8	11·2	Stouthamer, unpublished results
Bifidobacterium bifidum[4]	2·5–3·0	37·4	13·1	de Vries & Stouthamer, unpublished results
Actinomyces israeli[5]	2·0	24·7	12·3	Buchanan & Pine (1967)

1. Including one value for a respiratory deficient mutant of *S. cerevisiae*.
2. Including one value for a respiratory deficient mutant of *S. rosei*.
3. These organisms were grown in minimal medium. The ATP yield recorded for these organisms is for 1 mole of glucose, which is completely fermented. To calculate Y_{ATP}, a correction has been applied as only part of the glucose is fermented (see text).

4. This result is based on the fermentation balance

$$1 \text{ glucose} \rightarrow 1{\cdot}85 \text{ acetate} + 0{\cdot}21 \text{ lactate} + 0{\cdot}27 \text{ ethanol} + 0{\cdot}66 \text{ formate (C-recovery} = 92\%).$$

Fermentation takes place by the fructose 6-phosphate phosphoketolase pathway (de Vries *et al.*, 1967). This pathway gives a net ATP production/mole glucose of 3 moles ATP, when pyruvate is converted into acetate, formate and ethanol and of 2·5 moles when pyruvate is converted into lactate.

5. This result is obtained when the organism is grown in the absence of CO_2. In the presence of CO_2 much higher yields are obtained (see Section VII).

For several bacteria, the carbon content of the cells is higher than 40%. Values of 50% have been reported for *E. coli* (Roberts *et al.*, 1955). Under these circumstances, a correction for the higher carbon content of bacteria than that of glucose must be applied. This correction is relatively small however, and the error introduced by this factor is not very large.

Y_{ATP} has been calculated for several organisms. The results are given in Table I. In this Table, only results have been collected for organisms growing with glucose as energy source. In several cases, Y_{ATP} values for one organism have been determined by several authors. In this case the mean value of these determinations is given. From the results with these organisms we find a mean value of $10{\cdot}8 \pm 0{\cdot}4$ for Y_{ATP}, with a range of 8·5–13·1. It must be emphasized that the range of these values is not due to experimental errors but to the fact that 1 mole of ATP gives 13·4 g of dry weight in

TABLE II

Y_{ATP} values calculated from molar growth yields and ATP yields for several organisms growing with other substrates as energy source

Organism	Substrate	ATP yield	Y	Y_{ATP}	Ref.
St. faecalis	Gluconate	1	17·6	9·8	Sokatch & Gunsalus (1957)
	2-Ketogluconate	..	19·5	8·5	Goddard & Sokatch (1964)
	Ribose	1·67	21·0	21·6	Bauchop & Elsden (1960)
	Arginine	1	10·2	10·2	Bauchop & Elsden (1960)
	Pyruvate	1	10·4	10·4	Forrest (1965)
Z. mobilis	Fructose	1	9·2	9·2	Dawes *et al.* (1966)
A. aerogenes	Fructose	3	26·7	10·7	Hadjipetrou *et al.* (1964)
	Mannitol	2·5	21·8	10·0	Hadjipetrou & Stouthamer (1965)
	Gluconate	2·5	21·4	11·0	Hadjipetrou & Stouthamer, unpublished results

Bi. bifidum and only 8·5 g of dry weight in *Z. mobilis*. Y_{ATP} values calculated for the same organisms but growing with other substrates is given in Table II. If these values are included we find a mean value of 10·5 ± 0·3 for Y_{ATP} (range, 8·5–13·1) from 18 determinations.

It must be mentioned here that recently some indications have been obtained that in rumen micro-organisms growing in continuous culture at optimal growth rates, much higher Y_{ATP} values can be obtained (Hobson, 1965; Hobson and Summers, 1967). The reasons for this phenomenon are not known at this moment.

VI. ENERGY UNCOUPLING DURING GROWTH

To explain the constant value of Y_{ATP} it is necessary that there exists in growing cells an effective coupling between the energy-yielding metabolism and the energy-consuming reactions of cell biosynthesis. There is evidence for this in several cases. Herbert (1958) has shown that the Q_{O_2} of bacteria increases linearly with their growth rate. Similar findings have been reported by Neidhardt (1965) and by Tempest and Herbert (1965). Sometimes however, uncoupling during growth takes place (see also Section VIII B).

A. Influence of incubation temperature on molar growth yields

Senez (1962) has found that the molar growth yield of *A. aerogenes* remains practically constant between 22 and 37°C, although the exponential growth rate and the respiratory activity both increase. Above 37°C the growth rate and yield show a decline, although the respiratory activity still increases. Only above 42°C, when growth is completely inhibited, does the respiratory activity decline. This indicates that between 37 and 42°C, uncoupling occurs between oxidation and growth. Similar results have been reported by Senez (1962) for *Desulfovibrio desulfuricans*. MacKechnie (1966) has found that the growth yields and Y_O values for *P. aeruginosa* were higher at 30°C than at 37°C.

B. Nutrient limitations

Growth of *St. faecalis* in continuous culture with tryptophan as limiting nutrient has been studied by Rosenberger and Elsden (1960). It was found that catabolism of glucose proceeded at a rate unrelated to the growth rate. Under these conditions the molar growth yield increased linearly with the growth rate.

Senez and Belaïch (1965) reported a molar growth yield of 28·9 for *E. coli* growing with glucose, but after exhaustion of the inorganic phosphate linear growth is obtained with a molar growth yield of 10·9. Under the latter conditions uncoupled growth takes place. During growth the rate of heat production increases logarithmically until phosphate is exhausted and remains constant afterwards.

Anaerobic growth of a mutant of *A. aerogenes* blocked in α-oxoglutarate dehydrogenase activity was found to be inhibited by nitrate (Stouthamer, 1967) and the molar growth yield on glucose of this mutant was decreased from 45·5 to 29·9. In the presence of succinate, normal growth was obtained, indicating that under anaerobic conditions in the presence of nitrate a shortage of succinate results and uncoupling occurs. Pichinoty (1960) has studied the influence of the nitrogen source on aerobic growth of *A. aerogenes*. His results show that nitrate assimilation depresses the growth rate and the growth yield with glucose. The rate of glucose consumption was not affected by the change from growth with ammonia to growth with nitrate. In medium with ammonia the growth rates with glucose, pyruvate and citrate were different, but in medium with nitrate they were identical. This shows that during growth with nitrate the growth limiting step is the conversion of nitrate to ammonia. The result shows that when the growth rate is limited by the speed at which ammonia is made available, uncoupling occurs. A similar conclusion is reached from similar experiments with *D. desulfuricans* (Le Gall and Senez, 1960). They studied growth with ammonia or with N_2 as nitrogen source, and found that the growth yield with the first was about twice the value obtained with N_2. The rate of consumption of the carbon source was the same. The examples given above show that when growth is inhibited by the availability of one compound, uncoupling between growth and energy production occurs. Most likely a similar explanation must be given for the finding of Senez and Belaïch (1965) that the molar growth yield of *Z. mobilis* on glucose in minimal medium is only 4·11 g/mole. In complex medium it is 8·61 g/mole. It may be that the rate at which an unknown compound is synthesized from glucose is rate limiting, thus causing uncoupling of growth and energy production.

C. Influence of growth-inhibitory compounds

Hadjipetrou and Stouthamer (1965) have found that nitrite depresses the molar growth yield of *A. aerogenes*, growing on glucose. Nitrite acts as hydrogen acceptor as acetate production in the presence of nitrite is increased. Thus, although the ATP yield from glucose is increased, the molar growth yield is decreased, indicating that uncoupled growth takes place in the presence of nitrite. Similar results have been obtained from the study of Hadjipetrou *et al.* (1966) on the influence of ferricyanide on energy production by *E. coli*.

VII. APPLICATION OF Y_{ATP} FOR PREDICTING ATP YIELDS FROM UNKNOWN FERMENTATION PATHWAYS

The results given in Tables I and II show that Y_{ATP} is fairly constant, and this makes the utilization of Y_{ATP} for the prediction of ATP yields possible. If Y_{ATP} for the organism studied is not known, the mean value given in the previous Section may be used. As the value of Y_{ATP} for different organisms varies, we can only obtain an approximate estimate of the ATP yield in this case. This method for predicting ATP yields has been used in the following cases.

A. Propionic acid fermentation

Bauchop and Elsden (1960) have measured the molar growth yields of *Propionibacterium pentosaceum*, growing with glucose, glycerol and lactate as energy source. These values are included in Table III. The ATP yields that can be calculated from substrate phosphorylations (in the glycolytic system and from acetate formation) are much lower than obtained by dividing Y by the mean value of Y_{ATP}. For instance, for lactate only 1 mole of acetate is formed per 3 moles of lactate fermented, giving an ATP yield of 0·33 moles/mole. The molar growth yield is 7·6 predicting an ATP yield of about 0·7 moles/mole. It is thus evident that extra ATP is formed in this fermentation. Bauchop and Elsden supposed that the production of extra ATP is associated with the formation of succinate. The complete fermentation scheme for the propionic acid fermentation has been elucidated by Allen *et al.* (1964). In this scheme the extra ATP formation is coupled to the reduction of fumarate to succinate. The reducing agent in the fumarate reductase reaction is a reduced flavin. This reduction must compensate for the oxidation of triosephosphate, which generates NADH. Thus a transfer of hydrogen from NADH to flavin is necessary. This reaction is known to be coupled with oxidative phosphorylation in aerobic systems. Anaerobic reduction of fumarate in these systems by NADH has been shown to occur and to be coupled to ATP formation (Sanadi and Fluharty, 1963; Haas, 1964), and this may explain the extra ATP formation in *Prop. pentosaceum*. If the formation of 1 mole of succinate is associated with the formation of 1 mole of ATP (as expected) we may calculate ATP yields of 4 moles/mole glucose, of 2 moles/mole glycerol and of 1 mole/mole lactate. These values are in reasonable agreement with the molar growth yields found.

B. Fermentations with net CO_2 fixation

Actinomyces israeli requires substrate amounts of CO_2 for maximal growth and succinate is one of the major fermentation products from glucose (Pine and Howell, 1956; Buchanan and Pine, 1963). In the absence

TABLE III

Molar growth yields of organisms growing on substrates of which the exact ATP yields are not known

Organism	Substrate	Y	ATP yields from known reactions	Ref.
Propionibacterium pentoseceum	Glucose	37·5	2·7	Bauchop & Elsden (1960)
	Glycerol	20·0	1·0	
	Lactate	7·6	0·33	
Lipolytic bacterium 5s	Glycerol	22·0	..	Hobson (1965)
	Fructose[1]	60	..	Hobson & Summers (1967)
Clostridium tetanomorphum	Glutamate	6·8	0·62	Twarog & Wolfe (1963)
	Histidine	11·1	..	
Clostridium aminobutyricum	γ-Amino-butyrate	7·6	0·5	Hardman & Stadtman (1963)
	γ-Hydroxy-butyrate	8·9		
Ruminococcus albus	Cellobiose	90·2	8·9	Hungate (1963)
Actinomyces israeli	Glucose + CO_2	41·7–49·3[2]	2·8	Buchanan & Pine (1967)
	Glucose + malate	40·4	4·0	
Selenomonas ruminantium	Glucose[3]	17	..	Hobson (1965)
	Glucose[4]	62	..	
Bacteroides amylophilum	Maltose[5]	130	..	Hobson & Summers (1967)

1. In continuous culture at $D = 0 \cdot 1$.
2. The growth yield depends on the medium.
3. In batch culture.
4. In continuous culture with D between 0·2 and 0·45.
5. In continuous culture at $D = 0 \cdot 35$.

of CO_2, glucose is fermented to 2 moles of lactate and a molar growth yield of 24·7 g/mole is obtained (Buchanan and Pine, 1967). In the presence of CO_2 the anaerobic molar growth yields are much higher however (values of 41·1–51·0 are reported). From known reactions in which substrate phosphorylation occurs the formation of only 2·8 moles/mole ATP of glucose can be accounted for. Thus the extra formation of ATP must be associated with the change in the fermentation pattern due to the presence of CO_2. It seems in this case, however, that formation of the extra ATP might be associated with CO_2 fixation and not succinate formation. The exact fermentation scheme for this organism needs to be elucidated, as does its aerobic metabolism.

C. Butyric acid fermentation

Molar growth yields of *Clostridium tetanomorphum* growing with glutamate and histidine (Table III) have been reported by Twarog and Wolfe (1963). The ATP yield during this fermentation was calculated from the fermentation balance. During this fermentation ATP is produced from butyryl phosphate and acetyl phosphate. Not all the acetate formed in this fermentation is derived from acetyl phosphate however, as part of it is derived directly from mesaconate (Wachsman and Barker, 1955). The authors calculated a Y_{ATP} of 10·9, which is in very good agreement with the mean value of Y_{ATP} given in Section V. Hardman and Stadtman (1963) obtained higher yields of *Cl. aminobutyricum* with γ-aminobutyrate and γ-hydroxybutyrate as growth substrates (Table III), than can be accounted for by the well established pathways. The extra ATP might be generated by reactions associated with butyrate formation. In the same way as outlined in Section VII A, the extra ATP may be obtained during the transfer of hydrogen of NADH to flavins. The results of Twarog and Wolfe do not support this conclusion with *Cl. tetanomorphum*. It is very likely that ATP formation is indeed associated with butyrate formation in *Clostridium kluyveri*, which grows at the expense of the reaction—

$$\text{Ethanol} + \text{acetate} \rightarrow \text{butyrate.}$$

Work with cell-free extracts is needed to clarify these points.

D. Molar growth yield for some rumen micro-organisms

The growth yield of *Ruminococcus albus* growing on cellobiose has been measured by Hungate (1963). The interpretation of these results is, however, difficult. A cellular dry weight of 2·58 was formed from 9·79 mg of cellobiose, corresponding to a molar growth yield of 90·1. These cells contained a large amount of starch, however, and consequently the percentage of nitrogen in the cells was small. The formation of polysaccharide requires less ATP than the formation of cell material. Gunsalus and Shuster (1961) have pointed out that at least 80 g of polysaccharide may be formed from glucose per mole of ATP. The starch-containing cells had a nitrogen content of only 8·7%, whereas normal cells have a nitrogen content of 10·5%. Therefore Hungate corrected the observed yield to 74·89, by multiplying by 8·7/10·5. The organism was grown in a minimal medium and consequently some cellobiose was used as a carbon source, and for each mole of cellobiose utilized, only 342 − 74·89 g was fermented to provide ATP.

Hungate found about 4 moles of acetate/mole of cellobiose. If cellobiose is split by phosphorolytic cleavage an ATP yield of 9 moles/mole of cellobiose may be obtained. From the amount of cellobiose fermented 7·8 moles of ATP may be formed. This is about the same amount as obtained by dividing the corrected yield by the mean value of Y_{ATP} given in Section V.

Selenemonas ruminantium gave a molar growth yield on glucose of approximately 17 when grown in batch culture (Hobson, 1965). Glucose is mainly fermented to lactate under these conditions. Thus Y_{ATP} in this organism would be relatively low (about 8·5), unless the reported Y values are too low owing to lysis after complete glucose consumption. Further, the bacteria had a high carbohydrate content, indicating that part of the glucose had been assimilated.

In continuous culture, very high growth yields (between 60 and 70) were obtained at growth rates between 0·2 and 0·45. Hobson (1965) has shown that more volatile acid is formed under these conditions. However, an increase in ATP yield to such an extent that a growth yield of 60 to 70 can be obtained cannot as yet be explained.

Similarly the growth yield of "lipolytic bacterium 5s" on fructose increased to a maximum of 60 at dilution rates of about 0·1 and decreased at higher dilution rates (Hobson and Summers, 1967). No changes in pattern of fermentation products with growth rate were observed in this case. The growth yield of *Bacteroides amylophilus* on maltose increased to about 130 at a growth rate of 0·35. In this case the pattern of fermentation products was changed. Hobson and Summers (1967) suggest that from these results values of Y_{ATP} of about 20 are obtained for *Bact. amylophilus* or *Sel. ruminantium* and 15 for Bacterium 5s growing on fructose. A normal Y_{ATP} value during growth of Bacterium 5s on glycerol in continuous culture was, however, obtained. The figure for Bacterium 5s suggests that a normal Y_{ATP} is obtained during growth on glycerol in continuous culture. In these experiments, differences between growth yield determinations in batch and in continuous culture were nearly always found. An analysis of the cells produced under both conditions seems necessary—before an explanation can be provided.

E. Conclusion

The limited use that has been made of Y_{ATP} to predict ATP yields has indicated higher values than expected in several fermentations. The results suggest that oxidative phosphorylation occurs during hydrogen transfer to flavins, but only studies with cell-free extracts will verify this conclusion. Further; the results mentioned in Section VII D show that several organisms give high molar growth yields, which are difficult to interpret at this moment.

VIII. RESULTS AND SIGNIFICANCE OF Y_O DETERMINATION

The results of several determinations of Y_O in batch culture are given in Table IV. The first Y_O value to be determined was that of *S. cerevisiae* by Pasteur (1876) who reported that 414 ml O_2 are required to produce 1 g of dry yeast. This corresponds to a Y_O value of 26·9. The following conclusions can be reached from the results of Table IV.

TABLE IV

Y_O values of several organisms growing on various substrates

Substrate	*E. coli*	*A. aerogenes*	*P. fluorescens*	*P. oxalaticus*	*Ac. oxydans*	*S. cerevisiae*
Glucose	20·2	31·9	16·0†	..	4·9	25·0
Gluconate	..	27·5	17·1†	..	..	..
2-Keto-gluconate	..	..	17·9†	..	..	..
Citrate	..	18·0	13·2	..	..	..
Glutamate	11·7	7·3	16·7	..	..	..
Succinate	11·3	10·0	10·3	12·1	..	..
Glycerol	..	15·5	..	..	3·7	..
DL-lactate	11·7	5·9	12·2	..	..	..
Acetate	6·2	5·0	7·4	..	..	..
Oxalate	..	..	..	4·0	..	..
Formate	..	..	..	3·9	..	..

† These values are for *P. fluorescens* ATCC 9027 (MacKechnie, 1966). The other values are for *P. fluorescens* KB1. The values for *A. aerogenes* are from Hadjipetrou *et al.* (1964). The value for *S. cerevisiae* is the mean value of determinations by Pasteur (1876), Sperber (1945) and Bulder (1963). The values for *E. coli*, *P. fluorescens* KB1, *P. oxalaticus* and *Ac. oxydans* are from Whittaker and Elsden (1963).

A. Efficiency of oxidative phosphorylation

If we divide the Y_O value for glucose of Table IV by the mean value of Y_{ATP} given in Section V, we obtain P/O values that are very close to 2·0 for *E. coli* and *P. fluorescens* and that approach 3·0 for *A. aerogenes* and *S. cerevisiae*. High Y_O values have also been obtained in continuous culture (Table V). These results also point to P/O ratios of about 3 in *A. cloacae* and in *T. utilis*. Chen (1964) has also concluded from the molar growth yield on glucose and on the material-balance equation, that the supposition that 38 moles of ATP are produced per mole glucose oxidized is correct. From the growth yield and the ATP yield, calculated on the assumption of an ATP yield of 38 moles/mole glucose, he found a Y_{ATP} value of 10·96. This is in agreement with the mean Y_{ATP} value given in Section V. This indicates

TABLE V

Y_O values determined in continuous culture

Organism	D	Dry weight g/litre/h	O_2 mmole/litre/h	Y_O	Ref.
A. cloacae	0·25	1·0	20	25·0	Elsworth *et al.* (1958)
	0·50	2·1	40	26·2	Elsworth *et al.* (1958)
A. cloacae	0·30	0·83	24†	17·4‡	Pirt (1957)
	0·20	0·73	24†	15·1‡	Pirt (1957)
T. utilis	0·3	..	..	19·2	Button & Garver (1966)
	0·5	..	..	24·0	Button & Garver (1966)
	0·8	..	..	28·8	Button & Garver (1966)

† The O_2 value has been measured too large and must be divided by a factor of 1·4–2·0 (Pirt and Callow, 1958).
‡ The Y_O value must be multiplied by a factor of 1·4–2·0.

that in living bacterial cells the efficiency of oxidative phosphorylation is as high as in mitochondria. This is in contrast with the low P/O ratios found in cell-free extracts (Dolin, 1961). Hadjipetrou has measured P/O ratios of 0·4–0·7 in cell-free extracts of *A. aerogenes* for the oxidation of NADH. It is evident that the low P/O ratios in cell-free extracts are due to damage to the phosphorylative system. However, it does not seem that the efficiency of oxydative phosphorylation is the same in all organisms and for all oxidations.

The Y_O values for *P. fluorescens* on glucose, gluconate and 2-ketogluconate increase in this order (McKechnie, 1966). Glucose is partly metabolized in this organism by membrane-bound enzymes into 2-ketogluconate (Wood and Schwerdt, 1953; Ramakrishnan and Campbell, 1955; Hertlein and Wood, 1958). It has been shown with the solubilized enzymes that soluble coenzymes, such as NAD, NADP, FAD and FMN are not involved. The oxidases are directly linked to cytochromes. The 2-ketogluconate formed is further metabolized by phosphorylation and reduction to 6-phosphogluconate (Frampton and Wood, 1957). In the conversion of 6-phosphogluconate soluble coenzyme-linked enzymes are involved. It is thus likely that the hydrogen transfer in the oxidations of glucose and gluconate to 2-ketogluconate occurs in fewer steps than the oxidation of NADH. The lower Y_O values for glucose and gluconate further support this view. With *Acetobacter oxydans*, low Y_O values are obtained (Whitaker and Elsden, 1963). The oxidation of both glucose and glycerol in this organism takes place by membrane-bound enzymes (Arcus and Edson, 1956; King and Cheldelin, 1957). In these oxidations no soluble coenzymes are involved. The molar growth yields of *Gluconobacter liquefaciens* (= *Acetobacter aceti*) are very low (Stouthamer, 1962; Whitaker and Elsden, 1963). Also the P/O ratios in cell-free extracts are very low. It is very likely that at least in this organism the efficiency of oxidative phosphorylation is indeed low.

Under aerobic conditions, *Z. mobilis* metabolized glucose with an oxygen uptake of 1 mole/mole (Belaïch and Senez, 1965). However, the aerobic and anaerobic molar growth yields are the same, indicating that the aerobic and anaerobic ATP yields are the same and that oxidative phosphorylation does not occur in this organism.

B. Efficiency of growth on simple compounds

It is clear that growth on simple compounds leads to smaller Y_O values than growth on glucose (Table IV). These values are lower because these compounds must be converted into cellular monomers, e.g., carbohydrates, for which extra ATP is required. It is possible to calculate how much ATP is required for the assimilation of 1 g atom of C from these compounds from the data provided (Table IV). In this calculation it is assumed that the carbon content of the bacteria is 45%. The calculation is given in Table VI for succinate, acetate and formate. Two values are given corresponding to P/O ratios of 3 and 2. The value calculated from Y_{ATP} is given for comparison. From the Y_O values for glucose with *A. aerogenes* (Table IV), *S. cerevisiae* (Table IV), *A. cloacae* (Table V) and *T. utilis* (Table V) we may calculate that in these cases the amount of ATP needed to assimilate 1 g-atom of carbon is about the same as that calculated from Y_{ATP}, assuming a P/O ratio of 3. For the other substrates, the amount of ATP needed to

TABLE VI

ATP requirement for the assimilation of 1 g-atom of C from different substrates

Substrate	Y_O	ATP requirement for assimilation of 1 g-atom of C assuming a P/O ratio of 3†	2†
Glucose	31·9‡	2·5	1·6
Glucose	16·0§	5·0	3·3
Succinate	10·9	7·5	5·0
Acetate	6·2	12·9	8·6
Formate	3·9	20·4	13·6

† Value calculated from Y_{ATP}: 2·5 moles ATP/g-atom C. This value is calculated from the mean value of Y_{ATP} given in Section V.

‡ Value of *A. aerogenes* (Table IV).

§ Value for *P. fluorescens* (Table IV). The values for succinate and acetate are the mean of the Y_O values for all organisms recorded in Table IV.

assimilate 1 g-atom of carbon is very high even if we accept the assumption of a P/O ratio of 2 (assumption of a P/O ratio of 1 is ruled out by the observed Y_O values on glucose). These amounts of ATP are much higher than the amount we can calculate from the pathways by which these substrates are assimilated. For instance during growth of *Pseudomonas oxalaticus* on formate, assimilation occurs by CO_2 fixation in the Calvin cycle (Quayle and Keech, 1958, 1959). In this cycle 3 moles of ATP are required per mole of CO_2 assimilated. NADH is also involved in the assimilation of CO_2. In *P. oxalaticus*, formate is oxidized by a NAD-linked formate dehydrogenase (Johnson *et al.*, 1964). Thus in this case, in contrast to the chemoautrophic micro-organisms (see Section IX) no extra ATP is required for the formation of NADH. The expected amount of ATP for the assimilation of 1 g-atom of carbon from formate is $2{\cdot}5+3{\cdot}0=5{\cdot}5$ moles/mole. The observed amount is much larger. Similar calculations can be given for the other substrates. Two explanations may be given for these observations.

1. During growth on these simple compounds, the rate of conversion of these substrates to a component needed for growth is rate limiting. In this case ATP production might be higher than can be utilized for growth, which might lead to uncoupling of growth and ATP production. Part of the ATP is then hydrolysed by ATPase and the energy is dissipated as heat. Some examples of uncoupled growth have been described (see Section VI). They occur when one of the components of the medium is less assimilable as for instance the nitrogen source.
2. Growth on these simple compounds is sometimes slower than growth with glucose. Micro-organisms need a certain amount of energy for maintenance (Mallette, 1963; Marr *et al.*, 1963). At low growth rates the maintenance requirement will be relatively more important than at faster growth rates.

At this moment is looks as if the first explanation will be the more important.

IX. MOLAR GROWTH YIELDS OF CHEMOAUTOTROPHIC MICRO-ORGANISMS

In chemoautotrophic micro-organisms, the energy of oxidation of an inorganic substrate is utilized for synthesis of all cellular components by CO_2 fixation. The efficiency of this process is sometimes expressed in terms of the thermodynamic efficiency, which is in % (Baas-Becking and Parks, 1927)—

$$100\times\frac{\text{carbon atoms assimilated}\times 113}{\text{free energy change during substrate oxidation}}$$

The factor 113 is the heat of combustion of bacteria. The free-energy efficiency for a number of organisms has been given in reviews by Elsden

(1962) and Fromageot and Senez (1960). Another way of expressing these efficiencies is the relationship between the amount of carbon dioxide fixed and the amount of oxygen consumed. In most cases this relationship is determined by measuring the incorporation of $^{14}CO_2$ into cellular material. Generally this relationship has been measured with washed cell suspensions. Most of these results will be neglected in this article. The results of several experiments with growing cells show that the O_2/CO_2 ratios vary considerably (Table VII). As mentioned before (Section I C) Y_O values can be

TABLE VII

Relationship between oxygen consumed and carbon dioxide assimilated

Organism	Substrate	O_2/CO_2	Ref.
Thiobacillus thiooxydans	S	18	Waksman & Starkey (1922)
Thiobacillus thioparus	$S_2O_3^{2-}$	19	Starkey (1935)
Nitrobacter winogradskyi	NO_2^-	34	Fischer & Laudelout (1965)
Nitrobacter winogradskyi	NO_2^-	36	Schön (1965)
Thiobacillus str. C	$S_2O_3^{2-}$	11	Kelly & Syrett (1964)
Thiobacillus ferrooxydans	Fe^{2+}	27	Beck & Shafia (1964)
Thiobacillus ferrooxydans	S	7·5	Beck & Shafia (1964)
Thiobacillus neapolitans	$S_2O_3^{2-}$	5·0†	Hempfling & Visniac (1967)
Thiobacillus neapolitans	$S_2O_3^{2}$	2·4‡	Hempfling & Visniac (1967)
Hydrogenomonas str H.16	H_2	2·0§	Schlegel *et al.* (1961)

In several cases, these O_2/CO_2 have been calculated from growth yields in g/mole inorganic substrate oxidized, assuming a carbon content of the bacteria of 50%.

† This result is the maximal yield which may be obtained in continuous culture.
‡ This yield is corrected for maintenance requirement.
§ Value for non-growing cells.

calculated from O_2/CO_2 ratios. As an example this calculation will be given for *Thiobacillus thiooxydans* growing with sulphur as energy sources and an O_2/CO_2 ratio of 18. This means that 1 mole of CO_2 is assimilated/18 moles of O_2 taken up. Assuming a carbon content of 45%, we find $\frac{100}{45} \times 12$ g dry weight/18 moles of O_2 taken up. This gives a Y_O value of $\frac{100}{45} \times \frac{12}{18} \times \frac{1}{2} = 0.75$. In comparison with the values given in Table IV, the Y_O value for *Th. thiooxydans* is very low.

The smallest O_2/CO_2 ratio is that for the *Hydrogenomonas* strain. In this case the equation—

$$25H_2 + 8O_2 + 4CO_2 \rightarrow C_4H_6O_2 + 22H_2O$$

has been given (Schlegel *et al.*, 1961). The CO_2 fixed is completely converted into poly-β-hydroxybutyrate. It is known that in this organism NAD may be reduced directly to NADH by hydrogenase (Packer and Vishniac, 1955). The formation of β-hydroxybutyryl-CoA from CO_2 requires 9 NADH and 14 ATP. The ATP must be produced by oxidation of NADH. Assuming P/O ratios of 3, 2 and 1 for this oxidation we get the following equations for poly-β-hydroxybutyrate formation: P/O ratio 3, 13·7 H_2+ 2·35 O_2+4 CO_2; P/O ratio 2, 16 H_2+3·5 O_2+4·5 CO_2; and P/O ratio 1, 23 H_2+7 O_2+4 CO_2. The last equation gives the best fit to the experiment. It is unlikely, however, that in *Hydrogenomonas*, which is related to *Pseudomonas*, the P/O ratio is really 1. The Y_O values for *P. fluorescens* (Table IV) indicate P/O ratios of 2. It is thus very likely that uncoupling occurs during the experiment. With a P/O ratio of 2, 1 mole of CO_2 is fixed per 8 moles of ATP and with a P/O ratio of 3, fixation of 1 mole of CO_2/12 moles of ATP occurs (compare the results for *P. oxalaticus* during growth on formate in Table VI). The highest value is that for *Nitrobacter winogradskyi*. Assuming P/O ratios of 3, 2 and 1 we can calculate that 216, 144 and 72 moles of ATP, respectively, are required for the assimilation of 1 mole of carbon dioxide. These values are very much in excess of the amount required theoretically. What may be the reasons for these large differences? In the first place we have to consider the possibility that the respiratory chain is shorter than in heterotrophic bacteria. The redox potentials of the inorganic substrate are more electropositive than those of the pyridine nucleotides. Consequently, oxidation of the inorganic substrate is directly coupled to reduction of cytochrome *c*, without soluble cofactors. We have seen already (Section VII A) that when a similar situation exists in heterotrophic bacteria, low Y_O values and low growth yields are found. This is the case with *G. liquefaciens* (= *Ac. aceti*) (Stouthamer, 1962), *Ac. oxydans* (Whittaker and Elsden, 1963) and for glucose oxidation in *P. fluorescens* (MacKechnie, 1966). We may thus expect low P/O ratios for autotrophic bacteria.

Secondly we must mention one of the factors described in Section VII B for the explanation of the low Y_O values for growth of heterotrophic bacteria with simple carbon compounds, namely energy consumption for maintenance. Hempfling and Vishniac (1967) have measured the energy requirement for maintenance of *Thiobacillus neapolitans* and have found that it is the highest value recorded so far. It is seen from Table VII that the O_2/CO_2 ratio after correction for maintenance requirement is only half the uncorrected ratio. Thus the requirement for maintenance is very important, and we may expect that in the very slow growing *N. winogradskyi* the energy requirement for maintenance will be very high.

We have mentioned already that in the dehydrogenation of the inorganic substrate no soluble cofactors are involved. For the operation of the Calvin cycle, 2 moles of NADH are required, however, for the fixation of 1 mole of CO_2. This NADH must be formed by an energy-dependent reversion of the electron flow in the respiratory chain. Chance and Hollunger (1961) first demonstrated this reversion of electron flow in mitochondria. They found that 2 moles of ATP are required for reduction of NAD with succinate as hydrogen donor. Similar reactions have been demonstrated with cell-free preparations of *Nitrobacter*, *Ferrobacillus*, *Nitrosomonas* and *Thiobacillus* (Aleem *et al.*, 1963; Aleem, 1965, 1966). The amount of ATP required to obtain NADPH or NADH by hydrogen transfer from inorganic substrates in cell-free extracts varied from organism to organism. But the energy requirement for synthesis of reduced pyridine nucleotides in all chemo-autotrophic micro-organisms, except *Hydrogenomonas*, is certainly one of the factors involved in the explanation of the large O_2/CO_2 ratios.

The last factor that may be involved is the possibility that the oxygen uptake is not completely associated with phosphorylation. Hempfling and Vishniac (1967) suggest that oxygen uptake associated with sulphite oxidation in *Th. neapolitans* is not accompanied by phosphate esterification, because they do not find esterification during sulphite oxidation in cell-free extracts and because sulphite does not reduce cytochromes directly (Hempfling and Vishniac, 1965). They suggest that during sulphite oxidation only 2 moles of ATP are formed from phosphorylation on substrate level. They assume a P/O ratio of 3 for the other oxidation. The calculated ATP yield is then in accordance with the theoretical amount of ATP required for the fixation of 1 mole of CO_2, namely 3 moles in the Calvin cycle, 2 moles for NADH formation and 2·7 moles for the formation of cellular material (see Table VI). A weak point is still the assumption of a P/O ratio of 3 in a dehydrogenation in which no soluble cofactors participate. In several cases only one oxidative step is involved (*Nitrobacter*, *Ferrobacillus*). In these cases we cannot assume that only part of the oxygen taken up is associated with ATP formation and consequently the high O_2/CO_2 ratios must be ascribed completely to the reasons previously mentioned, i.e., uncoupling, high requirement for maintenance and ATP requirement for NADH formation.

X. INFLUENCE OF HYDROGEN ACCEPTORS ON MOLAR GROWTH YIELDS

A. Nitrate

The influence of nitrate on the growth of *A. aerogenes* has been studied by Hadjipetrou and Stouthamer (1965). The $Y_{NO_3^-}$ values calculated from this work are 103·4 for glucose and 54·4 for mannitol; these values are very high (even higher than the corresponding Y_O values) and also there is a large difference between them. The explanation for the difference between the $Y_{NO_3^-}$ values for glucose and mannitol must be sought in differences in the products of nitrate reduction. With glucose, nitrate is nearly completely reduced to ammonia; with mannitol, only about half of the nitrate is reduced to nitrite. The explanation for the large $Y_{NO_3^-}$ values is that the metabolism of glucose is still mainly fermentative in the presence of nitrate and the citric acid cycle does not function under these conditions (Forget and Pichinoty, 1964; Hadjipetrou and Stouthamer, 1965). Thus in the presence of nitrate, ATP is mainly produced from phosphorylation at the substrate level, and only small amounts of nitrate are used as hydrogen acceptor. For these two reasons it is evident that the $Y_{NO_3^-}$ values have not much significance. To calculate the amount of ATP produced during nitrate reduction, we must utilize three values: the molar growth yield; the amount of nitrate reduced; and the amount of acetate produced. From these results it has been calculated that per mole of nitrate reduced to nitrite, 2·6–3·0 moles ATP are produced. Thus respiration with oxygen and with nitrate seems to give the same ATP yield.

Further evidence for the formation of 3 moles of ATP during nitrate reduction comes from the study of a mutant of *A. aerogenes*, which is blocked at α-oxoglutarate dehydrogenase (Stouthamer, 1967). In this mutant the citric acid cycle does not function, and glucose is only oxidized to acetate. The molar growth yield for aerobic growth on glucose is about the same (44·4 g/mole) as that of the wild type under anaerobic conditions in the presence of nitrate (45·5 g/mole). It has already been mentioned that during anaerobic growth in the presence of nitrate, the citric acid cycle does not function. Thus under both conditions (aerobically in the mutant and anaerobically in the presence of nitrate in the wild type) the citric acid cycle does not function and energy production is the same. As the Y_O values indicate a P/O ratio of 3, the ATP yield during nitrate respiration must also be 3 moles/mole.

B. Sulphate

Molar growth yields of *D. desulfuricans* growing with lactate and pyruvate in minimal medium have been reported by Senez (1962). The molar growth yields on lactate and pyruvate are 9·9 and 9·4 g/mole, respectively. With lactate, 0·5 moles of sulphate mole of lactate were reduced, and with pyruvate 0·25 moles of sulphate mole of pyruvate. This corresponds to $Y_{SO_4^-}$ values of 19·8 and 37·2. Both substrates are converted into acetate, giving an ATP yield of 1 mole/mole from phosphorylation at the substrate level. It is known that reduction of sulphate is preceded by its conversion in the reaction sulphate + ATP→adenosine 5′-phosphosulphate + pyrophosphate (Peck, 1959). Thus the reduction of 1 mole of sulphate requires 2 moles of ATP, as the energy-rich bond in pyrophosphate is supposed to be lost. It is evident that during growth with lactate, ATP formation from acetylphosphate will be just enough for activation of the sulphate required as hydrogen acceptor. If no other sources of ATP formation were present, no growth with lactate would be possible. According to Senez (1962) the growth yields indicate an ATP yield of about 1 mole/mole for both lactate and pyruvate. Postgate (1956) has shown that cytochrome c_3 is involved in sulphate reduction. Senez (1962) has suggested that electron transfer during sulphate reduction via cytochrome c_3 yields 1 mole of ATP per electron pair, and that the ATP formed during electron transfer is needed completely for reduction of the hydrogen acceptor. This certainly explains the equal molar growth yields on lactate and in pyruvate, but to explain the complete loss of ATP from hydrogen transfer one has to consider that per electron pair, 0·5 mole of ATP is required for sulphate activation and, further, an additional 0·5 mole in the reduction of SO_3^{2-} to S^{2-}.

Although this explanation fits the data, there is no experimental evidence for an ATP involvement in sulphite reduction. Another unexplained feature is that growth took place in minimal medium, in which pyruvate is used as carbon and energy source. The conversion of pyruvate into cellular material requires more ATP than is needed for growth with other simple compounds (Table VI). This has not been taken into account in the above analysis, and a re-evaluation of the process is required.

REFERENCES

Aleem, M. I. H. (1965). *Biochim. biophys. Acta*, **107**, 14–28.

Aleem, M. I. H. (1966). *Biochim. biophys. Acta*, **113**, 216–224.

Aleem, M. I. H., Lees, H., and Nicholas, D. J. D. (1963). *Nature, Lond.*, **200**, 759–761.

Allen, S. H. G., Kellermeyer, R. W., Stjernholm, R. L., and Wood, H. G. (1964). *J. Bact.*, **87**, 171–187.

Arcus, A. C., and Edson, N. L. (1956). *Biochem. J.*, **67**, 385–394.

Baas-Becking, L. G. M., and Parks, G. S. (1927). *Physiol. Rev.* **7**, 85–106.

Battley, E. H. (1960). *Physiologia Pl.*, **13**, 192–203.

Bauchop, T., and Elsden, S. R. (1960). *J. gen. Microbiol.*, **23**, 457–469.

Beck, J. V., and Shafia, F. M. (1964). *J. Bact.*, **88**, 850–857.

Beck, R. W., and Shugart, L. R. (1966). *J. Bact.*, **92**, 802–803.

Belaïch, J. P., and Senez, J. C. (1965). *J. Bact.*, **89**, 1195–1200.

Buchanan, B. B., and Pine, L. (1963). *Sabouraudia*, **3**, 26–39.

Buchanan, B. B., and Pine, L., (1967). *J. gen. Microbiol.*, **46**, 225–236.

Bulder, C. J. E. A. (1960). *Experientia*, **16**, 565–566.

Bulder, C. J. E. A. (1963). Ph.D. Thesis, Technical University, Delft.

Bulder, C. J. E. A. (1966). *Arch. Mikrobiol.*, **53**, 189–194.

Button, D. K., and Garver, J. C. (1966). *J. gen. Microbiol.* ,**45**, 195–204.

Chance, B., and Hollunger, G. (1961). *J. biol. Chem.*, **236**, 1534–1543.

Chen, S. L. (1964). *Nature, Lond.*, **202**, 1135–1137.

Cooper, C. M., Fernstrom, G. A., and Miller, S. A. (1944). *Ind. Engng Chem. ind. Edn*, **36**, 504–509.
Czerniawski, E., Sedlaczek, L., and Zablocki, B. (1965). *Bull. Acad. pol.* Sci. *Cl II Sér. Sci. biol.*, **13**, 291–293.
Dawes, E. A., Ribbons, D. W., and Rees, D. A. (1966). *Biochem. J.*, **98**, 804–812.
DeMoss, R. D., Bard, R. C., and Gunsalus, I. C. (1951). *J. Bact.*, **62**, 499–511.
de Vries, W., Gerbrandy, S. J., and Stouthamer, A. H. (1967). *Biochim. biophys. Acta.*, **136**, 415–425.
Dolin, M. I. (1961). *In* "The Bacteria" (Ed. I. C. Gunsalus and R. Y. Stanier), Vol. 2, pp. 319–363. Academic Press, New York.
Elsden, S. R. (1962). *In* "the Bacteria" (Ed. I. C. Gunsalus and R. Y. Stanier), Vol. 3, pp. 1–40. Academic Press, New York.
Elsworth, R., Telling, R. C., East, D. N., and Ford, J. W. S. (1958). *Chemy Ind.*, 382.
Fischer, I., and Laudelout, F. H. (1965). *Biochim. biophys. Acta*, **110**, 259–264.
Forget, P., and Pichinoty, F. (1964). *Biochim. biophys. Acta.*, **82**, 441–444.
Forrest, W. W. (1965). *J. Bact.*, **90**, 1013–1018.
Forrest, W. W., and Walker, D. J. (1965). *J. Bact.*, **89**, 1448–1452.
Forrest, W. W., Walker, D. J., and Hopgood, M. F. (1961). *J. Bact.*, **82**, 685–690.
Frampton, E. W., and Wood, W. A. (1957). *Bact. Proc.*, **57**, 122.
Fromageot, C., and Senez, J. C. (1960). *In* "Comparative Biochemistry" (Ed. M. Florkin and H. S. Mason), Vol. 1, pp. 347–409. Academic Press, New York.
Goddard, J. L., and Sokatch, J. R. (1964). *J. Bact.*, **87**, 844–851.
Gunsalus, I. C., and Gibbs, M. (1952). *J. biol. Chem.*, **194**, 871–875.
Gunsalus, I. C., and Shuster, C. W. (1961). *In* "The Bacteria" (Ed. I. C. Gunsalus and R. Y. Stanier), Vol. 2, pp. 1–58. Academic Press, New York.
Haas, D. W. (1964). *Biochim. biophys. Acta.*, **92**, 433–439.
Hadjipetrou, L. P. (1965). Ph.D. Thesis, University of Utrecht.
Hadjipetrou, L. P., and Stouthamer, A. H. (1963). *Antonie van Leeuwenhoek.*, **29**, 256–260.
Hadjipetrou, L. P., and Stouthamer, A. H. (1965). *J. gen. Microbiol.*, **38**, 29–34.
Hadjipetrou, L. P., Gerrits, J. P., Teulings, F. A. G., and Stouthamer, A. H. (1964). *J. gen. Microbiol.*, **36**, 139–150.
Hadjipetrou, L. P., Gray-Young, T., and Lilly, M. D. (1966). *J. gen. Microbiol.*, **45**, 479–488.
Halliwell, G. (1960). *Biochem. J.*, **74**, 457–462.
Hardman, J. K., and Stadtman, T. C. (1963). *J. Bact.*, **85**, 1326–1333.
Hempfling, W. P., and Vishniac, W. (1965). *Biochem. Z.*, **342**, 272–287.
Hempfling, W. P., and Vishniac, W. (1967). *J. Bact.*, **93**, 874–878.
Herbert, D. (1958). *In* "Recent Progress in Microbiology" (Ed. G. Tunevall). Almqvist and Wiksell, Stockholm. pp. 381–396.
Herbert, D., Elsworth, R., and Telling, R. C. (1956). *J. gen. Microbiol.*, **14**, 601–622.
Hertlein, B. C., and Wood, W. A. (1958). *Bact. Proc.* **58**, 105.
Hobson, P. N. (1965). *J. gen. Microbiol.*, **38**, 167–180.
Hobson, P. N., and Summers, R. (1967). *J. gen. Microbiol.*, **47**, 53–65.
Hungate, R. E. (1963). *J. Bact.*, **86**, 848–854.
Hurwitz, J. (1958). *Biochim. biophys. Acta.*, **28**, 599–602.
Johnson, M. J. (1949). *J. biol. Chem.*, **181**, 707–711.
Johnson, M. J. (1967). *Science*, **155**, 1515–1519.
Johnson, P. A., Jones-Mortimer, M. C., and Quayle, J. R. (1964). *Biochim. biophys. Acta.*, **89**, 351–353.

Kelly, D. P., and Syrett, P. J. (1964). *J. gen. Microbiol.*, **34**, 307–317.
King, T. E., and Cheldelin, V. H. (1957). *J. biol. Chem.*, **224**, 579–590.
LeGall, J., and Senez, J. C. (1960). *C. r. hebd. Séanc. Acad. Sci., Paris*, **250**, 404–406.
MacKechnie, I. (1966). Ph.D. Thesis. University of Hull.
Mallette, M. F. (1963). *Ann. N.Y. Acad. Sci.*, **102**, 521–535.
Marr, A. G., Nilson, E. H., and Clark, D. J. (1963). *Ann. N.Y. Acad. Sci.*, **102**, 536–548.
Middleton, K. R. (1959). *J. Sci. Fd Agric.*, **10**, 218–224.
Monod, J. (1942). "Recherches sur la croissance des cultures bactériennes." Herman, Paris.
Monod, J. (1950). *Annls Inst. Pasteur*, **79**, 390–410.
Neidhardt, F. C. (1965). *Colloques int. Cent. natn Rech. scient.*, **124**, 329–336.
Nicholas, D. J. D., and Nason, A. (1957). *In* "Methods in Enzymology" (Ed. S. P. Colowick and N. O. Kaplan), Vol. 3, pp. 983–984. Academic Press, New York.
Novick, A., and Szilard, L. (1950). *Proc. natn Acad. Sci. U.S.A.*, **36**, 708–719.
Oxenburgh, M. S., and Snoswell, A. M. (1965). *J. Bact.*, **89**, 913–914.
Packer, L., and Vishniac, W. (1955). *J. Bact.*, **70**, 216–223.
Pardee, A. B. (1949). *J. Biol. Chem.*, **179**, 1085–1091.
Pasteur, L. (1876). *In* "Oeuvres de Pasteur" (Ed. P. Vallery-Radot), Vol. 5. Masson, Paris.
Peck, H. D. (1959). *Proc. natn. Acad. Sci. U.S.A.*, **45**, 701–708.
Perret, C. J. (1958). *J. gen. Microbiol.*, **18**, VII.
Pichinoty, F. (1960). *Folia Microbiol.*, Praha, **5**, 165–170.
Pine, L., and Howell, A. (1956). *J. gen. Microbiol.*, **15**, 428–445.
Pirt, S. J. (1957). *J. gen. Microbiol.*, **16**, 59–75.
Pirt, S. J., and Callow, D. S. (1958). *J. appl. Bact.*, **21**, 206–210.
Postgate, J. R. (1956). *J. gen. Microbiol.*, **14**, 545–572.
Quayle, J. R., and Keech, D. B. (1958). *Biochim. biophys. Acta.*, **29**, 223–225.
Quayle, J. R., and Keech, D. B. (1959). *Biochem. J.*, **72**, 623–630.
Ramakrishnan, T. and Campbell, J. J. R. (1955). *Biochim. biophys. Acta.*, **17**, 122–127.
Roberts, R. B., Abelson, P. H., Cowie, D. B., Bolton, E. T., and Britten, R. J. (1955). Carnegie Institute, Washington, Publication, 607, pp. 16.
Rosenberger, F., and Elsden, S. R. (1960). *J. gen. Microbiol.* **22**, 726–739.
Sanadi, D. R., and Fluharty, A. L. (1963). *Biochemistry*, **2**, 523–528.
Schlegel, H. G., Gottschalk, G., and von Bartha, R. (1961). *Nature, Lond.*, **191**, 463–465.
Schön, G. (1965). *Arch. Mikrobiol.*, **50**, 111–132.
Senez, J. C. (1962). *Bact. Rev.*, **26**, 95–107.
Senez, J. C., and Belaïch, J. P. (1965). *Colloques int. Cent. natn Rech. scient.*, **124**, 357–363.
Sokatch, J. T., and Gunsalus, I. C. (1957). *J. Bact.*, **73**, 452–460.
Sperber, E. (1945). *Ark. Kemi Miner. Geol.*, **21A**, 1–136.
Starkey, R. L. (1935). *J. gen. Physiol.*, **18**, 325–349.
Stouthamer, A. H. (1962). *Biochim. biophys. Acta.*, **56**, 19–32.
Stouthamer, A. H. (1967). *J. gen. Microbiol.*, **46**, 389–398.
Tempest, D. W., and Herbert, D. (1965). *J. gen. Microbiol.*, **41**, 143–150.
Twarog, R., and Wolfe, R. S. (1963). *J. Bact.*, **86**, 112–117.
Wachsman, J. T., and Barker, H. A. (1955). *J. Biol. Chem.*, **217**, 695–702.
Waksman, S., and Starkey, R. L. (1922). *J. gen. Physiol.* **5**, 285–310.

Whitaker, A. M., and Elsden, S. R. (1963). *J. gen. Microbiol.*, **31**, XXII.
Wood, W. A., and Schwerdt, R. F. (1953). *J. biol. Chem.*, **201**, 501–511.
Wood, W. A. (1961). *In* "The bacteria" (Ed. I. C. Gunsalus and R. Y. Stanier). Vol. 2, pp. 59–149. Academic Press, New York.

Author Index

Numbers in *italics* refer to the page on which references are listed at the end of each chapter

C

D

E

F

L

M

Q

R

S

U

V

W

Y

Z

Subject Index

B

E

I

M

R

T

U

V